IFIP Advances in Information and Communication Technology 510

IFIP – The International Federation for Information Processing

IFIP was founded in 1960 under the auspices of UNESCO, following the first World Computer Congress held in Paris the previous year. A federation for societies working in information processing, IFIP's aim is two-fold: to support information processing in the countries of its members and to encourage technology transfer to developing nations. As its mission statement clearly states:

IFIP is the global non-profit federation of societies of ICT professionals that aims at achieving a worldwide professional and socially responsible development and application of information and communication technologies.

IFIP is a non-profit-making organization, run almost solely by 2500 volunteers. It operates through a number of technical committees and working groups, which organize events and publications. IFIP's events range from large international open conferences to working conferences and local seminars.

The flagship event is the IFIP World Computer Congress, at which both invited and contributed papers are presented. Contributed papers are rigorously refereed and the rejection rate is high.

As with the Congress, participation in the open conferences is open to all and papers may be invited or submitted. Again, submitted papers are stringently refereed.

The working conferences are structured differently. They are usually run by a working group and attendance is generally smaller and occasionally by invitation only. Their purpose is to create an atmosphere conducive to innovation and development. Refereeing is also rigorous and papers are subjected to extensive group discussion.

Publications arising from IFIP events vary. The papers presented at the IFIP World Computer Congress and at open conferences are published as conference proceedings, while the results of the working conferences are often published as collections of selected and edited papers.

IFIP distinguishes three types of institutional membership: Country Representative Members, Members at Large, and Associate Members. The type of organization that can apply for membership is a wide variety and includes national or international societies of individual computer scientists/ICT professionals, associations or federations of such societies, government institutions/government related organizations, national or international research institutes or consortia, universities, academies of sciences, companies, national or international associations or federations of companies.

More information about this series at http://www.springer.com/series/6102

Zhongzhi Shi · Ben Goertzel
Jiali Feng (Eds.)

Intelligence Science I

Second IFIP TC 12 International Conference, ICIS 2017
Shanghai, China, October 25–28, 2017
Proceedings

Editors
Zhongzhi Shi
Chinese Academy of Sciences
Beijing
China

Jiali Feng
Shanghai Maritime University
Shanghai
China

Ben Goertzel
Machine Intelligence Research Institute
Rockville, MD
USA

ISSN 1868-4238 ISSN 1868-422X (electronic)
IFIP Advances in Information and Communication Technology
ISBN 978-3-319-88553-7 ISBN 978-3-319-68121-4 (eBook)
DOI 10.1007/978-3-319-68121-4

Printed on acid-free paper

This Springer imprint is published by Springer Nature
The registered company is Springer International Publishing AG
The registered company address is: Gewerbestrasse 11, 6330 Cham, Switzerland

Preface

This volume comprises the proceedings of the Second International Conference on Intelligence Science (ICIS). Artificial intelligence research has made substantial progress in special areas. However, deeper understanding of the essence of intelligence is far from sufficient and, therefore, many state-of-the-art intelligent systems are still not able to compete with human intelligence. To advance research in artificial intelligence, it is necessary to investigate intelligence, both artificial and natural, in an interdisciplinary context. The objective of this conference is to bring together researchers from brain science, cognitive science, and artificial intelligence to explore the essence of intelligence and related technologies. The conference provides a platform for the discussion of key issues related to intelligence science.

For the Second International Conference on Intelligence Science (ICIS 2017), we received more than 82 papers, of which 39 papers are included in this program as regular papers and nine as short papers. We are grateful for the dedicated work of both the authors and the referees, and we hope these proceedings will continue to bear fruit over the years to come. All papers submitted were reviewed by three referees.

A conference such as this cannot succeed without help from many individuals who contribute their valuable time and expertise. We want to express our sincere gratitude to the Program Committee members and referees, who invested many hours for reviews and deliberations. They provided detailed and constructive review reports that significantly improved the papers included in the program.

We are very grateful for the sponsorship of the following organizations: Chinese Association for Artificial Intelligence (CAAI), IFIP TC12, China Chapter of International Society for Information Studies, Shanghai Maritime University, Shanghai Society for Noetic Science. The event was co-sponsored by the Shanghai Association for Science and Technology (SAST), Shanghai Association for Artificial Intelligence (SAAI), Shanghai Logic Association (SLA), IEEE Shanghai Chapter, and Institute of Computing Technology, Chinese Academy of Sciences. Thanks to Professor Xiaofeng Wang for his role as chair of the Organizing Committee.

Finally, we hope you find this volume inspiring and informative.

August 2017

Zhongzhi Shi
Ben Goertzel
Jiali Feng

Organization

Sponsors

Chinese Association for Artificial Intelligence (CAAI)
China Chapter under International Society for Information Studies

Co-sponsors

IFIP Technical Committee 12
Shanghai Association for Science and Technology (SAST)
Shanghai Association for Artificial Intelligence (SAAI)
Shanghai Logic Association (SLA)
IEEE Shanghai Section Joint Chapter
Science and Technology on Information Systems Engineering Laboratory,
  National University of Defense Technology

Organizers

Shanghai Maritime University
Shanghai Association for Noetic Science

Supporters

Shanghai Municipal Commission of Economy and Informatization
Shanghai Association for Science and Technology (SAST)

Steering Committee

Muming Poo	Chinese Academy of Sciences, China
Xiongli Yang	Fudan University, China
Changjun Jiang	Donghua University, China
Aoying Zhou	East China Normal University, China
Guangjian Zhang	Chinese Association for Noetic Science, China

General Chairs

Yixin Zhong	Beijing University of Posts and Telecommunications, China
Youfang Huang	Shanghai Maritime University, China
Karlheinz Meier	Heidelberg University, Germany

Program Chairs

Zhongzhi Shi — Institute of Computing Technology, Chinese Academy of Sciences, China
Ben Goertzel — Machine Intelligence Research Institute, USA
Jiali Feng — Shanghai Maritime University, China

Program Committee

Agnar Aamodt, Norway
Ansguar Bernardi, Germany
Jean-Yves Beziau, Brazil
Guoqiang. Bi, China
Joseph E. Brenner, Switzerland
Jiang Cai, China
Wen Cai, China
Jinde Cao, China
Mihir Kr. Chakraborty, India
Dongsheng Chen, China
Huanhuan Chen, China
Weineng Chen, China
Zhicheng Chen, China
Zhihua Cui, China
Shifei. Ding, China
Gordana Dodig-Crnkovic, Sweden
Elias Ehlers, South Africa
Jiali Feng, China
Xiaolan Fu, China
Urich Furbach, Germany
Junbao Gao, China
Xiaozhi Gao, China
Ping Guo, China
Xiaokun Guo, China
Huacan He, China
Tzung-Pei Hong, Taiwan
Jincai Huang, China
Zhisheng Huang, The Netherlands
Changjun Jiang, China
Xiaoqian Jiang, USA
Shengming Jiang, China
Licheng Jiao, China
Yaochu Jin, UK
Glgun Kayakutlu, Turkey
Dehua Li, China
Gang Li, Australia
Jiuyong Li, Australia
Qingyong Li, China
Wu Li, China
Xiaoli Li, Singapore
Yu Li, France
Jiye. Liang, China
Chenglin Liu, China
Derong Liu, China
Zengliang Liu, China
Jiansu Luo, China
Ping Luo, China
Xudong Luo, China
Shaoping Ma, China
Pedro C. Marijuán, Spain
Shimin Meng, China
Zuqiang Meng, China
Eunika Mercier-Laurent, France
Duoqian Miao, China
Miec Owoc, Poland
He Ouyang, China
Gang. Pan, China
Yan Peng, China
Xi Peng, China
Germano Resconi, Italy
Fuji Ren, Japan
Paul S. Rosenbloom, USA
Qiangjin Shao, China
Yinghuan Shi, China
Yuhui Shi, China
Zhongzhi Shi, China
Katsunori Shimohara, Japan
Bailu. Si, China
Andreej Skowron, Poland
Fuchun Sun, China
Chaoli Sun, China
Huajin Tang, China
Lorna Uden, UK
S. Vadera, UK

Organizing Committee

Chairs

Xiaofeng Wang — Shanghai Maritime University and Shanghai Association for Noetic Science, China
Rigui Zhou — Shanghai Maritime University, China

Secretary General

Jin Liu — Shanghai Maritime University, China

Local Arrangements Chair

Miao Song — Shanghai Maritime University, China

Finance Chair

Han Zhang — Shanghai Maritime University, China
Email: hanzhang@shmtu.edu.cn

Publication Chair

Junbo Gao — Shanghai Maritime University, China

Publicity Chair

Min Liu — Shanghai Maritime University, China

International Liaison

Hong Zhu — Shanghai Maritime University, China
Email: hongzhu@shmtu.edu.cn

Keynote and Invited Presentations

Pattern Recognition by the Brain: Neural Circuit Mechanisms

Mu-ming Poo

Institute of Neuroscience, Chinese Academy of Sciences,
and CAS Center for Excellence in Brain Science and Intelligence Technology

Abstract. The brain acquires the ability of pattern recognition through learning. Understanding neural circuit mechanisms underlying learning and memory is thus essential for understanding how the brain recognizes patterns. Much progress has been made in this area of neuroscience during the past decades. In this lecture, I will summarize three distinct features of neural circuits that provide the basis of learning and memory of neural information, and pattern recognition. First, the architecture of the neural circuits is continuously modified by experience. This process of experience-induced sculpting (pruning) of connections is most prominent early in development and decreases gradually to a much more limited extent in the adult brain. Second, the efficacy of synaptic transmission could be modified by neural activities associated with experience, in a manner that depends on the pattern (frequency and timing) of spikes in the pre- and postsynaptic neurons. This activity-induced circuit alteration in the form of long-term potentiation (LTP) and long-term depression (LTD) of existing synaptic connections is the predominant mechanism underlying learning and memory of the adult brain. Third, learning and memory of information containing multiple modalities, e.g., visual, auditory, and tactile signals, involves processing of each type of signals by different circuits for different modalities, as well as binding of processed multimodal signals through mechanisms that remain to be elucidated. Two potential mechanisms for binding of multimodal signals will be discussed: binding of signals through converging connections to circuits specialized for integration of multimodal signals, and binding of signals through correlated firing of neuronal assemblies that are established in circuits for processing signals of different modalities. Incorporation of these features into artificial neural networks may help to achieve more efficient pattern recognition, especially for recognition of time-varying multimodal signals.

Interactive Granular Computing: Toward Computing Model for Turing Test

Andrzej Skowron

University of Warsaw and Systems Research Institute PAS
skowron@mimuw.edu.pl

The Turing test, as originally conceived, focused on language and reasoning; problems of perception and action were conspicuously absent.

— *Ch. L. Ortiz Jr. Why we need a physically embodied Turing test and what it might look like, AI Magazine 37(1) 55–62 (2016).*

Abstract. We extended Granular Computing (GrC) to Interactive Granular Computing (IGrC) by introducing complex granules (c-granules or granules, for short). They are grounded in the physical reality and are responsible for generation of the information systems (data tables) through interactions with the physical objects. These information systems are next aggregated as a part of networks of information systems (decision systems) in the search of relevant computational building blocks (patterns or classifiers) for initiating actions or plans or understanding behavioral patterns of swarms of c-granules (in particular, agents) satisfying the user's needs to a satisfactory degree. Agents performing computations based on interaction with the physical environment learn rules of behavior, representing knowledge not known a priori by agents. Numerous tasks of agents may be classified as control tasks performed by agents aiming at achieving the high quality computational trajectories relative to the considered quality measures over the trajectories. Here, e.g., new challenges are related to developing strategies for predicting and controlling the behavior of agents. We propose to investigate these challenges using the IGrC framework. The reasoning used for controlling of computations is based on adaptive judgment. The adaptive judgment is more than a mixture of reasoning based on deduction, induction and abduction. IGrC is based on perception of situations in the physical world with the use of experience. Hence, the theory of judgment has a place not only in logic but also in psychology and phenomenology. This reasoning deals with c-granules and computations over them. Due to the uncertainty the agents generally cannot predict exactly the results of actions (or plans). Moreover, the approximations of the complex vague concepts, e.g., initiating actions (or plans) are drifting with time. Hence, adaptive strategies for evolving approximations of concepts are needed. In particular, the adaptive judgment is very much needed in the efficiency management of granular computations, carried out by agents, for risk assessment, risk treatment, and cost/benefit analysis.

Keywords: (Interactive) Granular computing • Complex granule • Adaptive judgment • Interaction rule • Rough sets, Adaptive approximation of complex vague concepts and games • Risk management

Optimal Mass Transportation Theory Applied for Machine Learning

David Xianfeng Gu

Computer Science Department, State University of New York at Stony Brook, USA
gu@cs.stonybrook.edu

Abstract. Optimal mass transportation (OMT) theory bridges geometry and probability, it offers a powerful tool for modeling probability distributions and measuring the distance between distributions. Recently, the concepts and method of optimal mass transportation theory have been adapted into the field of Machine Learning. It interprets the principle of machine learning from different perspective and points out new direction for improving machine learning algorithms.

This work introduces the fundamental concepts and principles of optimal mass transportation, explains how to use OMT framework to represent probability distributions, measure the distance among the distributions, reduce dimensions, approximate the distributions. In general, the machine learning principles and characteristics are explained from the point of view of OMT.

Quantifying Your Brain and Identifying Brain Disease Roots

Jianfeng Feng[1,2]

[1] Warwick University, UK
[2] Fudan University, China

Abstract. With the available data of huge samples for the whole spectrum of scales both for healthy controls and patients including depression, autism and schizophrenia etc, we are in the position to quantify human brain activities such as creativity, happiness, IQ and EQ etc and search the roots of various mental disorders. With novel machine learning approaches, we first introduced functional entropy and entropy rate of resting state to characterize the dynamic behaivour of our brain. It is further found that the functional entropy is an increasing function of age, but a decreasing function of creativity and IQ. Its biological mechanisms are explored. With the brain wide associate study approach, for the first time in the literature we are able to identify the roots of a few mental disorders. For example, for depression, we found that the most altered regions are located in the lateral and medial orbitofrontal cortex for punishment and reward. Follow up rTMS at the lateral orbitofrontal cortex demonstrated the significant outcomes of the treatments.

Multi-objective Ensemble Learning and Its Applications

XinYao

Department of Computer Science and Engineering,
Southern University of Science and Technology, Shenzhen, 518055, China
xiny@sustc.edu.cn

Abstract. Multi-objective learning might be a strange concept as some people thought that the only objective in learning would be to maximise the generalisation ability of a learner. What else do you need, one might ask. It turns out that there are different perspectives and aspects to learning. First, machine learning is almost always formulated as an optimisation problem by defining a cost function, an energy function, an error function, or whatever-it-is-called function. Once this function is defined, the rest is the development (or the direct application) of an optimisation algorithm. Interestingly, most of such functions have more than one term. For example, it is not uncommon for such a function to include a loss term and a regularisation term, which then have to be balanced by a parameter (hyper-parameter). Much research work in the literature has been devoted to the setting and tuning of such a hyper-parameter. However, loss and regularisation are clearly two conflicting objectives from the perspective of multi-objective optimisation. Why not treat these two objectives separately so that we do not have to tune the hyper-parameter? The first part of this talk explains how this is done [1, 2] and what the potential advantages and disadvantages are. Second, measuring the error of a machine learner is not always as straightforward as one might think. Different people may use different metrics. It is not always clear how different metrics relate to each other and whether a learner performs well under one metric would look very poor according to a different metric. The second part of this talk gives an example in software effort estimation, where a more robust learner is trained using multi-objective learning, which performs well under different metrics [3]. The third and last part of this talk is devoted to two important areas in machine learning, i.e., online learning with concept drift [4] and class imbalance learning [5, 6]. Multi-objective learning in these two areas will be introduced.

References

1. Chandra, A., Yao, X.: Ensemble learning using multi-objective evolutionary algorithms. J. Math. Model. Algorithms **5**(4), 417–445 (2006)
2. Chen, H., Yao, X.: Multiobjective neural network ensembles based on regularized negative correlation learning. IEEE Trans. Knowl. Data Eng. **22**(12), 1738–1751 (2010)
3. Minku, L.L., Yao, X.: Software effort estimation as a multi-objective learning problem. ACM Trans. Softw. Eng. Methodol. **22**(4), 32 (2013). Article No. 35

4. Wang, S., Minku, L., Yao, X.; A multi-objective ensemble method for online class imbalance learning. In: Proceedings of the 2014 International Joint Conference on Neural Networks, IJCNN 2014, pp. 3311–3318. IEEE Press, July 2014
5. Wang, P., Emmerich, M., Li, R., Tang, K., Baeck, T., Yao, X.: Convex hull-based multi-objective genetic programming for maximizing receiver operating characteristic performance. IEEE Trans. Evol. Comput. **19**(2), 188–200 (2015)
6. Bhowan, U., Johnston, M., Zhang, M., Yao, X.: Reusing genetic programming for ensemble selection in classification of unbalanced data. IEEE Trans. Evol. Comput. **18**(6), 893–908 (2014)

On Intelligence: Symbiotic, Holonic, and Immunological Agents

Elizabeth Marie Ehlers

Academy of Computer Science and Software Engineering,
University of Johannesburg, South Africa
emehlers@uj.ac.za

Abstract. Currently, Artificial Intelligence has become a focal point of interest in the development of new and innovative applications in modern technology. The presentation will discuss and compare symbiotic, holonic, and immunological agents. Immunological agency will be considered in the context of biologically inspired Artificial Intelligence techniques as it is based on the advantageous adaptive properties of the vertebrate immune system. Furthermore, the discussion will explore how these types of agent can contribute to realizing intelligence in applications such as mentioned above.

Extreme Learning Machines (ELM) – Filling the Gap between Machine Learning and Biological Learning

Guang-Bin Huang

School of Electrical and Electronic Engineering,
Nanyang Technological University, Singapore

Abstract. One of the most curious in the world is how brains produce intelligence. Brains have been considered one of the most complicated things in the universe. Machine learning and biological learning are often considered separate topics in the years. The objectives of this talk are three-folds: (1) It will analyse the differences and relationships between artificial intelligence and machine learning, and advocates that artificial intelligence and machine learning tend to become different, they have different focus and techniques; (2) There exists some convergence between machine learning and biological learning; (3) Although there exist many different types of techniques for machine learning and also many different types of learning mechanism in brains, Extreme Learning Machines (ELM) as a common learning mechanism may fill the gap between machine learning and biological learning, in fact, ELM theories have been validated by more and more direct biological evidences recently. ELM theories actually show that brains may be globally ordered but may be locally random. ELM theories further prove that such a learning system happens to have regression, classification, sparse coding, clustering, compression and feature learning capabilities, which are fundamental to cognition and reasoning. This talk also shows how ELM unifies SVM, PCA, NMF and a few other learning algorithms which indeed provide suboptimal solutions compared to ELM.

How Can We Effectively Analyze Big Data in Terabytes or Even Petabytes?

Joshua Zhexue Huang

National Engineering Laboratory for Big Data System Computing Technology, Shenzhen University, Shenzhen, China
zx.huang@szu.edu.cn

Abstract. The big values in big data can only be dug out through deep analysis of data. In the era of big data, datasets with hundreds of millions objects and thousands of features become a phenomenon rather than an exceptional case. The internet and telecom companies in China have over hundred million customers. Such datasets are often in the size of terabytes and can easily exceed the size of the memory of the cluster system. Current big data analysis technologies are not scalable to such data sets because of the memory limitation. How can we effectively analyze such big data? In this talk, I will present a *statistical-aware strategy* to divide big data into data blocks which are distributed among the nodes of the cluster or even in different data centers. I will propose a *random sample partition data model* to represent a big data set as a set of distributed random sample data blocks. Each random sample data block is a random sample of the big data so it can be used to estimate the statistics of the big data and build a classification or prediction model for the big data. I will also introduce *an asymptotic ensemble learning framework* that stepwise builds ensemble models from selected random sample data blocks to model the big data. Using this set of new technologies, we will be able to analyze big data effectively without the memory limit. With this new architecture for big data, we are able to separate data analysis engines from data centers and make the data centers more accessible to big data analysis.

Cyborg Intelligent Systems

Gang Pan

Department of Computer Science, Zhejiang University, Hangzhou, China
gpan@zju.edu.cn

Abstract. Advances in multidisciplinary fields such as brain-machine interfaces, artificial intelligence, and computational neuroscience, signal a growing convergence between machines and biological beings. Especially, brain-machine interfaces enable a direct communication pathway between the brain and machines. It promotes the brain-in-loop computational paradigm. A biological-machine intelligent system consisting of both organic and computing components is emerging, which we called *cyborg intelligent systems*. This talk will introduce the concept, architectures, and applications of cyborg intelligent systems. It will also discuss issues and challenges.

Learning and Memory in Mind Model CAM

Zhongzhi Shi

Key Laboratory of Intelligent Information Processing, Institute of Computing Technology, Chinese Academy of Sciences, Beijing, 100190, China
shizz@ics.ict.ac.cn

Abstract. Intelligent science is the contemporary forefront interdisciplinary subject of brain science, cognitive science, artificial intelligence and other disciplines, which studies intelligence theory and technology. The one of its core issues is to build the mind model. Mind is the human spirit of all activities, is a series of cognitive abilities, which enable individuals to have consciousness, sense the outside world, think, make judgment, and remember things. Mind model CAM (Consciousness And Memory) is mainly composed of five parts, namely, memory, consciousness, high-level cognitive functions, perception input, behavior response. CAM is general intelligent system architecture. This lecture will mainly introduce the learning and memory mechanism of the mind model CAM, focusing on the physiological basis of memory, complementary learning system, learning and memory evolution and other issues. Human brain learning and memory is a comprehensive product of two complementary learning systems. One is the brain neocortex learning system, through the experience, slowly learning about knowledge and skills. The other is the hippocampus learning system, which memorizes specific experiences and allows these experiences to be replayed and thus effectively integrated with the new cortical learning system. Explore the complementary learning and memory of short - term memory and long - term memory in CAM. The essence of learning and memory evolution is through learning, not only to increase knowledge, but also to change the structure of memory.

Acknowledgements. This work is supported by the National Program on Key Basic Research Project (973) (No. 2013CB329502).

Factor Space and Artificial Intelligence

Peizhuang Wang

College of Intelligence Engineering and Math,
Technical University of Liaoning, China
peizhuangw@126.com

Abstract. Different from human intelligence, the subject of AI is not brain, but machine. How do machine emulate intelligence of brain? Is it possible to construct a brain-like machine? No matter how advanced science becomes, it is not possible to make a machine as a clone of brain. It is mystery that the insuperable barrier does not take away all belief from AI researchers. Even though the ebb of fifth generation computer in 1990s hints that computer must emulate from the structure of human brain, people still have confidence on AI facing the difficulty of structure-emulation. Indeed, we would cognize that brain is the cognition's subject, but not the very cognition. Is there a cognition theory keeping a little independence from brain? It concerns with the relationship between cognition information and ontological information. Even though brain has influence to ontology information, ontology information is independent from the subject of cognition essentially, and there exists inner cognition theory to guide artificial intelligence. There were theories arising in artificial intelligences, unfortunately, they are not deep and united but shallow and split. There have been no deep and united artificial intelligence theory yet. No a strong theory, no substantial practice! Therefore, we are going to build a strong theory of artificial intelligence. Factor space aims to build a mathematical theory of AI. Factor is a generalization of gene since mene was been called Mendelian factor. Gene is the root of a bunch of bio-attributes, factor is the root of a bunch of attributes for anything. Factor space is the generalization of Cartesian coordinate space with axes named by factors. Factor space provides united platform to do concept generation and implication, which is the mathematics of cognition. Which provides united theory for artificial intelligence, especially, for special information processing faced big data.

Contents

Theory of Intelligence Science

Ecological Methodology and Mechanism Approach

New Perspective to Intelligence Science

Yixin Zhong(✉)

School of Computer, Beijing University of Posts and Telecommunications, Beijing, China
zyx@bupt.edu.cn

Abstract. What would strongly appeal in the paper is that the ecological methodology and the mechanism approach, both are newly proposed by the author of the paper, be employed for the study of intelligence science. As presented, the unified theory of intelligence science is discovered and established due to the methodology and approach being adopted.

Keywords: Ecological methodology · Mechanism approach · Unified theory of intelligence

1 Introduction

The research on artificial intelligence, AI for short, has made a number of splendid progresses recently while also confronted with severe controversies. Some researchers declare that the great breakthrough in AI is coming in few decades and AI machines will take the domination over human beings. Other group of researchers, however, insists on the existence of huge difficulties in AI research in which many fundamental issues related to intelligence are still remained unclear, or even almost unknown [1]. This situation led the theme of intelligence science, instead of intelligence technology (or AI), to receive more and more attentions from academic circles worldwide.

It is well accepted that the first important issue one should take care in mind in the academic research is the question of whether the scientific methodology being adopted is appropriate or not. This is because of the fact that scientific methodology is the macroscopic guideline for research that will inspire the proper approach to the specific field of research. Therefore, whether or not the methodology is suitable for intelligence study will determine whether or not the research can be successful [2].

According to the principles mentioned above, we have noted that the methodology adopted in intelligence research has long been the reductionism featured with "divide and conquer" which had been an extremely powerful, and also successful, methodology in classical physics in modern times. This led AI research to the situation where a number of different, and mutually isolated, approaches being employed, namely the structuralism, the functionalism, and the behaviorism all in parallel and separate. As result, researchers employing different approaches have different understanding on

Z. Shi et al. (Eds.): ICIS 2017, IFIP AICT 510, pp. 3–9, 2017.
DOI: 10.1007/978-3-319-68121-4_1

what AI should be. In other words, reductionism is the major root that generates the controversies in AI [3].

In order to effectively improve the situation, we have carried out a long term investigations based on which we have summed up and would now like to propose a new methodology, named ecological methodology, for the intelligence science study from which a new approach to the intelligence research, termed mechanism approach, is also inspired [4].

The concepts of ecological methodology will be introduced in Sect. 2 and then the basic model of the typical intelligence process, based on ecological methodology, will be presented in Sect. 3. The mechanism approach to the study of intelligence science will be explained and compared with the other approaches in Sect. 4. The major results due to the new methodology and approach will be exposited in Sect. 5. Some conclusions will be given in the final section.

2 Ecological Methodology vs. Reductionism Methodology

2.1 Definition 1. Methodology

Methodology, scientific methodology more precisely, is not methods for dealing with specific problems but is a general term that refers to the general criterion, or the general guideline, for the proper selection of the specific methods to a field of study.

Therefore, the first, and also the most important, issue among other any things for scientific study should be making an effort in doing an assessment on whether the methodology to be employed is appropriate or not. Only the proper methodology employed could give guarantee to the success of the research.

One of the most influential, and also the most successful, methodologies in the history of natural sciences development is the "reductionism". It says that, a complex system can be divided into a number of sub-systems, which must be simpler and easier to handle, and that the solution for the system can then be obtained by summing up the ones of all sub-systems. So, this methodology can well be characterized as "divide and conquer".

Reductionism as a methodology has achieved numerous successes, particularly in the physical science study, or mechanical science study in modern times. The major characteristic for performing the reductionism, as was mentioned above, is that the complex problem must first be divided into a number of simpler, and mutually isolated, sub-problems and then each of the sub-problem be individually solved and finally all the sub-solutions are summed up. This caused no problem in classical physical science study.

However, if the system in consideration is a kind of information systems, the performing of reductionism will certainly cause serious problem. This is because of the fact that there exist complex information links among the sub-systems of the system and between the system and its environment and that all the information links are the 'life lines' of the information system. The 'life lines' of the information system may be cut-off when improperly dividing the system, hence leading the information system being "dead system".

Therefore, when dealing with the problems in information systems, intelligence systems in particular, which are usually complex systems, one should be very much careful on how to properly dividing on one hand and, on the other hand, should carefully seek for the new, and appropriate, methodology for recovering the life lines of the system.

2.2 Definition 2. Ecology and Ecosystem

As a kind of methodology, ecology is referred to the general guideline that emphasizes the study of interrelationships among the organisms of the system and between the system and their environment. All the interrelated organisms constitute the ecosystem.

Obviously, the methodology of ecology, or more conveniently the ecological methodology, is the methodology that is exactly what needed for the study of complex information systems, through which the life lines, or the interrelations, among all the parts of the ecosystem and between the ecosystem and its environment can hopefully be recovered. Consequently, the global properties and laws governing the complex information system can hence be recovered and studied.

It is known that intelligence is derived out from knowledge which in turn is from information. In other words, there will be no intelligence at all if there is no information provided.

Therefore, intelligence systems in all cases are complex information systems, or equivalently information processing systems, in nature. In other words, ecological methodology is the one needed, and suitable, for the study of intelligence science.

3 The Model of Intelligence Process in Perspective of Ecological Methodology

In accordance with the definition of ecological methodology and the investigation of various kinds of intelligence activities, it is found that the most typical intelligence process can be abstracted as the model in Fig. 1.

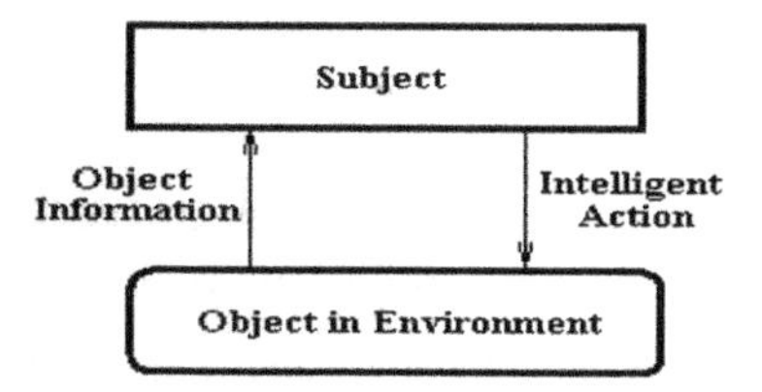

Fig. 1. Abstract model of typical intelligence process

As can be seen from Fig. 1 that the intelligence process must be within the framework of the subject-object interaction: First, the object in environment presents its object information to the subject. And then the subject handles the object information and produces, based on the results of the handling, an intelligence action that acts on the object in environment. The action must be intelligent enough. Otherwise the subject may suffer from risks of danger.

The process of producing intelligent action in Fig. 1 can be expressed in more detail. This leads to the model in Fig. 2.

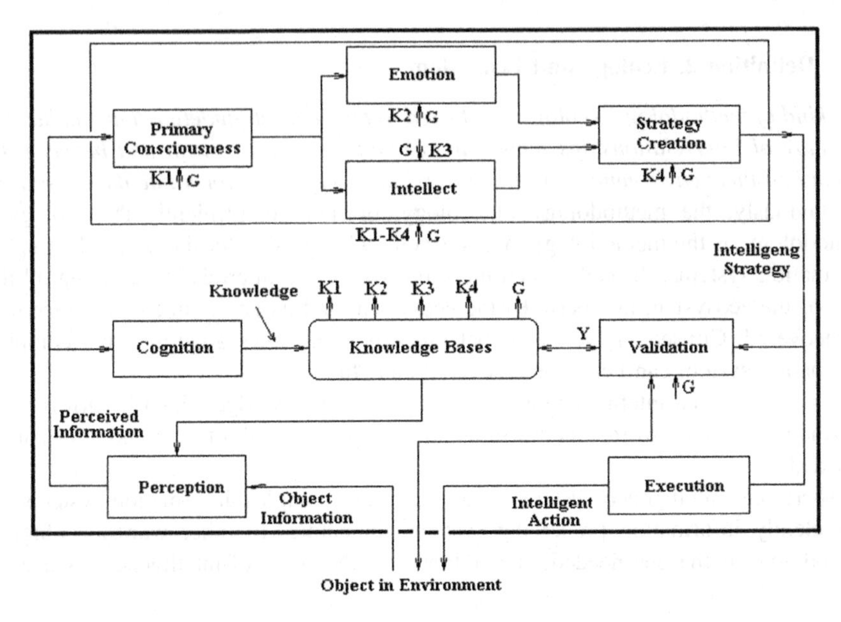

Fig. 2. The detailed version of Fig. 1

The entirety within the thick-line-box in Fig. 2 represents the subject who performs all the functions of intelligence: perception, cognition, strategy creation and strategy execution. The attributes that characterize the subject include various kinds of knowledge, K1 (innate and commonsense knowledge), K2 (empirical knowledge), K3 (regular knowledge), K4 (Art and other knowledge needed for strategy creation), and the goal for problem solving, G, all stored in the knowledge base at the center.

As can be seen from Fig. 2 that there exist various kinds of interrelations among the units of the subject and between the subject and the environment. The studies of the interrelations meet the requirements of ecological methodology.

More specifically, the major interrelations that the ecological methodology concerns the most include the followings (see in Fig. 2):

(1) The interrelation between the object in environment and the subject's unit of perception,
(2) The interrelation between the unit of perception and the unit of cognition,
(3) The interrelation between the unit of cognition and the unit of intelligence the latter of which consists of the unit of primary consciousness, the unit of emotion, the unit of intellect, and the unit of strategy creation,
(4) The interrelation between the unit of intelligence and the unit of execution, and
(5) The interrelation between the unit of execution and the object in environment.

4 Mechanism Approach vs. Other Approaches

The next issue is to develop the proper approach to the study of intelligence science based on the general guideline from the ecological methodology and the associated model. That means that we should understand and study these interrelations among the units of the system and their environment so that the laws and rules that govern the intelligence processes and activities can be established.

There have been many different approaches existed in history. Particularly in AI research, there are structuralism approach for dealing with neural networks, functionalism approach for dealing with physical symbol systems and expert systems, and behaviorism approach for dealing with sensor-motor systems. As mentioned above, these approaches are resulted from the reductionism methodology and can only explore some of the partial knowledge of the intelligence process.

Note that the mainstay in intelligence system is the series of interrelations among the units and their environment and that all the units of the subject cannot be treated separately from each other as mutually isolated ones. We should explore and discover the inherent mechanism that reasonably organizes all the units to fulfill the task of generating intelligence. This leads to the mechanism approach to intelligence science study.

To investigate the mechanism approach, let's look back the model in Fig. 2:

The function of interrelation (1) is to converse object information to perceived information; The function of interrelation (2) is to converse perceived information to knowledge; The function of interrelation (3) is to converse perceived information to intelligent strategy; The function of interrelation (4) is to converse the intelligent strategy to intelligent action; and The function of interrelation (5) is to apply intelligent action to the object in environment.

This can briefly be expressed as a series of conversion, that is,

(1) The conversion: object information → perceived information,
(2) The conversion: perceived information → knowledge,
(3) The conversion: knowledge → intelligent strategy,
(4) The conversion: intelligent strategy → intelligent action, and
(5) The conversion: intelligent action → object.

The series of conversion above can be expressed as "information → knowledge → intelligence", in which "object information" and "perceived information" are abstracted as "information" and similarly, "intelligent strategy" and "intelligent action" are abstracted as "intelligence".

Therefore, the most compact expression for the mechanism approach to intelligence science is the series of conversion:

$$\text{information} \rightarrow \text{knowledge} \rightarrow \text{intelligence} \tag{4.1}$$

The formula (4.1) means that, by employing mechanism approach to the study of intelligence science, the most important thing should be to have the information needed for the problem solving and to handle the series of the conversion from information to knowledge and further to intelligence, instead of the specific structure, functions, and behaviors of the system.

The formula (4.1) is the most important and significant result because of the fact that the system's structure, functions and behaviors must be determined by the mechanism of the intelligence system. In other words, mechanism of intelligence generation is the only dominating factor in the study of intelligence science whereas the system's structure, functions, and behaviors are the dominated factors.

5 Major Results Due to the Mechanism Approach

It is easy to see from the formula (4.1) that the essence of the mechanism approach can be named as "Information Conversion and intelligence Creation" provided that the following three conditions are given beforehand: (1) The object information that represents the problem to be solved, (2) The goal that indicates the final state for the problem solved, and (3) The knowledge that is needed for solving the problem in hand. This can also be clearly expressed in Fig. 3 below [4].

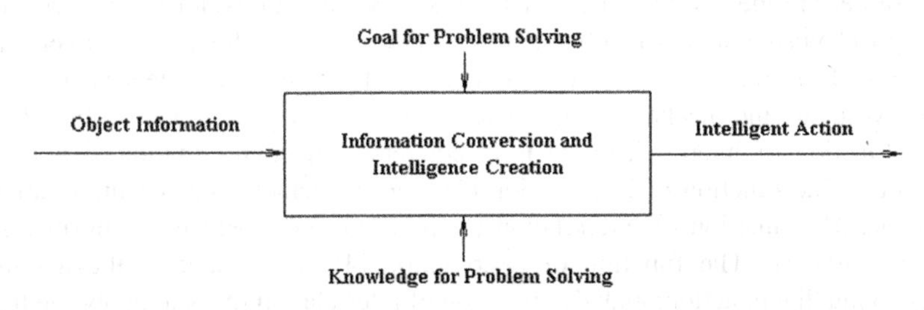

Fig. 3. Another expression for the formula (4.1)

The meaning one can see from Fig. 3 can be expressed in words that, as long as the three conditions are provided, the study of intelligence science is to perform the function of (4.1), the function of information conversion and intelligence creation, with the intelligent action being produced and executed.

If the result of the intelligent action is satisfied the problem is satisfactorily solved. Otherwise the error between the actual state and the goal state, as a kind of new information, should be feedback to the input of the system and the new knowledge should be learned and added for producing a new, and better, strategy. This "feedback-leaning-optimization" loop may be carried on a certain number of times until the result is satisfied.

This is the universal law as well as the unified theory of intelligence science.

6 Conclusions

(1) Scientific methodology is the first, and the most important, issue in scientific research. The reductionism methodology employed in intelligence research the past is not sufficient. The scientific methodology appropriate to the study of intelligence science should be the one of ecology, or more precisely the **methodology of information ecology**.

(2) The secondly important issue in any scientific research is to, under the guideline of the methodology employed, formulate an appropriate model representing the research. The model applied in AI in the past has been much too narrowly isolated. The appropriate models of typical intelligence process should be **the model of subject-object interaction** expressed in Figs. 1 and 2 respectively, instead of the model of brain only.

(3) The thirdly important issue in any scientific research is the approach selection, under the general guideline of the methodology employed and also based on the model selected. The structural approach, the functionalism approach, and the behaviorism approach applied in AI in the past are locally and shallowly valid. The universal approach to the study of intelligence science is the **mechanism approach expressed in the formula** (4.1).

(4) The mechanism approach to intelligence science study has been identified as the series of conversion: "object information → perceived information → knowledge → intelligent strategy → intelligent action", or more briefly "**information → knowledge → intelligence**" in which the conversions need to be supported by new logic [5] and new mathematics [6].

(5) The most interesting, and also most significant, result is the universal law governing the intelligence science that is the **Law of Information Conversion for Intelligence Creation** expressed in Figs. 2 and 3 which is derived from the methodology of information ecology, ecological model of intelligence process, and the mechanism approach. The law is universally valid, that is whenever the three conditions (the object information about the problem to be handled, the goal for the problem solving, and the knowledge needed for the problem solving) are provided, the task of intelligence science study is to perform the mechanism approach.

References

1. Zhong, Y.X.: AI: the unknown behind the hot wave. Science-Tech Review, No. 4 (2016)
2. Zhong, Y.X.: Principles of Information Science. BUPT Press, Beijing (1988)
3. Zhong, Y.X.: Principles of Advanced AI. Science Press, Beijing (2014)
4. Zhong, Y.X.: The Law of Information Conversion and Intelligence Creation. World Scientific Press, Singapore (2016)
5. He, H.C.: Principles of the Universal Logic. Science Press, Beijing (2005)
6. Wang, P.Z.: A factor space approach to knowledge representation. Fuzzy Sets Syst. **36**, 113–124 (1990)

Collaborative Model in Brain-Computer Integration

Zhongzhi Shi[1(✉)], Gang Ma[1,2], and Jianqing Li[1,3]

[1] Key Laboratory of Intelligent Information Processing, Institute of Computing Technology, Chinese Academy of Sciences, Beijing 100190, China
{shizz,mag,lijq}@ics.ict.ac.cn

[2] University of Chinese Academy of Sciences, Beijing 100049, China

[3] Shandong University, No. 27, Shan Da Nan Road, Jinan, Shandong, China

Abstract. Brain-computer integration is a new intelligent system based on brain-computer interface technology, which is integrated with biological intelligence and artificial intelligence. In order to make this integration effective and co-adaptive biological brain and machine should work collaboratively. ABGP-CNN based environment awareness and motivation driven collaboration will be presented in the paper. Motivation is the cause of action and plays important roles in collaboration. The motivation leaning method and algorithm will be explored in terms of event curiosity, which is useful for sharing common interest situations.

Keywords: Brain-computer integration · Environment awareness · ABGP-CNN model · Motivation driven collaboration · Motivation learning

1 Introduction

Brain-computer integration is a new intelligent system based on brain-computer interface technology, which is integrated with biological intelligence and artificial intelligence [1]. Brain-computer integration is an inevitable trend in the development of brain-computer interface technology. In the brain-computer integration system, the brain and the brain, the brain and the machine is not only the signal level of the brain machine interoperability, but also need to integrate the brain's cognitive ability with the computer's computing ability. But the cognitive unit of the brain has different relationship with the intelligent unit of the machine. Therefore, one of the key scientific issues of brain computer integration is how to establish the cognitive computing model of brain-computer integration.

At present, brain-computer integration is an active research area in intelligence science. In 2009, DiGiovanna developed the mutually adaptive brain computer interface system based on reinforcement learning [2], which regulates brain activity by the rewards and punishment mechanism. The machine adopts the reinforcement learning algorithm to adapt motion control of mechanical arm, and has the optimized performance of the manipulator motion control. In 2010, Fukayarna et al. control a mechanical car by extraction and analysis of mouse motor nerve signals [3]. In 2011,

Z. Shi et al. (Eds.): ICIS 2017, IFIP AICT 510, pp. 10–21, 2017.
DOI: 10.1007/978-3-319-68121-4_2

Nicolelis team developed a new brain-machine-brain information channel bidirectional closed-loop system reported in Nature [4], turn monkey's touch information into the electric stimulus signal to feedback the brain while decoding to the nerve information of monkey's brain, to realize the brain computer cooperation. In 2013, Zhaohui Wu team of Zhejiang University developed a visual enhanced rat robot [5]. Compared with the general robot, the rat robot has the advantage in the aspects of flexibility, stability and environmental adaptability.

Brain-computer integration system has three remarkable characteristics: (a) More comprehensive perception of organisms, including behavior understanding and decoding of neural signals; (b) Organisms also as a system of sensing, computation body and executive body, and information bidirectional exchange channel with the rest of the system; (c) Comprehensive utilization of organism and machine in the multi-level and multi-granularity will achieve system intelligence greatly enhanced.

The core of brain-computer integration is the cognitive computation model of brain computer collaboration. Cognitive process of brain computer collaboration is composed by the environment awareness, motivation analysis, intention understanding and action planning and so on, in support of the perception memory, episodic memory, semantic memory and working memory to complete brain machine group awareness and coordinated action. Supported by the National Program on Key Basic Research Project we are engaging in the research on Computational Theory and Method of Perception and Cognition of Brain-Computer Integration. The main goal of the project is the exploration of cyborg intelligence through brain-computer integration, enhancing strengths and compensating for weaknesses by combining the biological cognition capability with the computer computational capability. In order to make this integration effective and co-adaptive, brain and computer should work collaboratively. We mainly focus on four aspects, environment awareness, cognitive modeling, joint intention and action planning, to carry out the research of cognitive computational model.

(1) The environmental awareness of brain computer collaboration. For brain computer bidirectional information perception characteristics, the integration of visual features of the Marr visual theory and Gestalt whole perception theory in the wide range, research on the environment group awareness model and method by combination of brain and computer. The discriminative, generative and other methods are applied to analyze the features of environment perception information, mine perception information patterns and knowledge, generate high-level semantics, and understand well the environment awareness, build brain computer collaborative group awareness model.
(2) Cognitive modeling of brain computer collaboration. This involves combining the mutual cognitive characteristic of the brain computer, utilize and study the achievement made in intelligent science about consciousness, memory, study the cognitive cycle in brain computer collaborative information processing. According to the characteristic of the collaborative work of the brain computer, utilize the research results of physiological mechanism of brain information processing, study information representation method and reasoning mechanism in episodic memory and semantic memory of human brain, carry on cognitive modeling to information processing procedure in the brain computer collaboration.

(3) The joint intention driven by motivation. The essential characteristic of the brain computer collaborative work is that the agents have common goals, commitments, intentions, etc. that are jointly restrained. In order to describe the characteristics that should have for the brain computer system, study the joint intention theory to describe union restrictions of autonomous agents and reason balance of agent mental state; To the needing of brain computer collaborative work, study the essence of the behavior motivation and generation mechanism, put forward the intention model driven by motivation; Study the joint intention method for multi-agents, offer theory support for constructing the brain computer collaborative work of multi-agents.
(4) Action planning for brain computer collaboration. Under the support of ontology knowledge system, study action planning method of brain computer collaboration; using reinforcement learning and Markov decision theory, study the part of perception of the planning method, present the action planning theory with learnability and optimization methods.

Collaborations occur over time as organizations interact formally and informally through repetitive sequences of negotiation, development of commitments, and execution of those commitments. Both cooperation and coordination may occur as part of the early process of collaboration, collaboration represents a longer-term integrated process. Gray describes collaboration as a process through which parties who see different aspects of a problem can constructively explore their differences and search for solutions that go beyond their own limited vision of what is possible [6].

In this paper, a collaborative model for cyborg intelligence will be proposed in terms of external environment awareness and internal mental state. The collaboration between brain and computer is driven by motivation which is generated dynamically.

2 Conceptual Framework of Brain-Computer Integration

An effective approach to implementing engineering systems and exploring research problems in cyborg intelligence is based on brain-computer integration methods [7]. Using these methods, computers can record neural activity at multiple levels or scales, and thus decode brain representation of various functionalities, and precisely control artificial or biological actuators. In recent decades, there have been continuous scientific breakthroughs regarding the directed information pathway from the brain to computers. Meanwhile, besides ordinary sensory feedback such as visual, auditory, tactile, and olfactory input, computers can now encode neural feedback as optical or electrical stimulus to modulate neural circuits directly. This forms the directed information pathway from the computer to the brain. These bidirectional information pathways make it possible to investigate the key problems in cyborg intelligence.

How to interact between brain and computer at various is a critical problem in brain-computer integration. On the basis of the similarity between brain function partition and corresponding computing counterparts, a hierarchical and conceptual framework for brain-computer integration is proposed. The biological part and computing counterparts are interconnected through information exchange, and then cooperate to generate perception, awareness, memory, planning, and other cognitive functions.

For the brain part, abstracted the biological component of cyborg intelligence into three layers: perception and behavior, decision making, memory and consciousness (Fig. 1). We also divided the computer functional units into three corresponding layers: awareness and actuator, planning, motivation and belief layers. We also defined two basic interaction and cooperation operations: homogeneous interaction (homoraction) and heterogeneous interaction (heteraction). The former represents information exchange and function recalls occurring in a single biological or computing component, whereas the latter indicates the operations between the function units of both biological and computing parts. Homoraction is also modeled as the relationship between units within the same part. In the case of a single part in a brain-computer integration system, it will reduce to a biological body or computing device just with homoraction inside. Consequently, verifying the existence of heteraction is necessary for cyborg intelligent systems.

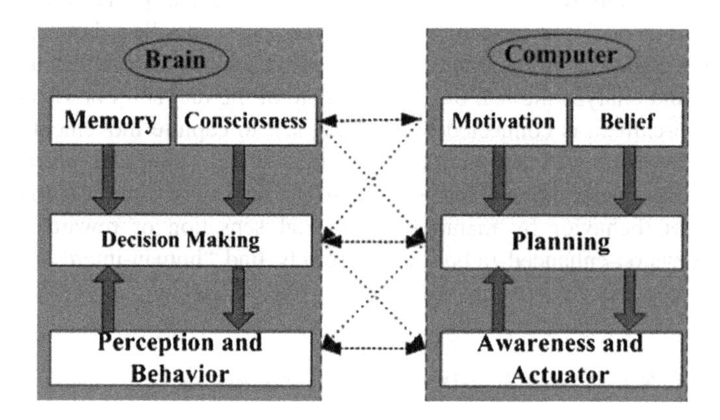

Fig. 1. Cognitive computational model

As typical brain-computer integration systems of “animal as the actuators”, rat cyborgs [8, 9], were developed to validate how the animals can be enhanced by the artificial intelligence. Ratbots are based on the biological platform of the rat with electrodes implanted in specific brain areas, such as the somatosensory cortex and reward area [10]. These electrodes are connected to a backpack fixed on the rat, which works as a stimulator to deliver electric stimuli to the rat brain. Figure 2 shows the physical implementation of the rat-robot navigation system [11]. In the automatic navigation of rats, five bipolar stimulating electrodes separately are implanted in medial forebrain bundle (MFB), somatosensory cortices (SI), and periaqueductal gray matter (PAG) of the rat brain. There is also a backpack fixed on the rat to receive the wireless commands.

There are two components which are necessary to implement the automatic navigation. Firstly, the communication between a computer and a rat needs to be solved. The stimulation signals are delivered by a wireless backpack stimulator which is

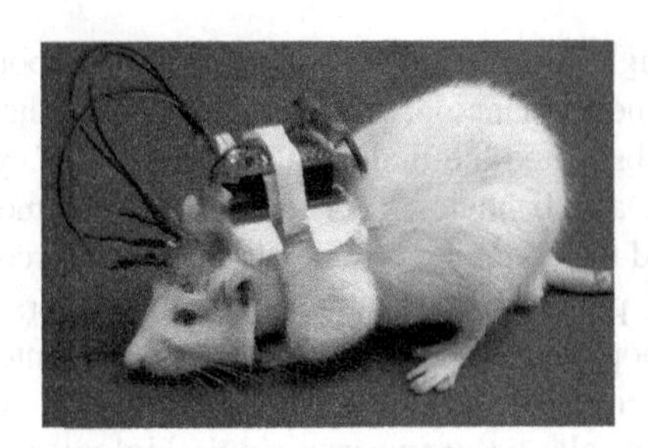

Fig. 2. Rat cyborg

comprised of stimulating circuit, control processor and Bluetooth transceivers. The control processor receives the computer instructions through the Bluetooth transceivers. Then it sends commands to the stimulator to control the rat behaviors. By receiving commands from the machine, the rat can perform a lot of navigation tasks, e.g. walking around mazes, climbing bridges, and stopping at a special place. Secondly, a video camera device used to capture the rat movement is installed above the scenario. With the video captured by the birdeye camera, the machine can establish a map of the environment and analyze the real time kinetic state of the rat [12]. For vison-enhanced ratbots, a mini-camera is connected to the backpack to capture movement or the surrounding environment. A computer analyzes video stream input and generates stimulation parameters that are then wirelessly sent to the backpack stimulator to control the rat's navigation behavior by manipulating virtual sensation or reward. Paper [13] reported that vision-enhanced ratbots can precisely find "human-interesting" objects, i.e., human faces and arrow signs, identified by object detection algorithms.

3 ABGP-CNN Based Environment Awareness

Agent can be viewed as perceiving its environment information through sensors and acting environment through effectors. As an internal mental model of agent, BDI model has been well recognized in philosophical and artificial intelligence area. As a practical agent existing in real world should consider external perception and internal mental state of agents. In terms of these considerations we propose a cognitive model through 4-tuple <Awareness, Belief, Goal, Plan>, and the cognitive model can be called ABGP model as shown in Fig. 3 [14].

Convolutional neural networks (CNN) is a multiple-stage of globally trainable artificial neural networks. CNN has a better performance in 2 dimensional pattern recognition problems than the multilayer perceptron, because the topology of the two-dimensional model is added into the CNN structure, and CNN employs three important structure features: local accepted field, shared weights, sub-sampling ensuring the invariance of the target translation, shrinkage and distortion for the input signal. CNN mainly consists of the feature extraction and the classifier. The feature extraction contains the multiple convolutional layers and sub-sampling layers. The classifier is consisted of one layer or two layers of full connected neural networks.

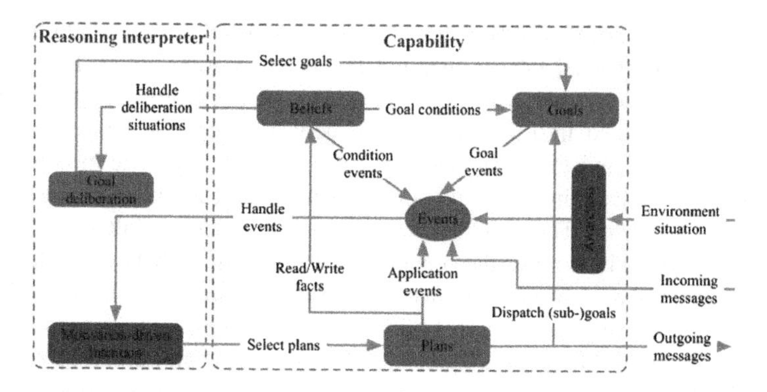

Fig. 3. Agent ABGP model

For the convolutional layer with the local accepted field and the sub-sampling layer with sub-sampling structure, they all have the character of sharing the weights

There are several methods developed for visual awareness [15]. Here we only describe CNN is used for visual awareness and construct agent model ABGP-CNN. For ABGP-CNN, the learning process of recognizing the natural scenes should mainly focus on how to train the CNN as its visual awareness module and how to build appropriate belief base, goals, and plans library. Training CNN includes what the multi-stages architecture is appropriate for the natural object recognition, what learning strategy is better.

4 Motivation Driven Collaboration

Motivation is an internal motive force and subjective reasons, which direct drive the individual activities to achieve a certain purpose, and the psychological state initiated and maintained by individual activities. Psychologists define motivation as the process that initiates, guides, and maintains goal-oriented behaviors. All kinds of behaviors and activities of the people can't be separated from the motivation.

Consider the dual nature of motivation, that is implicit and explicit, the motivation process is complexity. In general, implicit motivational processes are primary and more essential than explicit motivational processes. Here we only focus on explicit motivation and hypothesize that the explicit motivational representations consist mainly of explicit goals of an agent. Explicit goals provide specific and tangible motivations for actions. Explicit goals also allow more behavioral flexibility and formation of expectancies. In cyborg intelligent system we have developed two approaches for brain computer integration, that is, needs based motivation and curiosity based motivation.

4.1 Needs Based Motivation

In 1943, humanistic psychologist Maslow put forward the demand theory of motivation. Maslow's assumption that people in need, the sequence of human motivation, from the most basic physiological and safety needs, through a series of love and respect, the complex needs of self-realization, and need level has great intuitive appeal [16]. Green and others advocated that the theory of motivation is divided into 3 categories of physiology, behavior and social [17]. Merrick proposed that the motivation theory is divided into 4 categories, namely, the biological theory, cognitive theory, social theory and the combined motivation theory [18]. The biological motivation theory tries to explain the motivation of the work process based on the biological level of natural system. The mechanisms of these theories are often explained by energy and motion, which make the organism toward a certain behavior. The existing research on artificial system has been used to create the simulation of software agent and natural system using the theory of biological motivation.

Bach proposed the MicroPsi architecture of motivated cognition based on situated agents [19]. MicroPsi explores the combination of a neuro-symbolic cognitive architecture with a model of autonomous, polytelic motivation. The needs of MicoPsi cognitive system fall into three groups: physiological needs, social needs and cognitive [20]. Physiological needs regulate the basic survival of the organism and reflect demands of the metabolism and physiological well-being. Social needs direct the behavior towards other individuals and groups. They are satisfied and frustrated by social signals and corresponding mental representations. Cognitive needs give rise to open-ended problem solving, skill-acquisition, exploration, play and creativity. Urges reflect various physiological, social and cognitive needs. Cognitive processes are modulated in response to the strength and urgency of the needs.

According to brain computer integration requirements, a motivation could be represented as a 3-tuples $\{N, G, I\}$, where N means needs, G is goal, I means the motivation intensity [21].

A motivation is activated by motivational rules which structure has following format:

$$\mathrm{R} = (P, D, \ \mathrm{Strength}(P|D))$$

where, P indicates the conditions of rule activation; D is a set of actions for the motivation; Strength($P|D$) is a value within interval [0, 1].

4.2 Curiosity Based Motivation

Curiosity based motivation is through motivation learning algorithm to build a new motivation. Agent creates internal representations of observed sensory inputs and links them to learned actions that are useful for its operation. If the result of the machine's action is not relevant to its current goal, no motivation learning takes place. This screening of what to learn is very useful since it protects machine's memory from storing unimportant observations, even though they are not predictable by the machine and may be of sufficient interest for novelty based learning. Novelty based learning still can take place in such a system, when the system is not triggered by other motivations.

Motivation learning requires a mechanism for creating abstract motivations and related goals. Once implemented, such a mechanism manages motivations, as well as selects and supervises execution of goals. Motivations emerge from interaction with the environment, and at any given stage of development, their operation is influenced by competing event and attention switching signals.

The learning process for motivations to obtain the sensory states by observing, then the sensed states are transformed mutually by the events. Where to find novelty to motivate an agent's interestingness will play an important role. Once the interestingness is stimulated, the agent's attention may be selected and focused on one aspect of the environment. Therefore, it will be necessary to define observations, events, novelty, interestingness and attention before descripting the motivation learning algorithm.

Definition 1: ***Observation Functions***
Observation functions define the combinations of sensations from the sensed state that will motivate further reasoning. Observations containing fewer sensations affect an agent's attention focus by making it possible for the agent to restrict its attention to a subset of the state space. Where, a typical observation function can be given as:

$$\mathbf{O}_{S(t)} = \left\{ \left(o_{1(t)}, o_{2(t)}, \cdots, o_{L(t)}, \cdots \right) \middle| o_{L(t)} = s_{L(t)} (\forall L) \right\} \tag{1}$$

The equation defines observation function $\mathbf{O}_{S(t)}$ in which each observation focuses on every element of the sensed state at time t.

Definition 2: ***Difference Function***
A difference function Δ assigns a value to the difference between two sensations $S_{L(t)}$ and $S_{L(t')}$ in the sensed states $S_{(t)}$ and $S_{(t')}$ as follows:

$$\Delta\left(s_{L(t)}, s_{L(t')}\right) = \begin{cases} s_{L(t)}; & \text{if } \neg\exists s_{L(t')} \\ s_{L(t')}; & \text{if } \neg\exists s_{L(t)} \\ s_{L(t)} - s_{L(t')}; & \text{if } s_{L(t)} - s_{L(t')} \neq 0 \\ 0 & \text{otherwise} \end{cases} \tag{2}$$

Difference function offers the information about the change between successive sensations it calculates the magnitude of the change.

Definition 3: ***Event Function***
Event functions define which combinations of difference variables an agent recognizes as events, each of which contains only one non-zero difference variable. Event function can be defined as following formula:

$$\mathbf{E}_{S(t)} = \left\{ E_{L(t)} = \left(e_{1(t)}, e_{2(t)}, \cdots, e_{L(t)}, \cdots \right) \middle| e_{e(t)} \right\} \tag{3}$$

where,

$$e_{e(t)} = \begin{cases} \Delta\left(S_{e(t)}, S_{e(t')}\right); & \text{if } e = L \\ 0 & \text{otherwise} \end{cases} \tag{4}$$

Events may be of varying length or even empty, depending on the number of sensations to change.

Definition 4: ***Novelty Detection Function***
The novelty detection function, **N**, takes the conceptual state of the agent, $c \in \mathbf{C}$, and compares it with memories of previous experiences, $m \in \mathbf{M}$, constructed by long term memory to produce a novelty state, $n \in \mathbf{N}$:

$$\mathbf{N} : \mathbf{C} \times \mathbf{M} \rightarrow \mathbf{N} \tag{5}$$

Novelty can be detected by introspective search comparing the current conceptual state of an agent with memories of previous experiences [14].

Definition 5: ***Interestingness Function***
The interestingness function determines a value for the interestingness of a situation, $i \in \mathbf{I}$, basing on the novelty detected, $n \in \mathbf{N}$:

$$\mathbf{I} : \mathbf{N} \rightarrow \mathbf{I} \tag{6}$$

Definition 6: *Attention Selection*
Selective attention enables you to focus on an item while mentally identifying and distinguishing the non-relevant information. In cyborg we adopt maximal interestingness strategy to select attentions to create a motivation.

The following describes the basic steps of novelty based motivation learning and goal creation algorithm in the cyborg system.

Motivation learning algorithm

(1) Observe $\mathbf{O}_{S(t)}$ from $\mathbf{S}_{(t)}$ using the observation function
(2) Subtract $\mathbf{S}_{(t)} - \mathbf{S}_{(t')}$ using the difference function
(3) Compose $\mathbf{E}_{S(t)}$ using the event function
(4) Look for $\mathbf{N}_{(t)}$ using introspective search
(5) Repeat (for each $N_i(t) \in \mathbf{N}(t)$)
(6) Repeat (for each $I_j(t) \in \mathbf{I}(t)$)
(7) *Attention* = max $I_j(t)$
(8) Create a *Motivation* by *Attention*.

4.3 Motivation Execution

In cyborg system, the realization of the motivation module is through agent model ABGP. The current belief of the belief memory storage contains the agent motivation base. A desire is a goal or a desired final state. Intention is the need for the smart body to choose the current implementation of the goal. In agent, the goal is a directed acyclic graph by the sub goal composition, and realizes in step by step. According to a directed acyclic graph a sub goal is represented by a path to complete, the total goal will finish when all sub goals are completed.

4.4 Collaboration

In brain-computer integration rat brain should work with machine collaboratively. Here rat brain and machine can be abstracted as agent, so the collaboration can be viewed as joint intention. Joint intention is about what the team members want to achieve. Each team member knows the intention specifically and achieves it by collaboration [22].

In the joint intention theory, a team is defined as "a set of agents having a shared objective and a shared mental state". The team as a whole holds joint intentions, and each team member must inform others whenever it detects the goal state change, such as goal is achieved or the goal is no longer relevant.

For the joint intention, rat agent and machine agent have three basic knowledge: first, each one should select its intention; second, each one knows its cooperator who also select the same intention; and last, each one knows they are a team. They can know each other through agent communication.

5 Simulation Experiments

ABGP-CNN as the detailed implementation for the conceptual framework of brain-computer integration, here we give a simulation application to significantly demonstrate feasibility of conceptual framework of brain-computer integration based ABGP-CNN Agent model. The following will mainly represent the actual design of the rat agent based on ABGP-CNN supported by the conceptual framework of brain-computer integration.

Under belief knowledge conditions, the goals (here mainly visual information) constantly trigger the awareness module to capture environment visual information, and the event module converts the visual information into the unified internal motivation signal events which are transferred to action plan module. Then the action plan module will select proper actions to response the environment.

In simulation application, we construct a maze and design a rat agent based on ABGP-CNN to move in the maze depending on the guidepost of maze path in Fig. 4. The task of the rat agent is to start moving at the maze entrance (top-left of maze), and finally reach the maze exit (bottom right of maze) depending on all guideposts.

In order to fulfill the maze activity, the rat agent is implemented all the three basic modules, <Awareness>, <Motivation>, <Action Plan>. In the rat maze activity experiment, the rat agent is designed to have 3 basic motivation behaviors moving on, turning left and turning right in the maze. In order to guide rat's behaviors we construct a true traffic guidepost dataset means 3 different signals, moving on, turning left and turning right. The different signal corresponds to different guidepost images like in Fig. 5.

When rat agent moves on the path, its goals constantly drive awareness module to capture environment visual information (here guideposts in the maze) and generate the motivation signal events to drive its behaviors plan selection. In the experiment, there are 3 motivation signals, moving on, turning left and turning right according to the guideposts in the maze path. Which means the agent can response 3 types of action plans to finish the maze activities.

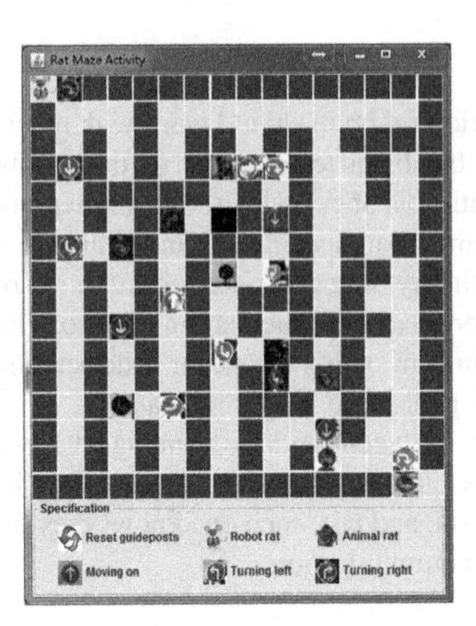

Fig. 4. Rat agent activities in maze

Fig. 5. Traffic guideposts in maze

6 Conclusions

This paper described the collaborative model of brain-computer integration, which is a new intelligent system based on brain-computer interface technology. In order to make this integration effective and co-adaptive biological brain and computer should work collaboratively. ABGP-CNN based environment awareness and motivation driven collaboration have been proposed in the paper. Motivation is the cause of action and plays important roles in collaboration. The motivation leaning method and algorithm has been explored in terms of event curiosity, which is useful for sharing common interest situations.

The future of brain-computer integration may lead towards many promising applications, such as neural intervention, medical treatment, and early diagnosis of some neurological and psychiatric disorders. The goal of artificial general intelligence (AGI) is the development and demonstration of systems that exhibit the broad range of general intelligence. The brain-computer integration is one approach to reach the AGI. A lot of basic issues of brain-inspired intelligence are explored in the book [1] in detail.

Acknowledgements. This work is supported by the National Program on Key Basic Research Project (973) (No. 2013CB329502), National Natural Science Foundation of China (No. 61035003), National Science and Technology Support Program (2012BA107B02).

References

1. Shi, Z.: Mind Computation. World Scientific Publishing Co. Pte. Ltd., Singapore (2017)
2. DiGiovanna, J., Mahmoudi, B., Fortes, J., et al.: Coadaptive brain-machine interface via reinforcement learning. IEEE Trans. Biomed. Eng. **56**(1), 54–64 (2009)
3. Fukuyama, O., Suzuki, T., Mabuchi, K.: RatCar: a vehicular neuro-robotic platform for a rat with a sustaining structure of the rat body under the vehicle. In: Annual International Conference of the IEEE Engineering in Medicine and Biology Society (2010)
4. O'Doherty, J.E., Lebedev, M.A., Ifft, P.J., et al.: Active tactile exploration using a brain-machine-brain interface. Nature **479**(7372), 228–231 (2011)
5. Wang, Y.M., Lu, M.L., Wu, Z.H., et al.: Ratbot: a rat "understanding" what humans see. In: International Workshop on Intelligence Science, in conjunction with IJCAI-2013, pp. 63–68 (2013)
6. Gray, B.: Collaborating: Finding Common Ground for Multiparty Problems. Jossey-Bass, San Francisco (1989)
7. Wu, Z., Zhou, Y., Shi, Z., Zhang, C., Li, G., Zheng, X., Zheng, N., Pan, G.: Cyborg intelligence: research progress and future directions. IEEE Intell. Syst. **31**(6), 44–50 (2016)
8. Berger, T.W., et al.: A cortical neural prosthesis for restoring and enhancing memory. J. Neural Eng. **8**(4) (2011). doi:10.1088/1741-2560/8/4/046017
9. Wu, Z., Pan, G., Zheng, N.: Cyborg intelligence. IEEE Intell. Syst. **28**(5), 31–33 (2013)
10. Feng, Z., et al.: A remote control training system for rat navigation in complicated environment. J. Zhejiang Univ. Sci. A **8**(2), 323–330 (2007)
11. Yu,Y., Zheng, N., Wu, Z., et al.: Automatic training of ratbot for navigation. In: International Workshop on Intelligence Science, in Conjunction with IJCAI 2013, Beijing, China (2013)
12. Wu, Z., Zheng, N., Zhang, S., Zheng, X., Gao, L., Su, L.: Maze learning by a hybrid brain-computer system. Sci. Rep. **6** (2016). Article no. 31746. doi:10.1038/srep31746
13. Wang, Y., et al.: Visual cue-guided rat cyborg for automatic navigation. IEEE Comput. Intell. **10**(2), 42–52 (2015)
14. Shi, Z., Zhang, J., Yue, J., Yang, X.: A cognitive model for multi-agent collaboration. Int. J. Intell. Sci. **4**(1), 1–6 (2014)
15. Ma, G., Yang, X., Lu, C., Zhang, B., Shi, Z.: A visual awareness pathway in cognitive model ABGP. High Technol. Lett. **22**(4), 395–403 (2016)
16. Maslow, A.H.: Motivation and Personality. Addison-Wesley, Boston (1954, 1970, 1987)
17. Green, R.G., Beatty, W.W., Arkin, R.M.: Human Motivation: Physiological, Behavioral and Social Approaches. Allyn and Bacon Inc., Boston (1984)
18. Merrick, K.E.: Modelling motivation for experience-based attention focus in reinforcement learning. Thesis, The University of Sydney (2007)
19. Bach, J.: Principles of Synthetic Intelligence – An architecture of motivated cognition. Oxford University Press, Oxford (2009)
20. Bach, J.: Modeling motivation in MicroPsi 2. In: Bieger, J., Goertzel, B., Potapov, A. (eds.) AGI 2015. LNCS, vol. 9205, pp. 3–13. Springer, Cham (2015). doi:10.1007/978-3-319-21365-1_1
21. Shi, Z., Zhang, J., Yue, J., Qi, B.: A motivational system for mind model CAM. In: AAAI Symposium on Integrated Cognition, Virginia, USA, pp. 79–86 (2013)
22. Shi, Z., Zhang, J., Yang, X., Ma, G., Qi, B., Yue, J.: Computational cognitive models for brain–machine collaborations. IEEE Intell. Syst. **29**(6), 24–31 (2014)

Entanglement of Inner Product, Topos Induced by Opposition and Transformation of Contradiction, and Tensor Flow

Jiali Feng(✉)

Information Engineering College, Shanghai Maritime University, Shanghai 201306, China
jlfeng@189.cn

Abstract. The law of unity of opposites, the mechanism of mutual change of quality and quantity, and the rule of dialectical transformation have become the key fundamental problems that need to be addressed in Intelligence Science. It is shown that the spatial-time position $x_t(u)$ of object u is the attribute describing where u is existing, the distance $d(x_t(u), y_t(v))$ between u and its contradiction v is the expressing relation distinguishing u from v. Based on the mechanism of distance vary with position change Δx_t was controlled by the law of unity of opposites, such that the description of the law can be transformed into a physical problem. By mean of three equivalent definition of distance, some of mathematical construction for describing physical move of u and v, such as Polarization Vector of Inner Product, Entangled Circle, Entangled Coordinates and Clifford Algebra can be induced, such that the Entangled relation both u and v can be transformed into a mathematical problem. The spatial-time position collection $\{z(x_t, y_t)\}$ with the collection of time arrows and the displacement arrows $(\Delta x_t, \Delta y_t, \Delta t)$ constructs a category E. A quantity x_t belong to a corresponding quality $q_v(u)$, during x_t varies with time change Δt in the qualitative criterion $[x_i, x_T)$, the mechanism that $q_v(u)$ is maintaining the same can be abstracted to be a Qualitative Mapping $\tau(x_t, [x_i, x_T))$ from a quantity x_t into quality $q_v(u)$, and the regulation of different quantity convert into different quality of new quality, can be represented by the Degree Function of Conversion $\eta(x_t)$, a Cartesian Closed Category can be gotten by $\tau(x_t, [x_i, x_T))$ and $\eta(x_t)$. A subobject classifier can be induced by the mechanism of a quality is changing to a simple (or non-essential) quality, so an Attribute Topos can be achieved by them. Tensor Flow and a Fixation Image Operator, and an approach for Image Thought has been presented, and their applications in Noetic Science and Intelligent Science are discussed too.

Keywords: Entanglement of inner product · Attribute Topos · Qualitative mapping function of conversion degree · Tensor Flow · Fixation Image Operator · Intelligence Science · Noetic Science · Meta Synthetic Wisdom

1 Introduction

The question: "Can Machines Think?" not only involves the basic contradiction between "spirit" and "substance" in philosophy, but also a chain of secondary contradictions induced by it, such that the law of unity of opposites between an object u

Z. Shi et al. (Eds.): ICIS 2017, IFIP AICT 510, pp. 22–36, 2017.
DOI: 10.1007/978-3-319-68121-4_3

and its contradiction object v, the mechanism of mutual change of quality and quantity, and the rule of dialectical transformation, i.e. so-called "law of unity of opposites and dialectic transformation" have become the key fundamental problems that need to be addressed in Noetic Science, Intelligence Science and Theory of Meta Synthetic Wisdom.

If the human brain is defined as an organ processing information, then two basic questions would be raised:

(1) What information is processed by human brain?
(2) What processes are taken by human brain on the information?

There are at least three answers for these two questions in cognition science:

(1) Symbol information and symbol operation;
(2) Neural signal information and neural or brain chemical operation;
(3) Stimuli information and feedback.

There are three school based on the three answers respectively in cognition science and artificial intelligence:

(1) Symbol School; (2) Connective or Structure School; (3) Behave or Control School.

Could the three approaches be united?

There are three answers:

(1) Minsky: they can't be unified [1];
(2) Prof. Zhong proposed that the various existing AI approaches can all be united within the framework of mechanism approach, and that Intelligence Science can be created [2]. Jiali Feng show that Zhong's Mechanism Approach can be implementedby using the Qualitative Mapping [3].
(3) Tsien, the Famous Scientist of China, not only proposed Noetic Science and Meta Synthetic Wisdom, but also suggested an implementing Scheme for MSW, noted by Quantity Wisdom ⊗ Image Wisdom ⇒ Quality Wisdom (MSW) as shown in Fig. 2. [4], as shown in Fig. 2(c). A Quantitative Method ⊗ Fixation Image Method ⇒ Qualitative Method (MSW) for MSW is proposed by Attribute Theory [5].

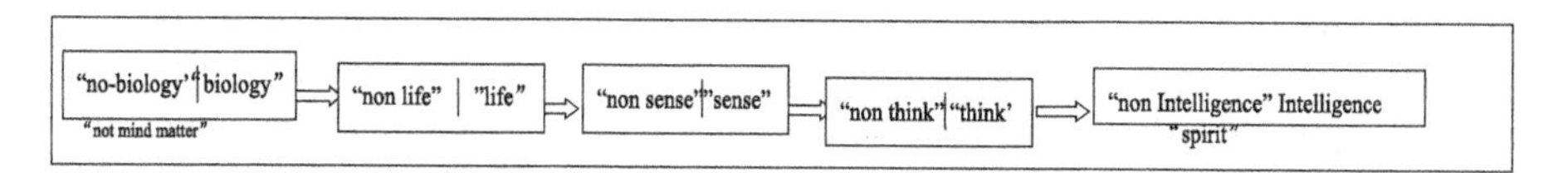

Fig. 1. Fig. 1 basic contradiction between "matter" and "spirit" and a chain of secondary contradictions

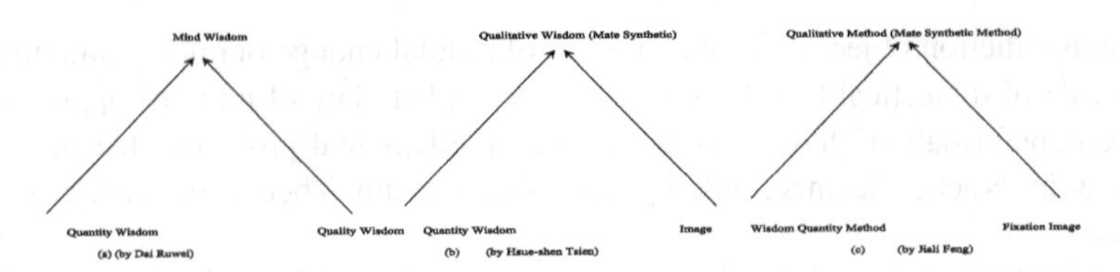

Fig. 2. (a) Quantity Wisdom ⊗ Quality Wisdom ⇒ Mind Wisdom (by Dai Ruwei) (b) Quantity Wisdom ⊗ Image Wisdom ⇒ Qualitative Wisdom (Mate Synthetic Wisdom) (by Hsue-shen Tsien) (c) Quantity Method ⊗ Fixation Image ⇒ Qualitative Method (Mate Synthetic Method (by Jiali Feng)

Based on the basic fact that the information received by human brain is, and only is the attributes of the objects, two basic hypothesis are proposed in this paper as follow:

Two Basic Hypothesis of Attribute Theory:

(1) Human thinking can be constructed by some of processing attributes, such as, receiving, interpretation, and coding of attributes.
(2) The mechanism of thinking construction and intelligent simulation can be considered as the mechanism of processing attribute.

If the two hypothesis are true, then we should study the following three basic questions:

(1) What are the attributes?
(2) How is an attribute and its transformation received, deciphered, interpreted and coded by brain?
(3) How can the operators of Human Brain be represented by Mathematics?

In philosophy, an attribute is defined to be as follow [6]:

Definition 1: An attribute is an expressing quality of an object when an interaction between the object and another object is happening.

As well-known that the spatial-time position of object u $z_1(t,u)$, is the attribute showing where u existing, the distance $d(z_1,z_2)$ between u and its contradiction v is the expressing relation distinguishing that u from v, based on the fact, the intrinsic quality $q_v(u)$ of u distinguish from its contradicted object v can be defined by the corresponding relation "$u \neq v \Leftrightarrow z_1(t,u) \neq z_2(t,v) \Leftrightarrow d(z_1,z_2) \neq 0$". Based on the mechanism of $d(z_1,z_2)$ varies with time change Δt was controlled by law of the law of unity of opposites about the contradictory movements, such that the philosophy question: "when the $d(z_1,z_2)$ vary with time change Δt, whether the intrinsic quality $q_v(u)$ of object u keeps itself or not?" can be transform into a physical problem.

By mean of three equivalent definition of distance $d(z_1,z_2)$, the Polarization Vector γ_c, γ_d and γ of Inner Product $z_1 \cdot z_2$, Entangled Vector, Entangled Circle $\odot O$, Entangled Coordinates and Clifford Algebra and so on for describing physical move of u and v, can be induced.

It is shown that if $\frac{z_1}{|z_1|}$ and $\frac{z_2}{|z_2|}$ are commeasurable each other, then Inner Product $z_1 \cdot z_2$ can be measured as an integer that can be decomposed a product of some of

Prime Factors, and a module can be deduced. But if $\frac{z_1}{|z_1|}$ and $\frac{z_2}{|z_2|}$ are non-commeasurable each other, then a measure system which value of unity continuously variable like the Fourier Orthogonal Base is necessary for the measure of Inner Product, and the requirement of value of distance $d(z_1, z_2)$ is an integer is the foundation for orbit quantization.

Let $\Delta t_{it} = t - i$ be the time change, $\Delta z(\Delta t_{it}) = (\Delta x(\Delta t_{it}), \Delta y(\Delta t_{it}))$ the displacement of u and v vary with Δt_{it}, then $\mathrm{z_t}(u,v) = (x_t(u), y_t(v)) = \mathrm{z_i}(u,v) + \Delta z(\Delta t_{it}) = (\Delta x(\Delta t_{it}), \Delta y(\Delta t_{it}))$, and a category E whose object collection is $[i, T) \times [z_i, z_T)$, the arrow collection is $\{(\Delta t_{it}, \Delta x_{it}, \Delta y_{it})\}$ can be constructed.

The mechanism that the quality $q_v(u)$ is maintaining the same, during the quantity varies with time change in its qualitative criterion, can be abstracted as the Qualitative Mapping $\tau(x_t, [x_i, x_T))$ from a quantity into its corresponding quality, and the regulation of different quantity convert into different quality of new quality, can be represented by the Degree Function of Conversion $\eta(x_t)$, such that a Cartesian Closed Category can be gotten by $\tau(x_t, [x_i, x_T))$ and $\eta(x_t)$. A subobject classifier can be induced by the mechanism of a quality is changing to a simple (or non-essential) quality, so an Attribute Topos can be achieved by them.

Because a Hilbert Space $\mathcal{H}$ can be expanded by the Family of Qualitative Mappings, such that not only the degree of conversion function can be represented by a linear combination of base, but also a Tensor Flow can be induced by a functor F from the Base of H to Base of H' with the time stream. A Fixation Image Operator, and an approach for Image Thought has been presented, some of applications in Noetic Science and Intelligent Science are discussed.

2 Entanglement Vectors Induced by Polarization Identity of Inner Product of Both Vectors z_1 and z_2

Let Z be the plan determined by vectors z_1 and z_2, and $z_1 = (x_1, y_1)$ and $z_2 = (x_2, y_2)$, the distance is defined the square root of inner product $z_1 \cdot z_2$ of z_1 and z_2

$$d(z_1, z_2) = \sqrt{z_1 \cdot z_2} = \sqrt{(x_2 - x_1)^2 + (y_2 - y_1)^2} = |z_1||z_2|cos\theta \tag{1}$$

But there is a Polarization Identity of inner product $z_1 \cdot z_2$

$$z_1 \cdot z_2 = \left(\frac{z_1 + z_2}{2}\right)^2 - \left(\frac{z_1 - z_2}{2}\right)^2 = \left(\frac{z_1}{2} + \frac{z_2}{2}\right)^2 - \left(\frac{z_1}{2} - \frac{z_2}{2}\right)^2 = \left(\frac{z_1}{2} + \frac{z_1}{2}\right)\left(\frac{z_2}{2} + \frac{z_2}{2}\right) \tag{2}$$

If let $\mathrm{A} = \mathrm{A}\left(\frac{x_1 + x_2}{4}, \frac{y_1 + y_2}{4}\right)$, $\mathrm{B} = \mathrm{B}\left(\frac{x_1 + x_2}{2}, \frac{y_1 + y_2}{2}\right)$, $\odot$A the circle which diameter equal to $d_{\odot \mathrm{A}} = |OB|$, $\odot$B the circle which center is B, its radius equal to $r_{\odot \mathrm{B}} = \frac{z_1 - z_2}{2}$. It is shown in Fig. 3(a) that there exist two Intersection Points of $\odot$A and of $\odot$B, $C(c_1, c_2)$ and $D(d_1, d_2)$, their c_1 and c_2 are the horizontal coordinate of C and the vertical coordinate of C, respectively, d_1 and d_2 of $D(d_1, d_2)$ as well. Since $BC \perp OC$, we have

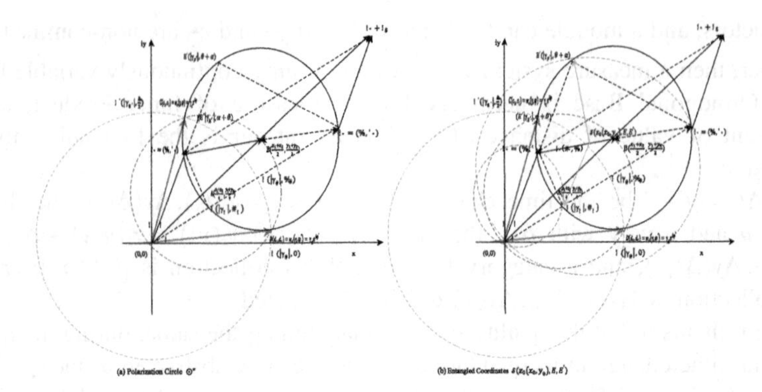

Fig. 3. (a) Polarization Circle $\odot O$ of Inner Product $z_1 \cdot z_2$ (b) Entangled Coordinates $\mathcal{S}(z_0(x_0, y_0), E, E')$

$$\begin{cases} \mathrm{OC}^2 = \mathrm{OB}^2 - \mathrm{BC}^2 = \left(\frac{z_1+z_2}{2}\right)^2 - \left(\frac{z_1-z_2}{2}\right)^2 = z_1 \cdot z_2 = |z_1||z_2|cos\theta \\ \mathrm{OD}^2 = \mathrm{OB}^2 - \mathrm{BD}^2 = \left(\frac{z_1+z_2}{2}\right)^2 - \left(\frac{z_1-z_2}{2}\right)^2 = z_1 \cdot z_2 = |z_1||z_2|cos\theta \end{cases} \tag{3}$$

But c_1 and c_2 of $C(c_1, c_2)$ are the solution of following two circles equations:

$$\begin{cases} \left(c_1 - \frac{x_1+x_2}{2}\right)^2 + \left(c_2 - \frac{y_1+y_2}{2}\right)^2 = \left(\frac{x_2-x_1}{2}\right)^2 + \left(\frac{y_2-y_1}{2}\right)^2 \\ \left(c_1 - \frac{x_1+x_2}{4}\right)^2 + \left(c_2 - \frac{y_1+y_2}{4}\right)^2 = \left(\frac{x_2-x_1}{4}\right)^2 + \left(\frac{y_2-y_1}{4}\right)^2 \end{cases} \tag{4}$$

And the equation of chord CD is the follow

$$c_1 = \frac{2(x_1x_2 + y_1y_2) - c_2(y_2 + y_1)}{x_1 + x_2} \tag{5}$$

By substituting (5) in (4), we get the solution of (4) as follow

$$\begin{cases} c_1 = \dfrac{2(x_1+x_2)(x_1x_2+y_1y_2) + (y_1+y_2)\sqrt{(x_1x_2+y_1y_2)[(x_2-x_1)^2+(y_2-y_1)^2]}}{(x_1+x_2)^2+(y_1+y_2)^2} \\ c_2 = \dfrac{2(x_1+x_2)(x_1x_2+y_1y_2) - (x_1+x_2)\sqrt{(x_1x_2+y_1y_2)[(x_2-x_1)^2+(y_2-y_1)^2]}}{(x_1+x_2)^2+(y_2+y_1)^2} \end{cases} \tag{6}$$

$$\begin{cases} d_1 = \dfrac{2(x_1+x_2)(x_1x_2+y_1y_2) - (y_1+y_2)\sqrt{(x_1x_2+y_1y_2)[(x_2-x_1)^2+(y_2-y_1)^2]}}{(x_1+x_2)^2+(y_1+y_2)^2} \\ d_2 = \dfrac{2(x_1+x_2)(x_1x_2+y_1y_2) + (x_1+x_2)\sqrt{(x_1x_2+y_1y_2)[(x_2-x_1)^2+(y_2-y_1)^2]}}{(x_1+x_2)^2+(y_2+y_1)^2} \end{cases} \quad (7)$$

Let $\odot O$ be the Circle whose center is the origin $O(0,0)$, its radius $r_{\odot O} = |\gamma|$, $|\gamma| = |\gamma_c| = \sqrt{c_1^2+c_2^2} = \sqrt{d_1^2+d_2^2} = |\gamma_d|$, then a Polar Coordinates $W(\gamma,\varphi)$ can be induced by Polarization Identity, such that the points $C(c_1,c_1)$ and $D(d_1,d_1)$ can be written as two vectors $w_c(\gamma,\varphi_c)$ and $w_d(\gamma,\varphi_d)$ satisfying the follow

$$\begin{cases} C(c_1,c_1) = w_c(\gamma,\varphi_c) = \gamma e^{i\varphi_c} = \gamma\cos\varphi_c + i\gamma\sin\varphi_c \\ D(d_1,d_1) = w_d(\gamma,\varphi_d) = \gamma e^{i\varphi_d} = \gamma\cos\varphi_d + i\gamma\sin\varphi_d \end{cases} \quad (8)$$

From the view of mathematical point, a Coordinates Transformation from product $Z \times Z$ to W-Plan has been constructed by (6), (7) and (8), $w: Z \times Z \to W$, for any $d_{ij}(z_i(x_i,y_i), z_j(x_j,y_j)) \in Z \times Z$, such that

$$w\big(d_{ij}(z_i(x_i,y_i), z_j(x_j,y_j))\big) = w(\gamma,\varphi)\begin{cases} \gamma^2 = z_1\cdot z_2 = |z_1||z_2|cos\theta \\ \varphi = \begin{cases} arctg\frac{c_2}{c_1} \ or \\ arctg\frac{d_2}{d_1} \end{cases} \end{cases} \quad (9)$$

(9) show us that both polarization vectors $w_c(\gamma,\varphi_c)$ and $w_d(\gamma,\varphi_d)$ is entangled together each other by identity: $\gamma^2 = \gamma_c^2 = \gamma_d^2$ among γ_c, γ_d and $d_{ij}(z_i(x_i,y_i), z_j(x_j,y_j))$.

Let $E(|\gamma_E|,\varphi_E)$ be the projection of z_2 on vector z_1, $|\gamma_E| = |z_2|cos\theta$, and $\varphi_E = \alpha$. $E'(|\gamma_{F'}|,\theta+\alpha)$ the projection of z_1 on vector z_2, $|\gamma_{E'}| = |z_1|cos\theta$, $\varphi_{E'} = \theta+\alpha$, as shown in Fig. 3(b), so we have $|z_1|^2 = |oE'|^2 + |z_1E'|^2$, $|z_2|^2 = |oE|^2 + |z_2E|^2$, addition that $|z_1 - z_2|^2 = |z_1|^2 + |z_2|^2 - 2|z_1||z_2|cos\theta$, then we have

$$|z_1 - z_2|^2 = |oE'|^2 + |z_1E'|^2 + |oE|^2 + |z_2E|^2 - 2|z_1||z_2|cos\theta \quad (10)$$

It is shown that between the pair contradicted object u and v, there is not only a polarization relation $\gamma^2 = \gamma_c^2 = \gamma_d^2$, but also there exist an entangled relation (10).

It is obvious to see that by using a coordinates transformation, the intersection of $\odot O$ and vector $z_1 - z_2$, note by $z_0 = x_0 + iy_0$, can be regarded to be the Origin point of respectively, and $m_{z_0} = \frac{m_{z_1}m_{z_2}}{m_{z_1}+m_{z_2}}$, the mass at the Electric Dipole Point z_0, because E and E' represent Force $f = |z_1||z_2|cos\theta$ action on m_{z_1} and m_{z_2} respectively, such that the Oscillation of EDP z_0 not only could be described by the, and some of Entangled reaction between u and v can be revealed out too.

In other hands, if let z_0 be the Harmonic Oscillation with mass m_{z_0}, then the law of Oscillation of u with mass m_{z_0} can be represented in. In special, when the distance

$d(z_1, z_2) \leq h_{pc}$, because we have not the unit for accurate measure position of $z_0(u)$, such that what the object u is become an uncertain problem, but the entangled relation between u and v is existing stile there.

In fact, we show from Fig. 3(a) that because $S_{\odot O} = \pi\gamma^2$, we get $\gamma^2 = z_1 \cdot z_2 = |z_1||z_2|cos\theta = \frac{S_{\odot O}}{\pi}$. And Fig. 3(b) show us that since $|\overline{Ez_2}| = |z_1||z_2|sin\theta$, the Exterior Product $z_1 \wedge z_2$ can be got by $z_1 \wedge z_2 = |z_1||z_2|sin\theta = |z_1||\overline{Ez_2}|$, and the Geometric Product $z_1 z_2$ can be defined by $z_1 z_2 = z_1 \cdot z_2 + z_1 \wedge z_2$ too, so a Geometric Algebra or Clifford Algebra as well.

From above analysis for Fig. 3, we have shown some of conceals information of Entangled Vector in bottom. One will naturally to ask: what more hidings are there the Entangled Vector? In order to get the answer, let us discuss it in detail in following.

Let E be the intersection point of $\odot O$ and vector z_1, $E(|\gamma_E|, \varphi_E)$ and $E(|\gamma_E|, \alpha)$ the E's W-Plan coordinate and Z-Plan coordinate, respectively, and $E(|\gamma_E|, \varphi_E) = E(|\gamma_E|, \alpha)$, then $\gamma_E = \sqrt{z_1 \cdot z_2} = \sqrt{|z_1||z_2|cos\theta}$, and $\varphi_E = \alpha$. Let E' be the intersection point of $\odot O$ and vector z_2, its W-Plan coordinate and Z-Plan coordinate are $E'(|\gamma_{E'}|, \varphi_{E'})$ and $E'(|\gamma_E|, \alpha + \theta)$, respectively, and $E'(|\gamma_{E'}|, \varphi_{E'}) = (E'|\gamma_{E'}|, \alpha + \theta)$, then $\gamma_{E'} = \sqrt{z_1 \cdot z_2} = \sqrt{|z_1||z_2|cos\theta}$, and $\varphi_{E'} = \alpha + \theta$.

Let F be the projection on vector z_1 of z_2, $F(|\gamma_F|, \varphi_F)$, $|\gamma_F| = |z_2|cos\theta$, and $\varphi_F = \alpha$. F' the projection on vector z_2 of z_1, $F'(|\gamma_{F'}|, \theta + \alpha)$, $|\gamma_{F'}| = |z_1|cos\theta$, $\varphi_{F'} = \theta + \alpha$.

Let G be the intersection point of $\odot A$ and x-axis, its W-Plan coordinate and Z-Plan coordinate are $G(|\gamma_G|, 0)$ and $G(|\gamma_G|, \alpha = 0)$, respectively, and $G(|\gamma_G|, \varphi_G) = G(|\gamma_G|, \alpha)$, $\varphi_G = \alpha$. G' be the intersection point of $\odot A$ and y-axis, its W-Plan coordinate and Z-Plan coordinate are $G'\left(|\gamma_{G'}|, \varphi_{G'} = \frac{\pi}{2}\right)$ and $G'\left(|\gamma_{G'}|, \frac{\pi}{2}\right)$, respectively, and $G'(|\gamma_{G'}|, \varphi_{G'})$, $\varphi_{G'} = \frac{\pi}{2}$. Since $|z_F| = |z_2|cos\theta$, we get the follow

$$|\gamma_E|^2 = |z_1||z_2|cos\theta = |z_F||z_1| = z_1 \cdot z_2 \tag{11}$$

Similarly,

$$|\gamma_{E'}|^2 = |z_1||z_2|cos\theta = |z_{F'}||z_2| = z_1 \cdot z_2 \tag{12}$$

Then we get

$$|z_F||z_1| = |z_{F'}||z_2| \tag{13}$$

We know that $\left|\frac{z_1}{|z_1|}\right|$ and $\left|\frac{z_2}{|z_2|}\right|$ are the unit of vector z_1 and z_2, respectively, note by $\left|\frac{z_1}{|z_1|}\right| \triangleq 1_{z_1}$ and $\left|\frac{z_2}{|z_2|}\right| \triangleq 1_{z_2}$, then we have from (13) that the norm $|z_F|$ of z_F in 1_{z_1} equal to the norm $|z_{F'}|$ of $z_{F'}$ in 1_{z_2}.

It is easy to see that because $\sqrt{|z_F|}$ possible are an integer number, a rational number or an irrational number, whether $|z_F|$ is commensurable by using the unit 1_{z_1}? Become a key problem. For example, $\sqrt{|z_F|} = \sqrt{2}$, because $\sqrt{2}$ can't be measured by using the unit 1_{z_1}, such that $|\gamma_E|$ can't be described in precision.

In order to discuss simple, first of all, let Z-Plan and W-Plan rotate an angle θ, such that the direct of vector z_1 turns into that of x_1, such that $1_{z_1} = 1_{x_1}$.

Second, assume $\sqrt{|z_F|}$ can be measured by 1_{x_1}, then there exits an unit system: $\left\{\frac{1_{x_1}}{m^k} = 1^k_{x_1}, k = 0, \ldots, N\right\}$, such that

$$|z_F| = n_0 1^0_{x_1} + n_1 1^1_{x_1} + \ldots + n_N 1^N_{x_1} = \sum\nolimits_{k=0}^{N} n_k 1^k_{x_1} = n_0.n_1 \ldots n_N \tag{14}$$

Since $1^k_{x_1} = 1^{k-1}_{x_1} \times m$, and $\frac{1_{x_1}}{m^N}$ is the minimum unit in this system, norm of $|z_F|$ can be written as follow:

$$|z_F| = n_0.n_1 \ldots n_N = n_0 \times 1 + n_1 \times m^1 \ldots n_N \times m^N = n \quad (\text{in} \frac{1_{x_1}}{m^N} = 1) \tag{15}$$

In other hands, if there are s prime factor of n, $p_l, l = 1, \ldots s$, such that $n = p_1 \times p_2 \times \ldots \times p_s$, then we get another unit system: $\left\{\frac{1_n}{p_l} = 1^{p_l}_n, l = 0, \ldots, s\right\}$, such that

$$|z_F| = n = (p_0 \times \ldots \times p_{l-1} \times \hat{p}_l \times p_{1+1} \times \ldots \times p_s) 1^{p_l}_n \tag{16}$$

here, symbol $\hat{p}_l$ represents the lack of the prime factor p_l. It is mean from (15) that by using prime factor p_l of n as unit for the measurement of the norm of $|z_F|$, its value equal to $(p_0 \cdot p_1 \cdot p_{l-1} \cdot \hat{p}_l \cdot p_{1+1} \cdot p_s)$. Therefore, prime factor p_l can be called an eigenvector of $|z_F|$, and $(p_0 \cdot p_1 \cdot p_{l-1} \cdot \hat{p}_l \cdot p_{1+1} \cdot p_s)$ is called the eigenvalues of the eigenvector p_l of $|z_F|$.

If $\sqrt{|z_F|}$ is an infinite rational number, for instant 1/3, then

If $\sqrt{|z_F|}$ can't be measured by 1_{x_1}, then we need an unit system whose value of unit is continuously variable, otherwise, the norm of $\sqrt{|z_F|}$ can't be measured.

We all know that

$$\int_0^x dt = t|_0^x = x \tag{17}$$

Because x is the norm of the interval $[0, x]$, the solution of (17) enlightens us that integration (17) is just the measure for norm of the interval $[0, x]$ using dt as the unity (or rule) of measure. In other words, dt can be considered to be the unit $1^k_{x_1}$, when $k \to \infty$, i.e. $dt = \lim\limits_{k \to \infty} 1^k_{x_1} = 1^\infty_{x_1}$.

In other hands, for the interval $[0, x]$, we have that $x(t) = (1-t)x_0 + tx_1$, let $\Delta x = x - x_0$, then $t = \frac{\Delta x}{x_1 - x_0}$, and $dx(t) = (x_1 - x_0)dt$, so we have

$$dt = \lim\nolimits_{\Delta x \to 0} \frac{\Delta x}{x_1 - x_0} = \frac{dx}{x_1 - x_0} \tag{18}$$

Integral for (18) we get

$$\int_0^{t_E} dt = \frac{1}{x_1 - x_0}\int_{x_0}^{x_E} dx = \frac{x}{x_1 - x_0}\Big|_{x_0}^{x_E} = \frac{x_E - x_0}{x_1 - x_0} = \frac{x_E}{x_1}\Big|_{x_0=0}^{x_E} \tag{19}$$

Since $|x_1| = |z_1| = 1_{z_1}$, we have the follow:

$$dt = \frac{dx}{x_1 - x_0}\Big|_{x_0=0}^{x_E} = \frac{dx}{|x_1|} = \frac{dx}{|z_1|} = \frac{dx}{|1_{x_1}|} \tag{20}$$

It is hiding from (18) and (19) that (20) could be assumed to being the minimum unit of measure.

Then we get

$$|x_F|\int_0^{t_E} dt = \frac{|x_F|}{|x_1|}\int_0^{x_E} dx = \frac{|x_F|}{|x_1|}x\Big|_0^{x_E} = \frac{|x_F||x_E|}{|x_1|}\ (\text{in}\frac{|x_F|dx}{|1_{x_1}|} = 1) \tag{21}$$

Let $\Delta x(\Delta t) = \lambda t$ then $x_t = x_0 + \Delta x(\Delta t) = x_0 + \lambda t$,

$$\varphi(x_t) = \int_0^{t_E} x_t dt = \int_0^{x_E}(x_0 + \lambda t)dx = \left(x_0 t + \frac{\lambda t^2}{2}\right)\Big|_0^{x_E} = \frac{\lambda x_E^2}{2} = \frac{\lambda \gamma^2}{2} \tag{22}$$

Let $\Delta y(\Delta t) = -i\lambda t$ then $y_t = y_0 + \Delta y(\Delta t) = y_0 - i\lambda t$,

$$\phi(y_t) = \int_0^{t_{E'}} y_t dt = \int_0^{y_E}(y_0 - i\lambda t)dx = \left(y_0 t - i\frac{\lambda t^2}{2}\right)\Big|_0^{y_{E'}} = -\frac{i\lambda y_{E'}^2}{2} = -\frac{i\lambda^2}{2} \tag{23}$$

Let $\Delta z(\Delta t) = \lambda(1-i)t$ and $z_t = z_0 + \Delta z(\Delta t) = (x_0 + iy_0) + \lambda(1-i)t$

Then we get

$$U(z_t) = \varphi(x_t) + i\phi(y_t) = \frac{\lambda x_E^2}{2} + i\left(-\frac{i\lambda y_{E'}^2}{2}\right) = \frac{\lambda}{2}\left(x_E^2 + y_{E'}^2\right) = \lambda\gamma^2 \tag{24}$$

$$U(z_t) = \varphi(z_t) + i\phi(z_t) = \varphi(x_0 + \lambda t) + i\phi(y_0 - i\lambda t) = \lambda^2\gamma^2 \tag{25}$$

Then we have a wave equation:

$$\frac{d^2U}{dt^2} - \frac{1}{\lambda^2}\frac{d^2U}{dz^2} = 0 \tag{26}$$

It is shown from above discuss that object u and v are entangled each other by (25). If $\Delta t < t_{pl}$, t_{pl} is Planck Time, the entangled relation (25) between u and v will be remained yet, then it could be called the quantum entanglement. An encryption and decryption algorithm based on the entangled vector γ_c and γ_d of z_1 and z_2, has been proposed by Jiali Feng [7]. The encryption and decryption algorithm based quantum entanglement among γ_c,γ_d and γ would be discussed in other paper.

3 The Attribute Topos Induced by Mechanism of Mutual Change Between Quality and Quantity

Because the position changes with time Δt_{it}, $\Delta z(\Delta t_{it}) = (\Delta x(\Delta t_{it}), \Delta y(\Delta t_{it}))$ induces a cone and a limit among arrows $\{(\Delta t_{it}, \Delta x_{it}), \Delta y_{it})\}$, as shown in Figs. 5, 6 and 7, such that a finitely complete category can be achieved.

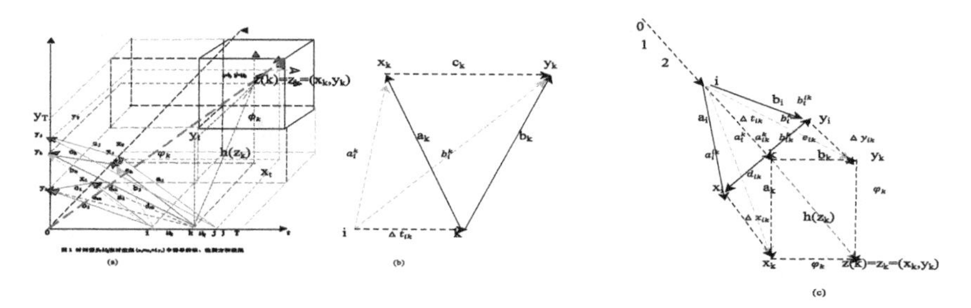

Fig. 4. (a) Category induced by arrows $\{(\Delta t, \Delta x, \Delta y)\}$ (b) Cone of arrows $F = \{(a_k, b_k, c_k)\}$, (c) Pullback (a_k, b_k) of position mapping (φ_k, φ_k)

Let $[i, T)$ and $[x_\mathrm{i}, x_\mathrm{T})$ be the time interval and position interval in which $q_v(u)$ maintains itself, then $q_v(u)$ must be fixed, as long as $x_\mathrm{t} = x_\mathrm{t} + \Delta x(\Delta t_{it})$ of u does not over out $[x_\mathrm{i}, x_\mathrm{T})$, i.e. $x_\mathrm{t} \in [x_\mathrm{i}, x_\mathrm{T})$, while the displacement $\Delta x(\Delta t_{it})$ vary with Δt_{it}. It is obvious that the mechanism of the conversion from quantity x_t into a quality $q_v(u)$ can be described by a Qualitative Mapping as follow

$$\tau(x_t, [x_0, x_\mathrm{T})) = \begin{cases} q_v(u) & x_t \in [x_\mathrm{i}, x_\mathrm{T}) \\ \neg q_v(u) & x_t \in [t_0, t_i] \end{cases} \tag{27}$$

Let $x_\mathrm{t}, x_\mathrm{s} \in [x_\mathrm{i}, x_\mathrm{T})$ be two quantities belonging to a same qualitative criterion $[x_\mathrm{i}, x_\mathrm{T})$, even they are conversing into the same quality $q_v(u)$, their degrees of conversion are different. To describe this case, a function of conversion degree $\eta(x_\mathrm{t}, [x_\mathrm{i}, x_\mathrm{T})) = \eta(x_\mathrm{t})$ has been proposed by Attribute Theory. In other words, let $[i, T) \times [x_i, x_T)$ be the product of $[i, T)$ and $[x_i, x_T)$, $[y_\mathrm{i}, y_\mathrm{T})$ codomain of $\eta(x_\mathrm{t})$, we get the follow

$$\eta\colon [i, T) \times [x_i, x_T) \to [\eta_\mathrm{i}, \eta_\mathrm{T}) \tag{28}$$

It is obvious that there is an equivalent relation, between the function of conversion degree η: $[i,T) \times [x_i,x_T) \to [\eta_i,\eta_T)$ and the η': $[i,T) \to [x_i,x_T) \times [\eta_i,\eta_T)$, this is said that in category

$$\hom([i,T) \times [x_i,x_T),[\eta_i,\eta_T)) \approx \hom([i,T),[x_i,x_T) \times [\eta_i,\eta_T)) \tag{29}$$

Therefore, we get a Cartesian Closed Category induced by the function $\eta(t,[x_i,x_T))$ and $\tau_i(x_k,[x_i,x_T))$. In other words, there is an evaluation morphism induced by $\eta(t,[x_0,x_T))$ and $\tau_i(x_k,[x_i,x_T))$, ev: $[i,T) \times [\eta_i,\eta_T)^{[i,T)} \to [\eta_i,\eta_T)$, such that ev$(t,\tau_i(x_k)) = \tau(x_k)(t) = \eta(t,x_k)$, here the exponential morphism is $\tau_i(x_k,[x_i,x_T))$, as shown in Fig. 5(a) [8].

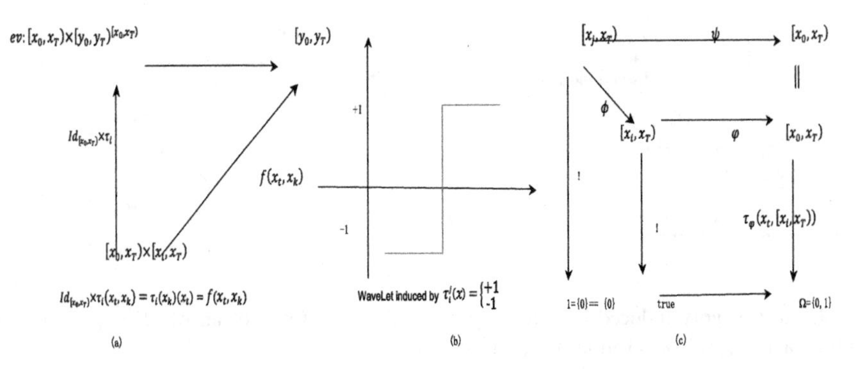

Fig. 5. (a) the Cartesian Closed Category induced by function $f(x_t,x_k)$ and $\tau_i(x_k)$ (b) Quality change $\tau_i^j(x) = \begin{cases} +1 \\ -1 \end{cases}$ of $\tau_j(x) - \tau_i(x)$ and its Harr wavelet (c) Pullback induced by Qualitative Mapping $\tau_\varphi(x_t,[x_i,x_T))$

If let $x_j \in [x_j,x_{j+1}) \subseteq [x_i,x_T)$ be the point that quality $q_v(u)$ transform to its non-essential quality $q_j(u)$, note by "$q_v \to q_j$" or q_v^j, $[x_j,x_{j+1})$ the qualitative criterion of $q_j(u)$, in which $q_j(u)$ will remain itself, as long as x_t varies with time change Δt_{it} but still in $[x_j,x_{j+1})$, and let $x_T \in (x_{T^-},x_{T^+})$ be the point at that the quality $q_v(u)$ transform to the quality $p_u(v)$ of v, the contradiction of u, or when the distance $|(x_{T^-},x_{T^+})| \le h_{pl}$, the Planck constant, a system $S_{(u,v)}$ can be constructed by u and v, i.e. $x_t \notin [x_i,x_T), x_t \in (x_{T^-},x_{T^+}) \wedge |(x_{T^-},x_{T^+})| \le h_{pl}$. Then we have a Qualitative Mapping as following

$$\tau(x_t,[x_0,x_T)) = \begin{cases} S_{(u,v)} & x_t \in [x_T,x_{T^+}] \wedge |[x_T,x_{T^+}]| \le h_{pl} \\ p_u(v) & x_t = [x_{T^-},x_T] \\ q_j(u) & x_t \in [x_j,x_{j+1}) \\ q_v(u) & x_t \in [x_i,x_T) \\ \neg q_v(u) & x_t \in [t_0,t_i] \end{cases} \tag{30}$$

The mechanism of a non-essential transforms from the quality $q_i(u)$ to its sub-quality $q_j(u)$, notation as

$$\tau_i^j(x) = \tau_j(x, [x_j, x_T)) - \tau_i(x, [x_i, x_j)) = \begin{cases} 0_j - 1_i = -1_i^j & x \in [x_i, x_j) \\ 1_j - 0_i = 1_i^j & x \in [x_j, x_T) \end{cases} \tag{31}$$

The (15) show that a Harr wavelet induced by the difference of $\tau_j(x, [x_j, x_T)) - \tau_i(x, [x_i, x_j))$ or $\tau_i^j(x)$, when $[x_i, x_j), [x_j, x_T) \subseteq [x_i, x_T)$, and $[x_i, x_j) \cap [x_j, x_T) = \emptyset$.

Let $\varphi : [x_i, x_j) \hookrightarrow [x_0, x_T)$ be the monic function induced by the inclusion relation $[x_i, x_j) \subseteq [x_0, x_T)$, the characteristic function $\tau_\varphi(x_t, [x_i, x_j))$ induced by φ is just the qualitative mapping which criterion is $[x_i, x_T)$. This means that, not only a truth map true: $\{0\} \to \Omega = \{0, 1\}$, but also a pullback square as shown in Fig. 4 can be induced by $\tau_\varphi(x_t, [x_i, x_T))$. In particular, if $\psi : [x_j, x_T) \hookrightarrow [x_0, x_T)$ is a monic morphism induced by the relation $[x_j, x_T) \subseteq [x_0, x_T)$, such that $\psi(x_k) \in [x_0, x_T)$, then there exists a function induced by $\psi\phi : [x_j, x_T) \hookrightarrow [x_i, x_T)$, for $\forall x_k \in [x_j, x_T)$, $\phi(x_k) \in [x_i, x_T)$, such that $\varphi \circ \phi(x_k) = \varphi(x_\ell) = \psi(x_k)$.

In other word, because $1 = \{0\}$, and there is the inclusion mapping $1 = \{0\} \subseteq \{0, 1\}$, for any subset $[x_j, x_T) \subseteq [x_0, x_T)$ and the morphism $[x_j, x_T) \to 1$, the pullback square can be gotten, but any subset $[x_j, x_T) \subseteq [x_0, x_T)$ is true, if and only if, $\tau_\phi(x_k, [x_j, x_T)) = 1$. This means that there is only one predication "for $\forall x_k \in [x_j, x_T) \subseteq [x_0, x_T)$, the true mapping $[x_j, x_T) \to 1 \to \Omega$", and Ω is a subobject classifier. Therefore, we get an Attribute Topos.

4 The Fixation Image Operator Induced by Orthogonal Expanded of Function

Let $[x_i, x_T) = \bigcup_{k=1}^{m} [x_i, x_{i+k})$, we get a collection of qualitative mapping $\{\tau_j(x_k, [x_j, x_{j+1})), j = i, i+1, \ldots, i+m,\}$, and for $\forall \tau_{j_1}(x_\ell, [x_{j_1}, x_{j_1+1})), \tau_{j_2}(x_k, [x_{j_2}, x_{j_2+1})) \in \{\tau_j(x_k, [x_j, x_{j+1}))\}$, $\exists x_\ell \in [x_{j_1}, x_{j_1+1})$, for $\forall x_k \in [x_{j_2}, x_{j_2+1})$, their inner product satisfying the following

$$\int_{j_2}^{j_2+1} \tau_{j_1}(x_\ell, [x_{j_1}, x_{j_1+1}))\tau_{j_2}(x_k, [x_{j_2}, x_{j_2+1}))dx_k = \begin{cases} 1 & j_1 = j_2 \\ 0 & j_1 \neq j_2 \end{cases} \tag{32}$$

This shows that not only the collection $\{\tau_j(x_k, [x_j, x_{j+1}))\}$ constructs an orthogonal base and by which a Hilbert Space $\mathcal{H}$ can be expanded, but also any function of conversion degree $f(x_t, [x_i, x_T))$ can be expanded into a linear combination of qualitative mappings as follow

$$f(x_t, [x_i, x_T)) = \sum\nolimits_{k=1}^{m} f_k(x_t, [x_i, x_{i+k}))\tau_j(x_k, [x_i, x_{i+k})) \tag{33}$$

In fact, because a generalized Fourier Transformation of $f(x_t, [x_i, x_T))F(f)$ can be defined by the inner product of $f(x_t, [x_i, x_T))$ and $\tau_i\left(x_\ell, \left[x_j, x_{j+w}\right)\right)$ as following:

$$F(f) = \left(f(x_t, [x_i, x_T)), \tau_i\left(x_\ell, \left[x_j, x_{j+w}\right)\right)\right) = \begin{cases} f_k(x_t, [x_i, x_{i+k})) & i = j \\ 0 & i \neq j \end{cases} \tag{34}$$

By (34) the degree of conversion function $f(x_t, [x_i, x_T))$ could be homomorphically (or similarly) mapped from the function space into a vector or the point $\left(f_1(x_t, [x_{i+1}, x_{i+2})), \ldots, f_m\left(x_t, \left[x_{i+(m-1)}, x_{i+m}\right)\right)\right)$ in the Hilbert space $\mathcal{H}$, so (34) is called the "Fixation-Image" operator of function $f(x_t, [x_i, x_T))$.

In other words, this shows that a Hilbert Space $\mathcal{H}$ is expanded by an Orthogonal Base $\left\{\tau_j\left(x_k, \left[x_j, x_{j+1}\right)\right)\right\}$ which is induced by the Subobject Classifier of Attribute Topos. In the next, let us discuss its application.

Let $\{f^s\}$ be the collection of the approximation f^s of function f, F the "Fixation Image" operator $F : \{f^s\} \rightarrow \mathcal{H}$, $F(\{f^s\}) = N(F(f), \delta(s))$ the F-Image of $\{f^s\}$ in $\mathcal{H}$, and is called the neighbors of F(f). It is obvious, under the "Fixation Image" operator F, a (classification or) recognition algorithm for the function f^t, i.e. $\tau_f(F(f^t), N(F(f), \delta(s)))$.

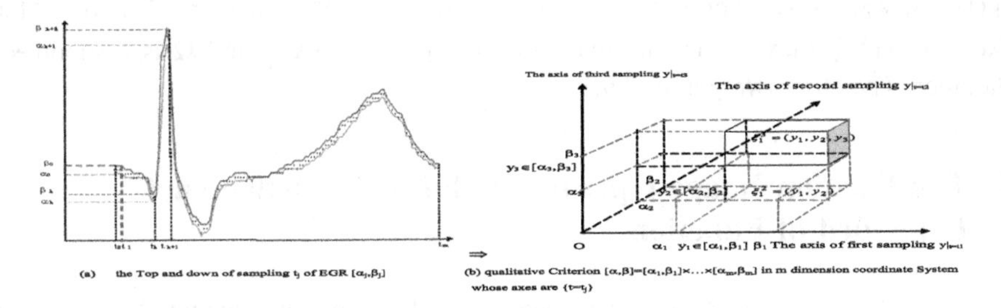

Fig. 6. (a) "Fixation-Image" operator of EGR. (b) Recognition algorithm of EGR based on Qualitative Mapping

A human faces recognition based on Attribute grid computer is developed by Gangxiao Lv, as shown in Fig. 7 [9].

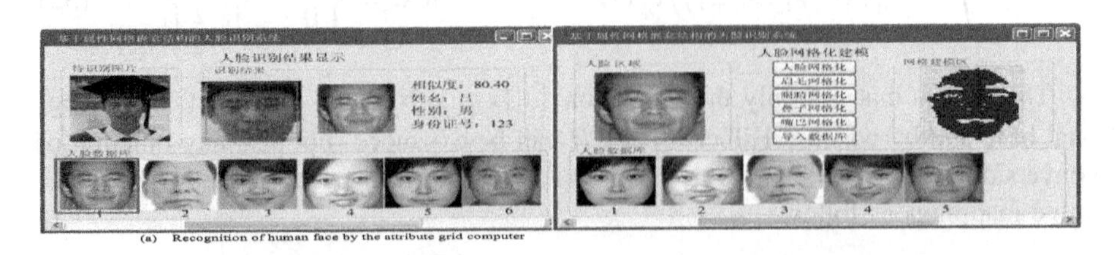

Fig. 7. Human face recognition based on qualitative mapping

It is obviously that, if the function f(x) instead by a n-dimensional pattern Min n-Hilbert space $\mathcal{H}$, then under the (18), M can be homomorphically (or similarly) mapped as a point F(M). in n-Hilbert space $\mathcal{H}$, and the approximation of the cluster with the model is mapped to the nearest neighbor $\mathrm{N}(\mathrm{F}(\mathrm{M}), \delta(\mathrm{s}))$, then a qualitative mapping $\tau_{\mathrm{M}}(\mathrm{M}^{\mathrm{t}}, \mathrm{N}(\mathrm{F}(\mathrm{M}), \delta(\mathrm{s}))$ for recognition of pattern M can be given.

4.1 The Tensor Flow Induced by Restriction Morphism F and Image Thinking

Let two Hilbert Space $\mathcal{H}$ and $\mathcal{H}'$ be expanded by base $\{\tau_{\mathrm{j}}(\mathrm{x}_{\mathrm{k}}, [\mathrm{x}_{\mathrm{i}}, \mathrm{x}_{\mathrm{i+k}}))\}$ and $\left\{\tau'_{j+1}(\mathrm{y}_{\mathrm{k+1}}, [\mathrm{y}_{\mathrm{i+1}}, \mathrm{y}_{\mathrm{i+k+1}}))\right\}$, respectively, $\mathrm{f}(\mathrm{x}_{\mathrm{t}}, [\mathrm{x}_{\mathrm{i}}, \mathrm{x}_{\mathrm{T}})) = \sum_{\mathrm{k}=1}^{\mathrm{m}} \mathrm{f}_{\mathrm{k}}(\mathrm{x}_{\mathrm{t}})\tau_{\mathrm{j}}(\mathrm{x}_{\mathrm{k}}, [\mathrm{x}_{\mathrm{i}}, \mathrm{x}_{\mathrm{i+k}}))$ the restriction function of function $g((\mathrm{y}_{\mathrm{t}}, [\mathrm{y}_{\mathrm{i}}, \mathrm{y}_{\mathrm{i+k}}))) = \sum_{\mathrm{k}=1}^{\mathrm{m}} \mathrm{g}_{\mathrm{k}}(\mathrm{y}_{\mathrm{t}})\tau'_{j}(\mathrm{y}_{\mathrm{k}}, [\mathrm{y}_{\mathrm{i}}, \mathrm{y}_{\mathrm{i+k}}))$, and $F: f(x_t) \to g(x_t)$ the restriction morphism between the function $f(x_t)$ and $g(x_t)$. Consider that time from t to t + 1, F could be dived tow partial, one is for the function $\mathrm{f}_{\mathrm{k}}(\mathrm{x}_{\mathrm{t}})F_f(\mathrm{f}_{\mathrm{k}}(\mathrm{x}_{\mathrm{t}})) \to \mathrm{g}_{\mathrm{k}}(\mathrm{y}_{\mathrm{t}})$, and second for the bases $F_\tau(\tau_{\mathrm{j}}(\mathrm{x}_{\mathrm{k}}, [\mathrm{x}_{\mathrm{i}}, \mathrm{x}_{\mathrm{i+k}}))) \to \tau'_{j+1}(\mathrm{y}_{\mathrm{k+1}}, [\mathrm{y}_{\mathrm{i+1}}, \mathrm{y}_{\mathrm{i+k+1}}))$, in this case, it should be satisfying as follow

$$\begin{aligned} F(\mathrm{f}(\mathrm{x}_{\mathrm{t}}, [\mathrm{x}_{\mathrm{i}}, \mathrm{x}_{\mathrm{T}}))) &= F\left(\sum_{\mathrm{k}=1}^{\mathrm{m}} \mathrm{f}_{\mathrm{k}}(\mathrm{x}_{\mathrm{t}})\tau_{\mathrm{j}}(\mathrm{x}_{\mathrm{k}}, [\mathrm{x}_{\mathrm{i}}, \mathrm{x}_{\mathrm{i+k}}))\right) = \sum_{\mathrm{k}=1}^{\mathrm{m}} F_f(\mathrm{f}_{\mathrm{k}}(\mathrm{x}_{\mathrm{t}}))F_\tau(\tau_{\mathrm{j}}(\mathrm{x}_{\mathrm{k}}, [\mathrm{x}_{\mathrm{i}}, \mathrm{x}_{\mathrm{i+k}}))) \\ &= \sum_{\mathrm{k}=1}^{\mathrm{m}} \mathrm{g}_{\mathrm{k+1}}(\mathrm{y}_{\mathrm{t+1}})\tau'_{j+1}(\mathrm{y}_{\mathrm{k+1}}, [\mathrm{y}_{\mathrm{i+1}}, \mathrm{y}_{\mathrm{i+k+1}})) = g((\mathrm{y}_{\mathrm{t+1}}, [\mathrm{y}_{\mathrm{i+1}}, \mathrm{y}_{\mathrm{i+k+1}}))) \end{aligned} \tag{35}$$

Let $\left(T^{j,(j+1)}_{k,(k+1)}\right)$ be the transformation tensor from $\tau_{\mathrm{j}}(\mathrm{x}_{\mathrm{k}}, [\mathrm{x}_{\mathrm{i}}, \mathrm{x}_{\mathrm{i+k}}))$ to $\tau'_{j+1}(\mathrm{y}_{\mathrm{k+1}}, [\mathrm{y}_{\mathrm{i+1}}, \mathrm{y}_{\mathrm{i+k+1}}))$, then the base morphism F_τ can be represented as follow

$$F_\tau(\tau_{\mathrm{j}}(\mathrm{x}_{\mathrm{k}}, [\mathrm{x}_{\mathrm{i}}, \mathrm{x}_{\mathrm{i+k}}))) = \tau'_{j+1}(\mathrm{y}_{\mathrm{k+1}}, [\mathrm{y}_{\mathrm{i+1}}, \mathrm{y}_{\mathrm{i+k+1}})) = \left(T^{j,(j+1)}_{k,(k+1)}\right)\tau_{\mathrm{j}}(\mathrm{x}_{\mathrm{k}}, [\mathrm{x}_{\mathrm{i}}, \mathrm{x}_{\mathrm{i+k}})) \tag{36}$$

Let $\left(\tau_{\mathrm{j}}(\mathrm{x}_{\mathrm{k}}, [\mathrm{x}_{\mathrm{i}}, \mathrm{x}_{\mathrm{i+k}})), \left(T^{j,(j+1)}_{k,(k+1)}\right), \tau'_{j+1}(\mathrm{y}_{\mathrm{k+1}}, [\mathrm{y}_{\mathrm{i+1}}, \mathrm{y}_{\mathrm{i+k+1}}))\right)$ be the triple of bases and tensor, since $\mathcal{H}$. and $\mathcal{H}'$ are orthogonal Hilbert space expanded by $\{\tau_{\mathrm{j}}(\mathrm{x}_{\mathrm{k}}, [\mathrm{x}_{\mathrm{i}}, \mathrm{x}_{\mathrm{i+k}}))\}$ and $\left\{\tau'_{j+1}(\mathrm{y}_{\mathrm{k+1}}, [\mathrm{y}_{\mathrm{i+1}}, \mathrm{y}_{\mathrm{i+k+1}}))\right\}$, respectively, and the series of coordination systems constructed by $\{[\mathrm{x}_{\mathrm{i}}, \mathrm{x}_{\mathrm{i+k}})\}, k = 1, \ldots, n$, when time is varying from t = i, i + 1 …, i + j, then a Tensor Flow with tow coordination systems and the transformation among them, called a (differential, if $\{\tau_{\mathrm{j}}(\mathrm{x}_{\mathrm{k}}, [\mathrm{x}_{\mathrm{i}}, \mathrm{x}_{\mathrm{i+k}}))\}$ is a differential coordination system) manifold can be constructed.

Because the problem how does the function $f(x_t)$ transform to function $g(x_{t+1})$, can be described by $\left(\tau_{\mathrm{j}}(\mathrm{x}_{\mathrm{k}}, [\mathrm{x}_{\mathrm{i}}, \mathrm{x}_{\mathrm{i+k}})), \left(T^{j,(j+1)}_{k,(k+1)}\right), \tau'_{j+1}(\mathrm{y}_{\mathrm{k+1}}, [\mathrm{y}_{\mathrm{i+1}}, \mathrm{y}_{\mathrm{i+k+1}}))\right)$, if let $\mathfrak{B}(\mathcal{H})$ be the think Category of Brian, then the (Image) Thinking about how does the function $f(x_t)$ transform to function $g(x_{t+1})$, can be described by the nature transformation between tow restriction morphism fonctors F and G, $\alpha : \mathcal{H} \rightarrow \mathfrak{B}(\mathcal{H})$, $\alpha(F \rightarrow G)$.

An application of the Tensor Flow model in Traditional Chinese Medicine as shown in Fig. 8 [8].

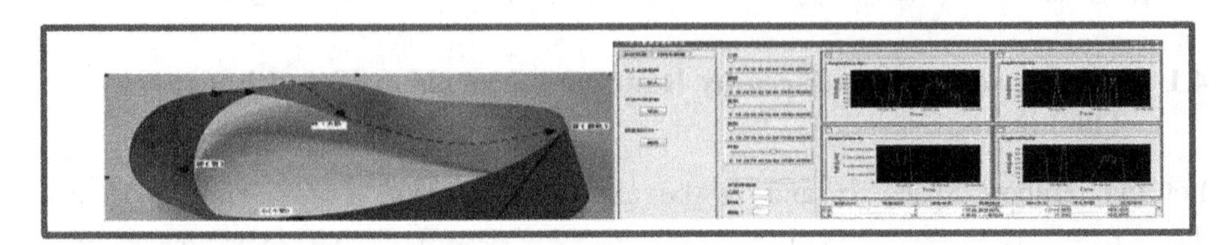

Fig. 8. "Möbius Strip" model of "five elements of life grams" in Chinese medicine

Acknowledgement. The authors specially thank Professor Pei-zhuang Wang for his introduction of the Factor Space Theory [10], Dr. He. Ouyang for the discussion in category and Topos and Prof. Zhongzhi Shi for his help in many time.

References

1. Minsky, M.: The Society of Mind. Simon & Schuster, New York (1985)
2. Zhong Y.: A unified model for emotion and intelligence in machine. In: Proceedings of Sino-Japan Symposium on Emotion Computing, 1–2 July 2003
3. Feng, J.: Qualitative mapping orthogonal system induced by subdivision transformation of qualitative criterion and biomimetic pattern recognition. Chin. J. Electron. **15**(4A), 850–856 (2006). Special Issue on Biomimetic Pattern Recognition
4. Tsien, H.: The letter to Dai Ruwei on 25, Janury, 1993. In: Lu, M. (ed.) Thought About Noetic Science of Hsue-sen Tsien, p. 252. Science Press (2012)
5. Feng, J., Wang, X.: Four-key inner product decomposition of inner product of a constant. In: The 10th China Conference on Machine Learning (2005)
6. Encyclopedia of Chinese Philosophy, vol. II, p. 1181, Beijing, August, 1987
7. Feng, J.: Attribute grid computer based on qualitative mapping and its application in pattern recognition. In: Lin, T.Y., Hu, X., Han, J. (eds.) 2009 IEEE International Conference on Granular Computing, pp. 154–161. IEEE Computer Society (2009)
8. Lane, S.M.: Category for The Working Mathematician, 2nd edn. Springer, Heidelberg (1998)
9. Feng, J.: The Attribute Theory Method in Noetic and Intelligence Science. Atom Energy Press, Beijing (1990). (in Chinese)
10. Wang, P.: Factor space, a mathematical preparing for the coming of big data tide (special talk). In: High-end Forum on Big Data. Chinese Academy of Sciences, Beijing, December 2014

Go Mapping Theory and Factor Space Theory Part I: An Outline

He Ouyang[1,2](✉)

[1] Sunbridge Grothendieck Institute, Pudong, Shanghai, People's Republic of China
908229188@qq.com

[2] Department of Mathematics, School of Science, National University of Defense Technology, Changsha, Hunan, People's Republic of China

Abstract. Inspired by Professor Wang, Peizhuang's Factor Space Theory (FST), we propose a new scheme, called GO Mapping Theory, or GMT in short, to formalize the concept formation knowledge representation of AI. This scheme can be viewed as an extension of Willie's Formal Concept Analysis (FCA), PZ Wang's Factor Space Theory, and it naturally includes Gouguen's L-Fuzzy Sets therefore it sounds a unified soft computing scheme. Potentially, GMT can be used for human-like knowledge representation and computation by modern computers. By deploying Grothendieck's topos theory (this is the origin for the name GO mapping), we developed a unified mathematical language understandable by robots which can represent human language: concepts and logic, which also unified the current learning techniques at a more abstract level, therefore can be used as a basis for AGI or Super AI.

By restating FST under category language, one can have a much more general setup for classical reductionist's view about multiple sensory system, we call it a cognitive frame. Then under the assumption of uncertainty of any measurement, we can naturally, in fact ontologically, obtain an L-fuzzy set by FST, then we can construct from this L-fuzzy set a L-presheaf by standard procedure, we call it the GO mapping, which by Barr's Embedding Theorem, can be viewed as the natural replacement for classical fuzzy sets. In fact, in topos theory, FST and GMT pairs a geometric morphism in Grothendieck's topos theory, which shows the amazing power of pure math in real life applications.

1 Factor Space Theory (FST)

Introduced by Prof Wang, PZ around 80's, [7], we have

Definition 1.1. *Let U be a set, called the universe, $f : U \to V_f$ is a map, we call f a factor for U. If $\{f : U \to V_f\}$ is a set of factors, we call $SS = \prod V_f$ the state space and $F : U \to SS$ the factor space.*

The basic idea is this: when we want to know U, or an object in U, we try to use several "detectors", or "sensors/measurements/instruments", to

Z. Shi et al. (Eds.): ICIS 2017, IFIP AICT 510, pp. 37–41, 2017.
DOI: 10.1007/978-3-319-68121-4_4

gauge/measure U, we call these detectors "the factors" (by Wang). This is in fact the reductionist/instrumentalists idea.

Afterwards, we try to analyze the result in the state space to see, to find out where these values falling. And from there, one can do logic operations and draw conclusions and extract information about U.

In this paper, we will assume all sets are topological spaces and we would like to modify the Definition 1.1 slightly as follows:

Before giving the definition, let's recall some terminologies from category theory.

Definition 1.2. *A diagram $\mathcal{D}$ in a category $\mathfrak{C}$ is a collection of objects as vertices and some morphisms among these objects as edges.*

A cone X is for a diagram $\mathcal{D}$ in $\mathfrak{C}$ is a collection of arrows $f_i : X \to D_i$ such that for any $g : D_i \to D_j$ in the diagram $\mathcal{D}$, following diagram

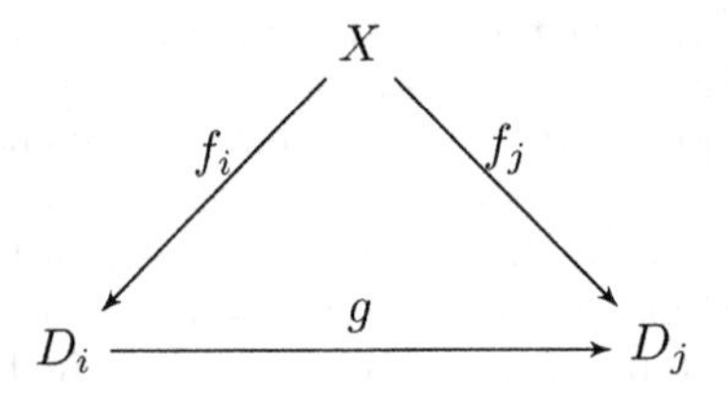

commutes. Cone is usually denoted by $\{f_i : X \to D_i\} = Cone$. Reference [5] for details.

This way, we can rewrite the factor space theory as follows.

From cognitive point of view, we can identify a cone as the most general format of factor space theory. Or we can view a cone as a cognitive frame for object X. $\mathcal{D}$ can be viewed the collection of instruments.

Let's remind what is the limit for a diagram. In a short sentence, the limit is a universal object satisfying

1. It is a cone for $\mathcal{D}$, denoted by $\lim \mathcal{D}$
2. It is a "minimum" cone among all cones over $\mathcal{D}$

which means for any cone K over $\mathcal{D}$, $\exists!$ arrow $F : K \to \lim \mathcal{D}$ such that

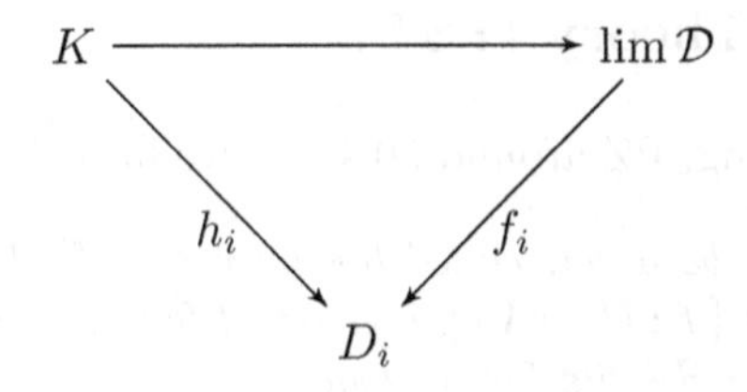

commutes.

After these preparation, we can give a more elegant definition for Factor Space Theory.

Definition 1.3. *Let* **Top** *be the category with topological spaces as objects and continuous maps as arrows. Then any diagram in* **Top** *has the limit. We call any cone* $\{f : U \to V_f\}$ *in* **Top** *the cognitive frame for the universe* U

We call $\lim V = \lim\{V_f\}$ *the state space and the unique map* $F : U \to \lim V$ *the cognitive map.*

2 Formal Concept Analysis (FCA)

In 80's, Wille developed a concept called Formal Concept Analysis, FCA in short. FCA has gained great applications in computer science.

In FCA, $\forall U, V \in$ **Sets**, $F \subset U \times V$, the triple (U, F, V) is called a formal context, if $A \subset U, B \subset V$ satisfy

$$FA = B, \quad F^{-1}B = A$$

We call A and B are Galois connected and (A, F, B) a formal concept.

This theory in some sense encodes the classic concept in classic logic for computer science. It is easily identified that set A is the extension for a concept, and B is the intension for this concept. Obviously, FCA (A, F, B) has best captured the classical usage of concepts for bots, if human mind can be determined by Boolean logic. In the work of [4], we prove that under Definition 1.3, we have

Theorem 2.1.

$$FST \supset FCA$$

Hence, from Sect. 1, we know that one cognitive from theory (CFT) is the most general set operation:

$$CFT \supset FST \supset FCA$$

3 Gouguen's L-Fuzzy Sets and Barr's Embedding

Let H be a Heytin algebra. Then Gouguen introduced a category of H-fuzzy sets by define:

- Obj(Fuz(H)) $= \{x : X \to H, X \in$ **Sets**$\}$
- Arrows $f : (X, x) \to (Y, \eta) \Leftrightarrow f : X \to Y$ in **Sets** and $x \leq \eta \circ f : X \to H$

Unfortunately, Fuz(H) is not a topos except H is a Boolean algebra. The great advantage for working math, AI, mind computing is that Topos behaves almost like **Sets**, except the law of excluding the middle (LEM).

In real life logic, LEM probably the most wrong law human follows, hence it is born for abandoning.

However, to proceed the basic logic deduction, subsets, power sets, products, limit, colimit and all basic ingredients. Topos provide all these needed goodies.

1980s Michael Barr [1] and Pitts [6] both showed that the Gouguen category is not a topos, but can be embedded in some way into a topos. Here we will give a short description on Barr's construction.

$\forall \alpha \in H$, we can define a "level" set $X_\alpha = \{y \in X : x(y) \geq \alpha\}$, then we call the collection $\mathcal{T}(X) = \tilde{X} = \{X_\alpha : \alpha \in H\}$ the tow of X. One can prove that $\mathcal{T}$: Fuz(H) $\rightarrow$ Tow(H) is an isomorphism of categories. Each element in fact looks like $G : H \rightarrow$ **Sets** satisfying the axioms for presheaf, i.e. G is a contravariant functor from partial order set considered as a category to the category set.

The Barr proved by providing an initial element z to H, one can "embed" the cat Tow(H) into a topos Sh(H^+), where $H^+ = H \cup \{z\}, \quad z < a \quad \forall a \in H$ and Sh(H^+) is the category of H^+-sheaves.

This is a very significant result. Because one can translate every "fuzzy" math problem into a problem about sheaf/presheaf then using the powerful tool sets since Grothendieck developed algebraic geometry.

This is analogue to embedding the rationals to reals!

From Barr and Pitts, we in fact showed how to consider every presheaf

$$G : H \rightarrow \mathbf{Sets}$$

to be some sort of "fuzzy sets" or "fuzzy concepts".

Using presheaves directly instead of the Zadeh/Gouguen fuzzy sets will provide powerful tools for soft computing & AI, from our perspective.

4 The GO Mapping Theory (GMT)

In general in a real world, any measurement can be viewed as a function

$$f : U \rightarrow V$$

Normally, V is a finitely dimensional vector space equipped with a positive metric C, most possibly a metric/a topology induced by an L^P-norm, $p = 1, 2, \infty$ would be the most popular ones. Then people will define/conceptualize a thing based on the values of the measurement. This is established since Aristotle's era and in 1980's formalised by FST, FCA and Feng's Property Mapping Theory (PMT).

Here we will develop a new scheme which can include all the theories above as a special case and naturally incorporate the intrinsic fuzziness of measurements and human mind computation.

Let's go back to the example above.

In reality, no measurement is precise (in face we do not have a god given correct precise value for any), hence in most cases what we know is that $\forall u \in U$, $f(u)$ is some neighborhood of a center value $x \in V$, i.e. $f(u) \in B(x)$ This way, one is more interested in $f^{-1}(B) \subset U$. For some open set B. Hence f induces a map

$$G : \mathcal{O}(V) \rightarrow \mathcal{P}(U)$$

by "defining"

$$G(B) = f^{-1}(B)$$

where $\mathscr{O}(V)$ is the collection of open sets of V. From [2,3], we know that any $f : U \to V$ induces a geometric morphism

- the direct image $f_* : PSh(U) \to PSh(V)$
- the inverse image $f^* : PSh(V) \to PSh(U)$

where PSh stands for the topos of presheaves.

This way, we can construct the GO mapping for a cognitive frame as follows: Let $F : U \to V = \lim V_f$ be the factor mapping. $\forall s \in PSh(U)$, We will call

$$F^* \circ s : \mathscr{O}(V) \to \mathbf{Sets}$$

a GO mapping by "Fuzzy" set s. In fact, $\forall B \in \mathscr{O}(V), F^* \circ s(B) = s(F^{-1}(B))$. Sometime we also call the functor F^* the GO mapping, denote it by G.

References

1. Barr, M.: Fuzzy set theory and topos theory. Canad. Math. Bull **29**(4), 501–508 (1986)
2. Johnstone, P.T.: Topos Theory. Courier Corporation (2014)
3. Johnstone, P.T.: Sketches of an Elephant: A Topos Theory Compendium, vol. 1 and 2. Oxford University Press, New York (2002)
4. Ouyang, H.: A unified theory for theories. In: A Presentation at Dalian International Conference on Oriental Thinking and Fuzzy Logic, August 2015
5. Pierce, B.C.: Basic Category Theory for Computer Scientists. MIT Press, Cambridge (1991)
6. Pitts, A.M.: Fuzzy sets do not form a Topos. Fuzzy Sets Syst. **8**(1), 101–104 (1982)
7. Wang, P.-Z., et al.: Factor space, the theoretical base of data science. Ann. Data Sci. **1**(2), 233–251 (2014)

Cognitive Computing

A Case-Based Approach for Modelling the Risk of Driver Fatigue

Qiaoting Zhong and Guangnan Zhang(✉)

Center for Studies of Hong Kong, Macao and Pearl River Delta,
Institute of Guangdong, Hong Kong and Macao Development Studies,
Sun Yat-Sen University, Guangzhou 510275, China
zsuzgn@hotmail.com

Abstract. Fatigue-related crashes are one of the major threats to road safety worldwide. Despite the substantial work in the domain of transportation science by both the industry and academia, there are few studies in applying case-based reasoning (CBR) approach to modelling the risk of driver fatigue. This research explores the potential for fatigued driving using a database of 16,459 traffic crashes reported from 21 cities in Guangdong province, China from 2006 to 2010. The CBR system under development differentiates between fatigued-driving and non-fatigued-driving cases based on various personal and environmental traffic characteristics. The advantage of using CBR in modelling fatigued driving has been demonstrated through empirical evaluation.

Keywords: Intelligent information processing · Intelligent decision making · Case-based reasoning · Traffic safety management · Driver fatigue

1 Introduction

Driver fatigue has already become a leading factor contributing to traffic crashes around the world [48]. In China, fatigued driving caused 887 (9.26%) of all highway crashes in 2011, resulting in 520 (8.1%) deaths and over RMB 37 million (10.82%) property losses [43]. To decrease the occurrences of traffic crashes and promote road safety, studying drivers' decision on driving under fatigue condition is urgent. Despite the substantial work in the domain of transportation science by both the industry and academia, there are few studies in applying case-based reasoning approach to modelling the risk of driver fatigue.

Case-based reasoning (CBR) has been used in both cognitive science and artificial intelligence [32], it makes use of the most similar previous cases to solve new problems [27,30,32]. Since the early 1980s, CBR has been successfully applied in various fields such as legal reasoning [5,6,42,45], planning [14,15,25, 28,34], E-commerce [18,41], medical diagnosis [4,9,19,40], incident management [24,49], and risk analysis [23,35]. CBR is the best fit for modelling fatigued driving because of its ability to simultaneously process a large number of highly

Z. Shi et al. (Eds.): ICIS 2017, IFIP AICT 510, pp. 45–56, 2017.
DOI: 10.1007/978-3-319-68121-4_5

interrelated variables to arrive at a decision [13,20]. Moreover, CBR models have significant merits as compared to statistical ones (*e.g.*, the widely employed matched case-control logistic regression [2,31]) and other artificial intelligence models (*e.g.*, neural network) regarding the comprehensibility of output [13].

The novelty of the research work presented lies in the following aspects. First of all, this work explores the potential for using CBR in dealing with driver fatigue management. Secondly, this research examines a highly comprehensive set of risk factors that relate to the driver fatigue. A database of 16,459 real traffic crashes reported from 21 cities in Guangdong province, China from 2006 to 2010 is used, and a total of 50 variables in 21 categories are examined in this study. Compared with the logistic regression model that uses the same variables, our proposed CBR model had better performance in modelling fatigued driving on AUC measures in all simulations. The outcome of our proposed CBR system can help targeting the group of drivers that may intend to drive under fatigue condition, which constitutes the crucial part of driver safety management in an intelligent transportation system.

This paper is organised as follows. Section 2 introduces previous relevant work from the literature. Section 3 describes the methodology. Section 4 presents the case study and empirical evaluation of our approach. Section 5 concludes the study.

2 Related Work

In recent years, there had been an increase in uptake of case-based reasoning (CBR) concepts in the research area of traffic control and management. For instance, Jagannathan et al. [22] developed a CBR prediction system that is capable of differentiating between the accident and non-accident cases. Mounce et al. [36] designed a CBR system to help selection of signal timing plans. Sadek et al. [39] developed a prototype CBR routing system. Li and Zhao [33] applied CBR to intersection control. Kofod-Petersen et al. [26] presented a prototype implementation of a CBR system that predicts traffic flow and calculates signal plans for urban intersections. CBR methods have also been used to model the vehicle control behaviour (*i.e.*, steer, throttle, and brake) of teen drivers [37]. Different from the previous ones, this work aims to explore the potential for using CBR in dealing with driver fatigue management.

The risk of driver fatigue is related to a combination of situational and individual factors. The increased risk may result from a mix of biological, personal, road and environmental related factors [29]. Previous studies on detecting driver's state of fatigue have used various biological and personal indicators, such as visual indicators (*e.g.*, face images and eye state [10,12,17]), physiological sensor signals (*e.g.*, EEG [46] and ECG [7]) and driving states (*e.g.*, accelerate, brake, shift and steer [21]). However, none of them (including the CBR approach of identifying the drivers' stress state [7]) considered road and environmental features that are crucial towards drivers' intention to drive under fatigue condition. By using CBR in this research, the risk of driver fatigue is modelled based on past similar circumstances considering not only personal but also road and environmental characteristics. Moreover, most previous studies verify their models through experiments with driving simulators, whereas we validate ours by using the reliable official source of traffic crash data.

3 System Design

Following the work of [27,33], we propose a case-based system of fatigued driving that consists of three main components: representation module, retrieve module, and reuse module. Figure 1 shows a schematic overview of the system architecture. The proposed system receives traffic information, through the *representation module*, a target case C_t and a case-base $\mathcal{C}$ are constructed based on such information. The *retrieve module* retrieves the most similar previous case from the case-base, and *reuse module* further manipulates the retrieved cases for the best outcome of the target case.

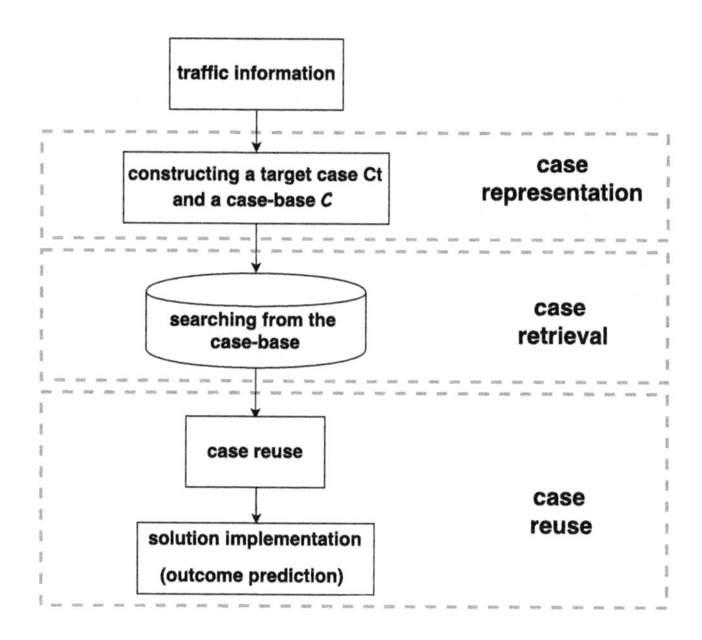

Fig. 1. Schematic diagram showing steps involved in modelling fatigued driving

3.1 Case Representation

A *case* in the system represents a previous traffic situation [33], and it is composed of a case number, a case description, and a case outcome. Formally, a case can be represented as follows:

- Case $C = \langle N, D, O \rangle$, where,
  - N is the case number;

- D is the case description;
- O is the case outcome (*i.e.*, fatigued-driving or non-fatigued-driving);

$\mathcal{C}$ is used to denote the case base that contains all possible cases.

The description of a case depicts the situation when the case appears [26,33]. Four categories of attributes are needed to describe the case. The first category contains the demographic information of the driver. The second one describes the vehicle information. The third category consists of road situation factors. The fourth category contains some environmental factors such as the weather situation. Formally, we have:

- Case description $D = \langle DD, VI, R, E \rangle$, where,
  - DD is the demographic information of the driver such as gender;
  - VI is the vehicle information such as vehicle type;
  - R is the road situation such as road type;
  - E is the environmental factors such as bad weather.

For example, a simple case (No. 0) that a crash occurred on the expressway at about 8 am on Saturday was caused mainly because the fatigued driving of the male motorcyclist would be represented as $\langle 0, \langle \{\text{male}\}, \{\text{motorcycle}\}, \{\text{expressway}\}, \{\text{moring, week-end}\} \rangle, \text{fatigued-driving} \rangle$.

3.2 Case Retrieval and Reuse

The crucial part of the retrieval module is the similarity measure, which calculates the similarity between the target case and the cases in the case-base [22]. For each target case, the similarity between the target case and the cases in the case-base is measured by calculating the Euclidean distance (the most widely used distance metric in CBR) between them. Given a target case t and a case c in the case-base, the similarity between both cases is:

$$S(t, c) = \sqrt{\sum_a (D_{t,a} - D_{c,a})^2}$$

where $D_{t,a}$ is the normalised value of attribute a $\in DD, VI, R, E$ in the target case t and $D_{c,a}$ is the normalised value of attribute a in the case c from the case base. Note that all the attributes are of equal importance in this paper for simplicity reasons.

With the above similarity measure, the most similar cases will be retrieved by using the K-nearest neighbour (KNN) algorithm [11], where K refers to the number of neighbours. The KNN algorithm, widely used in classification problems to assign objects to classes, matches each attribute in the target case to its corresponding attribute in the retrieved case [22]. The predicted outcome of the target case is then obtained by the majority vote of its K nearest neighbours, *i.e., the most common outcome of the K retrieved cases are reused and assigned to the target case directly.*

4 Case Study: Based on the Traffic Crash Data in China

In this study, we use the traffic crash data for the period 2006–2010 in Guangdong Province, China. These data are extracted from the Traffic Management Sector-Specific Incident Case Data Report, the Road Traffic Accident Database of China's Public Security Department. They are the only officially available source of traffic crash data in China. Data are recorded and reported by the traffic police on-scene who conducted assessments and provided feedback immediately to the headquarters of the Traffic Management Department. These reports include characteristics of drivers, vehicle features, road conditions, the time of crashes, the environmental context for each crash and the cause of crashes such as traffic violations like driving under fatigue condition [47]. To experiment with the proposed approach, we set up a case-based system in which each of the samples in our datasets was a case. Table 1 shows the case structure used

Table 1. Case structure

Case description		
Category	Feature	(Binary) attributes
Demography	Gender	Male
	Age	0–24, 25–44, 45 and above
	Residence	Urban
	Driving experience	0–2 years, 3–5 years, above 5 years
	Occupation	Clerk, migrant worker, farmer, self-employed, other
Vehicle	License condition	Valid
	Safety condition	Poor
	Insured or not	Insured
	Overloaded or not	Overloaded
	Vehicle type	Passenger car, truck, motorcycle
	Commercial or not	Commercial
Road	Road type	Express, first-class highway, second-class or below highway, urban expressway, urban ordinary highway, other
	Traffic lane	Isolated
	Road surface	Dry, wet, other
	Traffic control device	No device, signal, sign, other
Environment	Light condition	Daytime, with street lighting in the night, without street lighting in the night
	Weather	Rainy
	Weekend or not	Weekend
	Holiday or not	Holiday
	Time	00:00–06:59 (midnight to dawn), 07:00–08:59 (morning rush hours), 12:00–13:59 (noon), 17:00–19:59 (afternoon rush hours), other
	Season	Spring, summer, fall, winter
Case outcome: fatigued-driving, non-fatigued-driving		

in the experiment where relevant binary attributes are selected based on [48]. According to working time patterns and peoples' lifestyles in China, we classify five groups of time.

4.1 Evaluation Settings

To evaluate our proposed case-based system of drivers' risk of fatigued driving, the data set was partitioned into a training set and a test set. To be aligned with the literature (*e.g.*, [3]), we varied the percentage of the training and test data to study the possible variations of performance based on different partitions. In particular, ratios 10:90, 30:70, 50:50, 70:30 and 90:10 were used as the portions of training and test data. Each case in the test set was consecutively made the target case with the cases in the training set constituting the case-base. For each portion of training and test data, KNN algorithms with different numbers of K were run. Ties may occur in a binary classification problem if $K > 1$, to avoid ties, the most similar case ($K = 1$), the three most similar cases ($K = 3$), and the five most similar cases ($K = 5$) were retrieved in the current study. In comparison, the widely employed logistic regression was also conducted for each partition. The dependent variable indicated the occurrence of driving under fatigue condition in a crash, while the independent variables under consideration were the same as the CBR setting.

The performance of a computational intelligent system can be measured in many different ways. Typically, the models' predictive performances are measured regarding *overall accuracy*, *e.g.*, the percentage of fatigued-driving and non-fatigued-driving classified accurately in the current research [16]. However, other measures like the *true positive rate* (TPR, also called recall or sensitivity) tends to be the key concern when infrequent events (as fatigued driving in the current study) are to be predicted [44]. Complementary to TPR that measures the proportion of positives that are correctly identified as such, another primary evaluator is the *false positive rate* (FPR, also known as fall-out or false alarm ratio) that calculates the ratio of negatives wrongly categorised as positive. Based on pairs of TPR and FPR, a receiver operating characteristic (ROC) curve could be obtained to compare the performance of different models. We calculate the area under the ROC curve based on the work of Cantor and Kattan [8], and use IBM SPSS Statistics 23 to perform all the statistical analysis.

4.2 Results and Discussion

The available data set contained 16,459 traffic crashes, of which 384 (2.33%) were fatigue-related. Among all the fatigue-related crashes, 99% of drivers were male, 88% of them had more than two years of driving experience, and 72.4% came from urban areas. Drivers that were involved in fatigue-related crashes included self-employed workers (23.7%), migrant workers (18.5%), farmers (17%), clerks (8%), and employees from other professions (32.8%). Concerning vehicle information, trucks and motorcycles constituted 56% and 16.7%, respectively. For road situation, 65.9% of these fatigue-related crashes occurred in separate lanes,

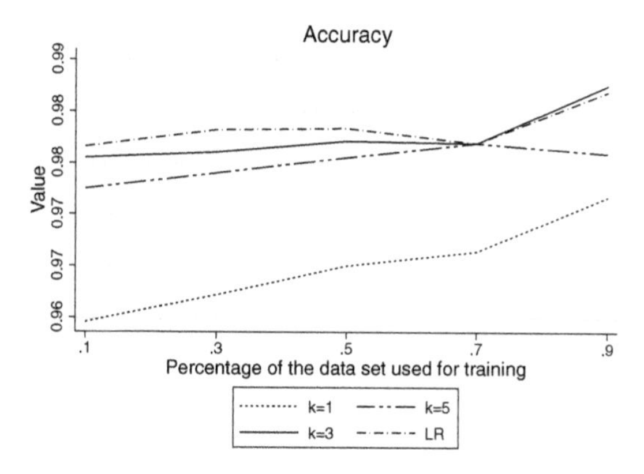

Fig. 2. Accuracy of the different models

Table 2. Evaluation results (TPR for true positive rate and FPR for false positive rate)

	10:90				30:70				50:50			
	k=1	k=3	k=5	LR	k=1	k=3	k=5	LR	k=1	k=3	k=5	LR
TPR	0.12	0.01	0	0	0.16	0.04	0.02	0	0.22	0.06	0.02	0
FPR	0.02	0	0	0	0.02	0.01	0	0	0.02	0	0	0

	70:30				90:10			
	k=1	k=3	k=5	LR	k=1	k=3	k=5	LR
TPR	0.26	0.11	0.04	0	0.23	0.17	0.1	0
FPR	0.02	0	0	0	0.01	0.01	0	0

and 38% occurred in expressway. Considering environmental features, 51.8% of the fatigue-related crashes occurred during 00:00–06:59, and 38% took place at daytime. Seasonal distribution of the fatigue-related crashes: 32.3%, 26.3%, 24.7%, and 16.7% occurred in the summer, winter, spring, and fall, respectively.

Figure 2 shows the comparison of the overall accuracy for all simulations. In overall, all simulations received very high accuracy in modelling fatigued driving, *i.e.*, more than 96%. Logistic regression (LR) had a better performance in comparison to the proposed CBR method (no matter what the value of K) for most training/test partitions. When 70% or above of the dataset was used for training, the CBR method with $K = 3$ achieved slightly higher accuracy than LR.

Table 2 summarises the results of true positive rate (TPR) and false positive rate (FPR) from all simulations. In all scenarios, we observed that the FPR is nearly zero, indicating that both LR and the proposed CBR method have excellent performance in correctly identifying non-fatigued-driving cases. However, LR received zero TPR for all the training/test partitions, which reflects that LR is relatively biased towards the majority class in our dataset (*i.e.*, the non-fatigued-driving case). Such bias may be caused by the imbalance situation of our dataset (*i.e.*, only 2.33% cases were fatigue-related). Recall that TPR

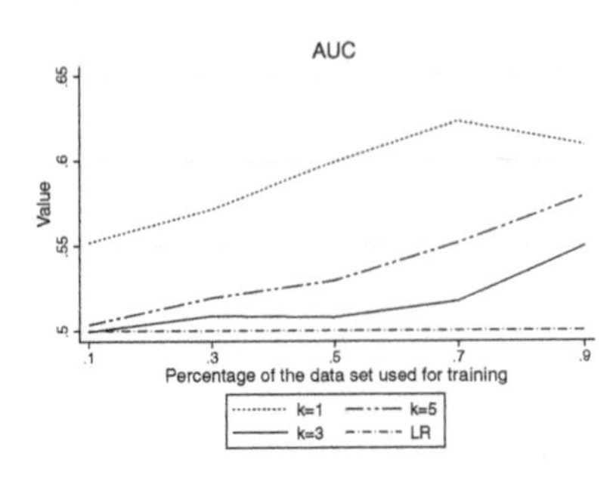

Fig. 3. AUC of the different models

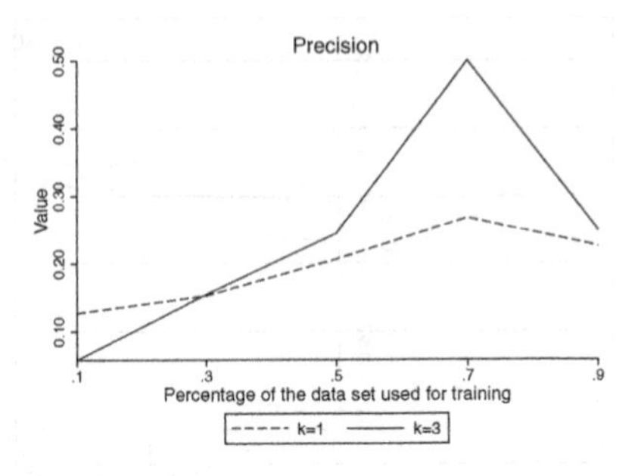

Fig. 4. Precision of the K = 1 and K = 3 models

reflects the hits of fatigued driving reality, in such circumstances, although LR got very high accuracy and very low FPR, it is meaningless [44]. In contrast, our proposed CBR models with K = 1 or K = 3 had shown their merits by receiving non-zero TPR in all the training/test partitions.

Figure 3 illustrates the trade-off between the TPR and the FPR of each simulation. Our proposed CBR models (no matter what the value of K) had better performance than LR in modelling fatigued driving on AUC measures. Simulation of K = 1 outperformed the rest as the percentage ratio of the training and test data were varied.

To sum up, we can see that the proposed CBR model with K = 1 and that with K = 3 have their strengths in modelling fatigued driving. To further differentiate them, we calculated the *precision* values in various partitions of training and test data. Precision measures the probability that a driver with a positive screening test indeed conducted fatigued driving. As shown in Fig. 4, when 10% of the data set were used for training, the K = 1 model had higher precision value, whereas the K = 3 model had better performance when more than 10% were used for training. Further empirical validation is required to study the optimal value of K for modelling fatigued driving.

4.3 Further Implications

Fatigue-related crashes are one of the major threats to road safety worldwide. The important implication of our research is: if driving under fatigue condition could be reduced or controlled successfully through the early detection, the occurrences of a crash would be reduced accordingly. When designing road traffic interventions to decrease incidences, it is well established that a change in the driver's attitude is of utmost importance. The outcome of our proposed CBR model can help targeting the group of drivers that may have an intention to drive under fatigue condition. In particular, our proposed CBR model can highlight the important features of such group of drivers in the aspects of various personal, vehicle, road, and environmental conditions. Specific countermeasures by integrating these highlighted features into a driver fatigue detection component of the intelligent transportation systems will be effective in improving traffic safety [38]. The results obtained in this study may likely be generalised to other provinces in China because the crash data in Guangdong Province and features considered are rather representative and comprehensive. The fact that Guangdong has a high percentage of residents migrating from other provinces, and is having one of the largest numbers of vehicles among all provinces in China reinforces the generalisability [48]. Furthermore, the findings of this study can also contribute as a reference to future road safety research for other countries.

5 Limitations and Future Work

A live case-based reasoning system has a "4 REs" cycle (*i.e.*, retrieve, reuse, revise, and retain) [1]. Since our proposed case-based system is not operating on live data, there is no component designed in the current study for the standard revise and retain process. Our proposed system could be extended to deal with live data by adding a new part of revise and retain in the future.

The advantage of using case-based reasoning in modelling fatigued driving has been demonstrated through comparing with logistic regression. In getting higher values of recall, precision and AUC, an important consideration is case attribute selection and weighting. The attributes we have chosen are based on previous work. However, it is likely that fatigued driving also occurs due to other conditions that have not yet been studied. Moreover, the contribution of each attribute to the similarity between the target case and the cases in case-base could differ as well. Thus, it is worth testing out different attributes and weighted assignments for attributes considered.

The empirical experiments show that the CBR system under development is capable of differentiating between fatigued-driving and non-fatigued-driving cases to some extent based on certain personal and environmental traffic characteristics. The evaluation of our proposed method is not exhaustive and could be expanded in several directions, for instance, to compare with other artificial intelligence techniques (*e.g.*, neural network). It would also be worth evaluating whether or not our proposed approach could be fruitfully applied to other risky behaviours rather than the kind we have considered.

Acknowledgements. This research was supported in part by the National Natural Science Foundation of China grant 71573286 and Ministry of Education Project for Humanities and Social Sciences Research (16JJDGAT006).

References

1. Aamodt, A., Plaza, E.: Case-based reasoning: foundational issues, methodological variations, and system approaches. AI Commun. **7**(1), 39–59 (1994)
2. Abdel-Aty, M., Pande, A., Lee, C., Santos, V.G.C.D.: Crash risk assessment using intelligent transportation systems data and real-time intervention strategies to improve safety on freeways. J. Intell. Transp. Syst. **11**, 107–120 (2007)

3. Adedoyin, A., Kapetanakis, S., Petridis, M., Panaousis, E.: Evaluating case-based reasoning knowledge discovery in fraud detection. In: Workshops Proceedings of the 24th International Conference on Case-Based Reasoning, pp. 182–191 (2016)
4. Althoff, K.D., Bergmann, R., Wess, S., Manago, M., Auriol, E., Larichev, O.I., Gurov, S.I.: Case-based reasoning for medical decision support tasks: the inreca approach. Artif. Intell. Med. **12**(1), 25–41 (1998)
5. Ashley, K.D.: Modeling Legal Arguments: Reasoning with Cases and Hypotheticals. MIT Press, Cambridge (1991)
6. Athakravi, D., Satoh, K., Broda, K., Russo, A.: Generating legal reasoning structure by answer set programming. In: Proceedings of the 8th International Workshop on Juris-informatics, pp. 24–37 (2014)
7. Begum, S., Ahmed, M.U., Funk, P., Filla, R.: Mental state monitoring system for the professional drivers based on heart rate variability analysis and case-based reasoning. In: Proceedings of Federated Conference on Computer Science and Information Systems, pp. 35–42 (2012)
8. Cantor, S.B., Kattan, M.W.: Determining the area under the ROC curve for a binary diagnostic test. Med. Decis. Making **20**(4), 468–470 (2000)
9. Chattopadhyay, S., Banerjee, S., Rabhi, F.A., Acharya, U.R.: A casebased reasoning system for complex medical diagnosis. Expert Syst. **30**(1), 12–20 (2013)
10. Chen, D., Wang, Z., Zhuo, Q., Wu, F.: A new algorithm for fatigue detection in driving. In: Wen, Z., Li, T. (eds.) Foundations of Intelligent Systems. AISC, vol. 277, pp. 1041–1049. Springer, Heidelberg (2014). doi:10.1007/978-3-642-54924-3_98
11. Cover, T., Hart, P.: Nearest neighbor pattern classification. IEEE Trans. Inf. Theory **13**(1), 21–27 (1967)
12. Devi, M.S., Bajaj, P.R.: Driver fatigue detection based on eye tracking. In: Proceedings of the First International Conference on Emerging Trends in Engineering and Technology, pp. 649–652 (2008)
13. Goh, Y.M., Chua, D.K.H.: Case-based reasoning for construction hazard identification: case representation and retrieval. J. Constr. Eng. Manag. **135**(11), 1181–1189 (2009)
14. Hammond, K.: CHEF: a model of case-based planning. In: Proceedings of the Fifth National Conference on Artificial Intelligence, pp. 267–271 (1986)
15. Hammond, K.J.: Case-Based Planning: Viewing Planning as a Memory Task. Elsevier, Amsterdam (2012)
16. Han, J., Kamber, M., Pei, J.: Data Mining: Concepts and Techniques. Morgan Kaufmann, Burlington (2011)
17. He, J., Roberson, S., Fields, B., Peng, J., Cielocha, S., Coltea, J.: Fatigue detection using smartphones. J. Ergon. **3**(3), 1–7 (2013)
18. Heras, S., Jordn, J., Botti, V., Julian, V.: Argue to agree: a case-based argumentation approach. Int. J. Approx. Reason. **54**(1), 82–108 (2013)
19. Holt, A., Bichindaritz, I., Schmidt, R., Perner, P.: Medical applications in case-based reasoning. Knowl. Eng. Rev. **20**(3), 289–292 (2005)
20. Hossain, M., Muromachi, Y.: A Bayesian network based framework for real-time crash prediction on the basic freeway segments of urban expressways. Accid. Anal. Prev. **45**, 373–381 (2012)
21. Wang, H., Liu, H., Song, Z.: Fatigue driving detection system design based on driving behavior. In: Proceedings of the 2010 International Conference on Optoelectronics and Image Processing, pp. 549–552 (2010)

22. Jagannathan, R., Petrovic, S., Powell, G., Roberts, M.: Predicting road accidents based on current and historical spatio-temporal traffic flow data. In: Proceedings of 2013 International Conference on Computational Logistics, pp. 83–97 (2013)
23. Jin, R., Han, S., Hyun, C., Cha, Y.: Application of case-based reasoning for estimating preliminary duration of building projects. J. Constr. Eng. Manag. **142**(2) (2016). http://dx.doi.org/10.1061/(ASCE)CO.1943--7862.0001072
24. Johnson, C.: Using case-based reasoning to support the indexing and retrieval of incident reports. In: Proceedings of European Safety and Reliability Conference: Foresight and Precaution, pp. 1387–1394 (2000)
25. Khattak, A., Kanafani, A.: Case-based reasoning: a planning tool for intelligent transportation systems. Transp. Res. Part C: Emerg. Technol. **4**(5), 267–288 (1996)
26. Kofod-Petersen, A., Andersen, O.J., Aamodt, A.: Case-based reasoning for improving traffic flow in urban intersections. In: Lamontagne, L., Plaza, E. (eds.) ICCBR 2014. LNCS, vol. 8765, pp. 215–229. Springer, Cham (2014). doi:10.1007/978-3-319-11209-1_16
27. Kolodner, J.: Case-Based Reasoning. Morgan Kaufmann, Burlington (1993)
28. Kolodner, J.L.: Retrieval and Organisation Strategies in Conceptual Memory: A Computer Model. Erlbaum, Mahwah (1984)
29. Lal, S.K., Craig, A.: A critical review of the psychophysiology of driver fatigue. Biol. Psychol. **55**(3), 173–194 (2001)
30. Leake, D.B.: CBR in context: the present and future. In: Leake, D.B. (ed.) Case-Based Reasoning, Experiences, Lessons, & Future Directions, p. 330 (1996)
31. Lee, C., Abdel-Aty, M., Hsia, L.: Potential real-time indicators of sideswipe crashes on freeways. Transp. Res. Rec.: J. Transp. Res. Board **1953**, 41–49 (2006)
32. Li, K., Waters, N.M.: Transportation networks, case-based reasoning and traffic collision analysis: a methodology for the 21st century. In: Methods and Models in Transport and Telecommunications: Cross Atlantic Perspectives, pp. 63–92 (2005)
33. Li, Z., Zhao, X.: A case-based reasoning approach to urban intersection control. In: Proceedings of the 7th World Congress on Intelligent Control and Automation, pp. 7113–7118 (2008)
34. Liu, W., Hu, G., Li, J.: Emergency resources demand prediction using case-based reasoning. Saf. Sci. **50**, 530–534 (2012)
35. Lu, Y., Li, Q., Xiao, W.: Case-based reasoning for automated safety risk analysis on subway operation: case representation and retrieval. Saf. Sci. **57**, 75–81 (2013)
36. Mounce, R., Hollier, G., Smith, M., Hodge, V.J., Jackson, T., Austin, J.: A metric for pattern-matching applications to traffic management. Transp. Res. Part C: Emerg. Technol. **29**, 148–155 (2013)
37. Ontañón, S., Lee, Y.-C., Snodgrass, S., Bonfiglio, D., Winston, F.K., McDonald, C., Gonzalez, A.J.: Case-based prediction of teen driver behavior and skill. In: Lamontagne, L., Plaza, E. (eds.) ICCBR 2014. LNCS, vol. 8765, pp. 375–389. Springer, Cham (2014). doi:10.1007/978-3-319-11209-1_27
38. Paaver, M., Eensoo, D., Kaasik, K., Vaht, M., Mäestu, J., Harro, J.: Preventing risky driving: a novel and efficient brief intervention focusing on acknowledgement of personal risk factors. Accid. Anal. Prev. **50**, 430–437 (2013)
39. Sadek, A.W., Demetsky, M.J., Smith, B.L.: Case-based reasoning for real-time traffic flow management. Comput.-Aided Civil Infrastruct. Eng. **14**(5), 347–356 (1999)
40. Sharaf-El-Deen, D.A., Moawad, I.F., Khalifa, M.E.: A new hybrid case-based reasoning approach for medical diagnosis systems. J. Med. Syst. **38**(2), 1–11 (2014)

41. Shimazu, H.: Expertclerk: a conversational case-based reasoning tool for developing salesclerk agents in e-commerce webshops. Artif. Intell. Rev. **18**(3–4), 223–244 (2002)
42. Simpson, R.L.: A computer model of case-based reasoning in problem solving: an investigation in the domain of dispute mediation. In: Georgia Institute of Technology, School of Information and Computer Science Technical report no. GIT-ICS-85/18 (1985)
43. Traffic Management Bureau, Ministry of Public Security. China Road Traffic Accidents Annual Statistical Report 2011, PRC, China (2011)
44. Weiss, G.M., Hirsh, H.: Learning to predict extremely rare events. In: Proceedings of the AAAI 2000 Workshop on Learning from Imbalanced Data Sets, pp. 64–68 (2000)
45. Yeow, W.L., Mahmud, R., Raj, R.G.: An application of case-based reasoning with machine learning for forensic autopsy. Expert Syst. Appl. **41**(7), 3497–3505 (2014)
46. Zhang, C., Wang, H., Fu, R.: Automated detection of driver fatigue based on entropy and complexity measures. IEEE Trans. Intell. Transp. Syst. **15**(1), 168–177 (2014)
47. Zhang, G., Yau, K.K., Chen, G.: Risk factors associated with traffic violations and accident severity in China. Accid. Anal. Prev. **59**, 18–25 (2013)
48. Zhang, G., Yau, K.K., Zhang, X., Li, Y.: Traffic accidents involving fatigue driving and their extent of casualties. Accid. Anal. Prev. **87**, 34–42 (2016)
49. Zubair, M., Khan, M.J., Awais, M.M.: Prediction and analysis of air incidents and accidents using case-based reasoning. In: Proceedings of the Third Global Congress on Intelligent Systems, pp. 315–318 (2012)

Gazes Induce Similar Sequential Effects as Arrows in a Target Discrimination Task

Qian Qian[1(✉)], Xiaoting Wang[1], Miao Song[2], and Feng Wang[1]

[1] Yunnan Key Laboratory of Computer Technology Applications, Kunming University of Science and Technology, 727 Jing Ming Nan Lu, Kunming 650500, Yunnan, China
qianqian_yn@126.com

[2] Information Engineering College, Shanghai Maritime University, Shanghai 200135, China

Abstract. Symbolic cueing paradigm has been widely used to investigate the attention orienting induced by centrally-presented gaze or arrow cues. Previous studies have found a sequence effect in this paradigm when arrows are used as central cues and simple detection tasks are included. The present study investigated the universality of the sequence effect with gaze cues and in discrimination tasks. It was found that sequence effects are not limited to specific cue types or specific tasks, and the sequence effect can even generalize across different cue types (from gaze to arrow, or from arrow to gaze). In addition, the sequence effect is not influenced by the repetition and switch of target identities (along with response keys). The results suggest that sequential processing is a common mechanism in attention orienting systems, and support the automatic retrieval hypothesis more than the strategy adjustment account.

Keywords: Attention orienting · Cueing effect · Sequence effect · Memory retrieval

1 Introduction

Orienting of attention refers to the ability for our attention system to be able to select pertinent input for further processing according to external cues [1]. It has been found that perception of a pointing arrow is enough to shift our attention to the pointed location reflexively [2]. Besides arrows, another people's gaze can also shift attention even when these symbolic cues are uninformative for our task in hand [3]. In a typical study of the symbolic cueing, observers were presented with a centrally-presented symbolic cue (e.g., a gaze or an arrow) indicating left or right, and after a certain time interval (stimulus onset asynchrony (SOA)) were instructed to respond to the appearance of a target to the left or right of the central cue. Although observers were told that the direction of the central cue did not predict where the target would occur, reaction time (RT) was reliably faster when the cue direction was toward (i.e., valid trials), rather than away from (i.e., invalid trials), the target. This facilitation of RT between valid and invalid trials is referred to as cueing effect, which is considered to be evidence of attention orienting.

Z. Shi et al. (Eds.): ICIS 2017, IFIP AICT 510, pp. 57–65, 2017.
DOI: 10.1007/978-3-319-68121-4_6

Several recent studies [4, 5] reported a sequence effect in symbolic cueing paradigm. That is, when using a predictive central arrow cue (i.e., the ratio of valid trials among all trials were 80%), the cueing effect in one trial is significant smaller when the previous trial is an invalid trial than when it is a valid trial. Since the cues of these studies successfully predict the target locations in most of the trials, the sequence effect has been attributed to the continuous updating of the predictive value that participants assign to the spatial cue. In other words, participants adapt their utilization of the cue depending on whether it is correctly or wrongly directed their attention on the previous trial. On the contrary, another explanation about the sequence effect observed in these studies is automatic memory processes in which information of previous trials is automatically retrieved from memory to facilitate performance on current trials. For example, when the previous trial type (valid or invalid) is consistent with the current trial type, performance will be facilitated, whereas when the previous and current trial types differ, performance is slowed due to the conflict between the two trial types. This automatic retrieval hypothesis is in line with the view from peripheral cueing studies [6] and further supported by the results of [7], in which the sequence effect was still found when the arrow cues did not predict the target location and the participants were explicitly asked to ignore the arrow cues.

In symbolic cueing studies, arrow and gaze cues are two representative attentional cues. However, so far as we know, the sequence effect of symbolic cueing is investigated only with arrow cues and only in a simple detection task [4, 7]. Therefore, there still have some questions for the sequence effect. First, whether sequence effects can be found in other cueing tasks. For example, a simple detection task only need participants to press one key whenever they detect the targets, but a discrimination task will ask participants to choose response keys according to the target identities. In the simple detection task, there are only one target and one response, but in the discrimination task, target identities and responses either repeated or switched between consecutive trials. We still don't know whether sequence effects exist in a discrimination task, and whether the change of target identities and responses between trials influences the sequence effect; Second, whether or not gaze cues induce sequence effects, this question is important because it can help to answer the question whether or not sequential processes in human attention orienting systems are common mechanisms responding to central symbolic cues; Third, if sequential processes are common mechanisms, whether or not the sequence processes of symbolic cueing can generalize across different cue types (e.g., from gaze to arrow cues), and whether there are differences between them.

In experiment 1, sequence effects were tested with a central gaze cue. If sequence effects are common mechanisms in symbolic cueing, significant sequence effects should also be observed in gaze cue condition. In addition, different from previous studies, a discrimination task, instead of a simple detection task, was used to test the flexibility of the sequence effect. If sequence effects are robust phenomenon in symbolic cueing, significant sequence effects should also be observed with a discrimination task. Considering that in the previous study [7], significant sequence effects were only found when the SOA of previous trials was relatively long, we expect that sequence effects of gaze cueing are significant, but only for long SOA conditions.

Another advantage for choosing a discrimination task instead of a simple detection task is that there are two kinds of targets, and participants need to press different buttons to respond to the different targets. Therefore, target identities and its corresponding responses will either repeated or switched between consecutive trials. By using a simple detection task, previous study [7] found a significant influence of target location alternation on sequence effects in arrow cueing. However, because participants only need to press one specific button to respond to the appearance of one specific target in a simple detection task, whether the sequence effect is also influenced by the repetition and switch of target identities (along with its corresponding responses) is still unknown. With a discrimination task, we can investigate how target locations and target identities affect sequence effects.

Experiment 2 was aimed to compare the sequence effects induced by gaze and arrow cues. During the experimental trials, gaze and arrow cues were randomly mixed to form four kinds of cue sequence conditions between consecutive trials: gaze-gaze sequence, gaze-arrow sequence, arrow-gaze sequence, and arrow-arrow sequence. This experimental design had several advantages. First, the sequence effects induced by gaze and arrow cues could be compared directly by comparing the magnitude of sequence effects between gaze-gaze sequence and arrow-arrow sequence conditions. Different from arrow cues, cueing effects by gazes are based on mechanisms specialized for gaze perception [8]. The biological-significance of the gaze cues should make it more meaningful and important for participants. Therefore, according to strategy adjustment account, participants may adapt their utilization of the gaze cues more strongly than arrow cues, leading to an enhanced sequence effect in gaze cueing. However, an automatic retrieval hypothesis will predict similar sequence effects for both gaze and arrow cues.

Second, the experimental design allowed us to investigate whether the sequence effect could generalize across cue types. According to the strategy adjustment account, sequence effects are originated from continuous updating of the predictive value that participants assign to the cue. Therefore, the change of cue types between consecutive trials should eliminate the sequential processing, because the last cue is new for participants and there is no reason for participants to update the predictive value of a new cue. On the contrary, an automatic retrieval hypothesis will predict undistinguishable sequence effects for four kinds of cue sequence conditions.

2 Experiment 1

2.1 Participants

A total of 33 students (with a mean age of 23 years, range 19 to 27 years, 13 females) consented to participate in this experiment. All participants were right-handed and reported normal or corrected-to-normal vision. All participants were naive as to the purpose of the experiment.

2.2 Apparatus and Stimuli

The stimuli were presented on a LCD display operating at a 75 Hz frame rate. The participants were seated approximately 60 cm away from the screen in a dimly-lit room.

A cross, subtending 1°, was placed at the center of the screen as a fixation point. Face photographs were included as central cues. The central stimulus was a photograph of a female face, about 4° wide and 7° height, displayed in eight-bit grayscale. The face photograph was manipulated to produce the left-gaze and right-gaze cues by cutting out the pupil/iris area of each eye and pasting it into the left and right corner of each eye, respectively, using Photoshop CS2 software. The target stimulus was a capital letter ('X' or 'O') measuring 1° wide, 1° high, and was presented 15° away from the fixation point on the left or right side.

2.3 Design and Procedure

The cue-target SOAs were 100 ms and 600 ms. On each trial, cue direction, target location, and SOA duration were selected randomly and equally. Therefore, the cue validity was 50%. For each participant, there were six blocks with 80 trials each. Including 20 training trials, there were in total 500 trials for each participant.

Participants were instructed to keep fixating on the center of the screen. First, a fixation cross appeared at the center of the screen for 1000 ms, and then the cue stimulus appeared. After a certain cue-target SOA, a target letter 'X' or 'O' appeared either at left or right until participants had responded or 1500 ms had elapsed. The cue stimulus was still remained on the screen after the appearance of the target. Participants were instructed to respond to the identity of the target letters by pressing the 'UP' key with their middle finger of the right hand or the 'DOWN' key with their index finger of the right hand, the mapping between the responses and the target letters were counter-balanced across the participants. Participants were also informed that the central stimuli did not predict the location in which target would appear, and that they should try to ignore the central cues.

2.4 Results

The participants missed or responded to a wrong target at about 2.6% of the all trials. Anticipations (RT of less than 100 ms) and outliers (RT over 1000 ms) were classified as errors and were excluded from analysis. After that, responses with RTs exceeding plus or minus two standard deviations of each participant's mean RT on each single cell of the design were also removed as errors. As a result, about 8.4% of all trials were removed as errors. The error rates did not vary systematically and no signs of any speed-accuracy trade-off were observed.

The mean RTs under different conditions can be seen in Fig. 1. A four-way ANOVA with previous SOA (pre-100 ms and pre-600 ms), SOA (100 ms and 500 ms), previous cue validity (pre-valid and pre-invalid), and cue validity (valid and invalid) as within-participants factors was conducted on the RTs. There was a significant main effect of previous SOA, $F(1, 32) = 9.122$, $p < .005$, with RTs becoming slower as the previous SOA was increased. The main effect of SOA was also significant, $F(1, 32) = 30.589$, $p < .0001$, with RTs becoming shorter as the current SOA was increased. The interaction between previous SOA and previous cue validity was

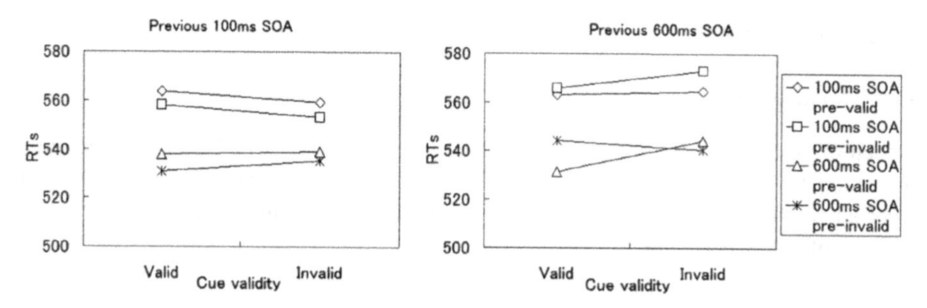

Fig. 1. Mean reaction times (RTs) under different previous and current cue validities and SOAs in Experiment 1.

significant, $F(1, 32) = 9.096$, $p < .005$. This interaction was qualified by a significant previous SOA × current SOA × previous cue validity × cue validity interaction, $F(1, 32) = 4.200$, $p < .049$. Further tests revealed that the previous cue validity × cue validity interaction (indicating the sequence effects) were only significant when both the previous and current SOAs were 600 ms, $F(1, 32) = 6.787$, $p < .014$, but not for other combinations of previous and current SOAs (ps > .4). The average sequence effects (calculated by cueing effects of pre-valid trials minus cueing effects of pre-invalid trials) for different SOA conditions are 0.3 ms (pre100–100), −3.2 ms (pre100–600), −6.1 ms (pre600–100), and 16.7 ms (pre600–600). No other effects or interactions reached significance.

In order to investigate the influence of target locations and target identities on the sequence effects, a four-way ANOVA with target location (repeated and switched), target identity (repeated and switched), previous cue validity (pre-valid and pre-invalid), and cue validity (valid and invalid) as within-participants factors was conducted on the RTs. The main effects of target location and target identity were both significant, $F(1, 32) = 15.044$, $p < .0001$, and $F(1, 32) = 37.151$, $p < .0001$, respectively. The interaction between target location and target identity was also significant, $F(1, 32) = 200.270$, $p < .0001$, reflecting that alternation of target identity between trials slowed RTs only when target location repeats between trials. The interaction between target location and previous cue validity was significant, $F(1, 32) = 12.734$, $p < .001$. No other effects or interactions reached significance.

The results of experiment 1 showed that significant sequence effects can be found in a discrimination task with gaze cues. Furthermore, similar to the finding of previous study [7], sequence effects were only significant under relatively long SOA conditions. However, contrary to the previous study [7], no significant influence of target location (or target identity) on the sequence effect was found. This is probably because that the sequence effects are not significant in most of trials when previous or current SOAs are short. An investigation that only involves the pre600–600 SOA condition will be infeasible for the small trial numbers. Therefore, the influence of target location and target identity will be investigated again in experiment 2.

3 Experiment 2

3.1 Participants

A total of 36 students (with a mean age of 23 years, range 19 to 27 years, 15 females) consented to participate in this experiment. All participants were right-handed and reported normal or corrected-to-normal vision. All participants were naive as to the purpose of the experiment.

3.2 Apparatus and Stimuli

The apparatus and the stimuli were the same as that in experiment 1 except that an arrow cue was included. For arrow cues, the central fixation stimulus was a horizontal line centered on the screen, 3° in length. An arrow head and an arrow tail appeared at the ends of the central line, both pointing left or both pointing right. The length of an arrow, from the tip of the arrow head to the ends of the tail, was 4°.

3.3 Design and Procedure

A relatively long SOA (600 ms) was used to ensure the capability of the design to induce sequence effects. On each trial, cue type (gaze cue or arrow cue), cue direction, and target location were selected randomly and equally. There were eight blocks with 80 trials each. Including 20 training trials, there were in total 660 trials for each participant. The procedure was the same as that in experiment 1.

3.4 Results

The participants missed or responded to wrong targets at an average of about 2.1% of all trials. Anticipations (RT of less than 100 ms) and outliers (RT over 1000 ms) were classified as errors and were excluded from analysis. After that, responses with RTs exceeding plus or minus two standard deviations of the participant's mean RT on each single cell of the design were also removed as errors. As a result, about 7.5% of all trials were removed as errors. The error rates did not vary systematically and no signs of any speed-accuracy trade-off were observed.

A four-way ANOVA with previous cue type (pre-gaze or pre-arrow), cue type (gaze or arrow), previous cue validity (pre-valid and pre-invalid), and cue validity (valid and invalid) as within-participants factors was conducted on the RTs. There was a significant main effect of cue validity, $F(1, 35) = 50.446$, $p < .0001$, demonstrating the cueing effect, and a significant interaction between previous validity and cue validity, $F(1, 35) = 14.06$, $p < .001$, indicating the significant sequence effect. As for the influence of cue types, the interaction between previous cue type and previous cue validity was significant, $F(1, 35) = 8.265$, $p < .007$, reflecting that RTs were shorter for pre-valid than for pre-invalid condition when the previous cue was an arrow, but this tendency was reversed when the previous cue was a gaze. The interaction between previous cue type and cue validity was also significant, $F(1, 35) = 4.548$, $p < .04$, indicating that cueing effects were larger for pre-arrow condition than for pre-gaze

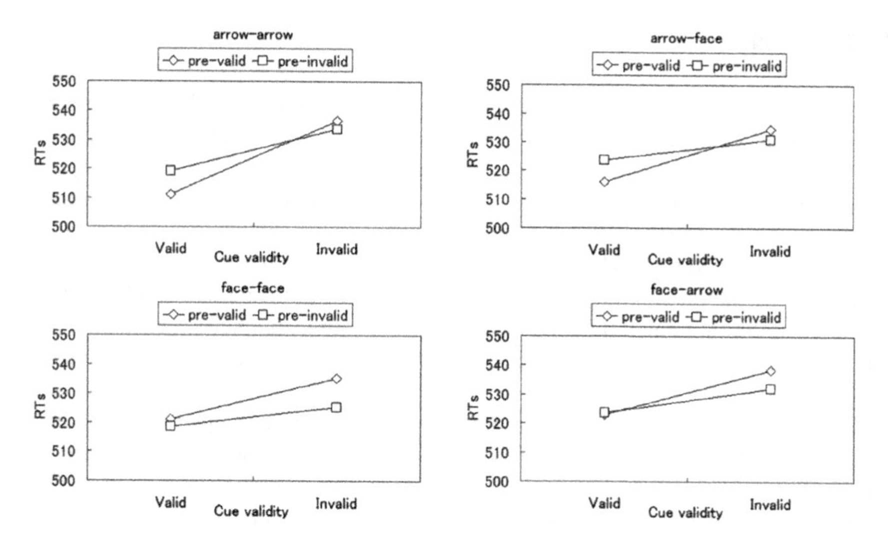

Fig. 2. Mean reaction times (RTs) under different previous and current cue validities and different combination of cue types between trials in Experiment 2.

condition. No other factors or interactions were significant. The average sequence effects for different cue sequence conditions are 11.1 ms (arrow-face sequence), 10.6 ms (arrow-arrow), 7.1 ms (face-arrow), and 7.2 ms (face-face). Paired-samples t tests confirmed that the magnitude of these four sequence effects is not significantly different from each other (ps > .59). The mean RTs under different cue sequence conditions can be seen in Fig. 2.

Similar to the experiment 1, a four-way ANOVA with target location (repeated and switched), target identity (repeated and switched), previous cue validity (pre-valid and pre-invalid), and cue validity (valid and invalid) as within-participants factors was conducted on the RTs. The main effects of target location and target identity were both significant, $F(1, 35) = 31.779$, $p < .0001$, and $F(1, 35) = 35.923$, $p < .0001$, respectively. Similar to that in experiment 1, the interaction between target location and target identity was significant, $F(1, 35) = 121.552$, $p < .0001$, demonstrating that alternation of target identity between trials slowed RTs only when target location repeats between trials. The interaction between target location and previous cue validity was also significant, $F(1, 35) = 25.232$, $p < .0001$. In addition, there was a significant main effect of cue validity, $F(1, 35) = 41.302$, $p < .0001$, demonstrating the cueing effect, and a significant interaction between previous validity and cue validity, $F(1, 35) = 12.774$, $p < .001$, indicating the sequence effect. Importantly, the target location × previous cue validity × cue validity interaction were marginally significant, $F(1, 35) = 3.989$, $p = .054$, replicating the influence of target locations on the sequence effect that was observed in the previous study. No other effects or interactions reached significance.

In all, significant sequence effects were found for all four combination conditions of two cue types. In addition, although the influence of target locations on the sequence effect was replicated, no significant influence of target identity alternation was found.

4 Discussion

The present study investigated the sequence effects (i.e., trial-by-trial influence) induced by gaze and arrow cues in cueing paradigm. In experiment 1, participants were asked to discriminate targets in a gaze cueing task, and it was found that gaze cues induced significant sequence effects in long SOA conditions. In addition, sequence effects were not influenced by the repetition and switch of target identities and its corresponding responses. When gaze and arrow cues were intermixed in experiment 2, the four combination conditions of cue types induced significant and similar sequence effects. In addition, though the influence of target locations on the sequence effects that was reported previously [7] was replicated, no significant influence of target identities was found.

Results from experiment 1 extend our knowledge about the sequence effects for the following aspects. First, sequence effects still exist when gazes are used as central cues. This finding suggests that sequential processing is a common phenomenon in symbolic cueing paradigm and is not limited to specific cue types, such as arrows.

Second, the appearance of sequence effects does not depend on specific task demands. So far as we know, all previous studies used the simple detection task to test the sequence effects. Before this study, it remains unknown whether or not the sequence effects can be found with different tasks.

Third, both experiment 1 and experiment 2 found that overall RTs were slowed down by the alternation of target identities (at least when target location repeated between trials), probably because participants need time to choose a different valid response key for a changed target. However, sequence effects were not influenced by the repetition and switch of target identities (along with its responses). From a memory perspective, there may be two major phases for the sequence processes: initial encoding phase in previous trials and later retrieval phase in current trials. In the former phase, the spatial organization information between cue directions and target locations needs to be encoded into memory; in the later phase, the information will be retrieved from memory to affect performance. Current results may reflect some memory mechanisms during the sequential processing. That is, the target identity information is not encoded and retrieved between trials.

Sequence effects by gaze and arrow cues were directly compared in experiment 2. Gaze and arrow cues were intermixed in a within-block design. Such design resulted four kinds of cue sequence conditions between consecutive trials: gaze-gaze sequence, gaze-arrow sequence, arrow-gaze sequence, arrow-arrow sequence. Interestingly, sequence effects are significant for all sequence conditions despite the alternation of cue types. That means gaze cues induce the same amount of sequence effects as arrow cues, and the sequence effect can generalize across different cue types (from gaze to arrow, or from arrow to gaze). This finding supports the automatic retrieval hypothesis and suggests that sequence effects induced by gaze and arrow cues are processed in the same system, probably the so-called implicit memory system in human brain [9, 10].

In all, the present study investigated the sequence effect in gaze cueing paradigm. Similar to the previous findings from arrow cueing, significant sequence effects were found. The results extend our knowledge by showing that sequence effects in cueing paradigm are not limited to a specific cue type or a specific task, and suggest that sequential processing is a common mechanism in attention orienting systems.

Acknowledgments. This research is supported by the NSFC (31300938, 61462053, and 61403251), and the Yunnan NSFC (2016FB107).

References

1. Posner, M.: Orienting of attention. Q. J. Exp. Psychol. **32**, 3–25 (1980)
2. Hommel, B., Pratt, J., Colzato, L., Godijn, R.: Symbolic control of visual attention. Psychol. Sci. **12**, 360–365 (2001)
3. Frischen, A., Bayliss, A., Tipper, S.: Gaze cueing of attention: visual attention, social cognition, and individual differences. Psychol. Bull. **133**, 694–724 (2007)
4. Jongen, E., Smulders, F.: Sequence effects in a spatial cueing task: endogenous orienting is sensitive to orienting in the preceding trial. Psychol. Res. **71**, 516–523 (2007)
5. Gómez, C.M., Flores, A., Digiacomo, M.R., Vázquez-Marrufo, M.: Sequential P3 effects in a Posner's spatial cueing paradigm: trial-by-trial learning of the predictive value of the cue. Acta Neurobiol. Exp. **69**, 155–167 (2009)
6. Dodd, M., Pratt, J.: The effect of previous trial type on inhibition of return. Psychol. Res. **71**, 411–417 (2007)
7. Qian, Q., Shinomori, K., Song, M.: Sequence effects by non-predictive arrow cues. Psychol. Res. **76**, 253–262 (2012)
8. Qian, Q., Song, M., Shinomori, K.: Gaze cueing as a function of perceived gaze direction. Jpn. Psychol. Res. **55**, 264–272 (2013)
9. Maljkovic, V., Nakayama, K.: Priming of pop-out: III. A short-term implicit memory system beneficial for rapid target selection. Vis. Cogn. **7**, 571–595 (2000)
10. Kristjánsson, A.: Rapid learning in attention shifts: a review. Vis. Cogn. **13**, 324–362 (2006)

Discrete Cuckoo Search with Local Search for Max-cut Problem

Yingying Xu, Zhihua Cui(✉), and Lifang Wang

Complex System and Computational Intelligence Laboratory,
Taiyuan University of Science and Technology, Taiyuan 030024, Shanxi, China
s_yingyingxu@163.com, zhihua.cui@hotmail.com,
wlf1001@163.com

Abstract. Max-cut problem is a well-known NP-hard problem, in this paper, a hybrid meta-heuristic algorithm is designed to solve it. In this algorithm, the discrete cuckoo search is employed to find the approximate satisfied solution, while the local optimal solution is used to further improve the performance. To test the validity, three other algorithms are used to compare, simulation results show our modification is effective.

Keywords: Max-cut problem · Discrete cuckoo search algorithm · Local search strategy

1 Introduction

Max-cut problem is a well-known NP-hard graph problem. For a graph $G = (V, E)$, $V = \{1, 2, \ldots, n\}$ is vertex set and E is the ordered set of undirected edges. Let w_{ij} be the weight associated with edge $\{i, j\} \in E$, then max-cut problem is to find an optimal partition (V_1, V_2) $(V_1 \cap V_2 = \phi,\ V_1 \cup V_2 = V)$, so that the total weights of the edges crossing different subsets is maximized, in other words, the objective function can be represented as:

$$\sum_{i \in V_1, j \in V_2} w_{ij} = \sum_{i<j} w_{ij} \cdot \frac{1 - x_i x_j}{2} \tag{1}$$

where $x_i \in \{1, -1\} (i = 1, 2, \cdots, n)$, $x_i = 1$ represents $x_i \in V_1$, as well as $x_i = -1$ denotes $x_i \in V_2$.

During the past years, many algorithms have been designed to solve it, including exact algorithm, approximate algorithm and heuristic algorithm [1, 2]. Heuristic algorithm [3–5] is an umbrella for all population-based stochastic optimization algorithm inspired by heuristic information [6], such as ant colony optimization [7, 8], fruit fly optimization [9], particle swarm optimization [10–12], artificial bee colony [13–16], social emotional optimisation algorithm [17], firefly algorithm [18–22] and bat algorithm [23–25].

Z. Shi et al. (Eds.): ICIS 2017, IFIP AICT 510, pp. 66–74, 2017.
DOI: 10.1007/978-3-319-68121-4_7

For max-cut problem, Laguna et al. [26] designed a hybrid version of cross entropy method, while Lin [27] proposed a discrete dynamic convexized method. Festa et al. [28] investigated several heuristics derived from greedy randomized adaptive search procedure and variable neighborhood search. In 2007, Wang [29] proposed a hybrid algorithm combining with chaotic discrete Hopfield neural network and genetic particle swarm optimization, while in 2011, Wang [30] designed another combination with tabu Hopfield neural network and estimation of distribution algorithm. Inspired by this work, Lin [31] designed an integrated method combined with particle swarm optimization and estimation of distribution algorithm, and a local search strategy is employed to improve the accuracy. Shylo [32] also employed global equilibrium search algorithms and tabu search to improve the accuracy.

In this paper, we propose a new heuristic algorithm to combining the cuckoo search algorithm and local search strategy, and apply it to solve max-cut problem. The rest of this paper is organized as follows: In Sect. 2, the details of our proposed hybrid algorithm are presented, as well as the simulation results are reported in Sect. 3.

2 Hybrid Algorithm

2.1 Discrete Cuckoo Search Algorithm

Cuckoo search algorithm was proposed in 2009 [33], up to now, many variants are proposed to improve the performance [34–36]. However, max-cut problem is a combination problem, and a discrete cuckoo search algorithm is designed to solve it.

In this discrete version, the position movement will use the following strategies [37]: Strategy 1:

$$x_{ijk}^{m+1} = \begin{cases} x_{ijk}^{m}, rand() \leq Sig(Step) \\ -1, other \end{cases} \tag{2}$$

where

$$Sig(Step) = 1/(1 + \exp(-Step)) \tag{3}$$

Strategy 2:

$$x_{ijk}^{m+1} = \begin{cases} -1, rand() \leq Sig(Step) \\ x_{ijk}^{m}, other \end{cases} \tag{4}$$

where

$$Sig(Step) = 1 - 2/(1 + \exp(-Step)) \tag{5}$$

Strategy 3:

$$x_{ijk}^{m+1} = \begin{cases} 1, rand() \leq Sig(Step) \\ x_{ijk}^{m}, other \end{cases} \tag{6}$$

where

$$Sig(Step) = 2/(1 + \exp(-Step)) - 1 \tag{7}$$

The most difference among three strategy is the sigma function $sig(step)$, and jump path $step$ is a random number with Levy distribution, to provide a deep insight, the Eqs. (3), (5) and (7) are plotted in Figs. 1, 2 and 3.

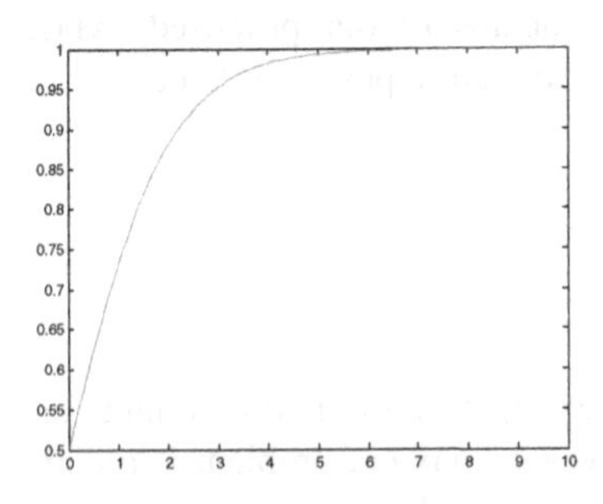

Fig. 1. Illustration for Eq. (3)

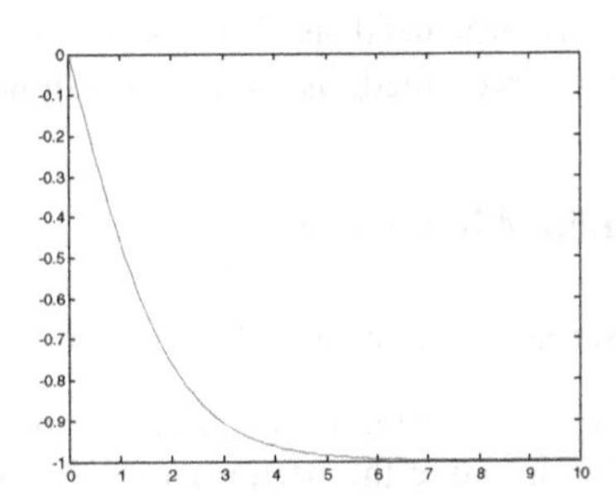

Fig. 2. Illustration for Eq. (5)

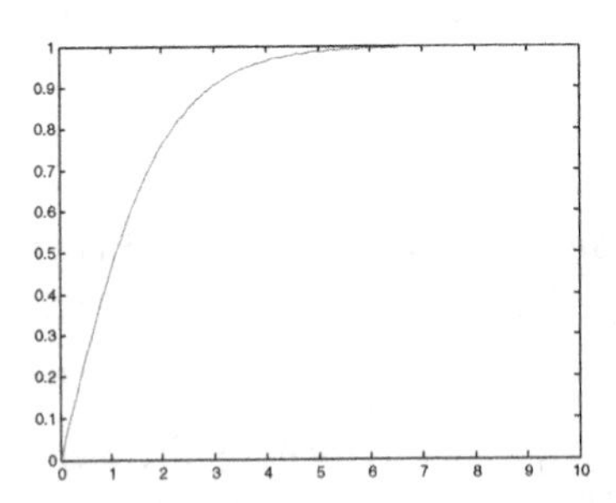

Fig. 3. Illustration for Eq. (7)

Algorithm 1 is the pseudocode of our discrete cuckoo search algorithm, where p_r is the predefined factor, and P_a is the probability of being discovered by the host bird.

Algorithm 1. Discrete cuckoo search algorithm

Begin
For each cuckoo, randomly initialize the position, the control factor p_r, the probability p_a of being discovered, Levy distribution *Step*;
Calculate the objective function values with Eq. (1) and Record the best position from swarm;
While (stop criterion is met)
    If $rand() \le p_r$
        Update the nest position for each cuckoo by strategy 1;
    Else
        If $Step \le 0$
            Update each cuckoo with strategy 2;
        Else
            Update each cuckoo with strategy 3;
        End
    End
    Evaluate the objective fitness for each cuckoo;
    If $rand() > p_r$
        Re-update the position of corresponding cuckoo with formulas:
        $x_{jk}^{t+1} = x_{jk}^{t'} + R \cdot (p_{gk}^{t'} - x_{uk}^{t'})$
        Evaluate the new cuckoos' finesses;
    End
    Record the best position
End
Output the best position
End

2.2 Local Search Strategy

For any partition (V_1, V_2), if the vertex j is moved from the current partition to another subset, the gain index g_j is defined as follows:

$$g_j = \begin{cases} \sum_{\{j,k\} \in E, k \in V_1} w_{jk} - \sum_{\{j,k\} \in E, k \in V_2} w_{jk}, j \in V_1 \\ \sum_{\{j,k\} \in E, k \in V_2} w_{jk} - \sum_{\{j,k\} \in E, k \in V_1} w_{jk}, j \in V_2 \end{cases} \quad (8)$$

Gain $g_j > 0$ means the vertex j should be moved with a lower objective function. With this manner, the following local search strategy is introduced:

Algorithm 2 Local Search Strategy

Begin
Use formula (8) to calculate the gain of each vertex
Descending all gains
Count the number (*sel*) of vertices with $g_j > 0$ is satisfied
If ($sel > 0$)
If $sel > 30$
$sel1 = \frac{1}{3} \cdot sel$;
Else
$sel1 = sel$;
Move the vertices of $sel1$, which are selected sequentially from sorted result.
Evaluate the fitness
End
End

2.3 Proposed Hybrid Algorithm

Our modification is a hybrid meta-heuristic method combining a discrete cuckoo search algorithm and local search strategy. The discrete cuckoo search is employed to find the approximate satisfied solution, while the local optimal solution is used to further improve the performance for the obtained approximate satisfied solution. Furthermore, to avoid the premature convergence, a mutation strategy is also employed to avoid the premature convergence. The pseudocode of our hybrid algorithm is listed in Algorithm 3. For each cuckoo, the mutation operation will randomly take 0.1% vertices to flip. We find that the algorithm will be improved after adding the mutation operation.

Algorithm 3 Cuckoo search with Local search

Begin
For the graph, initialise population *pop* , and record the best solution *gbest* ;
Find the approximate satisfied solution by the discrete cuckoo search (refer to Alg.1);
Where (Stop condition is not verified)
the local strategy (refer to Alg.2) is employed to improve the quality of the solution;
If a particle x in a group is not improved in two consecutive searches, a mutation strategy is employed to avoid the premature convergence;
End
End

3 Performance Evolution

To test the performance of our proposed hybrid algorithm, G-set graph benchmarks are employed, and compared with the following algorithms:

- Hybridizing the cross-entropy method (HCE, in briefly) [26];
- A new lagrangian net algorithm (LNA, in briefly) [38];
- Discrete Hopfield network with estimation of distribution algorithm (DHNN-EDA, in briefly) [14]
- Discrete cuckoo search with local search (DCSLS, in briefly)

The program is implemented with MATLAB. In this set of instances, the number of vertex range from 800 to 3000, the control factor p_r is 0.3, the probability p_a are both set to 0.5, the total generation is 500.

For chosen benchmarks, each instance will run 50 times, Table 1 provides the optimal value achieved by the three algorithms and our proposed algorithm. The last line noted as "*w/t/l*" is the comparison results between our proposed DCSLS and its competitors. "*w/t/l*" represents our algorithm wins in w functions, ties in t functions, and loses in l functions. It means DCSLS is better than HCE for six functions, while only worse than HCE for three functions. DCSLS are superior than DHNN-EDA and LNA for six functions too, while DHNN-EDA and LNA only better than DCSLS with two and three functions, respectively. In one word, DCSLS achieves the best performance when compared with HCE, DHNN-EDA and LNA.

Table 1. Comparison of the results

Instances	Best	HCE	DHNN-EDA	LNA	DCSLS
G1	11624	11584	11614	11490	11607
G2	11620	11595	11599	11505	11599
G3	11622	11574	11617	11511	11605
G11	564	552	494	560	530
G12	556	542	476	546	528
G13	582	564	520	572	552
G14	3064	3030	3027	3023	3035
G15	3050	3012	2988	2996	3016
G16	3052	3015	3001	2994	3018
w/t/l		**6/0/3**	**6/1/2**	**6/0/3**	

To provide a deep comparison, two non-parametric statistics tests: Friedman test and Wilcoxon test, are employed to show the differences among these four algorithms. In Table 2, the ranking value is: DCSLS < HCE < DHNN-EDA < LNA, DCSLS maintains the lowest ranking, it means the performance of DCSLS is more better. Table 3 implies there is significantly difference between DCSLS and DHNN-EDA.

Table 2. Friedman test

	Rank-value
HCE	2.33
DCSLS	1.94
LNA	2.89
DHNN-EDA	2.83

Table 3. Wilcoxon test

DCSLS vs	P-value
HCE	0.594
LNA	0.192
DHNN-EDA	0.050

4 Conclusion

In this paper, a new hybrid algorithm combining with discrete cuckoo search and local search strategy is designed. The cuckoo update manner of discrete cuckoo search is the same as [37], while the local search strategy is designed. Simulation results show our modification achieves the best performance when compared with other three algorithms.

Acknowledgments. This research is supported by the Natural Science Foundation of Shanxi Province under No. 201601D011045.

References

1. Xia, Y., Xu, Z.: An efficient lagrangian smoothing heuristic for max-cut. Indian J. Pure Appl. Math. **41**(5), 683–700 (2010)
2. Marti, R., Duarte, A., Laguna, M.: Advanced scatter search for the max-cut problem. Inf. J. Comput. **21**(1), 26–38 (2009)
3. Wang, G.G., Deb, S., Gao, X.Z., Coelho, L.: A new metaheuristic optimization algorithm motivated by elephant herding behavior. Int. J. Bio-Inspired Comput. **8**(6), 394–409 (2016)
4. Bilbao, M.N., Ser, J.D., Salcedo-Sanz, S., Casanova-Mateo, C.: On the application of multi-objective harmony search heuristics to the predictive deployment of firefighting aircrafts: a realistic case study. Int. J. Bio-Inspired Comput. **7**(5), 270–284 (2015)
5. Rajakumar, R., Dhavachelvan, P., Vengattaraman, T.: A survey on nature inspired meta-heuristic algorithms with its domain specifications. In: International Conference on Communication and Electronics Systems, pp. 550–555 (2016)
6. Xiao, R., Zhang, Y., Huang, Z.: Emergent computation of complex systems: a comprehensive review. Int. J. Bio-Inspired Comput. **7**(2), 75–97 (2015)
7. Dorigo, M., Gambardella, L.M., Middendorf, M., Stutzle, T.: Special section on ant colony optimization. IEEE Trans. Evol. Comput. **6**(4), 317–320 (2002)
8. Stodola, P., Mazal, J.: Applying the ant colony optimisation algorithm to the capacitated multi-depot vehicle routing problem. Int. J. Bio-Inspired Comput. **8**(4), 228–233 (2016)
9. Zhang, Y.W., Wu, J.T., Guo, X., Li, G.N.: Optimising web service composition based on differential fruit fly optimisation algorithm. Int. J. Comput. Sci. Math. **7**(1), 87–101 (2016)
10. Eberhart, R.C., Shi, Y.H.: Special issue on particle swarm optimization. IEEE Trans. Evol. Comput. **8**(3), 201–203 (2004)

11. Adewumi, A.O., Arasomwan, M.A.: On the performance of particle swarm optimisation with(out) some control parameters for global optimisation. Int. J. Bio-Inspired Comput. **8**(1), 14–32 (2016)
12. Grillo, H., Peidro, D., Alemany, M., Mula, J.: Application of particle swarm optimisation with backward calculation to solve a fuzzy multi-objective supply chain master planning model. Int. J. Bio-Inspired Comput. **7**(3), 157–169 (2015)
13. Lv, L., Wu, L.Y., Zhao, J., Wang, H., Wu, R.X., Fan, T.H., Hu, M., Xie, Z.F.: Improved multi-strategy artificial bee colony algorithm. Int. J. Comput. Sci. Math. **7**(5), 467–475 (2016)
14. Sun, H., Wang, K., Zhao, J., Yu, X.: Artificial bee colony algorithm with improved special centre. Int. J. Comput. Sci. Math. **7**(6), 548–553 (2016)
15. Lu, Y., Li, R.X., Li, S.M.: Artificial bee colony with bidirectional search. Int. J. Comput. Sci. Math. **7**(6), 586–593 (2016)
16. Yu, G.: A new multi-population-based artificial bee colony for numerical optimization. Int. J. Comput. Sci. Math. **7**(6), 509–515 (2016)
17. Guo, Z.L., Wang, S.W., Yue, X.Z., Yin, B.Y., Deng, C.S., Wu, Z.J.: Enhanced social emotional optimisation algorithm with elite multi-parent crossover. Int. J. Comput. Sci. Math. **7**(6), 568–574 (2016)
18. Wang, H., Wang, W.J., Zhou, X.Y.: Firefly algorithm with neighborhood attraction. Inf. Sci. **382**, 374–387 (2017)
19. Wang, H., Wang, W.J., Sun, H.: Firefly algorithm with random attraction. Int. J. Bio-Inspired Comput. **8**(1), 33–41 (2016)
20. Yu, G.: An improved firefly algorithm based on probabilistic attraction. Int. J. Comput. Sci. Math. **7**(6), 530–536 (2016)
21. Nasiri, B., Meybodi, M.R.: History-driven firefly algorithm for optimisation in dynamic and uncertain environments. Int. J. Bio-Inspired Comput. **8**(5), 326–339 (2016)
22. Fister, I., Fister, I., Yang, X.S., Brest, J.: A comprehensive review of firefly algorithms. Swarm Evol. Comput. **13**, 34–46 (2013)
23. Yang, X.S., Gandomi, A.H.: Bat algorithm: a novel approach for global engineering optimization. Eng. Comput. **29**(5–6), 464–483 (2012)
24. Cai, X., Gao, X.Z., Xue, Y.: Improved bat algorithm with optimal forage strategy and random disturbance strategy. Int. J. Bio-Inspired Comput. **8**(4), 205–214 (2016)
25. Xue, F., Cai, Y., Cao, Y., Cui, Z., Li, F.: Optimal parameter settings for bat algorithm. Int. J. Bio-Inspired Comput. **7**(2), 125–128 (2015)
26. Laguna, M., Duarte, A., Marti, R.: Hybridizing the cross-entropy method: an application to the max-cut problem. Comput. Oper. Res. **36**(2), 487–498 (2009)
27. Lin, G., Zhu, W.: A discrete dynamic convexized method for the max-cut problem. Ann. Oper. Res. **196**(1), 371–390 (2012)
28. Festa, P., Pardalos, P.M., Resende, M.G.C., Ribeiro, C.C.: Randomized heuristics for the max-cut problem. Optim. Method Softw. **17**(6), 1033–1058 (2002)
29. Wang, J.: A memetic algorithm with genetic particle swarm optimization and neural network for maximum cut problems. In: Li, K., Fei, M., Irwin, G.W., Ma, S. (eds.) LSMS 2007. LNCS, vol. 4688, pp. 297–306. Springer, Heidelberg (2007). doi:10.1007/978-3-540-74769-7_33
30. Wang, J., Zhou, Y., Yin, J.: Combining tabu hopfield network and estimation of distribution for unconstrained binary quadratic programming problem. Expert Syst. Appl. **38**(12), 14870–14881 (2011)
31. Lin, G., Guan, J.: An integrated method based on PSO and EDA for the max-cut problem. Comput. Intell. Neurosci. (2016). doi:10.1155/2016/3420671

32. Shylo, V.P., Shylo, O.V.: Solving the maxcut problem by the global equilibrium search. Cybern. Syst. Anal. **46**(5), 744–754 (2010)
33. Yang, X.S., Deb, S.: Cuckoo search via levy flights. In: World Congress on Nature and Biologically Inspired Computing, pp. 210–214 (2009)
34. Cui, Z.H., Sun, B., Wang, G.G., Xue, Y.: A novel oriented cuckoo search algorithm to improve DV-hop performance for cyber-physical systems. J. Parallel Distrib. Comput. **103**, 42–52 (2017)
35. Zhang, M.Q., Wang, H., Cui, Z.H., Chen, J.J.: Hybrid multi-objective cuckoo search with dynamical local search. Memetic Comp. (2017). doi:10.1007/s12293-017-0237-2
36. Li, F.X., Cui, Z.H., Sun, B.: DV-hop localisation algorithm with DDICS. Int. J. Comput. Sci. Math. **7**(3), 254–262 (2016)
37. Feng, D.K., Ruan, Q., Du, L.M.: Binary cuckoo search algorithm. J. Comput. Appl. **33**(6), 1566–1570 (2013). (in Chinese)
38. Xu, F.M., Ma, X.S., Chen, B.L.: A new lagrangian net algorithm for solving max-bisection problems. J. Comput. Appl. Math. **235**(13), 3718–3723 (2011)

A New Cuckoo Search

Zhigang Lian[1], Lihua Lu[1](✉), and Yangquan Chen[2]

[1] School of Electronic and Information Engineering, Shanghai DianJi University, Shanghai 200240, China
{lianzg,lulihua}@sdju.edu.cn

[2] School of Engineering, University of California, Merced, CA 95343, USA
ychen53@ucmerced.edu

Abstract. In this paper, we intend to formulate a new Cuckoo Search (NCS) for solving optimization problems. This algorithm is based on the obligate brood parasitic behavior of some cuckoo species in combination with the Lévy flight behavior of some birds and fruit flies, at the same time, combine particle swarm optimization (PSO), evolutionary computation technique. It is tested with a set of benchmark continuous functions and compared their optimization results, and we validate the proposed NCS against test functions with big size and then compare its performance with those of PSO and original Cuckoo search. Finally, we discuss the implication of the results and suggestion for further research.

Keywords: Cuckoo search · Lévy distribution · Particle swarm optimization

1 Introduction

In factual production and management, there are many complicated optimization problems. So many scientists have constantly proposed the new intelligent algorithms to solve them. For example, PSO was inspired by fish and bird swarm intelligence [1,2]. These nature-inspired metaheuristic algorithms have been used in a wide range of optimization problems including NP-hard problems such as the travelling salesman problem (TSP) and scheduling problems [3,4]. Based on the interesting breeding behavior such as brood parasitism of certain species of cuckoos, Yang and Deb [5] has formulated the Cuckoo Search (CS) algorithm. Yang and Deb in [6] review the fundamental ideas of CS and the latest developments as well as its applications. They analyze the algorithm, gain insight into its search mechanisms and find out why it is efficient.

In this paper, we intend to formulate a new Cuckoo Search (NCS) with different evolution mode, for solving function optimization problems. The NCS is based on the obligate brood parasitic behavior of some cuckoo species in combination with the Lévy flight behavior of some birds and fruit flies. Moreover

Z. Shi et al. (Eds.): ICIS 2017, IFIP AICT 510, pp. 75–83, 2017.
DOI: 10.1007/978-3-319-68121-4_8

it integrates the PSO with some evolutionary computation technique. The PSO has particularly gained prominence due to its relative ease of operation and capability to quickly arrive at an optimal/near-optimal solution. This algorithm is considered with the advantages of CS and PSO, avoid tendency to get stuck in a near optimal solution in reaching optimum solutions especially for middle or large size optimization problems. This work differs from existing ones at least in three aspects:

- it proposes the iterative formula of NCS, in which combines the iterative equations of CS and PSO.
- it finds the best combine parameters of NCS for different size optimization problems.
- by strictly analyzes the performance of NCS, we validate it against test functions and then compare its performance with those of PSO and CS with different random coefficient generate.

Finally, we discuss the implication of the results and suggestion for further research.

2 The New CS

2.1 The Model of New CS Algorithm

Yang [7] provides an overview of CS and firefly algorithm as well as their latest developments and applications. In this paper, we research on different random distribution number and their influence for algorithm. Furthermore, we also present a new CS with evolutionary pattern.

A New Cuckoo Search (NCS) combining CS and PSO is presented in this paper, in which is based on the obligate brood parasitic behavior of some cuckoo species and considered with the Lévy flight behavior of some birds and fruit flies. In the process of evolution, nests/particles/solutions of next generation share the historical and global best of the *ith* nests/particles/solutions. NCS improves upon the PSO and CS variation to increase accuracy of solution without sacrificing the speed of search solution significantly, and its detailed information will be given in following.

Suppose that the searching space is D-dimensional and m nests/particles/ solutions form the colony. The *ith* nest/particle represents a D-dimensional vector $X_i = (x_{i1}, x_{i2}, \ldots, x_{iD}), (i = 1, 2, \ldots, m)$, and it means that the *ith* nest/particle located at X_i in the searching is a potential solution. Calculate the nest/particle/solution fitness by putting it into a designated objective function. The historical best of the *ith* nests/particles/solutions denotes as $P_i = (p_{i1}, p_{i2}, \ldots, p_{iD})$, called $IBest$, and the best of the global as $P_g = (p_1, p_2, \ldots, p_D)$, called $GBest$ respectively. At the same time, some of the new nest/particle/solution should be generated combining Lévy walk around the best solution obtained so far. It will speed up the local search.

After finding above two best values, and with Lévy flight the nests/particles/solutions of NCS updates as formulas (1).

$$X_i^{(t+1)} = X_i^{(t)} + \alpha \oplus \delta * L'evy(\lambda) + (1-\delta)(R_1(P_i^{(t)} - X_i^{(t)}) + R_2(P_g^{(t)} - X_i^{(t)})); \quad (1)$$

In (1), Lévy follows a random walk Lévy distribution. A part from distant random position solution is far from the optimal solution, so it can make sure that the system does not fall into local optimal solution. Where the $\delta \in [0, 1]$ is weight index that is chosen according to different optimization problem. It reflects relatively important degree of the t generation Lévy fly, the best nests/particles/solutions of individual historical $P_i^{(t)}$ and the best nest/particle/solution of global $P_g^{(t)}$. In addition, the NCS evolution process, the global best nest/particle/solution $P_g^{(t)}$ may in $K\%$ forced preserved, $K \in (0, 100)$. Others parameters such as α and $R_1, R_2 \in (0, 1)$ are same as the ones in [2].

The search is a repeated process, and the stop criteria are that the maximum iteration number is reached or the minimum error condition is satisfied. The stop condition depends on the problem to be optimized. In the NCS evolution process, the nests/particles/solutions will be mainly updated through the three parts:

- Lévy walk;
- the distance between the best nests/particles/solutions of individual historical $P_i^{(t)}$ and its current nests/particles/solutions;
- the distance between the best nest/particle/solution of individual historical $P_i^{(t)}$ and its current nest/particle/solution.

There are some significant differences of NCS, CS and PSO. Firstly, their iterative equations are not the same. The NCS integrates the advantages of CS and PSO algorithm, which share the excellent information of nests/particles/solutions. It uses some sort of elitism and/or selection which is similar to the ones used in harmony search (HS). Secondly, the randomization is more efficient as the step length is heavy-tailed, and any large step is possible. Thirdly, the parameter δ is to be turned and easy to find the highest efficiency parameters which are adapted to a wider class of optimization problems. In addition, the NCS can thus be extended to the type of meta-population algorithm.

2.2 Pseudo Code of the NCS Algorithm

Algorithm 1. Pseudo code of NCS: Part 1

Require:
1: nests/particles/solutions population size: *PS*;
2: maximum of generation *Endgen*;
3: weight index δ;
4: forced preserved percent;
5: step size α;
Ensure: optimization results;
6: **procedure**
7: Generate stochastically PS size initial population;
8: Evaluate each nest/particle/solutions fitness/quality;
9: Generate initial global best population $P_g^{(t)}$ with the lowest fitness nest/particle/solution in the whole population;
10: Generate randomly initial individual historical best population $P_i^{(t)}$;
11: t:=0;
12: **while** (t ¡ *Endge*) or (stop criterion) **do**
13: t:=t+1;
14: Generate next nests/particle/solutions by Eq. (1);
15: $\beta = 3/2$;
16: $\sigma = (gamma(1+\beta) * sin(\pi * \beta/2)/(gamma(1+\beta)/2) * \beta * 2^{((\beta-1)/2)))^{(1/\beta)}}$;
17: $\mu = randn() * \sigma$, $v = randn()$;
18: $step = \mu./abs(\varrho).^{(1/beta)}$;
19: $stepsize = 0.01 * step. * (X_i^{(t)} - P_g^{(t)})$;
20: $X_i^{(t+1)} = X_i^{(t)} + \delta * stepsize. * randn() + (1-\delta) * (randn() * (P_g^{(t)} - X_i^{(t)}))$;
21: Evaluate nests/particles/solutions;
22: {Compute each nest/particle/solution's fitness F_i in the population;
23: Find new $P_g^{(t)}$ of nest/particle/solution by comparison, and update $P_g^{(t)}$;}
24: Keep the best nest/particle/solution;
25: Rank solutions and find the current best;
26: t:=t+1;
27: Choose a nest/particle/solution in population randomly;
28: **if** ($F_i > F_j$) **then**
29: Replace j by the new nest/particle/solution;
30: **end if**
31: A fraction p_α of worse nests/particles/solutions are abandoned and new ones are built;
32: The iteration calculation are as follows (X_{randn} is generated randomly):
33: $K = randn() > p_\alpha$;
34: $stepsize = randn() * (X_{randn} - X'_{randn})$;
35: $X_i^{t+1} = X_i^t + stepsize. * K$;
36: Evaluate nests/particles/solutions;
37: {Compute each nest/particle/solution's fitness F_i in the population;
38: Find new $P_g^{(t)}$ and $P_i^{(t)}$ by comparison, and update $P_g^{(t)}$ and $P_i^{(t)}$;}
39: Keep the best nest/particle/solution;
40: Rank solutions and find the current best;
41: Randomly selected nests/particles/solutions of populations K%, forced instead them by the highest quality nest/particle/solution $P_g^{(t)}$;
42: **end while**
43: **end procedure**

3 Numerical Simulation

3.1 Test Functions

To proof-test the effectiveness of NCS for optimization problems, 14 representative benchmark functions with different dimensions are employed to compare with CS and PSO described in Table 1, where the f_{10} is given as formula (2).

Table 1. Benchmark functions

$sphere : f_1(x)$	$Quadric : f_2(x)$	$Schwefe : f_3(x)$
$Rosenbrock : f_4(x)$	$Schwefel : f_5(x)$	$Rastrigin : f_6(x)$
$Ackley : f_7(x)$	$Griewank : f_8(x)$	$Generalized\ Penalized : f_9(x)$
$f_{10}(x)$	$Axis\ parallel\ hyper-ellipsoid : f_{11}(x)$	
$Sum\ of\ different\ power : f_{12}(x)$	$Michalewicz : f_{13}(x)$	$Schaffer\ f_7 : f_{14}(x)$

Table 2. The comparison results of the PSO, CS [6]and the NCS algorithm

In CS, the $p_\alpha = 0.25$, and in NCS the $p_\alpha = 0.25$ and $k = 10\%(e \sim n \quad is \times 10^n)$

$Fun = f_1$	NCS	δ	Min/Average/Std	$Fun = f_2$	NCS	δ	Min/Average/Std
Dim = 100		0.2	0.0407/5.3738/6.1344	Dim = 50		0.2	5.5837/136.355/104.5021
Best = 0		0.5	4.1881/4.1752/4.1881	Best = 0		0.5	**2.2119**/112.231/103.8505
PS = 200		**0.8**	**0.0161**/3.2514/2.9194	PS = 200		0.8	3.3954/109.8777/116.003
EG = 2000	PSO		4.2734e+4/6.4205e+4/1.7802e+4	EG = 2000	PSO		8.0679e+3/1.837e+4/5.6383e+3
	CS		1.4453e+3/ 1.9655e+3/311.6468		CS		9.8117e+3/1.1428e+4/938.2576
$Fun = f_3$	NCS	δ	Min/Average/Std	$Fun = f_4$	NCS	δ	Min/Average/Std
Dim = 50		0.2	**0.041**/0.2531/0.14356.1344	Dim = 50		0.2	**3.3818**/173.9222/140.0647
Best = 0		0.5	0.0439/0.2491/0.1356	Best = 0		0.5	22.6772/148.6759/88.6797
PS = 200		**0.8**	0.0678/0.2543/0.106	PS = 200		0.8	6.7335/147.4183/81.444
EG = 2000	PSO		2.8602e+3/1.7809e+4/7.2644e+3	EG = 2000	PSO		9.4946e+3/1.8978e+4/7.0683e+3
	CS		18.5934/20.3249/1.2497		CS		1.0000e+10/1.0000e+10/0
$Fun = f_5$	NCS	δ	Min/Average/Std	$Fun = f_6$	NCS	δ	Min/Average/Std
Dim = 30		0.2	−12569.5/−12565.5/4.8859	Dim = 50		0.2	0.0394/1.1456/1.0407
Best = −12569.5		0.5	**−12569.5**/−12567/2.7239	Best = 0		0.5	0.0244/0.9167/0.7278
PS = 200		**0.8**	−12569.5/−12566/4.0731	PS = 200		0.8	**0.0204**/1.1315/0.9799
EG = 2000	PSO		0/1.2656e+4/4.9802e+4	EG = 2000	PSO		1.0071e+5/1.0272e+5/1.1186e+3
	CS		−9.5134e+3/−8.9445e+3/199.3234		CS		166.4727/203.5985/16.2855
$Fun = f_7$	NCS	δ	Min/Average/Std	$Fun = f_8$	NCS	δ	Min/Average/Std
Dim = 100		0.2	2.5087/51.0446/29.1489	Dim = 100		0.2	0.0944/0.7814/0.3262
Best = 0		0.5	**0.0328**/0.4947/0.3140	Best = 0		0.5	0.2225/0.8737/0.2328
PS = 200		**0.8**	0.0437/0.4701/0.3093	PS = 200		0.8	**0.0777**/0.8240/0.2938
EG = 2000	PSO		6.6817e+4/8.8147e+4/9.6893e+3	EG = 2000	PSO		4.3883e+4/1.0700e+6/5.1800e+5
	CS		13.0822/16.9746/1.3364		CS		9.1896/11.6123/1.2995
$Fun = f_9$	NCS	δ	Min/Average/Std	$Fun = f_{10}$	NCS	δ	Min/Average/Std
Dim = 100		0.2	4.9369e−5/0.0045/0.0038	Dim = 100		0.2	2.1331e−4/0.0084/0.0061
Best = 0		0.5	1.0745e−4/0.0034/4.9369e−5	Best = 0		0.5	9.1625e−4/0.0075/0.006
PS = 200		**0.8**	**2.5771e−5**/0.0034/0.0037	PS = 200		**0.8**	**1.8272e−5**/0.0119/0.0156
EG = 2000	PSO		2.391e+4/5.0278e+4/1.5785e+4	EG = 2000	PSO		1.1415e+4/5.3205e+4/1.6434e+4
	CS		8.3128/10.159/0.8293		CS		24.833/31.9707/3.7447
$Fun = f_{11}$	NCS	δ	Min/Average/Std	$Fun = f_{12}$	NCS	δ	Min/Average/Std
Dim = 100		0.2	0.0129/0.6487/0.8072	Dim = 100		0.2	1.7951e−4/0.0069/0.0077
Best = 0		**0.5**	0.0057/0.4278/0.3422	Best = 0		0.5	6.4493e−5/0.0069/0.0114
PS = 200		**0.8**	0.0176/0.5174/0.4947	PS = 200		**0.8**	1.4869e−5/0.0064/0.0064
EG = 2000	PSO		2.1285e+5/2.1521e+5/1.7257e+3	EG = 2000	PSO		1.6815e+4/4.8188e+4/1.6373e+4
	CS		57.2611/80.8869/10.0869		CS		1.0000e+10/1.0000e+10/0
$Fun = f_{13}$	NCS	δ	Min/Average/Std	$Fun = f_{14}$	NCS	δ	Min/Average/Std
Dim = 100		**0.2**	**−39.6594**/−35.0528/1.7772	Dim = 100		0.2	0.0014/0.098/0.0852
Best = ?		0.5	−39.3823/−35.7054/1.9454	Best = 0		**0.5**	**6.1878e−4**/0.1044/0.0954
PS = 200		0.8	−39.5466/−35.3446/1.8935	PS = 200		**0.8**	0.0012/0.0822/0.0723
EG = 2000	PSO		2.2069e+5/2.2175e+5/493.2826	EG = 2000	PSO		3.7512e+4/7.1659e+4/1.6774e+4
	CS		−34.8234l/ − 32.8746/0.9454		CS		168.0519/199.9071/9.6655

$$f_{10}(x) = 0.1\{(sin(\pi y_1))^2 + \sum_{i=1}^{n-1}(y_i - 1)^2 \times [1 + 10(sin(\pi y_{i+1}))^2] + (y_n - 1)^2\} + \sum_{i=1}^{n} u(x_i, 10, 100, 4) \quad (2)$$

where

$$u(x_i, a, k, m) = \begin{cases} k(x_i - a)^m, & x_i > a \\ 0, & -a \leq x_i \leq a, y_i = 1 + \frac{1}{4}(x_i + 1), x_i \in [-50, 50] \\ k(-x_i - a)^m, & x_i < -a \end{cases} \quad (3)$$

3.2 Experimental Results and Comparison Used Against Test Function with Big Size

To more scientifically evaluate the proposed algorithm, we run the algorithm 32 times and compare their minimum values, mean values and standard deviations searched by NCS, CS and PSO. For each test function, the parameters and results used in experiments are listed in Table 2.

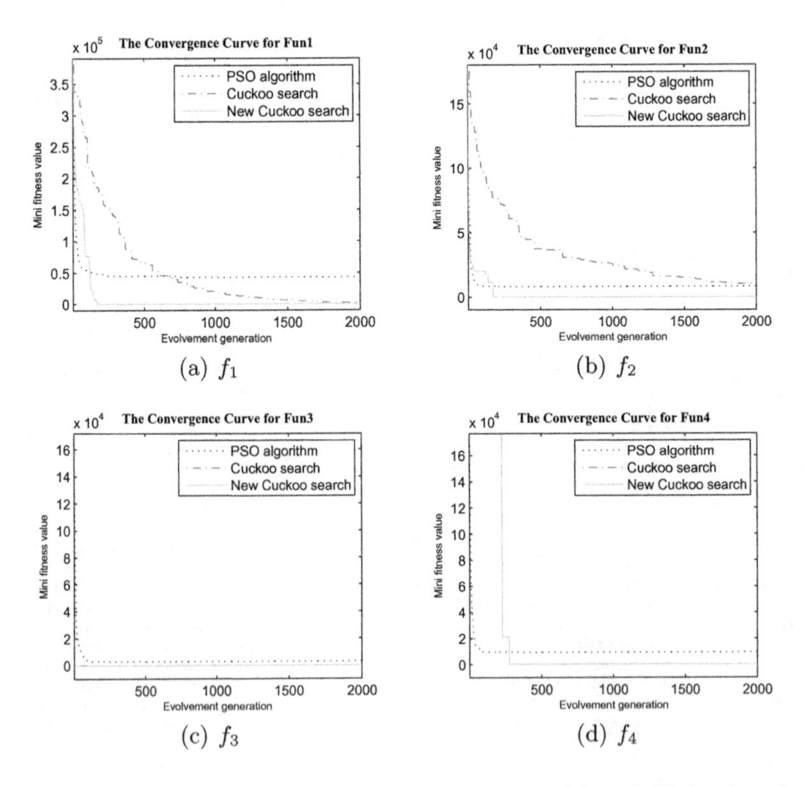

Fig. 1. Convergence figure of NCS comparing with PSO and CS for $f_1 - f_4$.

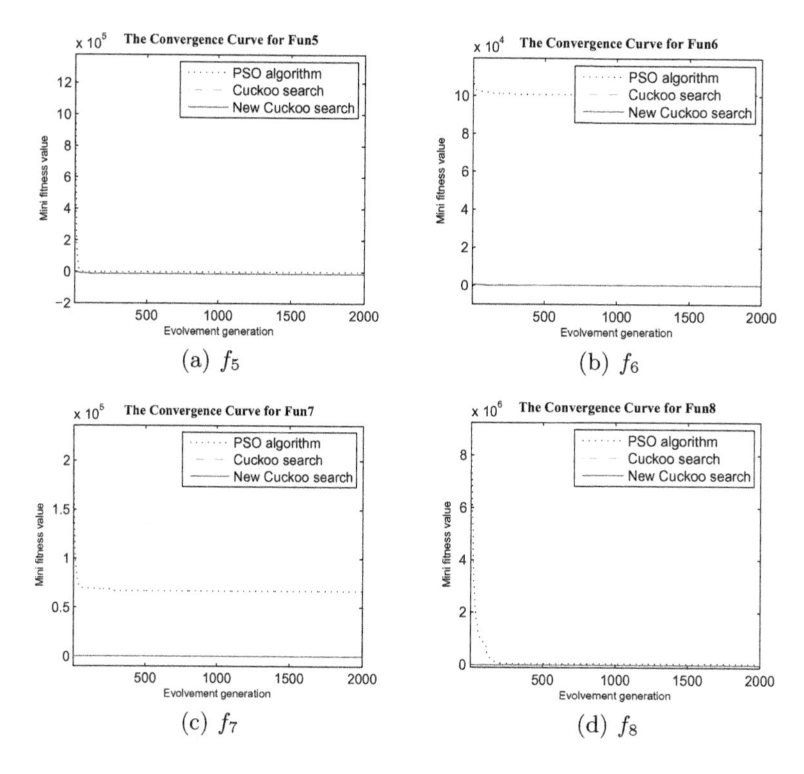

Fig. 2. Convergence figure of NCS comparing with PSO and CS for $f_5 - f_8$.

Remark 1: In Table 2, the parameters of *Fun* and *Dim* denote function and its dimension respectively. The *Best* is this functions optimum value. The *PS* and *EG* indicate algorithm population size and their terminate generation number. The better solutions and corresponding parameters found in NCS algorithm are illustrated with bold letters. The best minimum average and standard deviation are shown in italic and underline respectively.

From Table 2, in general it can observe that the $\delta \in [0.5, 0.8]$ has the highest performance since using them have smaller minimum and arithmetic mean in relation to solutions obtained by others. Especially the $\delta \approx 0.5$ has better search efficiency. In NCS, to optimize different problem should select various parameter δ, as a whole, it is better when $\delta \approx 0.5$. The CS optimization function f_4 and f_{12} is almost divergence. The NCS in all kinds of function optimization show excellent performance. In above function tests, the minimum value of NCS searched is NS and PSO searched 1/1000, or even 1/10000. Especially for large size problem, and function argument value in large range, optimal is strong search ability. NCS From simulation results we can obtain that the NCS is clearly better than PSO and CS for continuous non-linear function optimization problem. The NCS algorithm searches performance is strong, which can get better *the minimum*, *mean* and *standard deviation* relatively, but it computes

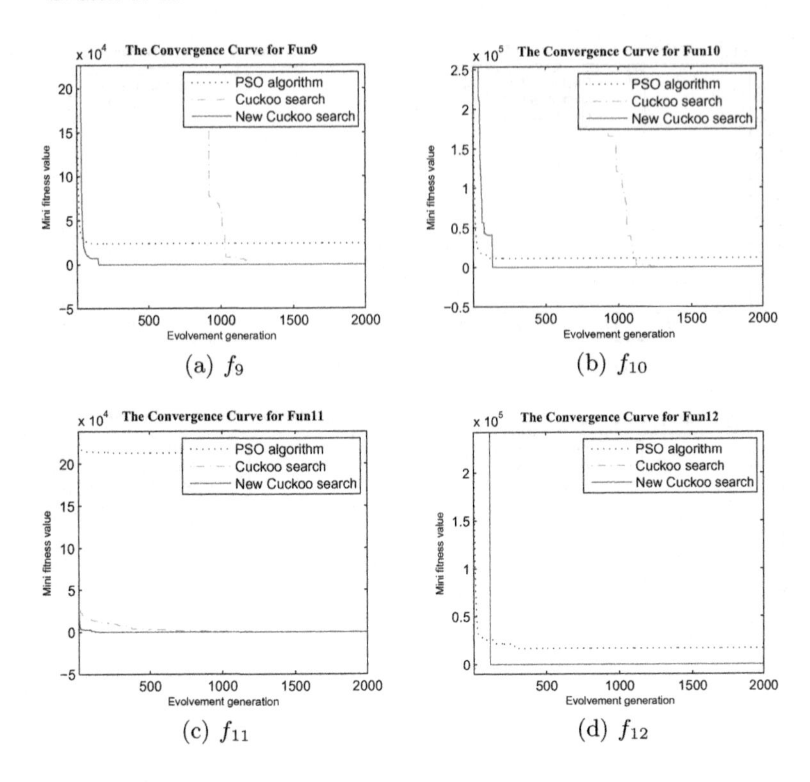

(a) f_9 (b) f_{10}

(c) f_{11} (d) f_{12}

Fig. 3. Convergence figure of NCS comparing with PSO and CS for $f_9 - f_{12}$.

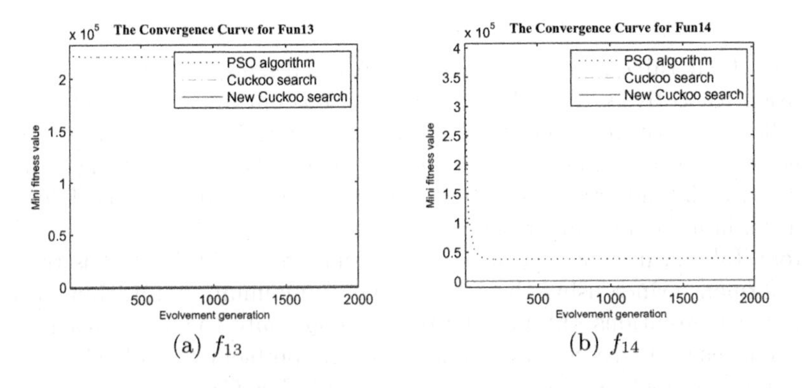

(a) f_{13} (b) f_{14}

Fig. 4. Convergence figure of NCS comparing with PSO and CS for $f_{13} - f_{14}$.

more part $(1 - \delta) * (randn() * (P_i^{(t)} - X_i^{(t)}) + randn() * (P_g^{(t)} - X_i^{(t)}))$, therefore more cost of hardware resources, although each running use a little more time, which is negligible. The convergence figures of most effective, distribution figures of run 32 times to search the optimal value of NCS comparing with PSO and CS for 14 instances are shown in Figs. 1, 2, 3 and 4.

From Figs. 1, 2, 3 and 4, it can discover that the convergence rate of NCS is clearly faster than the PSO and CS on every benchmark function. Especially it is more efficacious than PSO for middle and large size optimization problem. Accordingly, we can do state that the NCS is more effective than PSO and CS.

4 Conclusions and Perspectives

According to shortcoming of CS and PSO algorithm especially solving the middle or large size problem, we proposed NCS. Using 14 representative instances with different dimensions and compared with the PSO and CS, the performance of the NCS shows that is efficacious for solving optimization problems.

The proposed NCS algorithm can be considered as effective mechanisms from this point of view. There are a number of research directions which can be regarded as useful extensions of this research. Although this algorithm is tested with 14 representative instances, a more comprehensive computational study should be made to measure it. In the future it is maybe do experiments with different parameters and evaluate the performance of NCS. Furthermore, it finds the best parameters and usage scenarios such as TSP, scheduling, etc.

Acknowledgements. This work is supported by Academic Discipline Project of Shanghai Dianji University (Number: 16YSXK04) the Shanghai Natural Science Foundation (Number: 14ZR1417300).

References

1. Eberhart, R., Kennedy, J.: A new optimizer using particle swarm theory. In: Proceedings of the Sixth International Symposium on Micro Machine and Human Science, Nagoya, Japan, pp. 39–43 (1995)
2. Kennedy, J., Eberhart, R.: Particle swarm optimization. In: IEEE International Conferences on Neural Networks, Perth, Australia, pp. 1942–1948 (1995)
3. Lian, Z., Lin, W., Gao, Y., Jiao, B.: A discrete particle swarm optimization algorithm for job-shop scheduling problem to maximizing production. Int. J. Innov. Comput. Inf. Control **10**(2), 729–740 (2014)
4. Zhigang, L.: A united search particle swarm optimization algorithm for multi-objective scheduling problem. Appl. Math. Model. **34**, 3518–3526 (2010)
5. Yang, X.-S., Deb, S.: Cuckoo search via Lévy flights. In: World Congress on Nature and Biologically Inspired Computing, pp. 210–214. IEEE Publications (2009)
6. Yang, X.-S., Deb, S.: Cuckoo search: recent advances and applications. Neural Comput. Appl. **24**(1), 169–174 (2014)
7. Yang, X.-S.: Cuckoo search and firefly algorithm: overview and analysis. In: Yang, X.-S. (ed.) Cuckoo Search and Firefly Algorithm. SCI, vol. 516, pp. 1–26. Springer, Cham (2014). doi:10.1007/978-3-319-02141-6_1

Resting State fMRI Data Classification Method Based on K-means Algorithm Optimized by Rough Set

Xianzhe Li(✉), Weiming Zeng, Yuhu Shi, and Shaojun Huang

Lab of Digital Image and Intelligent Computation,
Shanghai Maritime University, 1550 Harbor Avenue,
Pudong 201306, Shanghai, China
lixianzhe1314@163.com

Abstract. With the development of brain science, a variety of new methods and techniques continue to emerge. Functional magnetic resonance imaging (fMRI) has become one of the important ways to study the brain functional connection and of brain functional connectivity detection because of its non-invasive and repeatability. However, there are still some issues in the fMRI researches such as the amounts of data and the interference noise in the data. Therefore, how to effectively reduce the fMRI data dimension and extract data features has become one of the core content of study. In this paper, a K-means algorithm based on rough set optimization is proposed to solve these problems. Firstly, the concept of important attributes is put forward according to the characteristics of Rough Set, and the attribute importance is calculated by observing the change of attribute positive domain. Then, the best attributes reduction is selected by the attribute importance, so that these important attributes are the best attributes reduction. Finally, the K-means algorithm is used to classify the important attributes. The experiments of two datasets are designed to evaluate the proposed algorithm, and the experimental results show that the K-means algorithm based on rough set optimization has more classification accuracy than the original K-means algorithm.

Keywords: Functional MRI · Rough set · K-means · Resting state

1 Introduction

The field of neuroinformatics mainly concludes: data collection, organization and analysis of neuroscience data, data calculation model and development of analytical tools, etc. Treated as a comprehensive subject of information science and neuroscience, neuroinformatics plays a vital role in information science and neuroscience research [1]. Functional magnetic resonance imaging (fMRI) technology is one of the most significant approaches to obtain the data of neuroinformatics. It has been widely used in

Fund Project: National Natural Science Foundation of China (Grants No. 31470954).

Z. Shi et al. (Eds.): ICIS 2017, IFIP AICT 510, pp. 84–92, 2017.
DOI: 10.1007/978-3-319-68121-4_9

human behavior experiment and pathology because of its noninvasive, repeatability and other advantages [2–8]. fMRI can be used to obtain high-resolution three-dimensional images of the brain through the BOLD (blood oxygen level dependent) effect, which can dynamically reflect changes in brain activity signals. However, there are some problems such as large amount of data and excessive interference noise in the research of fMRI. Therefore, how to effectively reduce and extract the feature of fMRI data has become the core contents of the research.

Rough set theory provides a new method that can extract attribute reduction set from fMRI data and obtain feature rules by subtracting the set. Rough set is a mathematical idea proposed by Polish mathematician Pawlak for dealing with uncertainty data in 1980 [9]. The main idea is to keep the classification ability under the premise of the same, get the problem of classification rules and decision rules through the knowledge reduction [10, 11]. The ultimate goal of rough set theory is to generate the final rule from the information (decision) system. There are two principles for the derivation of a feature rule: first, the rules should be used for the classification of database objects. That is, to predict the categories of unlabeled objects. Second, the rule should be used to develop a mathematical model in the field of research, and this knowledge should be presented in a way that can be understood by people. The main steps to process the data using rough set theory are as follows: (1) mapping information from the original database to the decision table (2) data preprocessing (3) calculated attribute reduction (4) from the data reduction derived rules (5) rules filtering. One of the most critical tasks is the reduction of the attributes of the data. In general, as a decision table elements, the real object often produces a large amount of data, and these data is not all valuable from the calculation point of view. Therefore, it would be meaningful to be able to extract the most valuable information from the large decision table effectively.

In this paper, a kind of K-means algorithm based on rough set optimization is proposed to apply the classification of fMRI data. First, use the rough set of ideas in the training set of fMRI data on the property reduction. Then, gain the best attribute reduction by calculating the importance of the property, and regard the best attribute reduction as an important attribute. Finally, treat the important attributes as data features, and classify the test fMRI data by k-means algorithm. Furthermore, various fMRI data experiments are used to demonstrate the effectiveness of the proposed method.

2 Knowledge of Rough Sets

Definition 1: Suppose $P \subseteq R, P \neq \varnothing$, $\cap P$ represents the intersection of all equivalence relations in P, as ind(P), and $[\mathrm{x}]_{ind(\mathrm{P})} = \bigcap_{P \subseteq R} [\mathrm{X}]_R$. ([x] is the non-distinguishable set of x).

Definition 2: Four-tuples = (U, A, V, f) is a knowledge representation system, U represents the domain of non-empty set of object, as $U = \{X_1, X_2, \cdots, X_n\}$; A represents non-empty finite set of the indicators, as $A = \{\mathrm{a}_1, \mathrm{a}_2, \cdots, \mathrm{a}_n\}$, (where $A = C \cup D$,

C represents the condition attribute and D represents the decision attribute); V is the range of indicator a, as $V = \cup V_a(\forall V_a \in V)$; $f = U \times A \to V$ is an information function which given the value of a message to each attribute, where $V = \cup V_a$ $(\forall V_a \in V)$.

Definition 3 (Pawlak approximation space): Let U be a finite nonempty universe of discourse. Let R be an equivalence relation over U. Then the ordered pair <U, R> is a Pawlak approximation space.

Approximation operators: Let <U, R> be a Pawlak approximation space U/R = {W1, W2…Wm} is a partition where W1, W2 … Wm are all equivalence classes of R. With each subset $X \subseteq U$, we associate two subsets:

$$\underline{R}(\mathrm{X}) = \cup\{\mathrm{W}_i \in \mathrm{U/R} : \mathrm{W}_i \subset \mathrm{X}\} \tag{1}$$

$$\overline{\mathrm{R}}(\mathrm{X}) = \cup\{\mathrm{W}_i \in \mathrm{U/R} : \mathrm{W}_i \cap \mathrm{X} \neq \varnothing\} \tag{2}$$

Called the R- upper and R- lower approximations of X, respectively.

$$\text{X's boundary of R} \qquad bn_R(\mathrm{X}) = \overline{R}X - \underline{R}X \tag{3}$$

$$\text{X's positive region of R} \qquad pos_R(\mathrm{X}) = \underline{R}X \tag{4}$$

$$\text{X's negative region of R} \qquad neg_R(\mathrm{X}) = U - \overline{R}X \tag{5}$$

Their relationship can be explained by Fig. 1.

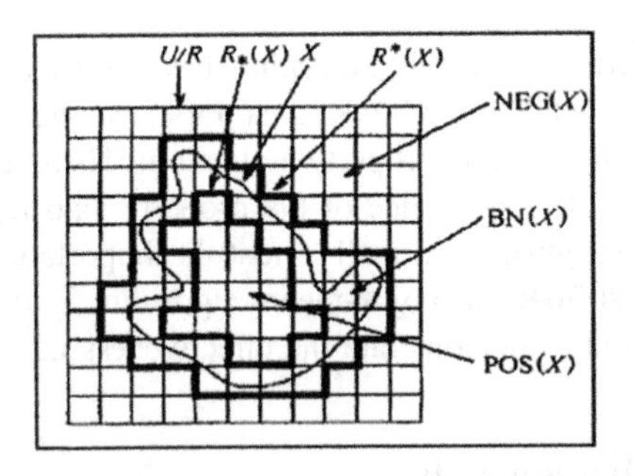

Fig. 1. Sketch map of concept of rough sets

Attribute positive field: Suppose P and Q are the two knowledge of the Domain of U. Definition positive field P of Q in fellow:

$$Pos_P(\mathrm{Q}) = \bigcup_{X \in U/Q} Pos_P(\mathrm{X}) = \{\mathrm{x}|\mathrm{x} \in \mathrm{U} \wedge [\mathrm{x}]_p \subseteq [\mathrm{x}]_Q\} \tag{6}$$

3 Experimental Methods

3.1 Data Acquisition and Data Preprocessing

(1) *Data of different eye states*

This study includes the following two behaviors: (1) Open eyes behavior. Subjects were asked to open their eyes, pegged to the top of the machine cross, and kept the frequency of blink as low as possible. (2) Close eyes behavior. Subjects were asked to close their eyes, keep less visual input. Data acquisition process requires subjects to keep their brain clear and keep away from the sleep state.

The BOLD fMRI data were acquired on a Siemens Trio 3.0 T of East China Normal University with a gradient echo EPI with 36 slices providing whole-brain coverage and 230 volumes, a TR of 2s and a scan resolution of 64 * 64. The in-plane resolution was 3.5 mm × 3.5 mm, and the slice thickness was 3.5 mm.

(2) *Data of Alzheimer's disease and health*

The data of 30 Alzheimer's disease data which was used in the current study was provided by the Alzheimer's disease Neuroimaging Initiative (ADNI) database (adni.loni.usc.edu). The ADNI is a public-private partnership guided by Michael W. Weiner, MD, which was founded in 2003. ADNI's initial target was to investigate and find a solution to predict the progression of early Alzheimer's disease and mild cognitive impairment by means of a combination of positron emission tomography, magnetic resonance imaging and other biological, neuropsychological and clinical evaluations.

The specific parameters for data acquisition are: Magnetic field strength 3.0 T Philips with a gradient echo EPI with 48 slices providing whole-brain coverage and 140 volumes, a TR of 3s and a scan resolution of 64 * 64. The in-plane resolution was 3.31 mm × 3.31 mm, and the slice thickness was 3.31 mm.

The 30 healthy subjects' data which was used in the current study, was provided by Common database of neuroimaging (http://www.nitrc.org/projects/fcon1000/). The data was published by Professor Yufeng Zang in NIFTI format.

The specific parameters for data acquisition are: Magnetic field strength 3.0T Philips with a gradient echo EPI with 33 slices providing whole-brain coverage and 215 volumes, a TR of 2s and a scan resolution of 64 * 64. The in-plane resolution was 3.13 mm × 3.13 mm, and the slice thickness was 3.6 mm.

3.2 Attribute Importance Calculation

Suppose that S = <U, A, V, f > is a knowledge expression system. $A = C \cup D$, $C \cap D = \varnothing$ C is the condition attribute set, D is the decision attribute. If $U/C = \{X_1, X_2, \cdots, X_n\}, U/D = \{Y_1, Y_2, \cdots, Y_n\}$ the degree of dependence between P and D can be computed as [12]:

$$k_c(\mathrm{D}) = \left[1 - \frac{cord(\mathrm{x})}{d}\right] \frac{1}{|U|} \sum_{i=1}^{m} |C\_(\mathrm{Y}_i)| = \frac{1}{|U|} \sum_{i=1}^{m} |pos_C(\mathrm{Y}_i)| \tag{7}$$

Thus, the significance of an attribute $a \in C$ can be calculated from the set of conditional attributes C as follows:

$$sig^{D}_{C-\{c\}}(c) = K_{C}(D) - K_{C-\{C\}}(D) \quad (8)$$

3.3 Best Attribute Reduction

The attribute significance of each condition attribute from each subject was calculated, and the attribute significance of the same condition attribute was added, the order table of attribute significance was obtained according to the order from large to small as well. We counted the attribute significances which were obtained for the combination of attribute reduction. In the premise of keeping the attribute table decision-making ability unchanged, we selected the least amount and the highest attribute significance from the combination of attribute reduction as the best attribute reduction.

3.4 K-means Algorithm

Step 1: From N data objects, K objects were selected as the initial clustering centers;
Step 2: According to the mean (central object) of each clustering object, calculate the distance between each object and the central object, and divide the corresponding object according to the minimum distance;
Step 3: Recalculate the mean of each cluster (central object);
Step 4: Calculate the standard measure function. When a certain condition is satisfied, such as meet the convergence of the function, the algorithm terminates; if the condition is not satisfied, then go back to step 2.

From the training set of data calculating the attribute significance, relying on the attribute significance to select the best attribute reduction of the rough set, the attribute of best attribute reduction was defined as an important attribute, and the rest of the attributes other than the best attribute were defined as non-significant attributes. The test sets for the two sets of data were tested separately; the K-means algorithm was used to classify the important and non-important attributes. Take $K = 2$, randomly generated the initial center of mass, the classification result was compared with the original data label to obtain the classification accuracy n. Considering that the correspondence between the label after K-means classification and the original label was uncertain, therefore, only values with a classification accuracy greater than 0.5 were selected. If n is less than 0.5, then $1 - n$ is the final classification accuracy. That is, the classification accuracy is at least 0.5, and the highest is 1.

4 Result Analyses

4.1 Reductions in fMRI Data for Different Eye States

We randomly selected 30 groups of subjects, a total of 13800 data as a training set, and 4600 data of the remaining 10 subjects as a test set. After acquiring the attribute reductions, the best attribute reduction is obtained by calculating the attribute significance. The best reduction in fMRI data for different eye states consists of 20 brain regions listed in Table 1. Rank in descending order according to attribute importance: $\{49, 4, 3, 10, 15, 40, 9, 47, 21, 46, 16, 42, 41, 50, 13, 22, 11, 44, 39, 35\}$.

Table 1. The best reduction in fMRI data for different eye states

ROI_name|ROI_Number	ROI_name|ROI_Number	ROI_name|ROI_Number
Frontal_Sup_Orb_L (3)	Frontal_Inf_Orb_R (16)	Occipital_Mid_R (42)
Frontal_Sup_Orb_R (4)	Olfactory_L (21)	Occipital_Inf_R (44)
Frontal_Mid_Orb_L (9)	Olfactory_R (22)	Temporal_Sup_R (46)
Frontal_Mid_Orb_R (10)	Paracentral_Lob_L (35)	Temporal_Mid_L (47)
Frontal_Inf_Oper_L (11)	Occipital_Sup_L (39)	Temporal_Inf_L (49)
Frontal_Inf_Tri_L (13)	Occipital_Sup_R (40)	Temporal_Inf_R (50)
Frontal_Inf_Orb_L (15)	Occipital_Mid_L (41)	

4.2 Reductions in fMRI Data for Alzheimer's Disease and Healthy Controls

We randomly selected 20 groups of subjects, a total of 7100 data as a training set, and 3550 data of the remaining 10 subjects as test set. After acquiring the attribute reductions, the best attribute reduction is obtained by calculating the attribute significance. The best reduction in fMRI data for Alzheimer's and healthy controls contains 17 brain regions listed in Table 2. Rank in descending order according to attribute importance: $\{19, 3, 16, 21, 50, 20, 23, 2, 48, 49, 43, 8, 40, 14, 35, 42, 10\}$.

Table 2. The best reduction in fMRI data for different eye states Alzheimer's disease

ROI_name|ROI_Number	ROI_name|ROI_Number	ROI_name|ROI_Number
Frontal_Sup_R (2)	Cingulum_Ant_R (20)	Temporal_Mid_R (48)
Frontal_Sup_Orb_L (3)	Olfactory_L (21)	Temporal_Inf_L (49)
Frontal_Mid_R (8)	Postcentral_L (23)	Temporal_Inf_R (50)
Frontal_Mid_Orb_R (10)	Paracentral_Lobule_L (35)	
Frontal_Inf_Tri_R (14)	Occipital_Sup_R (40)	
Frontal_Inf_Orb_R (16)	Occipital_Mid_R (42)	
Cingulum_Ant_L (19)	Occipital_Inf_L (43)	

4.3 Data Atlas

In order to more intuitively reflect the differences between important attributes and non-important attributes, the mean values of the different brain regions of each subject were obtained. The data maps are produced as follows:

(1) Different eye state data_Average data map (Line 1 to 230 is closed eyes, Line 231 to 460 is opened eyes) (Fig. 2).

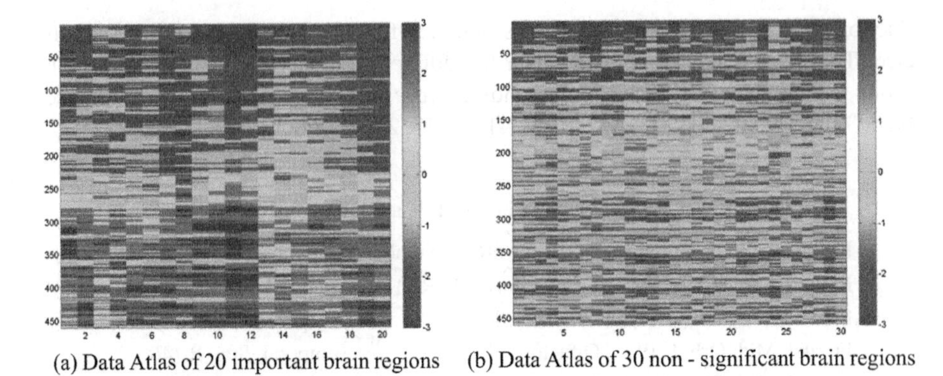

(a) Data Atlas of 20 important brain regions (b) Data Atlas of 30 non - significant brain regions

Fig. 2. Data Atlas of important and non-vital brain regions with different eye status data

(2) *Alzheimer's disease and normal control data _ Average data map* (Line 1 to 140 is Alzheimer's disease data, Line 141 to 355 is normal data) (Fig. 3).

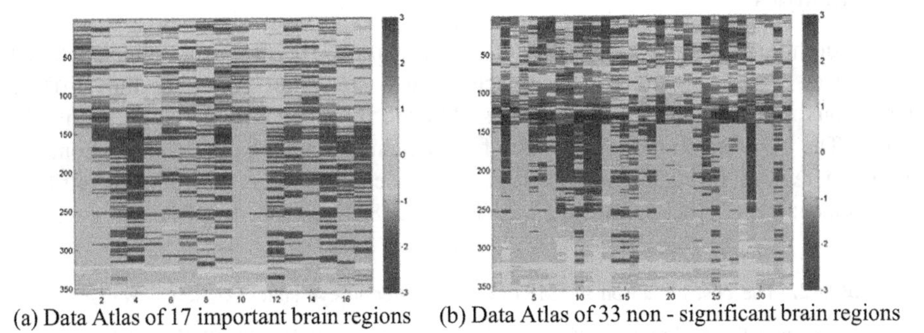

(a) Data Atlas of 17 important brain regions (b) Data Atlas of 33 non - significant brain regions

Fig. 3. Data Atlas of important and non-vital brain regions with Alzheimer's disease

We can clearly see from the two sets of data, in the data of the same control group, the distinction in the data of important attributes is greater, and different categories of data are more clearly distinguished. Therefore, the use of important attributes for classification can have a better effect. And the non-important attribute data is usually more disorganized and less distinct, so using it to classify is often ineffective and may even affect judgement.

4.4 Clustering Algorithm Based on Rough Set Optimization

The result of the attribute reduction is taken as an important attribute, and the data that has been subtracted is taken as a non-important attribute. Then the K-means algorithm is used to classify the raw data, the data containing only the important attributes, and the data containing only the non-significant attributes, respectively, where K = 2. The results are as follows (Table 3):

Table 3. Classification accuracy of K-means algorithms with different data

Type of data	All attribute classification accuracy %	Important attribute classification accuracy %	Non-important attribute classification accuracy %
Eye state	0.7159	0.8680	0.6239
Alzheimer's disease	0.8113	0.8592	0.7328

It can be found that, using the important attributes that obtained by the attribute reduction of the rough set to classify, can have a better effect. However, using the non-important attributes that has been subtracted to classify, its accuracy will be less than the raw data, and much less than the important attributes.

5 Discuss

In this paper, a K-means algorithm based on rough set optimization is proposed and applied to the hierarchical fMRI data classification. First, the concepts of important and non-important attributes were put forward. Applicate the rough set theory in the training data set for reduction of attributes, and select the best attribute reduction through the importance of the property. Regard the best attribute reduction of the attribute as an important attribute, the rest attributes as non-important attributes. The important attributes are the most significant cluster of data in the original attributes, and the rest of the attribute often plays a very weak role in the classification, and even affects the classification result. It can be seen from the experiment that the accuracy of classification using the important attributes will be higher than the original data, and the accuracy of the use of non-important attributes of the classification will be lower than the original data, and far below the important attributes. Therefore, the use of important attributes for classification calculation can greatly reduce the workload and can achieve better classification results. In the next step, the experiment can be improved by optimizing the attribute reduction algorithm, selecting the more reasonable strategy of superiority, dividing the more detailed brain region and so on, and obtain the scientific result.

References

1. Liu, H.B., Abraham, A., Zhang, W.S., Mcloone, S.: A swarm-based rough set approach for fMRI data analysis. Int. J. Innov. Comput. Inf. Control **7**, 3121–3132 (2011)
2. Taylor, M., Donner, E., Pang, E.: fMRI and MEG in the study of typical and atypical cognitive development. Neurophysiol. Clin. **42**, 19–25 (2012)
3. Shad, M.U., Keshavan, M.S.: Neurobiology of insight deficits in Schizophrenia: an fMRI study. Schizophr. Res. **165**, 220–226 (2015)
4. Huang, H., Wang, H.L., Zhou, Y.: Schizophrenia brain resting explore networks and perform network centers and to highlight the relationship between network. Chin. J. Psychiatry **48**, 175–181 (2015)

5. Li, H.J., Hou, X.H., Liu, H.H.: Toward systems neuroscience in mild cognitive impairment and Alzheimer's disease: a met analysis of 75 fMRI studies. Hum. Brain Mapp. **36**, 1217–1232 (2015)
6. Yang, J.J., Zeng, W.M.: Complex network-based connectivity study in patients with brain function. Chin. J. Med. Imaging **23**, 418–422 (2015)
7. Shi, Y.H., Zeng, W.M., Wang, N.Z., Chen, D.T.L.: A novel fMRI group data analysis method based on data-driven reference extracting from group subjects. Comput. Methods Programs Biomed. **122**, 362–371 (2015)
8. Shi, Y.H., Zeng, W., Wang, N., Zhao, L.: A new method for independent component analysis with priori information based on multi-objective optimization. J. Neurosci. Methods **283**, 72–82 (2017)
9. Pawlak, Z.: Rough set. Int. J. Comput. Inf. Sci. **11**, 341–356 (1982)
10. Wang, G.Y., Yao, Y.Y., Yu, H.: A review of research on rough sets theory and applications. J. Comput. Sci. 1229–1246 (2009)
11. Zhou, X.Z., Huang, B., Li, H.X., Wei, D.K.: Rough Sets Theory & Approaches for Knowledge Acquisition in Incomplete Information Systems. Nanjing University Press, Nanjing (2010)
12. Barnali, B., Swarnajyoti, P.: Hyper spectral image analysis using neighborhood rough set and mathematical morphology. In: Accessibility to Digital World (2016). ISBN 978-1-5090-4291

The Research of Attribute Granular Computing Model in Cognitive and Decision-Making

Ruqi Zhou[1,2](✉) and Yuepeng Zhou[2]

[1] Department of Computer Science, Guangdong University of Education, Guangzhou 510303, China
ruqzhou@163.com
[2] School of Data and Computer Science, Sun Yat-Sen University, Guangzhou 510006, China

Abstract. The cognitive activities of human beings are complicate and diversified. So far, there hasn't been a universal cognitive model. Each cognitive model generally only represents cognitive features in one or some aspects. Therefore, this paper aims to, based on the granular computing theory and principles, and with attributes and change laws as the main objects of the cognitive course, proposes a cognitive model which not only can describe the thing through attributes to represent the change law, but also can simulate the cognitive decision-making course with respect to the attribute change of the thing. Attribute granular computing, based on qualitative mapping, can simulate the cognitive functions of human brain, such as granulation, organization and causation. Petri net has asynchronous, concurrent and uncertainty characteristics, which is similar to the characteristics of some cognitive activities in human thinking process. Petri net is extended based on the basic concept and logic calculation rules of attribute granular computing in this paper. Some basic elements of a cognitive system, such as knowledge representation, reasoning, learning and memory mode are initially showed in the extended Petri net. The results show that this method can reflect the cognitive process of uncertainty identification and decision-making in a certain extent.

Keywords: Cognitive and decision-making · Attribute granular computing · Qualitative mapping · Fuzzy set · Petri net

1 Introduction

Neurophysiological studies indicate that human brain has a dual structure consisting of senses and consciousness, which are two independent brain functions respectively processed in two different parts in human brain [1]. According to the structure and physiological functions of the human brain, paper [2] propose a physiology-based cognitive logic theory which holds that the human brain is of a dual structure for senses and consciousness; the sensory memory gets information from the external world, and

Z. Shi et al. (Eds.): ICIS 2017, IFIP AICT 510, pp. 93–103, 2017.
DOI: 10.1007/978-3-319-68121-4_10

the consciousness memory gets information from the sensory system; the perception is the conjugation of the sensory information and the consciousness information; and the same transformation exists between the sensory information and the consciousness information. Paper [3] point out that the sense organs only respond to sensory attributes which are sensitive to the sense organs. The information collected by the human brain is the collection of sensory attributes, and the state, movement and change course of a thing is reflected through attributes and the change course of such attributes. The processing of the sensory attributes and their change laws is one of the core problems of the information processing in human brain. The cognitive course of human beings starts from these attributes received through sensory, and an attribute is essentially an information granule.

Paper [4] summarized the human's cognitive ability into: granulation, organization and causation. The concept of the information granule given by paper [5] can be interpreted with an attribute and its qualitative mapping operator, so that the granular computing can be used to simulate a part of the cognitive functions of the human brain.

Decision making is actually a cognitive process. In the decision-making process, people always make decision evaluation systematically from different perspectives in order to obtain more reliable decision results. Qian [7] proposed the multigranulation rough set theory, and proposed multigranulation decision rules which is consistent with human cognition. Song [6] determined the weights of the attributes for the decision objectives through mutual information and conditional entropy, and obtain the efficient sorting results. Yao [8] put forward three-way decision model, which is a common method of human cognitive decision-making. Based on the semantic model of Bayesian decision, a practical and effective decision-making rough set model was given [9, 10]. Three-way decision has been developed and applied in other areas [11, 12].

Feng [5, 13, 14] put forward a comprehensive decision and evaluation model of attribute coordinate. It is a multi-index factors and multi-attribute decision-making comprehensive evaluation model based on the qualitative mapping. The model is good at simulating the process of psychological cognition of decision maker and gives decision. Its characteristic is that its evaluation method is very close to the brain cognitive thinking mode, which can reflect the cognitive psychological preferences and their change curves of the decision maker. Based on this, this paper describes the brain cognition and decision-making process using the model of attribute granulation.

However, due to the lack of a formal mechanism for the reasoning process of attribute granule, Petri net is used to establish a cognitive decision-making model in this paper. The Petri net is a formalization method with good structure, and is characterized by concurrency, asynchrony, uncertainty and the like. It is similar to certain cognition activity in the thinking course of the human brain. If the places and transitions in the Petri net are simulated with attribute granule and qualitative mapping, the Petri net will initially show the characteristics of some basic elements necessary to a cognition system in terms of knowledge representation, knowledge reasoning, decision-making, learning mode, memory mode, etc. This is a new trying.

2 Qualitative Mapping Method in Cognition

Psychologists believe that human's senses reflect simple attributes of a thing while human's consciousness integrates such simple attributes into comprehensive attributes of the thing. Let $a(x)$ be sensory attributes of a thing x. The value of $a(x)$ generally can be divided into quantitative attribute value $d_{\mathrm{a(x)}}$ and qualitative attribute value $P_{a(x)}$. When x has qualitative attribute value $P_{a(x)}$, it is said that the x has the "properties" referred in $P_{a(x)}$. If $P_{a(x)}$ is the property by which the thing x can be distinguished from other things y, the $P_{a(x)}$ is called one property or characteristic of thing x [15].

Let $d(x)$ be a quantitative value of the property $p(x)$, and $[\alpha_i, \beta_i]$ be a certain qualitative criterion domain of $p(x)$. If $d(x) \in [\alpha_i, \beta_i], [\alpha_i, \beta_i]$ is a neighborhood of $d(x)$ [16].

If $[\alpha_i, \beta_i]$ and $[\alpha_j, \beta_j]$ are the qualitative criterions respectively for property $p_i(x)$ and property $p_j(x)$, we can perform the following logical operations on $[\alpha_i, \beta_i]$ and $[\alpha_j, \beta_j]$.

(1) Logic not: $\sim [\alpha_i, \beta_i], \sim [\alpha_j, \beta_j]$.

(2) Logic and: $[\alpha_i, \beta_i] \wedge [\alpha_j, \beta_j]$. The intersection $[\alpha_i, \beta_i] \cap [\alpha_j, \beta_j]$, which can be considered as the qualitative criterions for the conjunction property $q(x) = p_i(x) \wedge p_j(x)$.

Property $p_i(x)$ and property $p_j(x)$ are homogeneous if a criterion $[\alpha_{ij}, \beta_{ij}] \in [\alpha_i, \beta_i] \cap [\alpha_j, \beta_j]$ exists and makes $a(x) \in [\alpha_{ij}, \beta_{ij}] \wedge b(x) \in [\alpha_{ij}, \beta_{ij}]$ for any quantitative value $a(x)$ of $p_i(x)$ and any quantitative value $b(x)$ of $p_j(x)$. Or otherwise, $p_i(x)$ and $p_j(x)$ are heterogeneous.

(3) Logic or: $[\alpha_i, \beta_i] \vee [\alpha_j, \beta_j]$. The intersection $[\alpha_i, \beta_i] \cup [\alpha_j, \beta_j]$ can be considered as the qualitative criterions for the disjunction property $r(x) = p_i(x) \vee p_j(x)$.

(4) Logical implication: $[\alpha_i, \beta_i] \rightarrow [\alpha_j, \beta_j]$.

Let $\Gamma = \{[\alpha_i, \beta_i] | i \in I\}$ be the cluster of all qualitative criterions of a proposition p. The mapping $\Psi : \Gamma \rightarrow \Gamma$ is called as the criterion mapping of proposition p [10]. If $[\alpha_j, \beta_j] \in \Gamma$ exists for any $[\alpha_i, \beta_i] \in \Gamma$, then

$$\Psi([\alpha_i, \beta_i]) = [\alpha_j, \beta_j]$$

Ψ is called as the qualitative criterion transformation from $[\alpha_i, \beta_i]$ to $[\alpha_j, \beta_j]$, and noted formally as $[\alpha_i, \beta_i] \rightarrow [\alpha_j, \beta_j]$.

Obviously, homogeneous relations are equivalent based on the criterion mapping, while heterogeneous relations under the criterion mapping can be transformed when certain conditions are satisfied.

Philosophically, the law of quantitative change to qualitative change is represented as the transformation of quantitative properties to qualitative properties. Literatures [12] provide a mathematic model of this cognitive law:

Definition 1 [3]. Let $a(o) = \bigwedge_{i=1}^{n} a_i(o)$ be the integrated attributes of n factor attributes $a_i(o), i = 1, \ldots, n$, and $x = (x_1, \ldots, x_n)$ be the quantitative value of $a(o)$, where: $x_i \in X_i \subseteq R$ is the quantitative property value of $a_i(o), p_i(o) \in P_o$ is a certain attribute of

$a_i(o), \Gamma = \{[\alpha_i, \beta_i] | [\alpha_i, \beta_i]$ is the qualitative criterion of $p_i(o)\}$, and hypercube $[\alpha, \beta] = [\alpha_1, \beta_1] \times \ldots \times [\alpha_n, \beta_n]$ is the qualitative criterion of integrated attribute $p(o) = \overset{n}{\underset{i=1}{\wedge}} p_i(o)$. The mapping $\tau : X \times \Gamma \to \{0, 1\} \times P_o$ is called the qualitative mapping (QM) of $x = (x_1, \ldots, x_n)$ with n-dimensional hypercube $[\alpha, \beta]$ as a criterion. If a property $p(o) = \overset{n}{\underset{i=1}{\wedge}} p_i(o) \in P_o$ with $[\alpha, \beta] \in \Gamma$ and $[\alpha, \beta]$ as criterion exists for any $x \in X$, then:

$$\tau(x, [\alpha, \beta]) = \overset{n}{\underset{i=1}{\wedge}} (x_i \underset{?}{\in} [\alpha_i, \beta_i]) = \overset{n}{\underset{i=1}{\wedge}} \tau_{p_i(o)}(x_i)$$

where, $\tau_{p_i(o)}(x_i) = \begin{cases} 1 & x_i \in [\alpha_i, \beta_i] \\ 0 & x_i \notin [\alpha_i, \beta_i] \end{cases}$ is the real value of the property proposition $p_i(o)$.

The numerical domain $d_{a(x)}$ of attribute $a(x)$ of the thing x and the criterion domain cluster $\Gamma = \{[\alpha_i, \beta_i] | i \in I\}$ form a criterion topological space $T(d_{a(x)}, \Gamma)$. Therefore, the Definition 1 actually is a cognitive mode of the granular computing done in the criterion topological space.

3 Attribute Granules in Cognitive and Decision-Making

A granule is not only a cluster or a collection of entities, but also the abstraction of such cluster or collection [17, 18]. If all attributes of a thing is considered as a whole, a certain attribute can also be considered as a granule. The qualitative mapping, which creates maps from an attribute to a property, is an intuitive attribute computing but also an intuitive granular computing. Attributes are closely related to the qualitative criterion, or in other words, attributes are closely related to the qualitative mapping. When we say a thing x has the property $p(x)$ referred in $P_{a(x)}$ (the quantity property of an attribute $a(x)$), what we really mean is the thing has the property $p(x)$ under certain qualitative criterion. Therefore, an attribute with a bigger structure intuitively composed of attributes which contain qualitative criterion or qualitative mapping is called an attribute granule.

An attribute granule in the attribute topological space can be expressed as a 2-tuple:

$$(P, \Gamma) = (p_i(x), [\alpha_i, \beta_i]), \text{ or } (P, \Gamma) = (d_{a(x)}, [\alpha_i, \beta_i])$$

where, $a(x)$ is a sensory attribute of a thing x, $p_i(x)$ is the property of $a(x)$, $[\alpha_i, \beta_i]$ is a qualitative criterion domain, $d_{a(x)}$ is a quantitative attribute value of $a(x)$, and a granule $(p_i(x), [\alpha_i, \beta_i])$ is referred to as $\Gamma_i(p_i)$ for short. Formally, an attribute granule sometimes can also be equivalent to a first-order predicate formula.

Definition 2. Let $\Gamma(p_i)$ and $\Gamma(p_j)$ be two attribute granules. The computing formula of $\Gamma(p_i)$ and $\Gamma(p_j)$ with respect to logic connector can be defined as follows:

① $\sim \Gamma(p_i) = (\sim p_i, \Gamma')$, where, Γ' is the complementary operation of the qualitative criterion of Γ;

② $\Gamma_{\rm i}(p_i) \wedge \Gamma_j(p_j) = (p_i(x)\, \Delta p_j(x), [\alpha_{\rm i}, \beta_{\rm i}] \cap [\alpha_{\rm j}, \beta_{\rm j}], \Psi(\Gamma_{\rm i}) = \Gamma_j)$.
③ $\Gamma_{\rm i}(p_i) \vee \Gamma_j(p_j) = (p_i(x)\, \nabla p_j(x), [\alpha_{\rm i}, \beta_{\rm i}] \cup [\alpha_{\rm j}, \beta_{\rm j}], \Psi(\Gamma_{\rm i}) = \Gamma_j)$.
④ $\Gamma_{\rm i}(p_i) \rightarrow \Gamma_j(p_j) = (p_i(x)\, \Lambda p_j(x), [\alpha_{\rm i}, \beta_{\rm i}] \rightarrow [\alpha_{\rm j}, \beta_{\rm j}], \Psi(\Gamma_{\rm i}) = \Gamma_j)$.

Where, ∇ is attribute disjunction, Δ is attribute conjunction, Λ is attribute integration or reasoning, and Ψ is qualitative criterion transformation.

Definition 3. A formula $(\Gamma_i(p_i))$ is true if and only if the true value of its transformation function $\eta(\Gamma_i(p_i))$ is 1 or above 0.5.

Definition 4. The logical formula of an attribute granule can be interpreted semantically as:

A quadruple (4-tuple) D:

$$D = \{U, \Psi, \eta, \{\Gamma_i, i \in I\}\}$$

① U is a non-empty set called domain of interpretation (DOI);
② Ψ is a mapping and $P : U \rightarrow U$;
③ For any $i \in I, \eta : U \times \Gamma_i \rightarrow [0, 1]$ is a degree function;
④ $\Gamma_{\rm i}$ is the qualitative criterion domain.

In the interpretation D, each qualitative criterion transformation formula P corresponds to an element $v(P)$ in [0,1], which is called the true value of qualitative criterion transformation formula P in interpretation D, where:

① If P is the logical formula of the attribute granule $(\Gamma_{\rm i}(p_{\rm i}) = (p_{\rm i}(x), [\alpha_{\rm i}, \beta_{\rm i}])$, then:

$$v(p_i(x), [a_i, b_i]) = \begin{cases} 1 & \textit{iff} \quad x \in [\alpha_i, \beta_i] \\ \neg & \textit{iff} \quad x \notin [\alpha_i, \beta_i] \end{cases}$$

② If P is the logical formula of the attribute granule $\sim G$, then

$$v(\sim G) = (v(G))' = v(\sim p_i(\mathrm{x}), \Gamma - [\alpha_{\rm i}, \beta_{\rm i}]);$$

③ If P is the logical formula of the attribute granule $G \vee H$, then

$$v(G \vee H) = v(p_{\rm i}(x)\, \nabla p_{\rm j}(x), [\alpha_{\rm i}, \beta_{\rm i}] \cup [\alpha_{\rm j}, \beta_{\rm j}]);$$

④ If P is the logical formula of the attribute granule $G \wedge H$, then

$$v(G \wedge H) = v(p_{\rm i}(x)\, \Delta p_{\rm j}(x), [\alpha_{\rm i}, \beta_{\rm i}] \cap [\alpha_{\rm j}, \beta_{\rm j}]);$$

⑤ If P is the logical formula of the attribute granule $G \rightarrow H$, then

$$v(G \rightarrow H) = v(p_{\rm i}(\mathrm{x}) \Rightarrow p_{\rm j}(\mathrm{x}), [\alpha_{\rm j}, \beta_{\rm j}] = \Psi([\alpha_{\rm i}, \beta_{\rm i}]))$$

Where, $\Rightarrow$ is attribute reasoning.

Definition 5. G and H in the logical formula of the attribute granule are called equivalence. If the assignment of any interpretation D is $v(G) = v(H)$, then $G = H$.

Definition 6. $\Gamma_{\rm i}(p_{\rm i}) \rightarrow \Gamma_{\rm i}(p_{\rm i})$ is an interference in the logic of the attribute granule, which means that there is a mapping $\Psi_{ij}: \Gamma \rightarrow \Gamma$ to make $\Psi_{\rm ij}([\alpha_{\rm i}, \beta_{\rm i}]) = [\alpha_{\rm j}, \beta_{\rm j}]$ if $(p_{\rm i}(x) \Rightarrow p_j(x))$.

4 Cognitive Decision-Making and Fuzzy Attribute Granule

Definition 7 [5]. A mapping $\eta: X \times \Gamma \rightarrow [-1, 1]_i$ is a function representing the transformation degree of quality property set $p_i(\xi_i)$. If $\exists \eta_i(x) \in [-1,\ 1]_i$ for $\forall (x, N(\xi_i, \delta_i)) \in X \times \Gamma$, then:

$$\eta(x, \xi_I, \delta_i) = |x - \xi_i| \perp \delta_i = \eta_i(x)$$

where, the $N(\xi_i, \delta_i)$ is the qualitative criterion domain, δ_I is radius of the domain, ξ_I is core of the domain, $\eta_i(x)$ is diversity factor of $|x - \xi_i|$ and δ_i as to the mathematics nature.

Zadeh studied the size of divided class or granule and defined the information granularity as a proposition: value of x belongs to the fuzzy subset $G \subseteq U$ by membership λ. The x is a variable of U and the value of x is an entity of U, represented by g = x is G is λ, and noted formally as:

g = {$u \in U$: value of x ($v(x) = u$, v is assignment symbol on U) is calculated based on membership of λ in the fuzzy set $G \subseteq U$.}.

Obviously, in the formulation above, $0 \leq \lambda \leq 1$. Based on fuzzy set theory, λ is a fuzzy membership function. Based on logic, the λ is the fuzzy truth value or probability of the proposition.

It is easy to figure out by comparison that: under the qualitative mapping τ with transformation program function, a information granule ($p_i(x)$, $[\alpha_i, \beta_i]$ or ($d_{a(x)}$, $[\alpha_i, \beta_i]$) in Definition 1 shows that 'the fact attribute $a(x)$ having the property $p_{\rm i}(x)$ is calculated based on the membership of the attribute value $d_{a(x)}$ in the criterion $[\alpha_i, \beta_i]$'. They are the same in essence.

Definition 8. Let the criterion domain $N_i(o_i, r_i)$ be the core of a fuzzy set A, namely $N_i(o_i, r_i) = A_1$ = {$x \mid \mu_A(x) = 1$}. The $\mu_A(x)$ is membership of $x \in A$. λ cut set of A is A_λ = {$x \mid \mu_A(x) \geq \lambda$}, $\lambda \in (0,1]$, λ ∈ (0,1], consisting of element x of membership not less than λ. Ψ_λ is the λ criterion transformation of the fuzzy set A whose core is $N_i(o_i, r_i)$ if it meets following equation:

$$\Psi_\lambda(N_i) = A_\lambda$$

$$\Psi_\alpha([\alpha_i, \beta_i]) = N(T_\alpha(o_i), \Psi_\alpha(r_i)) = N_\alpha(o_k, r_k)$$

Definition 9. Let A and B be two fuzzy sets with criterion domains $[\alpha_i, \beta_i]$ and $[\alpha_k, \beta_k]$ as cores respectively. Ψ_α is the criterion transformation from $[\alpha_i, \beta_i]$ to $[\alpha_k, \beta_k]$ if it meets following equation:

$$\Psi_\alpha([\alpha_i, \beta_i]) = N(\Psi_\alpha(o_i), \Psi_\alpha(r_i)) = N_\alpha(o_k, r_k)$$

The following theorem can be obtained immediately:

Based on the Definition 10, criterion transformation $\Psi_{\alpha\lambda}$ as the composite transformation of Ψ_α and λ-criterion transformation if it meets conditions below:

$$\Psi_{\alpha\cdot\lambda}(N_i) = \Psi_\lambda \cdot \Psi_\alpha(N_i) = \Psi_\lambda(N_\alpha) = N_\lambda(o_\alpha, r_\alpha)$$

The mapping $\eta : X \times \Gamma \rightarrow [-1, 1]$ is the transformation degree function for $p_i(x)$ to reflect its nature characteristics; if $\forall(x, N(\xi_i, \delta_i)) \in X \times \Gamma, \exists\eta_i(x) \in [-1, 1]$, which makes

$$\eta(x, \xi_i, \delta_i) = |x - \xi_i| \perp \delta_i = \eta_i(x)$$

where, $\eta_i(x)$ is the variability between $|x - \xi_i|$ and δ_i.

5 Set Up Cognition Formal Model with Attribute Granule

The transition nodes in Petri net can be viewed as a qualitative mapping, and the quantity properties, quality properties and qualitative criterion domain mapping of the attributes of a thing care places in the Petri net. Therefore, if the place nodes are extended into an information granule, the transition nodes will naturally be the granular transformation mapping. The definition of ordinary Petri net is extended below, and is considered as a cognitive system model.

Definition 10. The form of a cognitive system can be defined as a nonuple (10-tuple) Petri net:

$$\text{CSPT} = \{P, T, F; M_0, S, O, N, W, \Psi, \Gamma\}$$

where, P is a finite set of place nodes of attribute granules; T is a finite set of transition nodes of attribute granules, equivalent to a granular transformation; and F is a marked relation on $P \times T$, representing the connections of place nodes to transition nodes, the rated input and the input strength calculation function S on connection line, as well as the corresponding connection strength. M_0 is a function defined on P with its value range being [0, K] (K is a bounded real number), indicating the initial marking of the place nodes at the beginning of the operation; N is the output strength function at a qualitative level; O is the operation of m-dimensional weighted qualitative operator; Ψ is the criterion transformation, Γ *is* qualitative criterion; and W is the attribute weight, a function of Γ.

If the number of input link in any place node is 0, the place node is called as the input node of RS. If the number of the output link n any place node is 0, the place node is called as the output node of CSPT.

CSPT is mainly characterized in that:

(1) The starting threshold of transition nodes of CSPT is the detection threshold of sensory nerves, and the starting course is the transformation of the qualitative mapping degree.
(2) Directed edges in CSPT can be divided into input link and output link. For any node, the link pointing to it is called its input link and that deviating from it is called its output link. The link strength is attached to the input and output links.
(3) The capacity of place node is a bounded real number and the start-up of transition node depends on whether value of the input of quantitative (or qualitative) property of attribute on the input link, connection strength and the degree function of qualitative mapping (which is called input strength) is higher than the starting threshold of the transition node.

This is an operable net. The logical relationship between place nodes can be represented as shown in Figs. 1, 2 and 3. Decision problems described in the qualitative mapping model can be represented by CSPT in a vivid and visual manner.

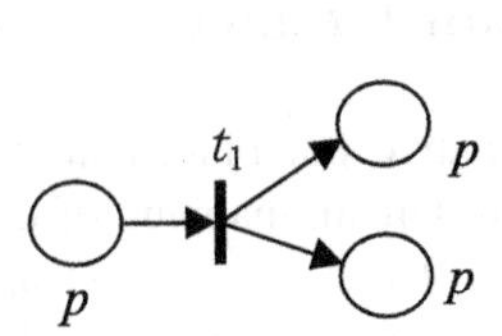

Fig. 1. One cause to multiple effects

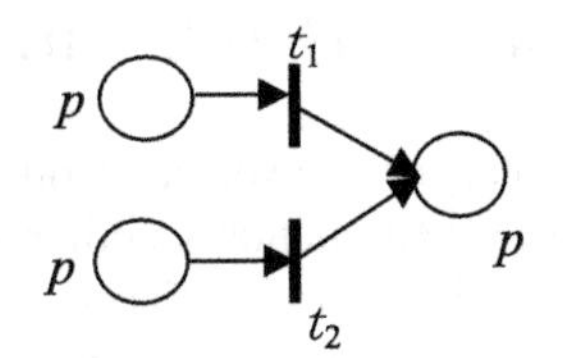

Fig. 2. Multiple causes correspond to the same effect

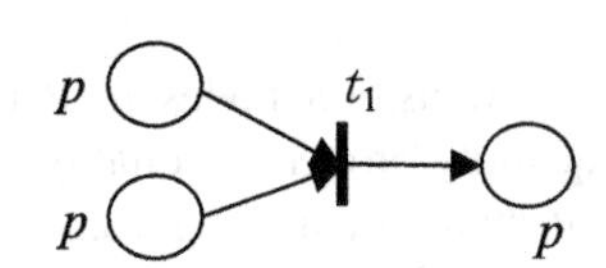

Fig. 3. Multiple causes to one effect

6 Decision-Making in CSPT

Paper [19, 20] proposed a logical structure of memory which holds that the sensory storage in the three-stage processing model for memory information is merely an input-oriented storage. In view of the whole process from input to output of the sensation, consciousness and the qualitative mapping, paper [21] indicate that the sensation is a test response to all simple attribute properties of an object, and that the consciousness is responsible for integrate them to form the overall consciousness of human brain to the object. The pattern recognition of the outer attribute information granules through sensation and consciousness can be represented with the following classification function:

$$s_y(p_i(\mathrm{x}), [\alpha_\mathrm{i}, \beta_\mathrm{i}]) = s_y \Delta\Gamma(p_i) = \begin{cases} s_y(x, y) & \text{If } s_y(x) \text{ is a simple factor of } p_i(x) \\ -s_y(x, y) & or \end{cases}$$

where, $(p_i(x), [\alpha_i, \beta_i])$ and Γ (p_{i}) represent the attribute granule of a thing x, s_y is sensory neuron which can be considered as the mapping of memory set $M(x, y)$ of the thing x from attribute set P_x to a subject y. Namely, s_y: $P_x \rightarrow M(x, y)$. Δ is an extraction operation.

The representation of attribute granules in CSPT is more natural, and the whole process of knowledge representation reflects the reasoning course of thinking operation in the human brain.

Assume the following thinking operation exists, $M(y)$ is the memory set of a subject y, and $s_i(y) \in M(y)$ is detecting neurons (or sensor or the like) of the attribute $s_i(x)$ of a thing x. Let $P_x = \{p_j(x) \,|\, j = 1, \ldots, n\}$ be the set of the attribute granules of x, and $p(x) = \wedge p_j(x)$ be the integration of r $(r \leq n)$ attributes of x, where, $\wedge$ is integration operator. If $r = n$ and $T(x) = \wedge p_j(x)$ is the integrated (comprehensive) attributes of x, then, the attribute detection on $T(x)$ by sensory neuron $s_i(y)$ can be represented as a mapping $s_i(y)$: $P_x \rightarrow M(y)$ from P_x to $M(y)$, and:

$$s_i(y)(T(x)) = s_i(y)\Delta(pi(x), [\alpha i, \beta i]) = s_i(x, y)$$

$s_i(x, y) \in M(y)$ is the sensory image of $s_i(x)$ in $M(y)$. The above formula indicates that: if $T(x)$ contains attributes $s_i(x)$ which can be detected by $s_i(y)$, then $s_i(y)$ will decompose and extract $s_i(x)$ from $T(x)$, and store the detection image $s_i(x, y)$ in $M(y)$. Or otherwise, $s_i(y)$ will tell the computer that x doesn't have attribute $s_i(x)$. The decision-making process of the above thinking operation can be represented in CSPT as shown in Fig. 4.

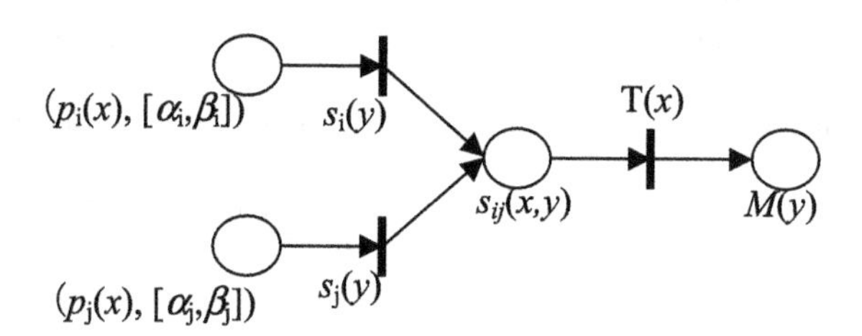

Fig. 4. Decision-making in CSPT

7 Conclusion

Sense organs of human being receive all kinds of sensory attribute information and respond accordingly. A thing represents its state, movement and change with its attributes and the attribute change courses. The cognitive process of human beings are closely related such attribute information received. Philosophical opinions hold that attribute not only expresses the quality property which needs to be expressed, but also has quantity property which needs to be limited and regulated. Based on such philosophical opinions in combination with the theories with respect to granular computing and qualitative criterion transformation, this paper provides the basic mathematical model of the cognition and the definition of Petri net. However, the cognitive process of human beings is very complicated. This paper gives merely the framework of a basic cognitive mode. More study work needs to be done.

Acknowledgments. This work is supported by the National High Technology Research and Development Program of China (863) sub project (2012AA101701-06), the Guangdong Science and Technology Plan Project (2012B010100049), the Characteristic Innovation Project of Guangdong Universities (natural science) (2014KTSCX195), open fund of Fujian Provincial Key Laboratory of Big Data Mining and Applications (Fujian University of Technology, 2017).

References

1. Badgaiyan, R.D.: Neuroanatomical organization of perceptual memory: an fMRI study of picture priming. Hum. Brain Mapp. **10**, 197–203 (2000)
2. Peking University Bio Intelligent Technology Research Group: The fifth generation computer and its cognitive logic methodology. Front. Sci. **1**(1), 18–23 (2007)
3. Feng, J., Dong, Z.: A mathematical model of perception schema generation and recognition based on attributive integration. Comput. Res. Dev. **34**(7), 487–491 (1997)
4. Zadeh, L.A.: Towards a theory of fuzzy information granulation and its centrality in human reasoning and fuzzy logic. Fuzzy Sets Syst. **19**, 111–127 (1997)
5. Feng, J.: Attribute network computing based on qualitative mapping and its applications in pattern recognition. J. Intell. Fuzzy Syst. **19**(2), 1–16 (2008)
6. Song, P., Liang, J.Y., Qian, Y.H.: A two-grade approach to ranking interval data. Knowl.-Based Syst. **27**, 234–244 (2012)
7. Qian, Y.H., Liang, J.Y., Yao, Y.Y., Dang, C.Y.: MGRS: a multigranulation rough set. Inf. Sci. **180**, 949–970 (2010)
8. Yao, Y.Y.: The superiority of three-way decisions in probabilistic rough set models. Inf. Sci. **181**(6), 1080–1096 (2011)
9. Yao, Y.Y.: Three-way decisions with probabilistic rough sets. Inf. Sci. **180**(3), 341–353 (2010)
10. Sun, B., Ma, W., Qian, Y.: Multigranulation fuzzy rough set over two universes and its application to decision making. Knowl.-Based Syst. **123**, 61–74 (2017)
11. Liu, D., Yao, Y.Y., Li, T.R.: Three-way investment decisions with decision theoretic rough sets. Int. J. Comput. Intell. Syst. **4**(1), 66–74 (2011)
12. Nauman, M., Azam, N., Yao, J.: A three-way decision making approach to malware analysis using probabilistic rough sets. Inf. Sci. **374**, 193–209 (2016)
13. Feng, J.: Qualitative mapping, linear transformation of qualitative criterion and artificial neuron. In: Proceedings of the IJCAI-2007, Workshop Theme: Complex Valued Neural Networks and Neuro-Computing: Novel Methods, Applications and Implementations, Hyderabad, India, 6–12 January 2007 (2007)
14. Feng, J., Wu, Z.: Conversion degree functions induced by qualitative mapping and artificial neurons. In: Proceedings of ICMLC 2003, IEEE 03EX693, pp. 1135–1140 (2003)
15. Feng, J., Dong, Z.: A mathematical model of sensation neuron detection based on attributive abstraction and integration. Comput. Res. Dev. **34**(7), 481–486 (1997)
16. Feng, J.: Granular transformation of qualitative criterion, orthogonality of qualitative mapping system and pattern recognition. In: Proceeding of 2006 IEEE International Conference on Granular Computing, Atlanta, Georgia, USA, pp. 10–12 (2006)

17. Zhong, C., Pedrycz, W., Wang, D.: Granular data imputation: a framework of granular computing. Appl. Soft Comput. **46**, 307–316 (2016)
18. Wolski, M., Gomolinska, A.: Rough granular computing in modal settings: generalised approximation spaces. Fundamenta Informaticae **148**, 157–172 (2016)
19. Yingxu, W., Ying, W.: Cognitive informatics models of the brain. IEEE Trans. Syst. Man Cybern. **36**(2), 203–207 (2006)
20. Wang, Y., Qi, Y.: Memory-based cognitive modeling for visual information processing. Pattern Recogn. Artif. Intell. **24**(2), 144–150 (2013)
21. Dong, Z., Feng, J.: An attributive coordinate representation of schematic memory. Comput. Res. Dev. **35**(8), 694–698 (1998)

Power Control in D2D Network Based on Game Theory

Kai Zhang(✉) and Xuan Geng(✉)

College of Information Engineering, Shanghai Maritime University, Shanghai, China
zhangkai8816@foxmail.com, xuangeng@shmtu.edu.cn

Abstract. This paper considers power control problem based on Nash equilibrium (NE) to eliminate interference in multi-cell device-to-device (D2D) network. The power control problem is modeled as a non-cooperative game model, and a user residual energy factor is introduced in the formulation. Based on the proof of the existence and uniqueness of Nash equilibrium, a distributed iterative game algorithm is proposed to realize power control. Simulation results show that the proposed algorithm can converge to Nash equilibrium quickly, and obtain a better equilibrium income by adjusting the residual energy factor.

Keywords: D2D · Nash equilibrium · Power control · Game theory

1 Introduction

With the requirement for high speed and efficiency of data transmission, the limited spectrum resource brings great challenges for mobile network communication. D2D communication is a new wireless technology, where two devices can communicate with each other without exchanging information from base station, so that it cannot only reduce the burden of base station, but also improve communication quality of cellular users [1]. However, the D2D users will be suffered from interference of other users in cellular system. Therefore, the interference elimination has been investigated in recent years [2].

In [3], the authors studied that the D2D users and the cellular users use the same channel resources by multiplexed mode. Although it can improve spectrum utilization, it will introduce a new kind of interference. The authors in [4] investigated a power control method for single cellular system containing one cellular link and one D2D link. The algorithm can reduce the interference significantly between D2D users and cellular users. In [5], the authors studied the interference elimination problem of multi-cell D2D network, where the cellular users are communicated by base station schedule and the D2D link communication is guaranteed by power control.

The previous work mainly focused on the centralized power control method for D2D network, while the distributed implementation is less concerned. Therefore, we studied distributed power control method in this paper by use of game theory. We first establish a static game model in hybrid multi-cell D2D network, and then prove the existence and uniqueness of the Nash equilibrium. Finally, the distributed power

Z. Shi et al. (Eds.): ICIS 2017, IFIP AICT 510, pp. 104–112, 2017.
DOI: 10.1007/978-3-319-68121-4_11

control method is iteratively implemented to obtain the optimal state of Nash equilibrium. Simulation results show that the game model designed in this paper can converge quickly to Nash equilibrium, and the system can get better balance by adjusting the residual energy factor.

2 System Model

The system model of D2D network is shown in Fig. 1. We consider a multi-cell system containing D2D links, and the adjacent cells use the same frequency band for multiplexing communication. Assume there are N cellular and D2D links in the system to use the same frequency resource. To eliminate the co-channel interference among the different cellular links and the interference between the cellular links and the D2D links, we propose a power control method based on game theory with pricing mechanism.

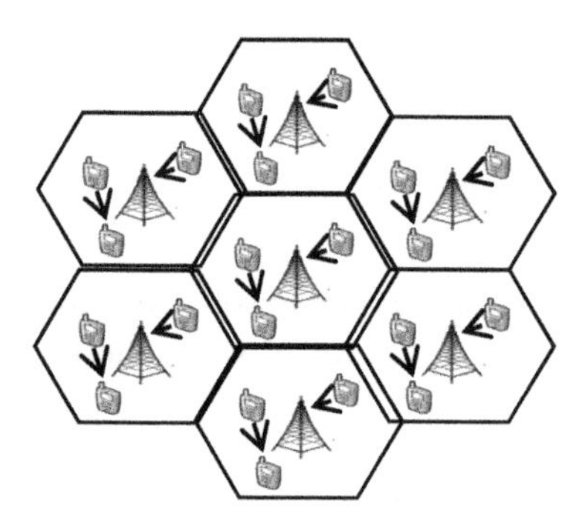

Fig. 1. D2D network system model

According to game theory, there are three elements should be considered, which include the player, the strategy and the utility function. We define the power control model as $G = [N, P, U]$ in multi-cell D2D network, and introduce the three elements:

(1) Player: Assume each cellular link and D2D link are the participants making decision of the game. We denote $N = \{1, 2, \ldots, N\}$ to be the participant set and each element represents one communication link.
(2) Strategy: Assume one communication link is denoted by $j \in N$, where the transmit power is $p_j \in P_j$. Here P_j is the available transmit power region of the link j, i.e. the strategy space of the game player. All the strategies constitute $P = (p_1, p_2, \ldots, p_N)$, and all the strategy space combinations can be expressed by $P = \times_{i \in N} P_i$. Besides, $P_{-j} \in \times_{i \in N, i \neq j} P_i$ represents all the left users' strategy space combination except the link j. Let $P_j = [0, p_{\max}]$ and $p_{\max}$ is the maximum available transmit power for the user.

(3) Utility function: Define $u_j(p_j, p_{-j}) = \log|1 + \frac{\varepsilon_j h_j p_j}{\sigma^2 + \sum_{k=1, k \neq j}^{N} h_k p_k}| - \varepsilon_j h_j p_j^2$, where u_j is the utility function of user j and it represents the payoff obtained by the game players after making decision. h_j is the channel gain of the link j and σ^2 is the variance of the additive white Gaussian noise. Unlike [6], we introduce a user residual energy factor ε in the utility function, which is defined as

$$\varepsilon = E_j^{\max} / \Delta E_j \tag{1}$$

where ΔE_j is the residual energy transmitted by link j, and its maximum valve is $E_j^{\max}$. The energy factor ε describes a price law. On one hand, when the supply exceeds the demand, the payment for the price of energy consumption is low and the user can consume more energy to obtain better performance. On the other hand, when the demand exceeds supply, the energy consumption should pay more prices. Therefore, adjusting ε can make the performance and energy consumption in a reasonable trade-off state. Note that ε is regarded as a control factor in the simulation.

Here we formulate the problem to establish a non-cooperative power control game model with pricing mechanism, which is

$$\max_{p_j \in P_j} u_j(p_j, p_{-j}), j \in N \tag{2}$$

In this paper, we use Nash equilibrium to solve the problem. We define the policy combination $p^* = (p_1^*, p_2^*, \ldots, p_N^*) \in p$ as a Nash equilibrium, and establish the below expression

$$u_j(p_j^*, p_{-j}^*) \geq u_j(p_j, p_{-j}^*), \forall p_j \in P_j, j \in N \tag{3}$$

If the game players adopt the strategy combination p^*, they cannot leave and change the strategy combination, so that the Nash equilibrium will be the optimal solution.

3 Non-cooperative Power Control Game Analysis

According to the model in Eq. (2), we prove the existence and uniqueness of Nash equilibrium in this section. After that, we use a distributed iterative power algorithm to solve the Nash equilibrium point.

3.1 Existence

Theorem 1: Nash equilibrium exists in the power control game model $G = [N, P, U]$.

Proof: Power strategy space is a non-empty, closed, and bounded convex set in Euclidean space. The utility function u_j is continuous for the strategy combination p, so that we obtain

$$\frac{\partial^2 u_j(p_j, p_{-j})}{\partial p_j^2} = \frac{-\varepsilon_j^2 h_j^2}{\left(\sigma^2 + \sum\limits_{k=1,\, k\neq j}^{N} h_k p_k + \varepsilon_j h_j p_j\right)^2} - 2\varepsilon_j h_j < 0 \tag{4}$$

Therefore, $u_j(p_j, p_{-j})$ is quasi-concave for p_j of the j th player [7]. According to the game existence theorem in [8], the game G has Nash equilibrium.

3.2 Uniqueness

Theorem 2: The iterative game algorithm can converge to the unique equilibrium point.

Proof: If we want to prove there is Nash equilibrium point in the game, the iterative function should meet the positive $r(p) \geq 0$, monotonic $r(p) \leq r(p')$, and expandability $r(Tp) > \frac{1}{T} r(p)$, where $r(p)$ is the optimal response strategy set.
Given the strategies p_{-j}, the optimal response strategy set of game players is

$$r_j(p_{-j}) = \left(p_j^* | p_j^* = \arg\max_{p_j} u_j(p_j, p_{-j})\right) \tag{5}$$

All the players $r_j(p_{-j})$ form the vector as

$$r(p) = (r_1(p_{-1}), r_2(p_{-2}), \cdots, r_j(p_{-j}), \cdots, r_N(p_{-N})) \tag{6}$$

According to the concave of utility function, we have $\arg\max_{p_j} u_j u_j(p_j, p_{-j}) = \min(\tilde{p}_j, p_{\max})$.
Let

$$\frac{\partial u_j(p_j, p_{-j})}{\partial p_j} = \frac{\varepsilon_j h_j}{\sigma^2 + \sum\limits_{k=1,\, k\neq j}^{N} h_k p_k + \varepsilon_j h_j p_j} - 2\varepsilon_j p_j h_j = 0 \tag{7}$$

and we obtain

$$\tilde{p}_j = \tilde{p}_j(p_{-j}) = \frac{-\left(\sigma^2 + \sum\limits_{k=1,\, k\neq j}^{N} h_k p_k\right) + \sqrt{\left(\sigma^2 + \sum\limits_{k=1,\, k\neq j}^{N} h_k p_k\right)^2 + 2\varepsilon_j h_j}}{2\varepsilon_j h_j} \tag{8}$$

(Negative solution)

Therefore, the optimal response strategy is

$$p_j^* = p_j^*(p_{-j}) = \min(\tilde{p}_j, p_{\max}) \tag{9}$$

The optimal response strategy vector is

$$r(p) = \left(p_1^*(p_{-1}), p_2^*(p_{-2}), \cdots, p_N^*(p_{-N})\right) \tag{10}$$

(1) Positive: If $p \geq 0$, then $r(p) \geq 0$. If all the elements of a vector are not less than the corresponding elements of another vector, the former one will be greater than or equal to the latter vector.
(2) Monotonicity: If $p \geq p'$, then $r(p) \leq r(p')$.

Proof: If $p \geq p'$, then

$$\sigma^2 + \sum_{k=1,\, k \neq j}^{N} h_k p_k \geq \sigma^2 + \sum_{k=1,\, k \neq j}^{N} h_k p'_k \tag{11}$$

Let $x = \sigma^2 + \sum_{k=1,\, k \neq j}^{N} h_k p_k$, we have

$$f(x) = \frac{-x + \sqrt{x^2 + 2\varepsilon_j h_j}}{2\varepsilon_j h_j} \tag{12}$$

and obtain

$$\frac{df(x)}{dx} = -\frac{1}{2\varepsilon_j h_j} + \frac{1}{2\varepsilon_j h_j} \frac{x}{\sqrt{x^2 + 2\varepsilon_j h_j}} < 0 \tag{13}$$

The function is a monotonically decreasing function, i.e. $r(p) \leq r(p')$, so that the optimal response strategy vector satisfies the monotonicity.

(3) Extensibility: If $\forall T > 1$, then $r(Tp) > \frac{1}{T} r(p)$.

Proof: We have

$$p_j(\tilde{T}p_{-j}) = \frac{-\left(\sigma^2 + T \sum_{k=1,\, k \neq j}^{N} h_k p_k\right) + \sqrt{\left(\sigma^2 + T \sum_{k=1,\, k \neq j}^{N} h_k p_k\right)^2 + 2\varepsilon_j h_j}}{2\varepsilon_j h_j} \tag{14}$$

$$\frac{1}{T}\tilde{p}_j(p_{-j}) = \frac{1}{T}\frac{-\left(\sigma^2 + \sum_{k=1,\,k\neq j}^{N} h_k p_k\right) + \sqrt{\left(\sigma^2 + \sum_{k=1,\,k\neq j}^{N} h_k p_k\right)^2 + 2\varepsilon_j h_j}}{2\varepsilon_j h_j} \tag{15}$$

so that

$$\begin{aligned}\frac{\tilde{p}_j(Tp_{-j})}{\frac{1}{T}\tilde{p}_{-j}(p_{-j})} &= T\frac{-\left(\sigma^2 + T\sum_{k=1,\,k\neq j}^{N} h_k p_k\right) + \sqrt{\left(\sigma^2 + T\sum_{k=1,\,k\neq j}^{N} h_k p_k\right)^2 + 2\varepsilon_j h_j}}{-\left(\sigma^2 + \sum_{k=1,\,k\neq j}^{N} h_k p_k\right) + \sqrt{\left(\sigma^2 + \sum_{k=1,\,k\neq j}^{N} h_k p_k\right)^2 + 2\varepsilon_j h_j}} \\ &= T\frac{\left(\sigma^2 + \sum_{k=1,\,k\neq j}^{N} h_k p_k\right) + \sqrt{\left(\sigma^2 + \sum_{k=1,\,k\neq j}^{N} h_k p_k\right)^2 + 2\varepsilon_j h_j}}{\left(\sigma^2 + T\sum_{k=1,\,k\neq j}^{N} h_k p_k\right) + \sqrt{\left(\sigma^2 + T\sum_{k=1,\,k\neq j}^{N} h_k p_k\right)^2 + 2\varepsilon_j h_j}} > 1\end{aligned} \tag{16}$$

and we obtain

$$r(Tp) > \frac{1}{T}r(p) \tag{17}$$

It means that the optimal response strategy vector can be extended.

According to the Ref. [9], we can conclude that the non-cooperative power control game has a unique Nash equilibrium point, which can be simply expressed by

$$p = \left(p_1^*(p_{-1}),\ p_2^*(p_{-2}), \cdots, p_N^*(p_{-N})\right) \tag{18}$$

3.3 Distributed Iterative Game Algorithm

Based on the proofs in Sect. 3.2, we present a distributed iterative algorithm to solve the Nash equilibrium. The realization is summarized below.

Step 1: Define the number of iteration M, and the stop criteria U.
Set $m = 0$.
Set the initial value of the strategy combination $p^{(0)} = 0$;

Step 2: Set $m = m + 1$.
Update the strategy via Eq. (9) to obtain new strategy $p^{(m)}$ by use of $p^{(m-1)}$.

Step 3: Repeat step 2 until $p^{(m)} - p^{(m-1)} \leq U$, and the algorithm ends. The output of Nash equilibrium point strategy is $p^{(m)}$.

Note that the players only know their own channel gain h_j and the corresponding energy factor ε_j when the Nash equilibrium point is calculated, so that the algorithm is implemented by distributed way.

4 Simulation Results

In this section, we perform the proposed non-cooperative power control game algorithm in D2D network by matlab simulation. The wireless system in simulation composes four cells c_1–c_4, and each cell has single cellular link and single D2D link, so that $N = 8$. The game player set is expressed as $N = \{1, 2, \ldots, 8\}$. Assume all the links use the same frequency band. The link gains h_j of the four cellular links are assumed to be 0.5, 0.7, 0.9, 1.1, and the other four D2D links gains are 1.2, 1.6, 2.0, 2. The variance of additive white Gaussian noise is unit. The maximum transmit power $p_{\max}$ of the D2D link and the cellular link is also unit. The compared algorithm comes from Ref. [9].

Figure 2 shows the valves of utility function varied with the iteration number. The control factor is set to be 2. Note the control factor in Ref. [9] represents power coefficient. From the simulation, it can be seen that our algorithm converges to the stable state within six times, while the compared algorithm in Ref. [9] needs eight times. Therefore, our method can converge more quickly than the compared one. Besides, we find that the valve of utility function of our method is bigger than the compared method, which means our method is more stable.

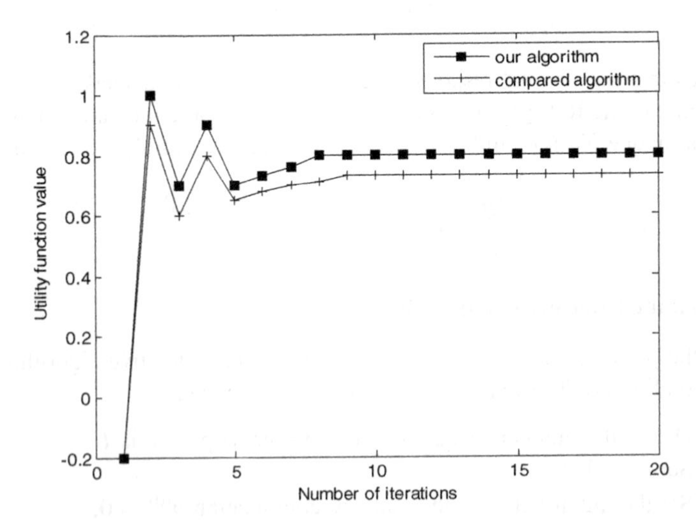

Fig. 2. Utility function value comparison for two algorithms

Figure 3 gives the bar chart of the iteration number varied with the control factor. It shows that our algorithm has smaller iteration number when the control factor is bigger than 1, which means the algorithm is more efficient.

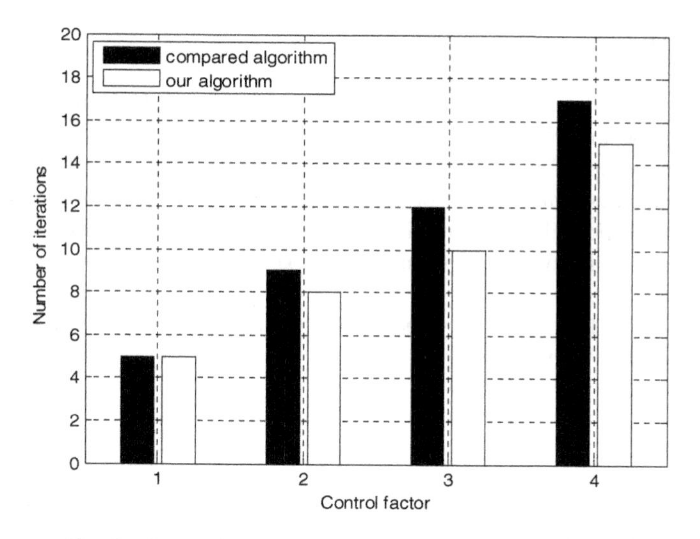

Fig. 3. Comparison of iteration number for two algorithms

Figure 4 gives the transmit power results for all of users in Nash equilibrium state. It observed that all of transmit power of users tend to be stable. Hence the proposed price function can make the system achieve the equilibrium stable state.

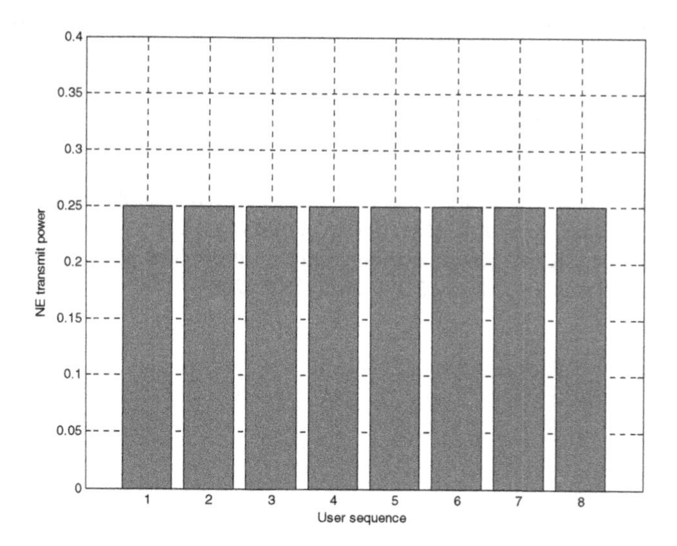

Fig. 4. Transmit power of each user in Nash equilibrium

From the simulations, we conclude that the power control algorithm based on pricing mechanism can effectively improve the performance of the system if pricing factor is adjusted reasonably.

5 Conclusion

In this paper, we investigate the power control problem in D2D network. A non-cooperative game model based on pricing mechanism is established to eliminate the interference in complex environment. We design a price function and prove its existence and uniqueness, and then propose a distributed iterative algorithm which can converge to Nash Equilibrium point. From the simulation, it can be observed that our algorithm improve the system performance compared with the conventional method.

Acknowledgement. This work was supported by the National Nature Science Foundation of China (Nos. 61401270, 61601283).

References

1. Fodor, G., Dahlman, E., Mildh, G., et al.: Design aspects of network assited device-to-device communications. IEEE Commun. Mag. **50**(3), 170–177 (2012)
2. Doppler, K., Rinne, M., Wijting, C., et al.: Device-to-device communication as an underly to LTE-advance networks. IEEE Commun. Mag. **47**(12), 42–49 (2009)
3. Janis, P., Yu, C.H., Doppler, K., et al.: Device-to-device communication underlaying cellular communications systems. Int. J. Commun. Netw. Syst. Sci. **2**(3), 169–247 (2009)
4. Yu, C.H., Tirkkonen, O., Doppler, K., et al.: Power optimization of device-to-device communication underlaying cellular communication. In: Proceedings of ICC 2009, Dresden, pp. 1–6 (2009)
5. Belleschi, M., Fodor, G., Abrardo, A.: Performance analysis of a distributed resource allocation scheme for D2D communications. In: Proceedings of Globecom 2011, Houston, pp. 1–6 (2011)
6. Chen, H., Wu, D., Tian, H., Wang, X., Du, J.: Non-cooperative game power control method in D2D network. Military Commun. Technol. **34**(4) (2013)
7. Boyd, S., Vandenberphe, L.: Convex Optimization. Cambridge University Press, California (2006)
8. Glecksberg, I.: A further generalization of the Kakutani fixed point theorem with application to Nash points points. Proc. Am. Math. Soc. **3**, 170–174 (1952)
9. Sung, C.W., Leung, K.: A generalized framework for distributed power control in wireless networks. IEEE Trans. Inf. Theory **51**(7), 2625–2635 (2005)

HCI Based on Gesture Recognition in an Augmented Reality System for Diagnosis Planning and Training

Qiming Li[1,2(✉)], Chen Huang[3], Zeyu Li[3,4], Yimin Chen[3], and Lizhuang Ma[1]

[1] Department of Computer Science and Engineering, Shanghai Jiaotong University, Shanghai 200240, China
[2] College of Information Engineering, Shanghai Maritime University, Shanghai 201306, China
qmli@shmtu.edu.cn
[3] Department of Computer Science and Technology, Shanghai University, Shanghai 200444, China
[4] Computer Center, Ruijin Hospital, Shanghai Jiaotong University School of Medicine, Shanghai 200025, China

Abstract. An Augmented Reality System for Coronary Artery Diagnosis Planning and Training (ARS-CADPT) is designed and realized in this paper. As the characteristic of ARS-CADPT, the algorithms of static gesture recognition and dynamic gesture spotting and recognition are presented to realize the real-time and friendly Human-Computer Interaction (HCI). The experimental results show that, with the use of ARS-CADPT, the HCI is natural and fluent, which improves the user's immersion and improves the diagnosis and training effects.

Keywords: Gesture recognition · Augmented reality · Human Computer Interaction

1 Introduction

Presently, 64 multi-slices computed tomographic coronary angiography technology has been considered as an effective way to diagnose coronary heart disease [1]. In the preoperative diagnosis planning process, the doctors are not accustomed to carry on the interactive diagnosis with computer by using the mouse and keyboard. 3D reconstruction based on Computed Tomography (CT) image sequence combined with augmented reality (AR) technology can effectively solve the above problems.

AR is a new technology that strengthens the user's perception of the real world by superimposing the virtual 3D information generated by the computer system onto the real scene. In fact, medicine is one of the earliest application fields of AR technology. State Andrei et al. [2] can draw a virtual 3D fetus on its abdomen position by ultrasonic scanning a pregnant woman. The doctor can understand the move and kick ability of

Z. Shi et al. (Eds.): ICIS 2017, IFIP AICT 510, pp. 113–123, 2017.
DOI: 10.1007/978-3-319-68121-4_12

the fetus through the Helmet-Mounted Displays (HMD) in 1994. AR technology can be used as an auxiliary means of surgical visualization. The 3D data of patient can be collected through Magnetic Resonance Imaging (MRI), CT or ultrasound images. According to the data, the corresponding virtual information can be rendered in real time. Combining with the actual situation of patient, the doctors can get more complete information, and improve the operation finally [3]. Wu [4] implements a spine surgery AR system, in which the surgeons can make use of 3D virtual model of preoperative patients to carry out spinal surgery simulation practice. In minimally invasive surgery, AR technology enables doctors to obtain the clairvoyant ability and improve the quality of surgery [5]. AR can also be used for medical training. According to statistics, over 50% of the augmented virtual reality application system are used in medical training, the most of which are realized based on virtual reality (VR) technology [6]. The amount of application system based on AR is relatively less. The AR based aid medical training system [7] is used to achieve medical training and examination through human body modeling.

The natural and real-time HCI is one of the three important features of AR system [8]. However, the traditional interaction mode such as using the mouse and keyboard cannot meet the application requirements. People are eager to realize the HCI in a very natural way. Gesture is just the most natural and intuitive way of interaction in human communication except language. Therefore, HCI based on gesture recognition has become a hot research topic. Gestures are usually defined as hand shapes and movements produced by the combination of palms, fingers, and even arms. The task of HCI based on gesture recognition is: firstly, recognize the meaning of the gesture correctly according to the data captured in real time, then trigger the corresponding instruction, and make the system feedback finally.

An AR system used for coronary artery diagnosis planning and training is designed and realized in this paper, which is called ARS-CADPT in the following paragraph. The system is very complicated, but the HCI based on gesture recognition is mainly discussed in this paper. The operating user or the lecturer can interact with the 3D model of the coronary arteries in a natural and intuitive manner with the defined gestures, and can perform simulation measurement of radius of vessels, and thus achieve a comprehensive and intuitive presentation and an accurate and detailed explanation of the patient's situation. The interns or students can understand and study the patient's coronary detail situation on a large tiled screen.

2 System Architecture

2.1 The Hardware Architecture of ARS-CADPT

As shown in Fig. 1, ARS-CADPT is constructed based on cluster architecture. It consists of several high-performance workstations, a parallel rendering and tiled display subsystem and a series of equipment for interaction data capture. Server 1 is used for 3D reconstruction of coronary artery based on CT images and storage the 3D coronary artery model database of all the previous patients. Server 2 is the surveillance and control center of the system. Server 3 is used for processing the lecturer's interaction data which is captured by the equipment such as Leap Motion, magnetic tracker, Microsoft Hololens, and so on. The display subsystem is consists of 5 parallel rendering nodes and a tiled screen, which is used for study, view and emulate for the student and intern users.

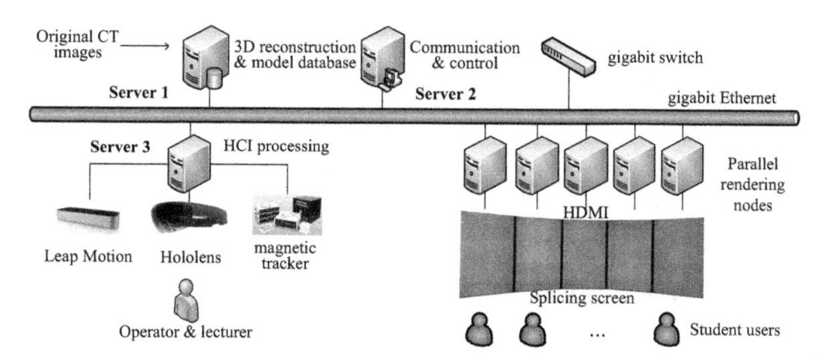

Fig. 1. The hardware of ARS-CADPT

2.2 The Software Framework of ARS-CADPT

The ARS-CADPT is realized based on multi-thread technique. The main thread is used for the diagnosis and training process, the HCI thread is used for the real-time interaction with the 3D coronary artery model, and the feedback is displayed to the users via the display thread. Therefore, the system mainly includes three function modules. 3D coronary artery reconstruction based on CT images, real-time HCI based on gesture recognition and synchronous display based on parallel rendering. Here into, the HCI module is the characteristic of the system. The interaction gestures used in the system are defined firstly. Then the algorithms for static gesture recognition, dynamic gesture spotting and recognition are proposed. The corresponding interaction operations are triggered according to the gesture recognition results finally.

3 Real-Time HCI

As shown in Fig. 2, the HCI module serves the main process of diagnosis and training. It is the bridge between the operating user and the system. The HCI in the system is accomplished based on the coordination of static and dynamic gesture recognition. The Leap Motion manufactured by Leap Company is used to capture the hand shape and motion trajectory.

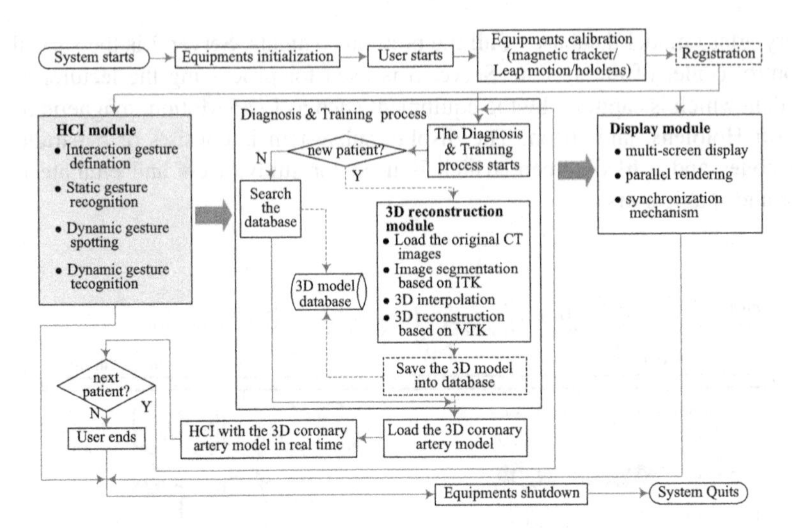

Fig. 2. The workflow and functions of the ARS-CADPT

3.1 Gesture Definition

Gesture includes static gesture and dynamic gesture. Considering a right-hander, the static gestures used in ARS-CADPT are defined in Table 1 and illustrated in Fig. 3.

Table 1. The static gestures and the corresponding HCI functions in ARS-CADPT

Legend	Class	HCI function	
Figure 3a	1	System input	Left mouse button down
Figure 3b	2		Right mouse button down
Figure 3c	3		Capture the hand trajectory as the trajectory of cursor
Figure 3d	4	Model control	Control 3D model in full synchronous mode according to the hand trajectory captured in real time
Figure 3e	5		Control 3D model in fine adjustment mode according to the recognition result of the real-time gesture trajectory
Figure 3f	6		Radius measurement of vessel stenosis
Figure 3g	7		In fine adjustment mode, control 3D model according to scale I
Figure 3h	8		In fine adjustment mode, control 3D model according to scale II
Figure 3i	9		In fine adjustment mode, control 3D model according to scale III
Figure 3j	10		In fine adjustment mode, control 3D model according to scale IV

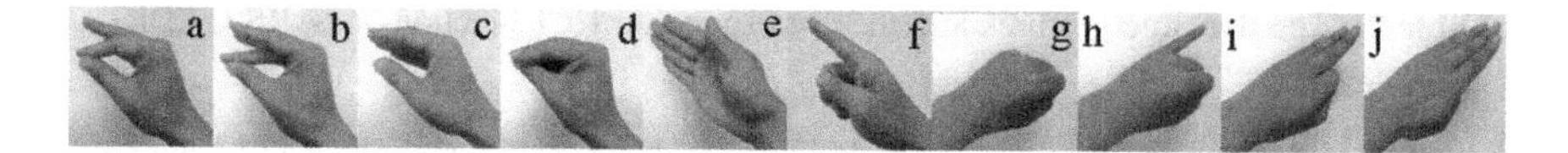

Fig. 3. The static gesture used in ARS-CADPT

The gestures can be divided into two categories: system input and model control. The former is used to achieve all kinds of system input instructions by gestures instead of mouse and keyboard, and the latter is used to manipulate and control the 3D models directly. Here into, the static gestures of class 1–3 are used for system input, and the static gestures of class 4–10 are used for model control.

In addition, the accurate measurement of vessel diameter, confirming the location and extent of stenosis and the situation of collateral circulation formation are objective gist for determining the diameter of stent during interventional treatment. However, in clinical practice, the measurement of vessel stenosis is mainly based on visual estimation, and its accuracy needs to be further improved. In ARS-CADPT, the operating user can interactively scale the virtual model freely and measure the radius of the blood vessel with the gesture shown in Fig. 3f.

There are two ways of interacting with the 3D model in ARS-CADPT. The full synchronization mode is to make the movement of 3D model completely consistent with the hand of which the static gesture is class 4 (shown in Fig. 3d), while the fine adjustment interaction mode is used for the slight, accurate and complex operations. The fine adjustment interaction mode is mainly realized by recognizing a series of dynamic gestures formed by the hand trajectory of which the static gesture is class 5 (shown in Fig. 3e). Therefore, the dynamic gestures needed in ARS-CADPT are defined in Table 2, and some examples are illustrated in Fig. 4. Here into, the translation gestures can be used for both system input and model control. The rotation and zoom gestures are used for model control only.

3.2 Static Gesture Recognition

The static gesture recognition algorithm based on rough sets theory was proposed. The static gesture recognition is considered as a decision table, denoted as $\mathrm{DT} = (U, C \cup D, V, f)$. Here into, U is a nonempty finite set of all the static gesture instances, called universe. C and D are also nonempty finite sets, C is called condition-attribute set, and D is called decision-attribute set. $V = \bigcup_{a \in C \cup D} V_a$, V_a is the range of attribute a. f: $U \times A \rightarrow V$ is called the information function, which assigns a value to each attribute. The data of static gestures are mainly captured by Leap Motion. The distance between the fingertips and the distance between the fingertips and the palms center are considered as the major factors influencing the static gestures, which are belong to C. The distance is discretized into five values, so $V_C = \{1, 2, 3, 4, 5\}$. There is only one decision attribute: the static gesture (denoted as d), i.e. $D = \{d\}$. According to Table 1, there is $V_d = \{1, 2, 3, 4, 5, 6, 7, 8, 9, 10, 11(\text{undefined})\}$. To sum up, the decision table of static gesture recognition can be modeled as shown in Table 3.

Table 2. The dynamic gestures and the corresponding HCI functions in ARS-CADPT

Gesture		Class	HCI function	
Translation	Left	11	Single-hand	Next menu | record | item | page | etc. Move the 3D model along the X axis in the positive direction
	Right	12		Previous menu | record | item | page | etc. Move the 3D model along the X axis in the negative direction
	Up	13		Next menu | record | item | page | etc. Move the 3D model along the Y axis in the positive direction
	Down	14		Previous menu | record | item | page | etc. Move the 3D model along the Y axis in the negative direction
	Forward	15		Previous menu | record | item | page | etc. Move the 3D model along the Z axis in the negative direction
	Backward	16		Next menu | record | item | page | etc. Move the 3D model along the Z axis in the positive direction
Rotation	Left	17	Double-hand	Rotate the 3D model clockwise in the top view
	Right	18		Rotate the 3D model anti-clockwise in the top view
	Up	19		Rotate the 3D model anti-clockwise in the elevation view
	Down	20		Rotate the 3D model clockwise in the elevation view
	Forward	21		Rotate the 3D model clockwise in the left view
	Backward	22		Rotate the 3D model anti-clockwise in the left view
Zoom	In	23		Enlarge the 3D model proportionally along three axes
	Out	24		Shrink the 3D model proportionally along three axes

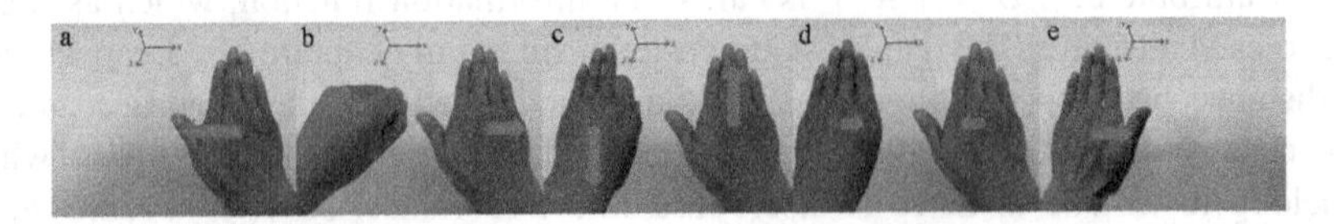

Fig. 4. Some examples of the dynamic gestures defined in ARS-CADPT: a. left translation (at default scale), b. right translation at scale I, c. right rotation at scale II, d. zoom out at scale IV, e. right translation by left hand (at default scale)

Table 3. Decision table of static gesture recognition

U	C								D
	1	2	...	10	i	ii	...	v	d
x_1	$v_{1,1}$	$v_{1,2}$	...	$v_{1,10}$	$v_{1,i}$	$v_{1,ii}$	...	$v_{1,v}$	$v_{1,d}$
x_2	$v_{2,1}$	$v_{2,2}$	...	$v_{2,10}$	$V_{2,i}$	$V_{2,ii}$	...	$V_{2,v}$	$v_{2,d}$
...	...	...	...	...	...	...	...	...	...
x_n	$v_{n,1}$	$v_{n,2}$	...	$v_{n,10}$	$V_{n,i}$	$V_{n,ii}$	...	$V_{n,v}$	$v_{n,d}$

Here into, $x_j(j = 1, 2, \ldots, n)$ is the j-th static instance, $U = \{x_1, x_2, \ldots, x_n\}$ is the set of static instance, and $v_{j,\ a}$ is the value of attribute $a(a \in C \cup D)$ in the j-th static instance.

The decision table is constructed according to the selected sample set. Then, the attribute reduction algorithm based on Skowron discernibility matrix and discernibility function is adopted: Firstly, construct the discernibility matrix; Secondly, construct the discernibility function; Thirdly, simplify the discernibility function using the absorption law; Finally, the conjunctive normal forms in the minimal disjunctive normal form of the discernibility function are all the D-reduct of C.

The classical reduction algorithm of attribute values is based on the value core concept. At first, calculate the value core of every instance in the decision table after attribute reduction; then get the minimal reduct from the value core table; finally, obtain the decision rules.

At last, the rules can be used to recognize the user's static gesture in real time.

3.3 Dynamic Gesture Spotting

Pavlovic et al. [9] divide the movements of the hand into two categories. One is the gesture that conveys the user's intention, and the other is meaningless action. Therefore, the starting point and termination point of each dynamic gesture must be located in the acquired continuous gesture data stream. It is the premise and foundation of dynamic gesture recognition. However, the existing dynamic gesture recognition methods usually assume either known spatial spotting or known temporal spotting, or both [10], which is unrealistic in the practical applications.

According to the data captured by Leap Motion, a segment of right hand motion trajectory is drawn in Fig. 5a. It can be seen the intervals of points are different. That means the speed is changing during the gestures. The curve shown in Fig. 5b is the speed variation during the gestures in Fig. 5a. It is clearly illustrated that the speed climbs up and then declines for several times. Each speed jump corresponds to a wave crest on the speed curve. There are five obvious wave crests which exactly correspond to five gestures. So, a simple method is to set a threshold. If speed is above the threshold, a gesture is detected. But this method would arouse some problems. One is that some noise points exist. Another is that the speed of dynamic gestures varies from person to person, and setting a threshold is not-so-flexible. In fact, the dynamic gestures defined in our system are all completed in a speed jump. Therefore, we could think that a dynamic gesture is generated only by judging an upward tendency of speed. Thirty data points are enough to represent the tendency from the experiment. We define the

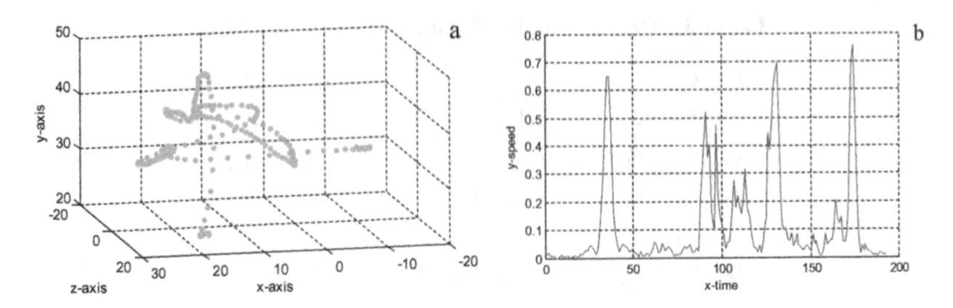

Fig. 5. A segment of dynamic gesture: a. Trajectory, b. Speed curve

upward tendency, which is the speed at any moment is higher than before. The fuzzy set of the standard upward tendency is denoted as $\underline{A}$, and a new fuzzy set $\underline{B}$ is obtained every time, and compute the close degree between $\underline{A}$ and $\underline{B}$ using Hamming close degree. The equation is:

$$N_H(\underline{A},\underline{B}) = 1 - d(\underline{A},\underline{B})/n = 1 - \delta(\underline{A},\underline{B}). \tag{1}$$

where $d(\underline{A},\underline{B}) = \sum_{i=1}^{n} \left|\mu_{\underline{A}}(x_i) - \mu_{\underline{B}}(x_i)\right|$, $\mu(x_i) = \begin{cases} 1 & y_i > Max \\ 0 & y_i \le Max \end{cases}$, $Max = \begin{cases} y_i & y_i > Max \; or \; i = 1 \\ Max & y_i \le Max \end{cases}$.

The lower the close degree is, the greater likelihood it is an upward tendency. If the close degree is less than 0.3, we believe the set $\underline{B}$ represents the upward tendency.

3.4 Dynamic Gesture Recognition

After the gesture spotting, a series of independent dynamic gesture trajectories can be obtained in real time.

1. Single-hand gesture

Let $C = \{c_1, c_2, \ldots, c_n\}$ be the set composed of the n classes dynamic gestures, and $A = \{a_1, a_2, \ldots, a_m\}$ be the set composed of the m attributes influencing dynamic gestures. s_k is the k-th gesture sample to be recognized. After s_k was recognized by all the attribute classifiers, a decision matrix is obtained and denoted as:

$$\boldsymbol{DM}(s_k) = \left(\left(f_{11}^k, f_{21}^k, \ldots, f_{m1}^k\right)^{\mathrm{T}}, \left(f_{12}^k, f_{22}^k, \ldots, f_{m2}^k\right)^{\mathrm{T}}, \ldots, \left(f_{1n}^k, f_{2n}^k, \ldots, f_{mn}^k\right)^{\mathrm{T}}\right). \tag{2}$$

Here into, the row vector $\boldsymbol{f}_i = \left(f_{i1}^k, f_{i2}^k, \ldots, f_{in}^k\right)(i = 1, 2, \ldots, m)$ is the recognition results of s_k by attribute classifier a_i with respect to the m classes, while the column vector $\boldsymbol{f}_i = \left(f_{1j}^k, f_{2j}^k, \ldots, f_{mj}^k\right)^{\mathrm{T}} (j = 1, 2, \ldots, n)$ is the recognition results of s_k by all the attribute classifiers with respect to the c_j-th class. then the intersection of the function f_j: $A \to [0, 1]$ (if the output of classifier is not in the interval [0, 1], it can be satisfied according to normalization) and the vector f_i, f_{ij}^k, indicates the degree of certainty that s_k is recognized as the c_j-th class by the attribute classifier a_i.

Let g be the fuzzy measure defined over $P(A)$, the power set of A. The fuzzy measure on the single-point set, i.e. fuzzy density $g_i = g(\{a_i\})(i = 1, 2, \ldots, m)$ represents the degree of credibility that the attribute classifier a_i makes decision. If $X \in P(A)$, $g(X)$ represents the degree of credibility that the attribute classifier X makes decision. However, as the single attribute classifiers are designed for a certain attribute feature of dynamic gestures, they should have different degrees of credibility for different gestures, namely, the single attribute classifiers should have different fuzzy densities. Let $g_j = (g_{1j}, g_{2j}, \ldots, g_{ij} \ldots, g_{mj})$ be the fuzzy density vector of class c_j, where g_{ij} represents the degree of credibility of the attribute classifier a_i with respect to class c_j. Then the fuzzy integral over A of the function f_j^k with respect to the fuzzy measure g_j is the overall objective estimate for s_k belonging to class c_j. In the way, for a certain gesture sample s_k, the system gives an integral value for every class, and the class that the greatest integral value corresponds with will be adopted as the recognition result.

2. Two-handed gesture

As for the two-handed dynamic gestures, the positions of the palm center at the beginning and end of the gesture are captured. Let $B^l = (b_x^l, b_y^l, b_z^l)$, $B^r = (b_x^r, b_y^r, b_z^r)$, $E^l = (e_x^l, e_y^l, e_z^l)$ and $E^r = (e_x^r, e_y^r, e_z^r)$ be the coordinates of both hands at the beginning and end of the gesture respectively, then the lengths of line segment B^lB^r and line segment E^lE^r (denoted as d_b and d_e respectively)and the angle between them (denoted as φ) can be calculated. By projecting φ to plane YOZ, XOZ and XOY respectively, the three direction angles (denoted as α, β and γ) can be calculated. At last, the current gesture can be recognized by the following rules:

IF $\varphi < \theta_1$ AND $(d_e - d_b) > \sigma$ ($< \sigma$), THEN *class*(g) = *zoom in* (*out*)

IF $\varphi > \theta_2$ AND $max(\alpha, \beta, \gamma) = \alpha \mid \beta \mid \gamma$, THEN *class*($g$) = *Rotate around the X|Y|Z axis*

Where θ_1, $\theta_2$and σ are thresholds predefined.

4 Experimental Results

The related experiments are carried out based on the system platform. Here into, the recognition rate for static gesture achieves an average performance of 97.3%. After the dynamic gesture spotting, the continuous dynamic gesture stream is divided into a set of isolated dynamic gestures. The system achieves an average performance of 92.4% for the dynamic gestures according to the algorithm proposed in Sect. 3.4.

The system is used for coronary artery diagnosis planning and teaching. As shown in Fig. 6, the lecturer is controlling the 3D coronary artery model in a natural and intuitive manner with the defined gestures, and thus achieves a comprehensive and intuitive presentation and an accurate and detailed explanation of the patient's situation.

At the same time, the student users or interns can study and understand the patient's coronary detail situation on a large tiled screen.

Fig. 6. System running instance: the lecturer is interacting with the 3D coronary artery model.

5 Conclusion and the Future Work

This paper presented an augmented reality system for coronary artery diagnosis planning and training. One of its advantages is to realize the real-time and friendly HCI by using the algorithms of static gesture recognition and dynamic gesture spotting and recognition. It can be concluded that the proposed solutions make the HCI more natural and convenient, make the explanation clearer and more intuitive, and finally achieve a better effect for the preoperative diagnosis planning and training.

It also can be concluded that the AR technology has great potential to apply to the computer-aided medical system. Some examples of AR-based surgical applications have been presented in the literatures [11, 12]. Meanwhile, there are still some technical challenges for further research and exploration. For example, the gesture set defined in ARS-CADPT is just a little subset of the human gesture set, and the gestures people used in daily life are much more complicated. This puts forward higher requirements to the gesture recognition algorithms. Moreover, there is still a lot of work to do before ARS-CADPT can be applied to real-time surgery.

Acknowledgement. This work is supported by Natural Science Foundation of China (Grant No.: 61472245) and Shanghai Municipal Natural Science Foundation (Grant Nos. 14ZR1419700 and 13ZR1455600).

References

1. Miller, J.M., Rochitte, C.E., Dewey, M., et al.: Diagnostic performance of coronary angiography by 64-row CT. N. Engl. J. Med. **359**(359), 2324–2336 (2008)

2. State, A., Chen, D.T., Tector, C., et al.: Case study: observing a volume rendered fetus within a pregnant patient. In: IEEE Conference on Visualization, pp. 364–368 (1994)
3. Tang, S.L., Kwoh, C.K., Teo, M.Y., et al.: Augmented reality systems for medical applications: Improving surgical procedures by enhancing the surgeon's 'view' of the patient. IEEE Eng. Med. Biol. Mag. **17**(3), 49–58 (1998)
4. Wu, J.R., Wang, M.L., Liu, K.C., et al.: Real-time advanced spinal surgery via visible patient model and augmented reality system. Comput. Methods Programs Biomed. **113**(3), 869–881 (2014)
5. De Paolis, L.T., Aloisio, G.: Augmented reality in minimally invasive surgery. In: Mukhopadhyay, S.C., Lay-Ekuakille, A. (eds.) Advances in Biomedical Sensing, Measurements, Instrumentation and Systems. Lecture Notes in Electrical Engineering, vol. 55, 305–320. Springer, Heidelberg (2010). doi:10.1007/978-3-642-05167-8_17
6. Alexandrova, I.V., Rall, M., Breidt, M., et al.: Animations of medical training scenarios in immersive virtual environments. In: Workshop on Digital Media and Digital Content Management, pp. 9–12 (2011)
7. Oliveira, A.C.M.T.G., Tori, R., Brito, W., et al.: Realistic simulation of deformation for medical training applications. In: 15th Symposium on Virtual and Augmented Reality, pp. 272–275 (2013)
8. Azuma, R.T.: A survey of augmented reality. Presence-teleoperators Virtual Environ. **6**(4), 355–385 (1997)
9. Pavlovic, V.I., Sharma, R., Huang, T.S.: Visual interpretation of hand gestures for human-computer interaction: a review. IEEE Trans. Pattern Anal. Mach. Intell. **19**(7), 677–695 (1997)
10. Jonathan, A., Vassilis, A., Quan, Y., et al.: A unified framework for gesture recognition and spatiotemporal gesture segmentation. IEEE Trans. Pattern Anal. Mach. Intell. **31**(9), 1685–1699 (2009)
11. Chen, X., Xua, L., Wang, Y., et al.: Development of a surgical navigation system based on augmented reality using an optical see-through head-mounted display. J. Biomed. Inf. **55**(C), 124–131 (2015)
12. Chen, X., Xu, L., Wang, H., et al.: Development of a surgical navigation system based on 3D Slicer for intraoperative implant placement surgery. Med. Eng. Phy. **41**, 81–89 (2017)

The Effect of Expression Geometry and Facial Identity on the Expression Aftereffect

Miao Song[1(✉)], Qian Qian[2], and Shinomori Keizo[3]

[1] Shanghai Maritime University, 1550 Haigang Ave, Shanghai 201306, China
miaosong@shmtu.edu.cn

[2] Kunming University of Science and Technology, Kunming 650500, China
qianqian_yn@126.com

[3] Kochi University of Technology, Kami-Shi 782-8502, Japan
shinomorikeizo@kochi-tech.ac.jp

Abstract. Few studies have systematically examined the effect of expression geometry and facial identity on this cross-identity expression aftereffect. This issue is, however, critical for understanding the nature of expression adaptation. We measured expression aftereffects using a cross-identity/cross-expression geometry factorial design. The results show that expression aftereffect would reduce if adaptor and tests have different identities or expression geometries, and that identity-independent expression aftereffect is relatively more robust to variance in expression geometry in comparison to identity-dependent expression aftereffect. Based on these results, we discussed their psychophysical and physiological implication for understanding the nature of identity-dependent and identity-independent expression representations.

Keywords: Facial expression · Visual adaptation · Neural representation

1 Introduction

Adaptation is a universal phenomenon in the human visual system, which refers to that the previous sensory experiences would affect the subsequent perception and response properties of visual system [1]. For instance, an observer adapts to a smiling face for a minute, and then looks at a middle expression within a series of ambiguous expression images morphing between smiling and angry expressions, this middle expression is tended to be perceived as an angry expression. This illusion is called expression aftereffect induced by the visual adaptation [2]. As visual adaptation can isolate and/or temporarily reduce the contribution of specific neural populations, visual adaptation is a time-honored tool for researchers to investigate the neural representation of human sensory system.

Previous study has found expression aftereffect in different person condition was weaker than that in same person condition, which indicated that there are two different neural representations in facial expression system, i.e., identity-dependent neural representation and identity-independent neural representation [3]. However, two issues should be further considered. First, although variance in identity impaired expression aftereffect, when adapting faces and test faces are of different person, not only facial

Z. Shi et al. (Eds.): ICIS 2017, IFIP AICT 510, pp. 124–129, 2017.
DOI: 10.1007/978-3-319-68121-4_13

identity but also expression geometry is also changed. It raises a possibility that the reduction of expression aftereffect in different person condition is simply induced by the difference of expression geometry instead of the difference of face identity between adaptor and test faces. Second, as the similarity between the adaptor and the tests is an important factor to modulate the aftereffect in the low-level visual adaptation, it is necessary to examine whether the similarity of facial identities also affects the magnitude of expression aftereffect.

To clarify these two issues and systematically examine the effects of expression geometry and facial identity on the expression aftereffect, five experimental conditions were tested in current study. The interaction of facial identity and expression geometry were considered in 2 × 2 combinations to constitute four conditions (same or different: facial identity and expression geometry between adaptor and test face): same identity/same geometry, same identity/different geometry, different identity/same geometry, different identity/different geometry. In fifth condition, to examine the effect of identity similarity on the expression aftereffect, we used the computer-morphing technology to generate the artificial faces of different similarity, and tested on the same ambiguous images.

2 Method

There were five adapting conditions in the experiment, in which test faces were always the same among conditions, but the identity and/or the expression geometry of adaptor were manipulated. The five conditions are respectively termed as SI/SG, SI/DG, DI/SG, DI/DG, and DIDS, where the first two letters indicates whether adaptor and test face are of the same identity (SI: Same Identity) or not (DI: Different Identity), and last two letters indicates whether the expression geometry of adaptor are the same with that of test face (SG: Same Geometry) or not (DG: Different Geometry). The last DIDS refers to the condition in which the expression aftereffect is measured with the face adaptor of different similarity.

2.1 Subject

The subjects were 8 paid students with normal or corrected-to-normal vision (5 from Kochi University of Technology and 3 from Shanghai Maritime University, mean age: 20.2, SD = 3.7). All 8 subjects participated SI/SG, SI/DG, DI/SG, DI/DG conditions, and 5 subjects from Kochi University of Technology participated the DIDS condition.

2.2 Stimuli and Apparatus

The face stimuli were selected from the affiliated image set of the Facial Action Coding System [4] and the Cohn-Kanade AU-Coded Facial Expression Database [5],which are coded by Face Action Coding System and enables us to select same expression configuration for different photographic subjects.

In SI/SG, SI/DG, DI/SG, and DI/DG conditions, happy, angry, surprised and disgusted expressions constituted two expression pairs, i.e., happy-angry and surprised-disgusted expression pairs. Illustrated by the case of surprised-disgusted expression pair, we selected two female photographic subjects (F01 and F02) depicting two expression configurations (C01 and C02) of surprise and disgusted expressions, resulting in four combinations of identity and expression configuration (F01 with C01, F01 with C02, F02 with C01, and F02 and C02) (see Fig. 1a). The same method was used to select the adaptors of happy-angry expression pair, except that the two photographic subjects showing happy and angry expressions were male.

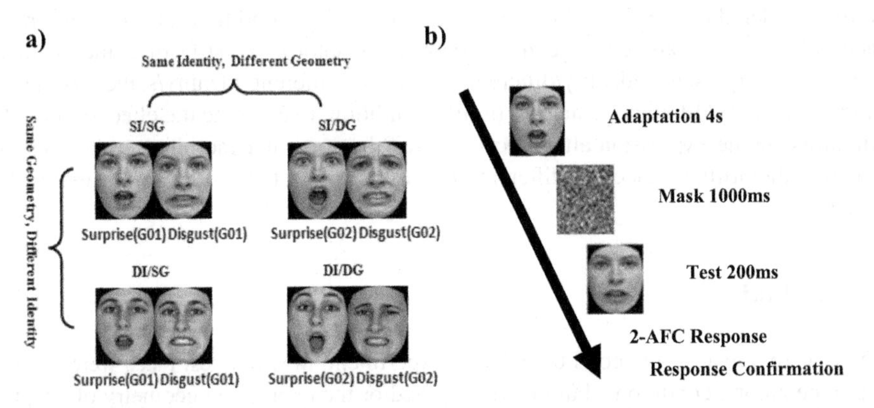

Fig. 1. The adaptors of surprise-disgusted expression pair used in four conditions (a), and experimental procedure (b).

Using two expression images of the same photographic subject, we morphed a series of test faces using Abrosoft FantaMorph 5.0 for surprised-disgusted and happy-angry expression pairs, respectively, with the nine middle ambiguous images, which varied from 30% to 70%, served as test stimuli.

The test faces in SI/SG, SI/DG, DI/SG, and DI/DG conditions were always the ambiguous expression images morphed between F01 with surprised expression (C01) and F01 with the disgusted expression (C01). We used the same test faces but different adaptors in the four experimental conditions. For instance, In the same identity/same configuration condition, the same images used to construct the test faces were used as the adaptors (i.e., F01 with surprised expression C01, and F01 with disgusted expression C01). In different identity/different configuration condition, the adaptors were the face images with different identities and expression configuration that differed from test faces (i.e., F02 with surprise expression C02, and F02 with disgusted expression C02).

In DIDS condition, we used computer-morphing technology to create the similar faces, because the morphing faces resembled two original faces when the features of two faces are blended together. The similar faces were created by morphing the test face between F01 with G01 and F02 with G01 on smiling and angry expression pair using the Fanta morphing software. We created the nine similar faces with identity proportion of test face from zero to 100% in steps of 10% in terms of the scale of the morphing software, with zero refer to that the created adapting face is dissimilar to the test face, while one denote the adapting face is absolutely same with test face, we select those face with proportion equal with 30%, 50% serves as adapting face. The reason not to select the face images with similarity strength greater than 50% as adapting face, because these faces are too similar with test face so that subject may consider them as the same person rather than similar person.

2.3 Procedure

The experiment consisted of 10 blocks, each for one of eight experimental conditions (2 expression pair × 5 experimental conditions), performed in an order that was randomized across subjects. Each block includes 2 adapting images and each adapting image was presented 20 times in 9 test images, resulting in a total of 360 trials for each block. The trials for different test in a block were also randomized. The duration for one block is approximately 40 min. The subject participated one block every other day and finished all experiments within 20 days.

Each subject was tested individually. They learned the experimental task through oral instruction and a short training session. In each trial, adaptor and test image were sequentially presented in the center of screen (Fig. 1b). After the presentation of test image, subjects perform a two-alternative forced choice (2-AFC) task to classify presented image into one of two categories (i.e., two expression images used to create the morphed series). The subject was instructed to attend to the face stimuli but no fixation point was given. This is to prevent subjects from overly attending to local facial features near fixation point. The duration of adapt stimuli and test stimuli were determined to 4000 and 200 ms, respectively, with 100 ms noise mask between adaptation and test stage to minimize the possible apparent motion effects.

3 Result

All adapting conditions generated significant after effect (Fig. 2), confirming the expression aftereffect reported in the previous literatures [2, 6]. A two-way analysis of variance (ANOVA) has been performed on SI/SG, SI/DG, DI/SG, and DI/DG conditions. There was a significant main effect for adapting condition ($F(3, 18) = 10.19$, $P < 0.001$), indicating that the expression aftereffect were significantly different for four different adapting conditions. There was no significant main effect for expression pairs ($F(1, 36) = 0.11$, $p = 0.742$) and no interaction between expression pair and adapting condition, thus, the data from different expression pair was merged for the further analysis.

The size of expression aftereffect in DI/SG condition was weak relative to that in SI/SG condition ($t(17) = 6.57$, $p < 0.001$), indicating that the reduction of expression aftereffect across identity is held even when the adaptor and test face has the same expression geometry. This result suggests that only variance in expression geometry cannot well explain the reduction of expression aftereffect, thus confirming the interference from identity system to the expression system. On other hand, the size of

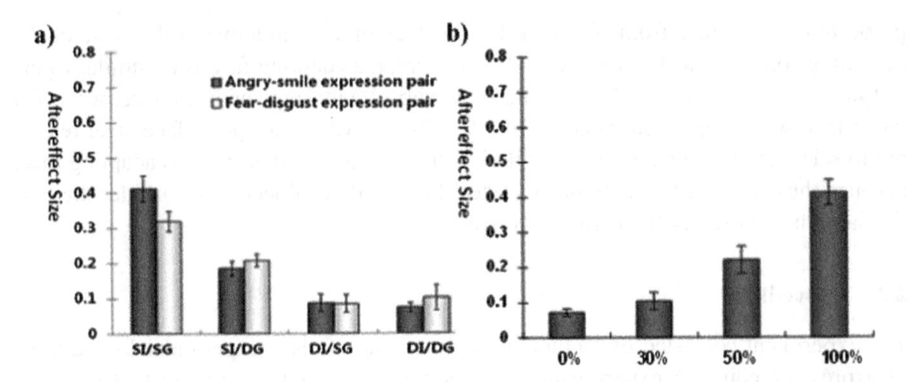

Fig. 2. The aftereffect sizes in SI/SG, SI/DG, DI/SG, DI/DG (a) and DIDS (b) conditions. Error bars denote SEM.

expression aftereffect in SI/DG condition was weaker than that in SI/SG condition ($t(17) = 5.84$, $p < 0.001$). In contrast, the aftereffect size in DI/SG condition was approximately the same with that in DI/DG condition ($t(17) = 0.19$, $p = 0.851$). These are new findings suggesting that variance in expression geometry impairs the identity dependent expression aftereffect, but not the identity independent expression aftereffect.

For DIDS condition, the result reveals that identity similarity between the adapting images and test could significantly increase the expression aftereffect. The adapting effects by 30% and 50% similar faces are 10.4% (S.E.M. = 2.3%) and 22.1% (S.E.M. = 3.9%), and the adapting effect by the 50% similar faces is significantly stronger than that by the 30% similar faces ($t8 = 2.843$, $p < 0.02$). It is worth noting that the adapting effect by the similar faces is still weaker than that in the SI/SG adapting condition (100% similar faces), but stronger than that in the DI/SG adapting condition (0% similar faces) (Fig. 2b). It suggests that the expression aftereffect would increase as the function of identity strength, thus confirming that the expression aftereffect is identity-dependent.

4 Discussion

We firstly found that the expression aftereffect is still much reduced when adaptors and test face have the identical expression geometry. Such finding is consistent with Ellamil et al.'s work [7], which used the artificial faces of anger and surprise expression as adaptors and found the reduction of adaptation effect when adaptor and test faces have same expression morphing prototype but different facial texture and contour. Together, our and Ellamil et al.'s observation suggests that the reduced aftereffect cannot be simply attributed to the dissimilarity of expression geometry between adaptor and test faces, it does reflect a functional interference from identity system on expression system. This observation consolidates Fox's proposal of the identity dependent neural representation. We next found that identity similarity between the adapting images and tests could significantly increase the expression aftereffect, thus indicating that the expression aftereffect is identity-dependent in specific range of identity strength.

Our data fits with the hypothesis of partially overlapping neural representations for the perception of expression and identity [8, 9]. These two systems depend on the different facial component to extract the information and process the expression. As the structural reference hypothesis stated [10], the face structure information is not only important in the identity discrimination, but also used by observers as a reference to compute and recognize expressions. The identity dependent expression system may more rely on the facial shape and/or structure information (e.g., local edge, facial contour) to perform the expression geometry analysis, while the identity independent expression system seems to depend on the abstract emotional information. It would be interesting for the future researches to locate the cortex of these two different neural representations.

Acknowledgement. This work is supported by the National Natural Science Foundation of China under Grants No. 61403251, and Shanghai Municipal Natural Science Foundation under Grants No. 14ZR1419300.

References

1. Mather, G., Verstraten, F., Anstis, S.: The Motion Aftereffect: A Modern Perspective. MIT Press, Cambridge (1998)
2. Webster, M.A., Kaping, D., Mizokami, Y., Duhamel, P.: Adaptation to natural facial categories. Nature **428**, 557–561 (2004)
3. Fox, C.J., Barton, J.J.S.: What is adapted in face adaptation? The neural representations of expression in the human visual system. Brain Res. **1127**, 80–89 (2007)
4. Ekman, P., Friesen, W.V., Hager, J.C.: Facial action coding system: the manual on CD ROM. In: A Human Face, Salt Lake City (2002)
5. Kanade, T., Cohn, J.F., Tian, Y.: Comprehensive database for facial expression analysis. In: Proceedings of the Fourth IEEE International Conference on Automatic Face and Gesture Recognition, Grenoble, France, pp. 46–53 (2000)
6. Butler, A., Oruc, I., Fox, C.J., Barton, J.J.S.: Factors contributing to the adaptation aftereffects of facial expression. Brain Res. **1191**, 116–126 (2008)
7. Ellamil, M., Susskind, J.M., Anderson, A.K.: Examinations of identity invariance in facial expression adaptation. Cogn. Affect. Behav Neurosci. **8**(3), 273–281 (2008)
8. Baudouin, J.Y., Gilibert, D., Sansone, S., Tiberghien, G.: When the smile is a cue to familiarity. Memory **8**, 285–292 (2008)
9. Calder, A.J., Young, A.W.: Understanding the recognition of facial identity and facial expression. Nat. Rev. Neurosci. **6**, 641–651 (2005)
10. Ganel, T.: Goshen-Gottstein, Y.: Effects of familiarity on the perceptual integrality of the identity and expression of faces: the parallel route hypothesis revisited. Q. J. Exp. Psychol. Hum. Percept. Perform. **30**(3), 583–597 (2004)

Big Data Analysis and Machine Learning

A Dynamic Mining Algorithm for Multi-granularity User's Learning Preference Based on Ant Colony Optimization

Shengjun Liu[1(✉)], Shengbing Chen[2], and Hu Meng[3]

[1] Anhui USTC-GZ Information Technology Co., Ltd., Hefei 230031, China
liusj@ustc-it.com

[2] Key Lab of Network and Intelligent Information Processing, Department of Computer Science and Technology, Hefei University, Hefei 230601, China

[3] HEFEI City Cloud Data Center Co., Ltd., Hefei 230094, China

Abstract. Mining user's learning preference is one of the key issues in the personalized online learning system, which is of great significance technology for modern educational. In this paper, using the hierarchical characteristics of the knowledge points in the course domain, we defined the equivalence relation and equivalence of knowledge points, and defined the structure of the knowledge points quotient space. Then, the functions of support, pheromone concentration and preference were defined on various levels, and an improved ant colony optimization was proposed to handle the multi granularity data structure of quotient space. An algorithm of multi-granularity Learning Preference Mining based on Ant Colony Optimization (ACO-LPM) was proposed to address the problems about too many learning knowledge points and too few user's test data in the online personalized learning system. The pheromone has the characteristic of dynamic evaporation, so, the preference patterns mined by ACO-LPM can be changed with the change of user interest in real time. The experimental results show that the algorithm can mining the user's learning preferences in online learning system effectively and efficiently.

Keywords: Quotient space · Preference Mining · Ant Colony Optimization · Personalized learning · Granular computing

Online learning, which breakthrough the constraints of time and space, provides a convenient and efficient learning platform for learners. It has become an important means of modern education for its three characteristics: the various learning modalities, the multiple teachers and students' role, and the rich learning resources. Personalized learning is the hotspot of online learning, which makes learners achieve the best learning effect under the minimum time and the best learning experience according to learners' learning characteristics and preference model. Researchers have done a lot of work in this domain. Jiunn used neural network to analyze students' online browsing behavior and get students' learning styles and learning preferences [1]. Du studied the personality traits of learners and the association between learning behaviors, and used the data mining algorithm to obtain the learner behavior model [2]. Qiu employed

Z. Shi et al. (Eds.): ICIS 2017, IFIP AICT 510, pp. 133–142, 2017.
DOI: 10.1007/978-3-319-68121-4_14

solomon learning style scale for pre-test, and obtained user interest model through the data mining of user learning history [3]. Lin and Yan studied the news recommendation in the mobile network environment, constructed the keyword vector by using the spatial model, and clustered the document according to the similarity degree, to obtain the gravity vector of each document cluster and build the user preference model [4, 5]. Ren put forward a U-I-C user interest model, which obtained scenario user preferences by adding the scene information in the user-project matrix [6]. Wang put forward the idea and method of user preference based on ontology and label [7]. Wolfgang studied the scenario information of the users in the mobile environment and found the user preference information by using the collaborative filtering algorithm [8]. Chen presented logistic curve model and hyperbolic model to analyze the user behavior, and proposed a user preference model based on multi-vector tree [9]. Pazzan used the expected information gain to analysis the annotations of users when they were browsing the pages and get user interest preferences [10]. Adomavicius mined the user's individual access records to construct the user model by using the associated association rules and user's personal information [11]. These methods have greatly improved the efficiency of the users' preferences in different backgrounds and applications. However, because of the characteristics of the massive knowledge points and the few test data, the problems of learning preferences on knowledge points has not been solved yet, and becomes a hot issue.

Granular computation can reduce the complexity of the solution, which inspired by mankind who can solve complex problems at different levels and solve them at the appropriate size [12]. In recent years, Granular computation (such as quotient space, rough set and fuzzy set) have been successfully applied to complex problems in many fields such as industrial control, transportation, graphic image processing, decision support, and biological information [13]. With the consideration of all the facts (the universe, structure, projection, etc.), quotient space can meet all needs of online learning system, such as the domain of knowledge and the dependence analysis. In this paper, we used quotient space theory to explore a dynamic mining algorithm for user's learning preference based on Ant Colony Optimization.

1 Quotient Space Structure of Knowledge Points

In online learning system, each knowledge point corresponds to a concept, which comes from the domain knowledge ontology, the knowledge points have a certain hierarchical structure and complex dependencies. As shown in Fig. 1, there is a inclusion relation between the knowledge points in different level. At same level, the knowledge has three relationships: pre-order, brotherhood, and equivalent. The knowledge point KP is defined as follows:

Definition 1: Knowledge point K is a triple (C, T, f), where C is the corresponding concept of knowledge point K; T is the topology of various relationships between knowledge points. The online learning system mainly has inclusion and pre-order structure; f is the property of knowledge.

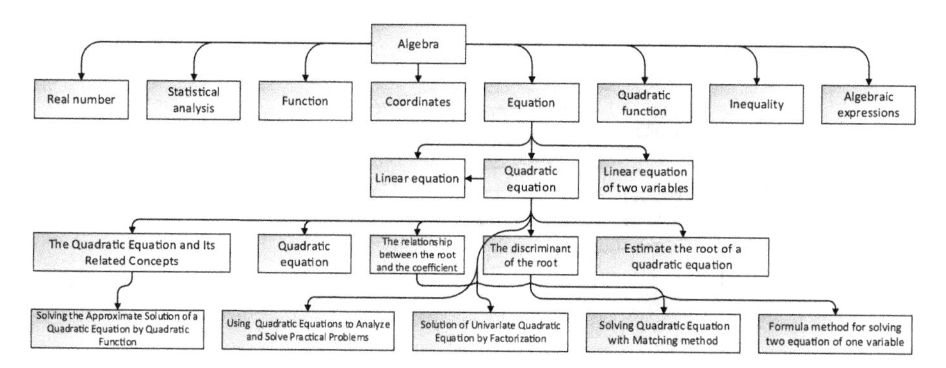

Fig. 1. Topological relationship of knowledge points in junior middle school

Definition 2: If the knowledge of the learning point K_i need to use the knowledge point K_j, then K_j is K_i's pre-order knowledge (also known as preparatory knowledge), expressed as $K_j < K_i$.

Definition 3: If K_u is the upper level knowledge point of K_i and K_j, and $C_u \supseteq C_i + C_j$, $T_u \supseteq T_i + T_j$, then K_u contains K_i and K_j, K_i and K_j are brotherhood relations, K_u is K_i and K_j's Father knowledge points.

Since K_u is not the same level as K_i and K_j, the internal relationship between K_i and K_j disappears naturally in the upper level, so $T_u \supseteq T_i + T_j$ is the operation of the parent node K_u level.

From the above definition we can see that if the brotherhood of the knowledge points is denoted by R, then R has both reflexivity, symmetry and transitivity, that is, R is the equivalence relation. Using the brotherhood relationship R constructs the equivalence class, the new triples ($[C]$, $[T]$, $[f]$) form a larger granularity of knowledge space, which is identified by the parent knowledge point.

Constructing the equivalence class mapping p: $(C, T) \rightarrow ([C], [T])$ which is the continuous natural projection of the knowledge point concept space. So, it meet the conditions of false warranty principle and fidelity principle define as follows:

1. False warranty principle: If the problem in the quotient space has no solution, it must also has no solution in any finer space.
2. Fidelity principle: Assuming that the problem has solution in the semaphore quotient space {C1, T1, f1}, {C2, T2, f2}, then it has a solution in its synthetic quotient space {C3, T3, f3}.

Using the false warranty principle and the fidelity principle, we can mining users' preferences on knowledge points in different granularity. The number of knowledge points were reduced by the equivalence. At the same time, the data sparse problems have also been cut down because the equivalent knowledge points have larger size.

The knowledge point space constructed by the brotherhood relationship R is shown in Fig. 2, the top is the root node, and its child node is called the inner node (corresponding to the equivalence class which R constructed). The inner node can also contain other inner nodes (finer equivalence), the bottom level is the leaf node,

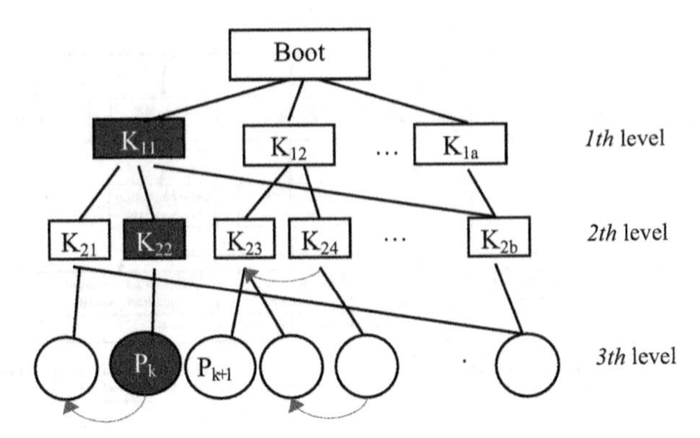

Fig. 2. The structure of knowledge points quotient space with brotherhood relationship

corresponding to the specific knowledge point. We call the sub-nodes of the root for the *1th* level, the sub-nodes of the *1th* level node for the *2th* level, and so on, the specific knowledge points for the *n-th* level.

In the structure of quotient space shown in Fig. 2, each level is a set of equivalence of a certain granularity. When a user study a leaf node (a specific knowledge point), each inner node on the path is considered to be accessed.

2 Functions Definition of Multi-granularity Ant Colony Optimization

2.1 Ant Colony Optimization

Ant Colony Optimization (ACO) proposed by Dr. Marco Dorigo who inspired by the natural ant colony foraging process. Ants can find the shortest path from the food source to the nest in un-visual conditions. During the foraging process, the ants release pheromone, which is proportional to the quality of the food source, and sense the pheromone. Then ants tend to move in the direction of high pheromone concentration. Thus, the group behavior of the ant colony shows a positive feedback: the more ants travel on a path, the more probability the latter choose the path, so, the path with good quality and short distance will attract more ants, and the pheromone concentration grows faster [14].

2.2 Functions Definition of ACO in Multi-granularity Data

In order to apply the ACO to the mining of multi-granularity knowledge points of hierarchical structure, several functions such as support degree, pheromone concentration and preference function in ACO are defined as follows:

Definition 4: Support η_l represents the probability that the user accesses path l, expressed as the access frequency of the path.

Let K_{ij}, K_{im} be the two knowledge points (equivalence class) nodes of the *i-th* level, the number of nodes in *i-th* level is n_i, path $l(i, j, m)$ represents a certain path from node K_{ij} to node K_{im}, the number of times the user visits the path $l(i, j, m)$ is expressed as $C_l(i, j, m)$. Then for the hierarchical interest pattern, its support $\eta_l(i, j, m)$ is:

$$\eta_l(i,j,m) = \frac{C_l(i,j,m)}{\sum_{k=1}^{n_i} C_l(i,j,k)} \tag{1}$$

Definition 5: The pheromone concentration τ_l (t) represents the user's interest in access to a path l, fades with time and increase with the user accesses the path l.

$$\tau_l(t+1) = (1-\rho)\tau_l(t) + \delta\Delta\tau_l \tag{2}$$

In formula (2), ρ is pheromone volatilization coefficient, δ is pheromone concentration increment adjustment coefficient, $\Delta\tau_l$ is pheromone increased value, the formula is as follows:

$$\Delta\tau_l = \frac{\tau_l(t) \times t + F_b}{t-1} - \tau_l(t) \tag{3}$$

In formula (3), F_b is the feedback value of the user's access to the node at time t + 1.

Definition 6: The preference function P_l (t) represents the user's preference for path l, including two factors which are pheromone and support. Let $\tau_w(t)$, η_w represent the pheromone and support of w-th path. Then the preference function P_l (t) is defined as:

$$P_l(t) = \frac{[\tau_l(t)]^\alpha \cdot [\eta_l]^\beta}{\sum_{w=1}^{n} [\tau_w(t)]^\alpha \cdot [\eta_w]^\beta} \tag{4}$$

In formula (4), n is the total number of paths in this level. If the threshold of the preference function is given and the value of the preference function P of all the path nodes is checked by (4), then the user's learning preference on multi-granularity knowledge points can be obtained.

3 Dynamic Mining Algorithm for Multi-granularity Learning Preferences

In this paper, we use the dynamic volatility of pheromone to dynamically mining the learning preferences on multi-granularity knowledge points. The main idea is that the user's knowledge learning activity corresponds to the ant's foraging behavior, the process of the users' learning knowledge points corresponds to the ants' foraging activity cycle. All of the users' learning actions are recorded in log files. Based on these log files, we can use ACO to find preferred function values of path nodes that formed by all knowledge points, and dynamically mine learning preferences for each knowledge point of the users, and then offer the users needed content for further study. Multi-granularity Learning Preference Mining algorithm based on Ant Colony Optimization (ACO-LPM) described as follows:

Input: Pre-processed *m* learning records, preference function threshold T_p
Output: learning preferences *H* of all levels in Multi-granularity knowledge space
Step 1: Initialize the ACO parameter, t=0, $\tau_l(t)$=0, η_l=0, C_l=0;
Initialize the knowledge point space tree, *Tree=NULL*;
Initialize the learning preference set, $H = (H_1, \ldots, H_n) = NULL$;
Step 2: According to the domain knowledge ontology, construct the quotient space tree of domain knowledge *Tree*;
Step 3: FOR (i=1 to m)
Find the first *i* log records corresponding to the path *l* in the tree *Tree*;
Calculate F_b of all nodes on path *l*, update C_l, $\tau_l(t)$.
END FOR
Step 4: Calculate η_l of all path nodes from equation (1);
Step 5: Initialize α, β;
FOR (i=1 to *Tree.High*)
Calculate $P_l(t)$ of all path nodes in the i-th level;
If $P_l(t) > T_p$, then $P_l(t) \rightarrow H_i$;
END FOR
Step 6: Return *H*.

In step 4 of the above algorithm, *Tree.High* represents the height of the tree *Tree*.

The time complexity of the algorithm: according to the definition of the quotient space structure, if the average of each equivalence class is composed of *m* elements, then the number of nodes in the upper lever is m^2 times less than the number of nodes in the lower level in the multi-granularity knowledge quotient space structure. While the time complexity of standard ACO is $O(I * N^2 * k)$, in which *I* is the number of iterations, *N* is the number of vertices, and *k* is the number of ants in the ant colony. So the time complexity of the i-th level is $O(I * (N/m^{2i})^2 * k)$, which has a greater reduction compared to the original space. Meanwhile, because the data of user online learning is the same, so the training data of each node in i-level increased m^{2i} times, which can effectively solve the problem of sparse data in the online learning system.

4 Experiment and Result

This section verified the effectiveness of the multi-granularity mining algorithm from two aspects: dynamic mining process display and practical application system. The experimental data was got from the online learning system, and the user learning behavior is recorded in log files. In order to record the user learning behavior, we added knowledge points information on the basis of the W3C extended log ExLF, which format is: "c-ip date time cs-username cs-method cs-uri-stem cs-user-agent sc-knowledgepoint cs-iscorrect sc-status sc-bytes". For example: "205.12.15.179 [01/Jul/2015:09:10:12] utest1 'GET/index/course/?courseid=10379 HTTP/4.0' M050311 0 200 598". Table 1 listed the log identification and description:

Table 1. The log format definition based on W3C extended standard

Field identifier	Description
c-ip	IP address of client that accessed the server
date	The date when the transaction was completed
time	The time when the transaction was completed(use UTC standard by default)
cs-username	Authenticated user name (anonymous user with '-')
cs-method	Action method that the client performed(GET and POST)
cs-uri-stem	Full URL of the data source that been accessed
cs(Referer)	Previous URL that the user browsed, the current page is linked from the URL
cs-user-agent	The browser version information used by the user client
sc-knowledge point	The knowledge points that the user visited
cs-iscorrect	Answer correctness (1 is correct, 0 is wrong)
sc-status	Returned status after the execution, described by HTTP or FTP
sc-bytes	The number of bytes sent by the server to the client

4.1 Dynamic Change Process Experiment of User Learning Preference

Dynamic mining is one of the advantages of the ACO-LPM algorithm. We took some log information to do simulation test on the situation of user's interest change. After the pre-processing of log records, knowledge that involved in one learning action combined with the case of right or wrong answer (*F*: wrong, *T*: right) as a record, such as: "*K*01*F*" represents the answer of knowledge *K*01 is wrong. Selected part of the knowledge of learning behavior data as follows (Table 2):

Table 2. The data of user learning behavior

Learning record	Knowledge point learning situation list
S001	K01F, K02F, K08T
S002	K01F, K02F, K06T, K09T, K12T
S003	K01T, K02F
S004	K01T, K02T, K07F
S005	K01T, K04T, K05F, K07F, K10T
S006	K03F, K07F, K08F
S007	K03T, K05T, K07F, K08F
S008	K02T, K03T, K07F, K08T, K15T
S009	K01T, K03T, K07T, K11T

Assuming that the minimum support is 2, the traditional mining algorithm will get frequent itemsets in consideration of the correctness of the answer{{K01K02: 3}, {K03K07: 2}}(In order to facilitate the expression, we call item {K01K02} for itemset1, item set {K03K07} for itemset2). However, if you carefully analyze the data flow, you will find two facts: First, In the long run, the user has the preferences of knowledge point K01K02, itemset1 coincidence occurs 3 times, and the error times with the correct times of 2: 1; Second, recent user learning has changed, itemset1 did not appear in the last 5 learning, but the itemset2 appeared 2 times, indicating that the user was concern about itemset2 recently (Comprehensive practice of knowledge points K03 and K07). In order to show the changes of frequent item sets more clearly and intuitive, we added value of pheromone volatilization coefficient ρ and concentration incremental coefficient δ, the parameters of ACO are defined as: $\rho = 0.15$, $\delta = 0.3$, $\alpha = 1$, $\beta = 0.5$. The preference function curve of each item is shown in Fig. 3. It can be seen that at the beginning of the training, the preference function value of the node is exponentially increasing due to the user's continuous error at the knowledge point K01K02, and the preference function value of the other knowledge points is zero; In the second half of the training, the knowledge point K01K02 is gradually reduced because it has not been accessed, but the preference value of the knowledge point K03K07 is gradually increased with the access, and even more than the knowledge point K01K02. It can be concluded that this algorithm can dynamically capture the current user's preference when the user's preference changes.

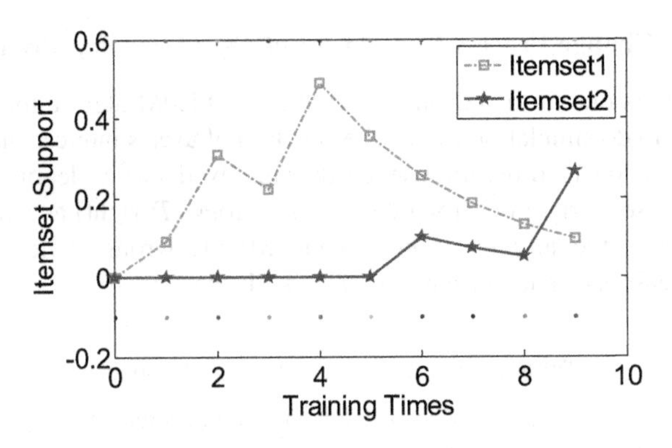

Fig. 3. The curve of user learning preference function

4.2 Experiment Data Analysis of Practical Application System

In order to verify the effectiveness of this method, we use the learning resources and system logs of the online education online system of Anhui Education Publishing Network Company (http://www.timeep.com/cms/index.html) on line in 2015. We have extracted the user learning logs of junior high school mathematics, physics, chemistry, a total of 12,000 log information, and recorded them in Math, Physics, Chemistry three data sets.

After preprocessing above data, we used the current mainstream mining algorithms: BP neural network (referred to as the BP-NN) and FP-growth algorithm for mining frequent items (referred to as FP-growth), and multi-size paper ACO-LPM learning preferences mining Algorithm, respectively, to test and verify the accuracy rate. In order to unify the format, ACO-LPM algorithm in this paper only mine for the first three granularity level. We set the ACO parameters as: $\rho = 0.05$, $\delta = 0.5$, $\alpha = 1$,

$\beta = 0.5$. The accuracy of each learning preferences mining algorithms is shown in the following table:

It can be seen from Table 3 that the ACO-LPM mining algorithm in this paper is superior to BP-NN and FP-growth algorithm as a whole. And with the level of the study increased, the accuracy of mining increases. This situation is not random, but by the character of hierarchical structure user interest, the higher the level, the user interest model contains more content, and the more vague concept, the higher the hit rate.

Table 3. Accuracy comparison of the ACO-LPM with other algorithms

Knowledge field	BP-NN	FP-growth	ACO-LPM		
			Original space	Level-2	Level-1
Math	69.5%	72.3%	73.7%	80.2%	85.7%
Physics	66.8%	68.6%	70.1%	75.5%	80.9%
Chemistry	66.2%	67.0%	68.7%	76.1%	81.7%

5 Conclusion

In online learning system, it is a great challenge to mining learning preferences on knowledge points because of the massive knowledge points, the few single user test data, and the change of user's learning preferences. Based on the Multi-granularity feature of knowledge points, this paper defines a quotient space structure of the knowledge points. On this basis, the ACO is introduced into the quotient space, and the multi-granularity dynamic mining algorithm is proposed by using the characteristics of pheromone dynamic volatility. Experiments show the effectiveness of the method.

Acknowledgments. This work was supported by the National Natural Science Foundation of China (Grant No. 61672204), Natural Science Foundations of Higher Education Institutions of Anhui Province (Grant Nos. KJ2012B149, KJ2013A226, KJ2015A229, KJ2015A257), Key Projects of Domestic Visiting and Training for Middle-aged and Young Scholar in Anhui Province (Grant No. gxfxZD2016211).

References

1. Lo, J.-J., Shu, P.-C.: Identifying learning styles by observing learners' browsing behavior. Br. J. Educ. Technol. (2008, in press)
2. Jin, D., Qinghua, Z.: The research of mining association rules between personality and behavior of learner under web-based learning environment. In: The 4th International Conference on Web-Based Learning (ICWL 2005), Hong Kong, China, 31 July–3 August 2005
3. Baishuang, Q.: Research on User Model in Adaptive Learning System Based on Semantic Web. Northeast Normal University, Changchun (2008). (in Chinese)
4. Hongfei, L., Yuansheng, Y.: The user table and update mechanism. J. Comput. Res. Dev. **39**(7), 844–847 (2002). (in Chinese)

5. Shukui, Y.: Design and Implementation of User Interest Extraction System for Mobile Network News. Beijing University of Posts and Telecommunications, Beijing (2012). (in Chinese)
6. Ziting, R.: Research on Personalized Recommendation User Interest Modeling in Mobile Environment. University of Electronic Science and Technology of China, Chengdu (2015). (in Chinese)
7. Hongming, W.: Design and Implementation of User Preference Extraction System Based on Ontology and Tag. Beijing University of Posts and Telecommunications, Beijing (2011). (in Chinese)
8. Woemd, W., Schueiler, C., Wojtech, R.: A hybrid recommender system for context-aware recommendations of mobile applications. In: Proceedings of the International Conference on Data Engineering, Workshops in Conjunction with the International Conference on Data Engineering, ICDE 2007, pp. 871–878 (2007)
9. Shuran, C.: Research on Personalized Service Oriented User Interest Modeling and Application. Chongqing University (2007). (in Chinese)
10. Pazzani, M., Billsus, D.: Learning and revising user profiles: the identification of interesting web sites. Mach. Learn. **27**, 313–331 (1997)
11. Adomavicius, G., Tuzhilin, A.: Using data mining methods to build customer profiles. IEEE Comput. **34**, 74–82 (2001)
12. Zhang, L., Zhang, B.: Dynamic quotient space model and its basic properties. Pattern Recogn. Artif. Intell. **25**(2), 181–185 (2012). (in Chinese)
13. Wang, G.Y., Zhang, Q.H., Jun, H.U.: An overview of granular computing. CAAI Trans. Intell. Syst. **2**(6), 8–26 (2007). (in Chinese)
14. Chen, S., Lv, G., Wang, X.: Offensive strategy in the 2D soccer simulation league using multi-group ant colony optimization. Int. J. Adv. Rob. Syst. **13**, 1 (2016)

Driver Fatigue Detection Using Multitask Cascaded Convolutional Networks

Xiaoshuang Liu[1], Zhijun Fang[1(✉)], Xiang Liu[1], Xiangxiang Zhang[1], Jianrong Gu[2], and Qi Xu[3]

[1] School of Electronic and Electric Engineering, Shanghai University of Engineering Science, Shanghai, China
{xsliu, zjfang, xliu, M020215124}@sues.edu.cn

[2] Information Center, Shanghai University of Engineering Science, Shanghai, China
rong154@sues.edu.cn

[3] College of Information Engineering, Shanghai Maritime University, Shanghai, China
qixu@shmtu.edu.cn

Abstract. Driving fatigue is one of the main reasons of traffic accidents. In this paper, we apply the multitask cascaded convolutional networks to face detection and alignment in order to ensure the accuracy and real-time of the algorithm. Afterwards another convolution neural network (CNN) is used for eye state recognition. Finally, we calculate the percentage of eyelid closure (PERCLOS) to detect the fatigue. The experimental results show that the proposed method has high recognition accuracy of eye state and can detect the fatigue effectively in real- time.

Keywords: Fatigue detection · Cascaded convolutional neural network · Face alignment · PERCLOS

1 Introduction

Along with the development of the auto industry and the transportation industry, traffic accidents have caused great loss in the property and damage to the society. Amongst these traffic accidents more than 20% of these traffic accidents are caused by fatigue driving. Safe driving has become a hot issue in today's society, Therefore, it is of great significance to develop a real-time and accurate fatigue detection system to send fatigue warning information when the driver is tired, which can effectively reduce the occurrence of traffic accidents.

At present, fatigue testing contains three main directions. First, fatigue detection based on the vehicle state detection method, mainly through the turning angle, vehicle driving speed to detect whether the driver fatigue, this method is subject to external interference, the detection accuracy has a greater impact. Second, based on driver's physiological information [7], mainly by detecting the driver's heart rate, pulse and other physiological signals to determine whether the driver is in a state of fatigue, This method requires the driver to carry a lot of testing equipment, very cumbersome, and the driver

Z. Shi et al. (Eds.): ICIS 2017, IFIP AICT 510, pp. 143–152, 2017.
DOI: 10.1007/978-3-319-68121-4_15

has a great interference. Third, fatigue detection methods based on computer vision [6, 8–10], this method is a non-intrusive way, the facial features can be calculated by analyzing the changes of facial expression, such as eye closure duration, yawning and so on.

In the fatigue detection, driver face detection and alignment are important. The multitask cascaded convolutional networks to face detection and alignment [1] has proven to be an effective method. Another very important step is the detection of human eye state. Compared to the traditional active infrared radiation method [2], normal camera image employs a safer passive way. To detect the state of eyes, There are many methods, such as AdaBoost classifier [3], SVM classifier [4] and so on. However, their ability of expressing features is limited. Recently, convolutional neural network (CNN) achieve remarkable progresses in a variety of computer vision tasks. In our paper, we design a driver fatigue detection system using multitask cascaded convolutional networks. As shown in Fig. 1, the method mainly includes five parts: Joint face detection and alignment using multitask cascaded convolutional networks, normalize the current image and ground truth shape according to the scaled mean shape, extract the area of eye, state of eye recognition, fatigue detection.

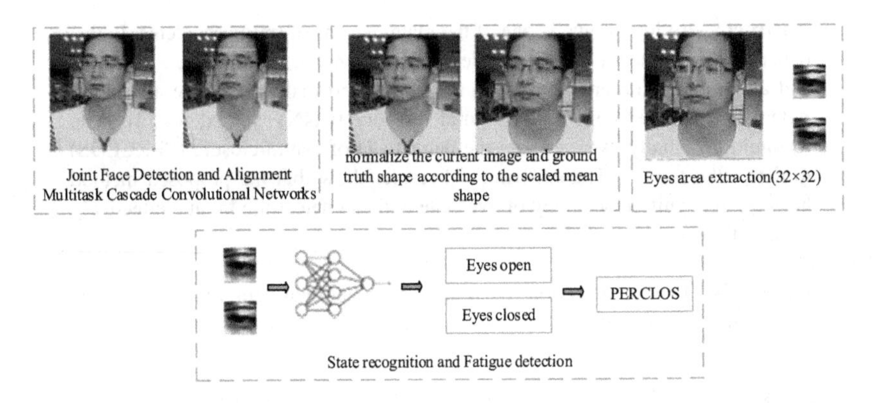

Fig. 1. Algorithm block diagrams.

2 Joint Face Detection and Alignment Using Multitask Cascaded Convolutional Networks

Fatigue detection system should have high recognition accuracy and can detect the fatigue effectively in real-time. How to quickly and accurately detect the face of the driver and the eye alignment and overcome the impact of a certain light are the difficulties of fatigue detection system. Kaipeng et al. [1] propose a new cascaded CNNs-based framework for joint face detection and alignment, and carefully design lightweight CNN architecture for real-time performance. The overall pipeline is shown in Fig. 2, which is the input of the following three-stage cascaded framework.

Stage 1: Exploit a fully convolutional network, called proposal network (P-Net), to obtain the candidate facial windows and their bounding box regression vectors. Then candidates are calibrated based on the estimated bounding box regression vectors.

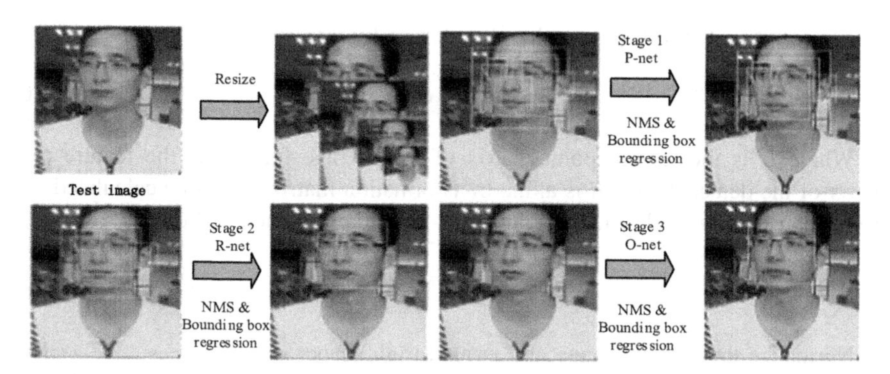

Fig. 2. Joint face detection and alignment using multitask cascaded convolutional networks.

After that, employ nonmaximum suppression (NMS) to merge highly overlapped candidates.

Stage 2: All candidates are fed to another CNN, called refine network (R-Net), which further rejects a large number of false candidates, performs calibration with bounding box regression, and conducts NMS.

Stage 3: This stage is similar to the second stage, but in this stage we aim to identify face regions with more supervision. In particular, the network will output five facial landmarks' positions.

3 Extraction Area Eye

3.1 Face Normalization

In order to accurately extract the eye areas, we need to calculate the average face. Then normalize the current image and ground truth shape according to the scaled mean shape, this process is 2D affine transformation. The 2D affine transformation is a method used to change the rotation angle, the scale, and the location of a shape. The transformation can be represented as Eq. (1).

$$\begin{cases} x = ax' + by' + c \\ y = dx' + ey' + f \end{cases} \tag{1}$$

Where $(x_i, y_i)^T$ is the coordinate of the *i*th feature point on the average face, $(x_i', y_i')^T$ is the coordinate of the *i*th feature point on the detected face. It has a matrix representation shown as Eq. (2).

$$\begin{bmatrix} x_i \\ y_i \\ 1 \end{bmatrix} = \begin{bmatrix} a & b & c \\ d & e & f \\ 0 & 0 & 1 \end{bmatrix} \begin{bmatrix} x_i' \\ y_i' \\ 1 \end{bmatrix} = M \begin{bmatrix} x_i' \\ y_i' \\ 1 \end{bmatrix} \tag{2}$$

For convenience, Eq. (2) can be rewritten as Eq. (3).

$$U = Kh \tag{3}$$

Where U is the feature point matrix of the average face, K is the feature point matrix of the detected face. h is affine transformation matrix. It can be calculated with least squares solution. Then, the solution of h can be obtained as Eq. (4).

$$h = \left(K^T K\right)^{-1} K^T U \tag{4}$$

Normalize the current image and ground truth shape.

According to the scaled mean shape aimed at change the detected faces' rotation angle, the scale, and the location of a shape. As shown in Fig. 3.

Fig. 3. Normalize the current image and ground truth shape according to the scaled mean

3.2 Eye Area Extraction

In this paper, we extract the area of eyes based on the facial landmarks after normalization as shown in Fig. 4. The eye area has a size of 32 × 32.

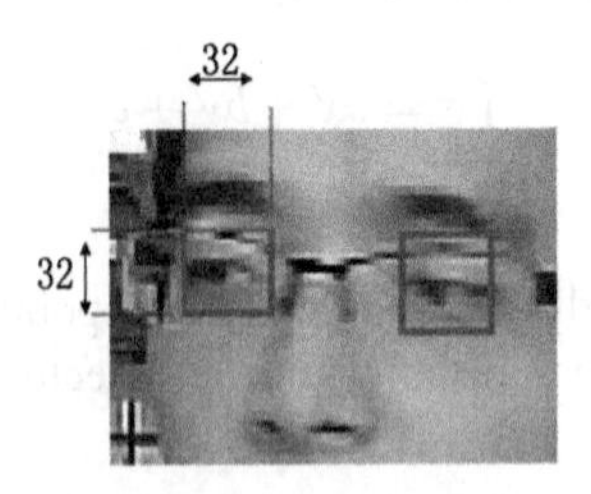

Fig. 4. The extraction result of the eye area.

4 Eye State Recognition

CNN expresses features more better, avoiding the manual feature selection. So we used convolutional neural network to detect the state of eyes.

4.1 Convolutional Neural Network

To have high recognition accuracy of state of eyes and can detect the fatigue effectively real-time, three convolutional layers are used in our proposed network as shown in Fig. 5. Each convolution layer connects a pooling layer, the first convolution layer is connected with a max pooling, the last two convolutions are connected with average pooling. The ReLU layers add non-linear constraints and the Dropout layers prevents overfitting in the networks.

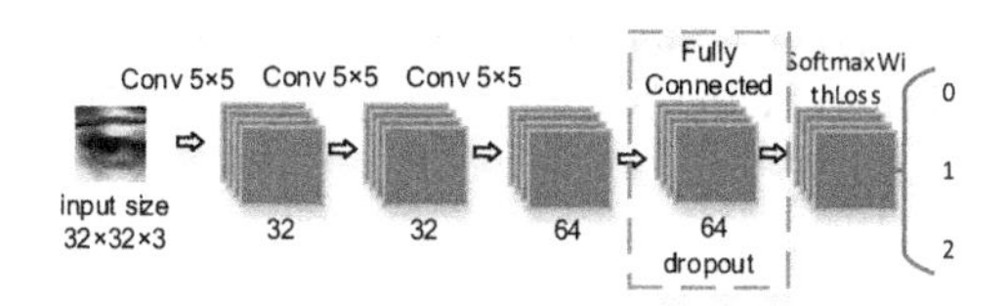

Fig. 5. Structure of the convolution neural network.

4.2 Activation Functions

Sigmoid function and tanh function are commonly used non-linear activation functions, but these functions exist the gradient vanishing, So we use the ReLU function (Rectified linear unit) which is defined as Eq. (5).

$$f(x) = \begin{cases} x & if\, x \geq 0 \\ 0 & if\, x < 0 \end{cases} \tag{5}$$

ReLU can effectively alleviate the problem of gradient vanishing, So as to train the deep neural network directly in a supervised manner. The network can get sparse expression after the ReLU function, with the advantage of unilateral suppression.

5 Fatigue Detection Based on PERCLOS

After eye area extraction, the next step is to detect driver fatigue based on PERCLOS (percentage of eyelid closure over the pupil over time). PERCLOS is an established parameter to detect the level of drowsiness. Level of drowsiness can be judged based on the PERCLOS threshold value, PERCLOS is a parameter that is used to detect driver fatigue [5]. It is calculated as (6).

$$f_{PERCLOS} = \frac{n_{close}}{N_{total}} \times 100\% \quad (6)$$

Let n_{close} be the number of eye-close frames over a period time. N_{total} is the total number of frames over a period time. When the driver is in a state of fatigue, the driver's PERCLOS value will be higher than normal. We set the PERCLOS threshold, when the driver's PERCLOS value is higher than this threshold, then the current driver is considered fatigue.

6 Experiment and Results

VS2013, running on a Win7 system with Intel (R) Core(TM) i7-6700HQ, CPUs (3.40 GHz), 32 GB memory, GPU NVNID GeForce GTX 1070.

6.1 Train

In order to overcome the influence of light on image, the training data must contain data for different light intensities to enhance the robustness of the network, as shown in Fig. 6.

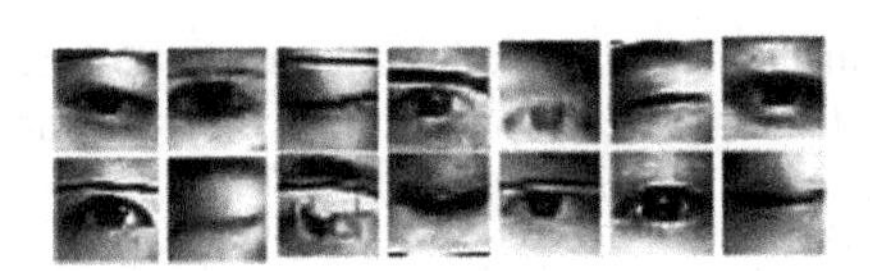

Fig. 6. Different light intensities of Parts of the training samples.

Since we perform eye state recognition, here we use the following two different kinds of data annotation in our training process:

(1) negatives: 36 × 36 sample area was randomly intercepted near the eye area, regions whose the intersection-over-union (IoU) ratio is less than 0.4 to any ground-truth eyes as shown in Fig. 7.

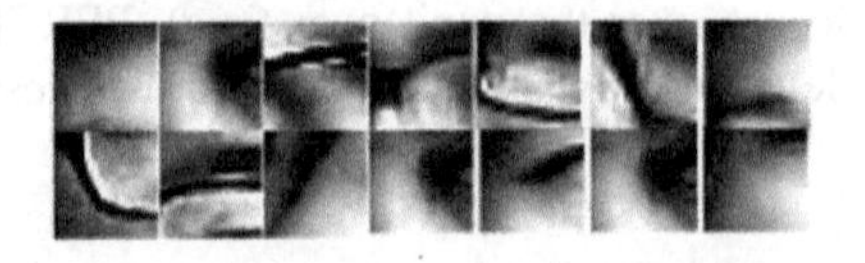

Fig. 7. Negatives training samples.

(2) positives: Positive samples are divided into two types, open eyes samples and closed eyes samples, their IoU above 0.6 to a ground truth face, as shown in Fig. 8.

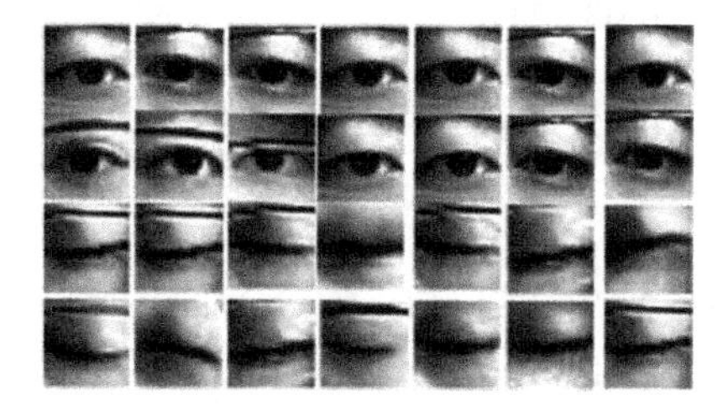

Fig. 8. Positives training samples.

6.2 Training Results

We select images including eye images of open and closed as positives samples, and randomly crop several patches to collect negatives samples. We select 120000 images as training samples. The eye state recognition rate of the network has an increase in the number of iterations when training the samples, the result is shown in Fig. 9.

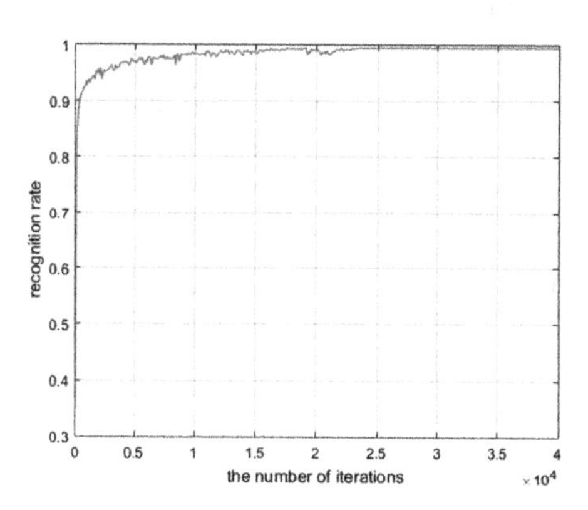

Fig. 9. The result of recognition rate.

With the increase of the iteration number, the accuracy rate gradually increased, the final accuracy rate between 0.995 to 0.996 fluctuations. In order to test the performance of the network, we collected three sections of video data, respectively, the accuracy rate shown in the Table 1.

Through statistical 5 tests videos includes 1239 frames of 320 * 240 images, computing the average time-consuming of the method include each module and overall time. Table 2 is the time-consuming result. The method complies with the requirement of real-time.

Table 1. The test result of eye state.

Video	Number of frames	Accuracy
Testl	243	97.11%
Test2	239	95.39%
Test3	249	98.39%
Testl	243	97.11%
Test2	239	95.39%

Table 2. The test of time consuming.

Video	Face detection, alignment and Eye area extraction (ms)	State recognition (ms)	Total (ms)
Testl	23.41	6.57	34.92
Test2	24.50	7.10	36.63
Test3	23.12	7.09	34.76
Test4	23.42	6.51	34.09
Test5	22.13	6.98	33.46

6.3 Fatigue Detection Based on PERCLOS

When the driver is in a state of fatigue, the driver's PERCLOS value will be higher than normal, by setting the PERCLOS threshold, when the driver's PERCLOS value is higher than this threshold, then the current driver is considered fatigue. In this paper,

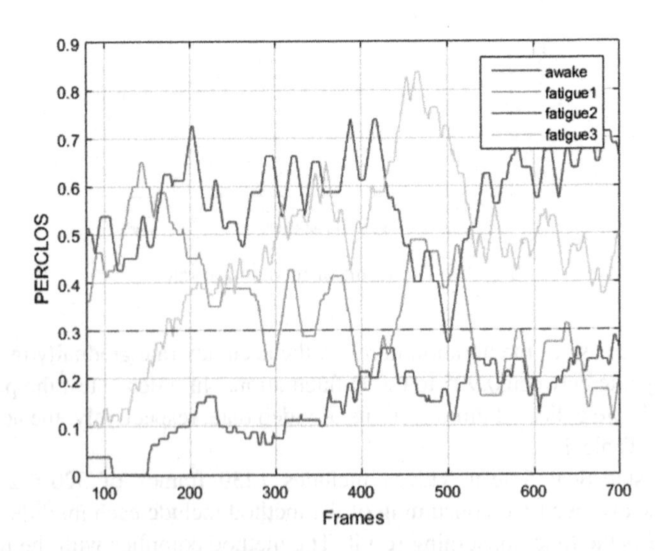

Fig. 10. The PERCLOS value.

the PERCLOS threshold is set to 0.30, when the driver is fatigue, the PERCLOS value is bigger than 0.30, Fig. 10 shows PERCLOS result.

Figure 11 shows the Sample images of detection results.

Fig. 11. Driver Fatigue Detection system.

7 Conclusion

In this paper we propose a driver fatigue detection system. This system uses the multitask cascaded convolutional networks to face detection and alignment. And then use another convolution neural network (CNN) for eye state recognition. Finally we calculate the percentage of eyelid closure (PERCLOS) to detect the fatigue. The method of eye state recognition provides high accuracy and can detect the fatigue effectively in real-time. Tests show that the system implementation is successful and the system does indeed infer fatigue reliably.

Acknowledgments. This research was partially supported by the National Nature Science Foundation of China (No. 61461021) and Local Colleges Faculty Construction of Shanghai MSTC (No. 15590501300).

References

1. Zhang, K., Zhang, Z., Li, Z., Qiao, Y.: Joint face detection and alignment using multitask cascaded convolutional networks. IEEE Sig. Process. Lett. **10**, 1499–1503 (2016)
2. Zhang, F., Su, J., Geng, L., Xiao, Z.: Driver fatigue detection based on eye state recognition. In: 2017 International Conference on Machine Vision and Information Technology (CMVIT), Singapore, pp. 105–110 (2017)
3. Shi-Feng, X.U., Zeng, Y.: Eyes state detection based on adaboost algorithm. Comput. Simul. **7**, 214–217 (2007)
4. Yao, S., Li, X., Zhang, W., Zhou, J.: Eyes state detection method based on LBP. Appl. Res. Comput. **6**, 1897–1901 (2015)
5. Wierwille, W.W., Ellsworth, L.A., Wreggit, S.S., Fairbanks, R.J., Kim, C.L.: Research on vehicle-based driver status performance monitoring: development, validation, and refinement of algorithms for detection of driver drowsiness. National Highway Traffic Safety Administration (1994)

6. Liu, A., Li, Z., Wang, L., Zhao, Y.: A practical driver fatigue detection algorithm based on eye state. In: 2010 Asia Pacific Conference on Postgraduate Research in Microelectronics and Electronics (PrimeAsia), Shanghai, pp. 235–238 (2010)
7. Wang. Y., Liu, X., Zhang, Y., Zhu, Z., Liu, D., Sun, J.: Driving fatigue detection based on EEG signal. In: 2015 Fifth International Conference on Instrumentation and Measurement, Computer, Communication and Control (IMCCC), Qinhuangdao, pp. 715–718 (2015)
8. Zhao, G.F., Han, A.X.: Method of detecting logistics driver's fatigue state based on computer vision. In: 2015 International Conference on Computer Science and Applications (CSA), Wuhan, pp. 60–63 (2015)
9. Li, X., Wu, Q., Kou, Y., Hou, L., Xie, H.: Driver's eyes state detection based on adaboost algorithm and image complexity. In: 2015 Sixth International Conference on Intelligent Systems Design and Engineering Applications (ISDEA), Guiyang, pp. 349–352 (2015)
10. Guohui, H., Wanying, W.: An algorithm for fatigue driving face detection and location. In: 2015 8th International Conference on Intelligent Computation Technology and Automation (ICICTA), Nanchang, pp. 130–132 (2015)

A Fast Granular Method for Classifying Incomplete Inconsistent Data

Zuqiang Meng(✉) and Hongli Li

College of Computer, Electronics and Information, Guangxi University, Nanning 530004, People's Republic of China
zqmeng@126.com

Abstract. Today extracting knowledge from "inferior quality" data that is characterized by incompleteness and inconsistency is an unavoidable and challenging topic in the field of data mining. In this paper, we propose a fast granular method to classify incomplete inconsistent data using attribute-value block technique. Firstly, a granulation model is constructed to provide a foundation for efficient computation. Secondly, an algorithm of acquiring classification rules is proposed and then an algorithm of minimizing rule sets is proposed, and with these proposed algorithms, a classification algorithm is designed to construct a rule-based classifier. Finally, we use the experiment results to illustrate the effectiveness of the proposed algorithms.

Keywords: Attribute-value blocks · Data classification · Classifier · Incomplete inconsistent data · Granulation model

1 Introduction

Data classification is an important task in the field of data mining. Related classification methods and techniques are increasingly extensively studied and some of them have been successfully used to solve practical problems [1–3]. In the era of big data, the volume and variety of data is growing and growing. Big data shows features of not only large volume but also "inferior quality". "Inferior quality" of data embodies in many aspects, two of which are incompleteness and inconsistency. Data's incompleteness means that there are missing values in data while the inconsistency refers to that data contains conflicting descriptions. There are many reasons that cause the incompleteness and inconsistency, such as objective and subjective factors, noisy data, variety of data. In fact, problems caused by the incompleteness and inconsistency are unavoidable when extracting knowledge from big data [4]. The incompleteness usually enhances the degree of inconsistency. Actually the problems of incompleteness and inconsistency are interwoven and can not totally separated, which makes the problem of knowledge extraction more complicated. Therefore, it is difficult and meaningful to solve classification problems oriented to "inferior quality" data that is characterized mainly by incompleteness and inconsistency.

This work is supported by the National Natural Science Foundation of China (No. 61363027, 61762009), the Guangxi Natural Science Foundation (No. 2015GXNSFAA139292).

Z. Shi et al. (Eds.): ICIS 2017, IFIP AICT 510, pp. 153–163, 2017.
DOI: 10.1007/978-3-319-68121-4_16

There are many ways to handle missing values. Stefanowski [5]. distinguished two different semantics for missing values: the "absent value" semantics and the "missing value" semantics. Grzymala-Busse [6]. further divided such missing values into three categories according to their comparison range: "do not care" conditions, restricted "do not care" conditions, and attribute-concept values. In 2009 we utilized sorting technique to design a fast approach to compute tolerance classes [7]. But this approach has the problem of data fragmentation when the degree of missing values increases to a certain level. Recently, we presented a new method called index method to quickly compute attribute-value blocks [8], which can completely eliminate data fragmentations. In this paper, we studied on a fast classification problem of incomplete inconsistent data by using and improving such index methods.

2 Preliminaries

2.1 Incomplete Decision Systems (IDSs) and Attribute-Value Blocks

A decision system that contains missing values is called an incomplete decision system (IDS), which can be described as 4-tuple: $IDS = (U, A = C \cup D, V = \bigcup_{a \in A} V_a, \{f_a\})_{a \in A}$, where U is a finite nonempty set of objects, indicating a given universe; both C and D are finite nonempty set of attributes (features), called condition attribute set and decision attribute set, respectively, where $C \cap D = \phi$; V_a is the domain of attribute $a \in A$, and $|V_a|$ denotes the number of elements in V_a, i.e., the cardinality of V_a; f_a: $U \rightarrow V$ is an information function from U to V, which maps an object in U to a value in V_a. Sometimes $(U, A = C \cup D, V = \bigcup_{a \in A} V_a, \{f_a\})_{a \in A}$ is expressed as $(U, C \cup D)$ for simplicity if V and f_a are understood. Without loss of generality, we suppose $D = \{d\}$ in this paper; that is, D is supposed to be composed of only one attribute. A missing value is usually denoted as "*". That is, if there exists $a \in C$ such that $* \in V_a$, then the decision system $(U, C \cup D)$ is an incomplete decision system.

For an incomplete decision system $IDS = (U, C \cup D)$, we define the concept of **missing value degree** (the degree of missing values) [8], denoted as ***MD(IDS)***, for *IDS*:

$$MD(IDS) = \frac{\text{the number of missing attribute values}}{|U||C|}.$$

Obviously, missing value degrees have great effects on classification performance, but related studies on this problem are little reported.

Attribute-value blocks were presented by Grzymala-Busse to analyze incomplete data [8]. Here we first introduce some concepts related to attribute-value blocks.

In an incomplete system $(U, C \cup D)$, for any $a \in C$ and $v \in V_a$, (a, v) is said to be an **attribute-value pair**, which is an atomic formula of decision logic [9]. Let $[(a, v)]$ denote all the objects from U which can be matched with (a, v), and $[(a, v)]$ is the so-called **attribute-value (pair) block** [10]. According to the semantics of "do not care" conditions, we have:

$$[(a,v)] = \begin{cases} \{y \in U | f_a(y) = v \,\text{or}\, f_a(y) = *\}, & \text{if } v \neq *; \\ U, & \text{else.} \end{cases}$$

Actually, $v = f_a(x)$ for some $x \in U$, and therefore block $[(a, v)]$ is usually denoted by $K_a(x)$ or $S_a(x)$ in many studies, i.e., $K_a(x) = S_a(x) = [(a, f_a(x))]$. For $B \subseteq C$, the attribute-value block with respect to B, $K_B(x)$, is defined as follows:

$$K_B(x) = \bigcap_{a \in B} [(a, f_a(x))] = \bigcap_{a \in B} K_a(x).$$

Property 1. For $B', B'' \subseteq C$, if $B' \subseteq B''$, $K_{B''}(x) \subseteq K_{B''}(x)$.

2.2 Incomplete Inconsistent Decision Systems (IIDSs)

In an incomplete system $(U, C \cup D)$, since V_D does not contain missing values, D can partition U into a family of equivalence classes, which are called decision classes. In this paper, we let D_x denote the decision class that contain object x, where $x \in U$.

Definition 1. For an object $x \in U$, let $m_B(x) = \frac{|K_B(x) \cap D_x|}{|K_B(x)|}$, denoting the degree to which object x belongs to decision class D_x with respect to B, and then $\mu_B(x)$ is called the **consistency degree** of object x with respect to B.

Obviously, $0 < \mu_B(x) \leq 1$. If $\mu_C(x) = 1$, object x is said to be a **consistent object**, otherwise an **inconsistent object**. It is not difficult to find that an inconsistent object x means that block $K_C(x)$ intersects at least two different decision classes, i.e., $K_C(x) \not\subseteq D_x$.

For an incomplete decision system $(U, C \cup D)$, if U contains inconsistent objects, i.e., there exists $y \in U$ such that $\mu_C(y) < 1$, then the decision system is said to be an **incomplete inconsistent decision system** (IIDS), denoted as $IIDS = (U, C \cup D)$.

Definition 2. Let $id(IIDS)$ denote the ratio of the number of inconsistent objects to the number of all objects in U, i.e., $id(IIDS) = |\{x \in U \mid \mu_C(x) < 1\}| / |U|$, and then $id(IIDS)$ is called **inconsistency degree** of the decision system $IIDS$.

Obviously, $0 \leq id(IIDS) \leq 1$. In order to judge if an object x is consistent, we need to compute its consistency degree $\mu_B(x)$, which is possibly a time-consuming computational process because it involves set operations. The following property can make preparation for efficiently computing $\mu_B(x)$.

Property 2. For incomplete inconsistent decision system $(U, C \cup D)$ and any $x \in U$, $K_B(x) \cap D_x = \{y \in U | y \in K_B(x) \,\text{and}\, f_d(y) = f_d(x)\}$, and therefore $\mu_B(x) = \frac{|\{y \in U | y \in K_B(x) \wedge f_d(y) = f_d(x)\}|}{|K_B(x)|}$, where $B \subseteq C$ and $D = \{d\}$.

3 A Granulation Model Based on IIDSs

In order to compute $K_B(x)$, we generally need to traverse all objects in U, and therefore it takes the computation time of $O(|U|^2|B|)$ to compute $K_B(x)$ for all $x \in U$, which is a time-consuming computation process. However, we notice that when considering only one attribute, we can derive some useful properties to accelerate the computation process.

Definition 3. In an *IIDS* $(U, C \cup D)$, for a given attribute $a \in C$, let U_a^* denote the set of all objects that $f_a(x) = *$, i.e., $U_a^* = \{x \in U | f_a(x) = *\}$; attribute a can partition $U - U_a^*$ into a family of equivalence classes, which are pairwise disjoint, and let $[x]_a$ denote an equivalence class containing x, i.e., $[x]_a = \{y \in U - U_a^* | f_a(y) = f_a(x)\}$, where $x \in U$, and let Γ_a denote such a family, i.e., $\Gamma_a = \{[x]_a | x \in U - U_a^*\}$.

Each element in Γ_a is an equivalence class. Those objects are drawn together in an equivalence class due to that they have the same attribute value on corresponding attribute. Sometimes, in order to emphasize the attribute value, we let $\Gamma_a(v)$ denote an equivalence class in Γ_a where all objects have attribute value v, i.e., $\Gamma_a(v) = \{y \in U - U_a^* | f_a(y) = v\}$.

It is not difficult to find that $\Gamma_a \cup \{U_a^*\}$ is a coverage of U; each element in $\Gamma_a \cup \{U_a^*\}$ is a subset of U, and they are also pairwise disjoint.

Property 3. Given Γ_a and U_a^*, for any $B \subseteq C$ and $x \in U$, we have

$$K_a(x) = \begin{cases} [x]_a \cup U_a^* & \textit{if } f_a(x) \neq * \\ U & \textit{else} \end{cases}$$

where $[x]_a \in \Gamma_a$.

Property 3 provides a method for us to use Γ_a and U_a^* to compute $K_a(x)$. We notice that $|V_a|$ almost does not increase with $|U|$, with which we can design an efficient algorithm to compute Γ_a and U_a^* for all $a \in C$. The algorithm is described as follows.

Algorithm 1: compute Γ_a and U_a^* for all $a \in C$
Input: $(U, C \cup D)$, where $U = \{x_1, x_2, ..., x_n\}$
Output: Γ_a and U_a^* for all $a \in C$
Begin
(1) For each $a \in C$ do
(2) {
(3) Let $U_a^* = \emptyset$ and $\Gamma_a = \emptyset$;
(4) For $i = 1$ to n do // $n = |U|$
(5) {
(6) If $f_a(x_i) = *$ then {let $U_a^* = U_a^* \cup \{x_i\}$; continue;}
//Suppose $\Gamma_a = \{\Gamma_a(v_1), \ldots, \Gamma_a(v_t)\}$ at this moment, where $t = |\Gamma_a| \leq |V_a|$
(7) Let flag = 0;
(8) For $j = 1$ to t do
(9) If $f_a(x_i) = v_j$ then { $\Gamma_a(v_j) = \Gamma_a(v_j) \cup \{x_i\}$; let flag = 1; break;}
(10) If flag = 0 then { let $\Gamma_a(v_{t+1}) = \{x_i\}$; $\Gamma_a = \Gamma_a \cup \{\Gamma_a(v_{t+1})\}$ };
(11) }
(12) }
(13) Return Γ_a and U_a^*;
End.

From Algorithm 1, we can find that its time complexity is $O(|C||U|t) \leq O(|C||U||V_a|)$. As mentioned above, $|V_a|$ almost does not increase with $|U|$, so $|V_a|$ can be regard as a constant generally. Therefore, the time complexity of this algorithm almost approaches linear complexity $O(|C||U|)$.

In fact, Algorithm1 is to granulate each "column" for an *IIDS* and therefore to construct a granulation model for the *IIDS*. Let $\Gamma_a^* = \Gamma_a \cup \{U_a^*\}$, and such a granulation model is denoted as $\mathfrak{R} = (U, \{\Gamma_a^*\}_{a \in C}, D)$ in this paper.

With the granulation model, we can compute any block $K_B(x)$ by using formula $K_B(x) = \bigcap_{a \in B} K_a(x)$. In order to quickly compute $K_B(x)$, we should know "where $K_a(x)$ is". So we construct an index structure to store the addresses of $K_a(x)$ for all $a \in C$ and $x \in U$. Such an index structure is expressed as a matrix $\psi = U \times C = [m(x, a)]_{x \in U, a \in C}$, where $m(x, a)$ is the index or address of $\Gamma_a(v)$ and $v = f_a(x)$. The algorithm of constructing matrix ψ is described as follows.

Algorithm 2: construct matrix ψ for granulation model $\mathfrak{R}$
Input: $\mathfrak{R} = (U, \{\Gamma_a^*\}_{a \in C}, D)$
Output: $\psi = U \times C = [m(x,a)]_{x \in U,\, a \in C}$,
Begin
(1) For each $a \in C$ do
(2) {
(3) For each $\theta \in \Gamma_a^*$ do
(4) {
(5) For each $x \in \theta$ do
(6) {
(7) If $f_a(x) = *$ then let $m(x,a) = null$;// in this case, $\theta = U_a^*$
(8) Else let $m(x,a) = loc(\Gamma_a(v))$; // the index or address of $\Gamma_a(v)$
(9) }
(10) }
(11) }
End.

Actually, Algorithm 2 is to traverse all objects in Γ_a^* for all $a \in C$. We notice that $\cup \Gamma_a^* = U$ and the elements (subset) in Γ_a^* are pairwise disjoint, therefore the complexity of this algorithm is exactly equal to $O(|U||C|)$, which is linear complexity.

Both granulation model $\mathfrak{R}$ and index matrix ψ are denoted as ordered pair $[\mathfrak{R}, \psi]$. If there is no confusion, $[\mathfrak{R}, \psi]$ is also called a granulation model. The purpose of constructing $[\mathfrak{R}, \psi]$ is to provide a way to quickly compute block $K_B(x)$ for any $x \in U$, with complexity of about $O(|K_B(x)||B|)$.

4 A Granulation-Model-Based Method for Constructing Classifier

4.1 An Attribute-Value Block Based Method of Acquiring Classification Rules

A classification rule can viewed as an implication relation between different granular worlds, and each object x can derive a classification rule, which is a "bridge" between such two worlds. Firstly, let's consider the following inclusion relation: $K_B(x) \subseteq D_x$. $K_B(x)$ and D_x has their own descriptions, which are formulae of decision logic [9]. Suppose their descriptions are ρ and φ, respectively. Then, object x can derive rule $\rho \to \varphi$. According to Property 1, when removing attribute from B, $K_B(x)$ would enlarge, and therefore the generalization ability of rule $\rho \to \varphi$ would be strengthened. But its consistency degree $\mu_B(x)$ may decrease and then increase its uncertainty. Therefore the operation of removing attributes from B must be done under a certain limited condition. Now we give the concept of object reduction, which is used to acquire classification rules.

Definition 4. In an *IIDS* = $(U, C \cup D)$, for any object $x \in U$, $B \subseteq C$ is said to be a **reduct** of object x, if the following conditions can be satisfied: (a) $\mu_B(x) \geq \mu_C(x)$, and (b) for any $B' \subset B$, $\mu_{B'}(x) < \mu_C(x)$.

Actually, it is time-consuming to find a reduce for an object, because it needs take too much time to search each subset of B so as to satisfy condition (b). A usual method is to select some attribute from C to constitute B such that $\mu_B(x) \geq \mu_C(x)$. Such a method is known as **feature selection**, with which we can easily obtain corresponding classification rule. In the following, we give an algorithm to perform feature selection for all objects $x \in U$ and derive corresponding classification rules.

Algorithm 3: construct a classification rule set
Input: $[\mathfrak{R}, \psi]$ and $(U, C \cup D)$ // suppose $U = \{x_1, x_2, \ldots, x_n\}$ and $C = \{a_1, a_2, \ldots, a_m\}$
Output: S // a rule classification set
Begin
(1) Let $S = \varnothing$;
(2) For $i = 1$ to n do //$n = |U|$
(3) {
(4) Let $B = C$;
(5) For $j = 1$ to m do //$m = |C|$ if $\mu_{B-\{a_j\}}(x_i) \geq \mu_C(x_i)$ then let $B = B$-$\{a_j\}$;
(6) Use B to construct $\rho \to \varphi$;
(7) Let $S = S \cup \{\rho \to \varphi\}$;
(8) }
(9) Return S;
End.

In Algorithm 3, step (5) is to remove redundant attributes in C, which is actually to perform feature selection. Let $B_j = B-\{a_j\}$. According to Property 2, with $[\mathfrak{R}, \psi]$,

$\mu_{B_j}(x_i)$ can be computed in the complexity of $O\left(\sum_{j=1}^{m}|K_{B_j}(x_i)||B_j|\right)$. Therefore, the complexity of Algorithm 3 is $O\left(\sum_{i=1}^{n}\sum_{j=1}^{m}|K_{B_j}(x_i)||B_j|\right)$. Suppose t is the average size of blocks $K_{B-\{a_j\}}(x_i)$ for all $a_j \in C$ and $x_i \in U$ and h is the average length of B_j, then $O\left(\sum_{i=1}^{n}\sum_{j=1}^{m}|K_{B_j}(x_i)||B_j|\right) = O(|U||C| \cdot t \cdot h)$. Generally, $t << |U|$ and $h << |C|$, so $O(|U||C| \cdot t \cdot h) << O(|U|^2|C|^2)$.

It should be pointed that Algorithm 3 can not guarantee each generated attribute subset is a reduct, but it does remove some redundant attributes from C and therefore can finish the task of feature selection.

4.2 Rule Set Minimum

Since each rule in S is induced by an object in U, these rules and objects are one to one correspondence. Usually, we let r_i denote the rule that is induced by object x_i, i.e., r_i corresponds to x_i. We notice that $|S| = |U|$ and there are many redundant rules in it. Therefore we need to further remove those redundant rules, and this process is so-called rule set minimum.

Definition 5. For rule r: $\rho \rightarrow \varphi$, let **coverage(r)** denote all objects which can match rule r, i.e., coverage(r) = $\{x \in U \mid x| = r\}$.

For two rules r_x: $\rho_x \rightarrow \varphi_x$ and r_y: $\rho_y \rightarrow \varphi_y$, if coverage($r_x$) $\subseteq$ coverage(r_y), then rule r_x is redundant and should be removed. Based on this consideration, we design the following algorithm to minimize the rule set S.

Algorithm 4: minimize a rule set
Input: $[\mathfrak{R}, \psi]$ and S // S is a rule set which is generated by Algorithm 3
Output: MS // MS is a minimized rule set
Begin
(1) Compute |coverage(r)| for all $r \in S$;
(2) Sort all rules from S in a descending order by |coverage(r)| and suppose $S = \{r_1, r_2, \ldots, r_n\}$ after sorting;
(3) Let $MS = \{r_1\}$;
(4) For $i = 2$ to n do //$n = |U|$
(5) {
(6) If $x_i \notin$ coverage(r_{i-1}) then let $MS = MS \cup \{r_i\}$; // r_i is induced by x_i
(7) }
(8) Return MS;
End.

In Algorithm 4, the key operation is to compute coverage(r_{i-1}). Suppose B_{i-1} is a set of all attributes which are contained in rule r_{i-1}, and then computing coverage(r_{i-1}) is equivalent to computing block $K_{B_{i-1}}(x_{i-1})$, whose complexity is $O(|K_{B_{i-1}}(x_{i-1})||B_{i-1}|)$ by using $[\mathfrak{R}, \psi]$. Suppose the average size of blocks $K_{B_{i-1}}(x_{i-1})$ is p and the average length of $B_{i\text{-}1}$ is o, then the complexity is $O(|K_{B_1}(x_1)||B_1| + |K_{B_2}(x_2)||B_2| + \ldots + |K_{B_n}(x_n)||B_n|) = O(|U| \cdot p \cdot o)$. Generally, $p << |U|$ and $o << |C|$. Therefore $O(|U| \cdot p \cdot o) << O(|U|^2|C|)$.

4.3 A Classification Algorithm for Constructing Rule-Based Classifier

Using the above four provided algorithms, we here give a complete algorithm to acquire a rule set, which is used as a classifier to classify incomplete inconsistent data. The complete algorithm is described as follows.

Algorithm 5: construct a rule-based classifier
Input: $(U, C \cup D)$
Output: *MS* // *MS* is a rule-based classifier
Begin
(1) Use Algorithms 1 and 2 to construct granulation model $[\mathfrak{R}, \psi]$;
(2) Use Algorithm 3 to construct rule set *S* by using $[\mathfrak{R}, \psi]$;
(3) Use Algorithm 4 to minimize rule set *S* to be *MS* by using $[\mathfrak{R}, \psi]$ and *S*;
(4) Return *MS*;
End.

As analyzed above, the complexities of Algorithms 1 and 2 are all $O(|C||U|)$, and those of Algorithms 3 and 4 are $O(|U||C| \cdot t \cdot h)$ and $O(|U| \cdot p \cdot o)$, respectively, which are much less than $O(|U|^2|C|^2)$ and $O(|U|^2|C|)$, respectively. Therefore, it can be seen that the time-consuming step is step (3), and then the complexity of Algorithm 5 is $O(|U||C| \cdot t \cdot h)$, which is much less than $O(|U|^2|C|^2)$.

5 Experimental Analysis

In order to verify the effectiveness of the proposed methods, we conduct several experiments using UCI data sets (http://archive.ics.uci.edu/ml/datasets.html). These experiments ran on a PC equipped with a Windows 7, Intel(R) Xeon(R), CPU E5-1620v3,and 8 GB memory. The data sets are outlined in Table 1, where |U|, |C| and |Vd| stand for the numbers of samples, condition attributes, and decision classes, respectively.

For there exist missing values in incomplete decision system and the relation between objects are tolerance relation, instead of equivalence relation, we can not use sorting technique to accelerate the process of computing blocks and therefore need to compare

Table 1. Description of the four data sets.

No.	Data sets	$\|U\|$	$\|C\|$	$\|V_d\|$	*MD(IIDS)*	*id(IIDS)*
1	Voting-records	435	17	2	0.0563	0.6736
2	Tic-Tac-Toe	958	9	2	0	0
3	Mushroom	8124	22	2	0.0139	0
4	Nursery	12960	8	5	0	0

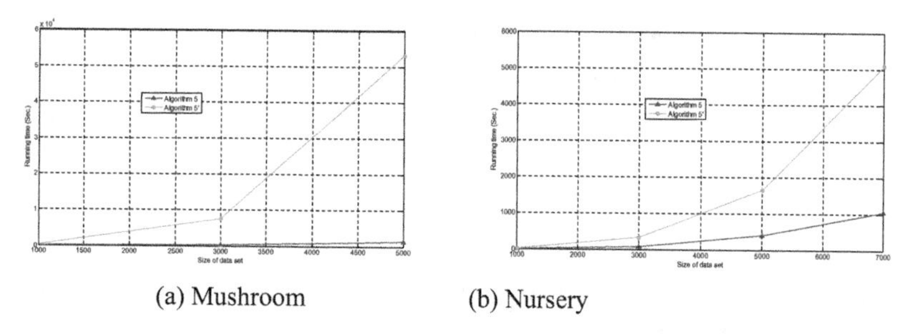

(a) Mushroom (b) Nursery

Fig. 1. Running times of Algorithm 5 and Algorithm 5′ for mushroom and nursery

x with all other objects in U when computing block $K_B(x)$, where $x \in U$. Replace the method of computing attribute-value blocks in Algorithm 5, which is based on the granulation model, with such a method of computing blocks and keep other parts uncharged. Thus we would obtain another algorithm, denoted as **Algorithm 5′**. To compare the running times for a varying number of data records, we execute Algorithm 5 and Algorithm 5′ on two data sets, Mushroom and Nursery, for four times, with randomly extracting different objects at each time, and the results are shown in Fig. 1.

Form Fig. 1 we can find that the running times of Algorithm 5′ increase much more rapidly than that of Algorithm 5. Therefore, the constructed granulation model can greatly improve computational efficiency for Algorithm 5.

Algorithm 5 consists of Algorithms 1, 2, 3 and 4. We count the time for each algorithm when they are executed on Mushroom and Nursery, and the results are shown in Table 2.

It can be seen from Table 2 that, comparing with Algorithms 3 and 4, it only takes a little time for Algorithms 1 and 2 to construct granulation model. This means that constructing granulation model cost very little but it forms the foundation for fast feature selection and building classifiers. Additionally, Table 2 also shows that Algorithm 3 is the most time-consuming algorithm.

Table 2. Running times of Algorithms 1–4 for mushroom and nursery

Data sets and the used algorithms		Size of data set			
		1000	3000	5000	7000
Mushroom	Algo. 1 + Algo. 2	0.062	0.109	0.548	0.710
	Algo. 3	20.524	293.039	1148.922	2748.864
	Algo. 4	1.602	49.749	233.802	545.356
Nursery	Algo. 1 + Algo. 2	0.042	0.058	0.164	0.358
	Algo. 3	1.955	28.818	119.987	278.489
	Algo. 4	1.382	53.047	287.059	771.875

To verify the classification performances of Algorithm 5, we utilize Voting-records, Tic-Tac-Toe, and Nursery to test Algorithm 5 using 10-fold cross-validation. The results are shown in Table 3.

Table 3. Precision and recall of Algorithm 5

Data set	Precision	Recall
Voting-records	0.9103	0.8950
Tic-Tac-Toe	0.9958	0.9969
Nursery	0.9850	0.7634

From Table 3, we can find that Algorithm 5 can have relatively high precision and recall on these data sets. This shows that the proposed algorithm has better application values. Additionally, Table 3 also shows that Algorithm 5 is suitable not only for incomplete inconsistent data but also for complete consistent data. Of course, it has better classification performances on complete consistent data than on incomplete inconsistent data.

6 Conclusion

Extracting rules from data sets and then using a rule set as a classifier to classify data is one of our purposes recent years. In this paper, oriented to incomplete inconsistent data, we first used attribute-value block technique to construct a granulation model, which actually consists of a block-based model and a index matrix; secondly, based on the constructed granulation model, an algorithm of acquiring classification rules is presented and then an algorithm of minimizing rule sets is proposed; with the proposed algorithms, we designed a classification algorithm for constructing a rule-based classifier; finally, we conducted some experiments to verify the effectiveness of the proposed algorithm. The experiment results are consistent with our theoretical analysis. Therefore, the work in this paper has a certain theoretical value and application value, and provides a new idea to classify incomplete inconsistent data.

References

1. Liu, Y., Bi, J.W., Fan, Z.P.: A method for multi-class sentiment classification based on an improved one-vs-one (OVO) strategy and the support vector machine (SVM) algorithm. Inf. Sci. **394**, 38–52 (2017)
2. Nakashima, T., Schaefer, G., Yokota, Y., et al.: A weighted fuzzy classifier and its application to image processing tasks. Fuzzy Sets Syst. **158**, 284–294 (2007)
3. Stimpfling, T., Bélanger, N., Cherkaoui, O., et al.: Extensions to decision-tree based packet classification algorithms to address new classification paradigms. Comput. Netw. **122**, 83–95 (2017)
4. Conde-Clemente, P., Trivino, G., Alonso, J.M.: Generating automatic linguistic descriptions with big data. Inf. Sci. **380**, 12–30 (2017)
5. Stefanowski, J., Tsoukiàs, A.: Incomplete information tables and rough classification. Comput. Intell. **17**, 545–566 (2001)
6. Grzymala-Busse, J.W.: A rough set approach to data with missing attribute values. In: Wang, G.-Y., Peters, J.F., Skowron, A., Yao, Y. (eds.) RSKT 2006. LNCS, vol. 4062, pp. 58–67. Springer, Heidelberg (2006). doi:10.1007/11795131_10

7. Meng, Z.Q., Shi, Z.Z.: A fast approach to attribute reduction in incomplete decision systems with tolerance relation-based rough sets. Inf. Sci. **179**, 2774–2793 (2009)
8. Meng, Z.Q., Gan, Q.L., Shi, Z.Z.: On efficient methods of computing attribute-value blocks in incomplete decision systems. Knowl.-Based Syst. **113**, 171–185 (2016)
9. Meng, Z., Gan, Q.: An attribute-value block based method of acquiring minimum rule sets: a granulation method to construct classifier. In: Shi, Z., Vadera, S., Li, G. (eds.) IIP 2016. IAICT, vol. 486, pp. 3–11. Springer, Cham (2016). doi:10.1007/978-3-319-48390-0_1
10. Grzymala-Busse, J.W., Clarka, P.G., Kuehnhausen, M.: Generalized probabilistic approximations of incomplete data. Int. J. Approx. Reas. **55**, 180–196 (2014)

Sentiment Analysis of Movie Reviews Based on CNN-BLSTM

Qianzi Shen, Zijian Wang, and Yaoru Sun(✉)

Department of Computer Science, Tongji University, Shanghai, China
{sqz,1410482,yaoru}@tongji.edu.cn

Abstract. Sentiment analysis has been a hot area in the research field of language understanding, but complex deep neural network used in it is still lacked. In this study, we combine convolutional neural networks (CNNs) and BLSTM (bidirectional Long Short-Term Memory) as a complex model to analyze the sentiment orientation of text. First, we design an appropriate structure to combine CNN and BLSTM to find out the most optimal one layer, and then conduct six experiments, including single CNN and single LSTM, for the test and accuracy comparison. Specially, we pre-process the data to transform the words into word vectors to improve the accuracy of the classification result. The classification accuracy of 89.7% resulted from CNN-BLSTM is much better than single CNN or single LSTM. Moreover, CNN with one convolution layer and one pooling layer also performs better than CNN with more layers.

Keywords: Natural language processing · CNN · LSTM · Sentiment analysis

1 Introduction

As the significant carries of emotion, texts are in an irreplaceable position on the Internet. The sentiment analysis is usually used to measure the emotional attitude of texts and related studies grow quickly. In the literature, Hazivassiloglou and McKeown investigated semantic orientation of adjectives based on cluster methods [1]. Turney and Littman chose seven pairs of words that have strong orientation and then set SO-PMI (semantic orientation-pointwise mutual information) to judge a word with a standard word. According to SO-PMI, they chose two groups of words in which one is positive (P word) and the other is negative (N word) as standard words [2]. Kim chose a method based on sentiment knowledge and summed up the weighing of phrases and describing words [3]. Pang studied sentiment analysis of English texts with machine learning [4]. With the development of artificial intelligence and deep learning, more and more related networks and methods have been involved in natural processing. In a variety of networks, CNN [5] is widely used for its generalization. Because of its advantage in learning the higher level features, CNN has achieved a great breakthrough [6] and being applied to many areas, such as image classification, face recognition and other tasks related to image [7]. Among them, LSTM [8] is a kind of recurrent neural networks and is able to keep the information for a long time because it has memory cells. In other words, LSTM is a time-order network [9], while BLSTM (Bidirectional Long Short Term Memory) is a kind of network combining LSTM with BRNN [10],

Z. Shi et al. (Eds.): ICIS 2017, IFIP AICT 510, pp. 164–171, 2017.
DOI: 10.1007/978-3-319-68121-4_17

which has two directions to input data and two hiding layers to save the information in both directions and sharing the same output layer. In this paper, we designed a more complex model containing both CNN and BLSTM for sentiment analysis.

The rest of the paper is organized as follows: Sect. 2 reviews the related work to CNN and LSTM. Section 3 describes the model we used in this work. Section 4 shows the experiments and their results. Section 5 presents the conclusion and future works.

2 Methods and Models

2.1 CNN

Usually, convolutional neural networks [5] are used in image classification. In this study, CNN was used as a feature important part in our sentiment analysis model.

Convolutional neural networks provide a learning model which is end to end and the parameters of the model can be trained with gradient descent algorithm. A typical convolutional neural network consists of an input layer, convolution layers, pooling layers, a full connection layer and an output layer. Its architecture is shown in Fig. 1.

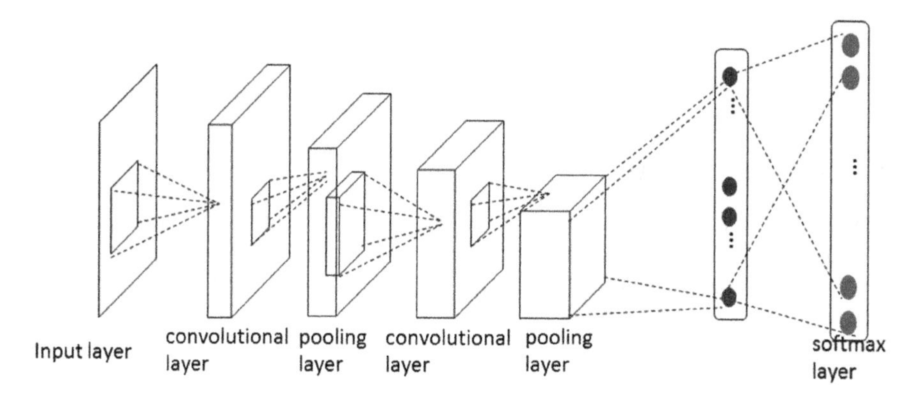

Fig. 1. Structure of typical CNN.

Both images and texts can be presented as a matrix. The matrix V is the input of a CNN. We suppose that H_i is the i layer of the network $(V = H_0)$.

If H_i is a convolutional layer:

$$H_i = f(H_{i-1} \odot W_i + b_i) \tag{1}$$

In formula (1) W_i is the weight vector of H_i layer. Operator $\odot$ is the convolution operation of the convolution kernel and the H_{i-1} layer. The output of the convolution will be added with the offset b_i. And finally, H_i could be calculated by a nonlinear excitation $f(x)$.

If H_i is a pooling layer:

$$H_i = subsampling(H_{i-1}) \tag{2}$$

The purpose of this layer is to reduce dimensionality and keep the features stable to some extent.

What we want is a probability distribution Y. With several convolution layers and pooling layers, we get a model:

$$Y(i) = P(L = l_i | H_0; (W, b)) \tag{3}$$

Gradient descent algorithm is used to train the parameters (W and b) of each layer.

$$E(W, b) = L(W, b) + \frac{\lambda}{2} W^T W \tag{4}$$

$$W_i = W_i - \eta \frac{\partial E(W, b)}{\partial W_i} \tag{5}$$

$$b_i = b_i - \eta \frac{\partial E(W, b)}{\partial b_i} \tag{6}$$

λ is the parameter to control the overfitting and η is the learning rate.

2.2 LSTM and BLSTM

LSTM [8] is a kind of recurrent neural networks (RNN). Because of its ability to make use of long term information, it is suitable for sequential data processing in texts. LSTM unit is shown in Fig. 2. And BLSTM [10] is included in LSTM, which is bidirectional.

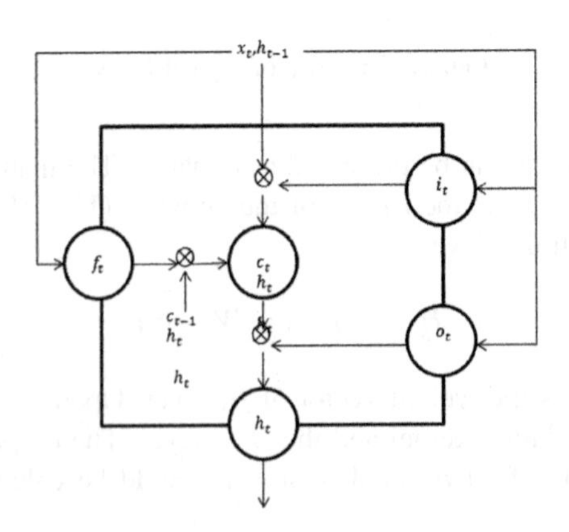

Fig. 2. LSTM memory block.

A LSTM memory block unit contains three gates, input gate, forget gate and output gate. These gates facilitate saving, reading, resetting and updating the long term information. We supposed x as the input, c as the memory cell and the h as the output. We calculate the candidate cell $\tilde{c}_t$, W_{xc} and W_{hc} are weights of input and the output from previous time.

$$H_i = f(H_{i-1} \odot W_i + b_i) \tag{7}$$

Input gate value i_t is used to control the impact of memory cell. Forget gate value f_t is used to control the impact from history information to current memory. And output gate value is used to control the output of the unit.

$$i_t = \sigma(W_{xi}x_t + W_{hi}h_{t-1} + W_{ci}c_{t-1} + b_i) \tag{8}$$

$$f_t = \sigma\left(W_{xf}x_t + W_{hf}h_{t-1} + W_{cf}c_{t-1} + b_f\right) \tag{9}$$

$$o_t = \sigma(W_{xo}x_t + W_{ho}h_{t-1} + W_{co}c_{t-1} + b_o) \tag{10}$$

x_t is the current input value. h_{t-1} is the previous output value. And c_{t-1} is the previous memory cell. σ is an activation function that we usually choose logistic sigmoid algorithm for it.

$$c_t = f_t \otimes c_{t-1} + i_t \otimes \tilde{c}_t \tag{11}$$

$$h_t = o_t \otimes \tanh(c_t) \tag{12}$$

3 CNN-BLSTM Model

3.1 Word Embedding

Word embedding is a significant technology in NLP when referring to deep learning. It turns each word into a feature embedding containing semantic information. In this study, word embedding was calculated with GloVe algorithm (Global Vector for word Representation) which is represented by Stanford in 2014 [11].

It's the loss function of GloVe:

$$J(\theta) = \frac{1}{2}\sum\nolimits_{i,j=1}^{W} f\left(P_{ij}\right)\left(u_i^T v_j - logP_{ij}\right)^2 \tag{13}$$

In this formula, $f(x)$ is a truncation function to reduce the disturbance of frequently used words.

3.2 CNN-BLSTM Model

In this paper, our model combined one-layer CNN and BLSTM. Each word in the texts has been processed to be a 50-dimensional vector in the preprocessing of the model. The embedding processing is shown in Fig. 3.

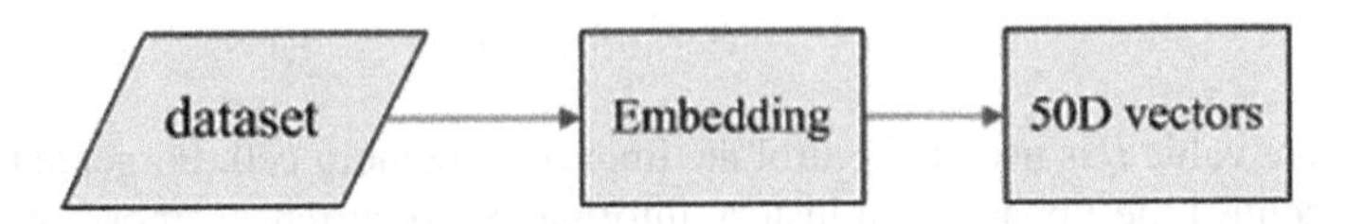

Fig. 3. Word embedding.

The pretrained words in the dataset were projected into word vectors in input part of the our model. A convolution layer, a maxpolling layer, a dropout layer and a BLSTM layer was followed sequentially. The whole structure is shown in Fig. 4.

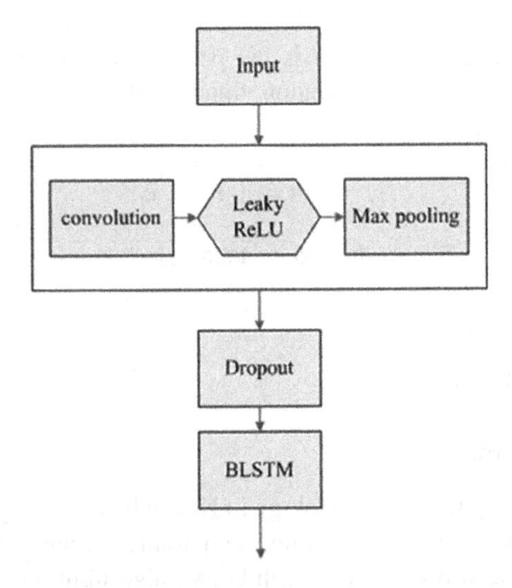

Fig. 4. CNN & BLSTM model structure.

The input is the result from word embedding process. The input layer receive a maximum of 1000 word vectors, each of them has 50 dimensions. We used convolution kernel size of 5 in CNN layer with activation function of LeakyRelU, a max pooling layer followed it with pool size of 4. In order to prevent model overfitting, we used a dropout layer after the last max pooling layer. The last part of the model is a BLSTM layer and two dense in which the last dense used softmax as activation and output 2 dimension vector refer to probabilities of the positive and negative classes.

4 Experiments and Results

4.1 Dataset

We used the dataset about the sentiment classifications of movie reviews in IMDB. The labeled dataset consists of 50000 IMDB movie reviews. They only have two types of reviews, positive or negative. The whole dataset was divided into two parts. 40000 labeled reviews were training set and the other 10000 were test set.

4.2 Experiments

4.2.1 Experiment 1

We compared simple BLSTM and CNN-BLSTM model for sentiment classifications. BLSTM model used a BLSTM layer with 128 unit outputs. CNN-BLSTM model connected a CNN layer with the size of 5-unit kernel, a max pooling layer with pool size 4 and an BLSTM layer with 128-unit output. The maximum feature numbers are 20000, maximum lengths are 1000 and the loss function is categorical cross entropy in the both models. The results are shown in Table 1.

Table 1. Comparison of BLSTM and CNN-LSTM.

	BLSTM	CNN-BLSTM
Accuracy	82.5%	85.3%

The accuracy of each single model is respectively 82.5% and 85.3%. CNN-BILSTM performed better than BLSTM model. The local semantic information in a sentence extracted by the convolution layer could lead to the better performance of CNN-BLSTM.

4.2.2 Experiment 2

We compared six different models. Three of them are models with respectively 1, 2 or 3 CNN layers with a BLSTM layer and a full connected layer with pre-trained word embedding (using GloVe, dimension: 50), and the others are models without pre-trained word embedding. In every model, each CNN layer used 64 filters with kernel size 5, each CNN layer was followed by a max-pooling layer with pool size 4, the last pool layer was followed by a dropout layer and a BLSTM layer with 128 unit outputs, the last layers were two full connected layers with respectively 128 unit outputs/activation function of LeakyReLU and 2-unit outputs/activation function of softmax. Every model was trained with 10 epochs, the training batch size is 128 with loss of categorical cross entropy and RMSProp optimizer. The comparison results are shown in Table 2.

The results showed that the 1CNN-BLSTM model with pretrained word embedding got the best accuracy 89.7%, and the 3CNN-BLSTM model with pretrained word embedding got the lowest accuracy 77.7%.

Table 2. Comparison of 6 models.

	1CNN-BLSTM	2CNN-BLSTM	3CNN-BLSTM	Average
With pretrained word embedding	89.7%	86.2%	77.7%	84.5%
Without pretrained word embedding	85.3%	81.6%	79.1%	82.0%
Accuracy	87.9%	83.9%	78.4%	

At first we could compare the models with pretrained word embedding and models without those, the result showed that average accuracy of the former is 84.5% while average accuracy of the latter is 82.0%. It's obviously that the pretrain word embedding enhanced the performance of CNN-BLSTM, which could be caused by the extra semantic information brought by word embedding.

We could also compared models with different numbers of CNN layer. It was kind of weird that the models with more CNN layer got worse performance. This consequence might be on the contrary of the cognition of us that the model with more complex structure should perform better. But it's known that IMDB datasets only have 50000 samples (40000 for training and 10000 for testing). The consequence could be the results of under-fitting of the complex model.

5 Conclusion and Future Work

In this work, we described a series of experiments with CNN and BLSTM and obtained a satisfactory accuracy that could reach up to 89.7%. The whole process contained two steps. The first step was the data pre-processing and each word was turned into a 50D word vector. The result was obviously better if we did word embedding before the data was put into our model directly.

The second step was to accomplish the sentiment analysis with CNN and BLSTM. In this step, we changed the numbers of layers of convolution and pooling in CNN to find a simple CNN with one layer of convolution and one layer of pooling which performed best by comparison. The accuracy of two-layer CNN which contained two convolution layers and two pooling layers was 86.2% and the accuracy of three-layer CNN which contained three convolution layers and three pooling layers was 77.7%.

In the model, the dropout function was also added to make the result reliable because of its reduction of over fitting. We also investigated whether the additional pre-processing of the data could give the result an advantage like adding POS (part of speech) tagging to the experiment, because not all of the words make sense. In general, adjectives and adverbs describe the feelings of the authors and therefore in the future work, we will devote to exploring more significant methods to do pre-processing of the natural language data.

Acknowledgment. This work was supported by the Grant from the National Natural Science Foundation of China (61173116).

References

1. Hatzivassiloglou, V., McKeown, K.R.: Predicting the semantic orientation of adjectives. In: Proceedings of the Eighth Conference on European Chapter of the Association for Computational Linguistics, pp. 174–181. Association for Computational Linguistics (1997)
2. Turney, P.D.: Thumbs up or down semantic orientation applied to unsupervised of reviews. In: Proceedings of the 40th Annual Meeting of the Association for Computation Linguistics, pp. 417–424. ACL, Somerset (2002)
3. Kim, S.M., Hovy, E.: Automatic detection of opinion bearing words and sentences. In: Proceedings of the Companion Volume to the International Joint Conference on Natural Language Processing (IJCNLP), vol. 8 (2005)
4. Pang, B., Lee, L., Vaithyanathan, S.: Thumbs up?: sentiment classification using machine learning techniques, pp. 79–86 (2002)
5. Bouvrie, J.: Notes on convolutional neural networks (2006)
6. LeCun, Y., Bengio, Y., Hinton, G.: Deep learning. Nature **521**(7553), 436–444 (2015)
7. Gu, J., Wang, Z., Kuen, J., Ma, L., Shahroudy, A., Shuai, B., Wang, G.: Recent advances in convolutional neural networks. arXiv Preprint arXiv:1512.07108 (2015)
8. Hochreiter, S., Schmidhuber, J.: Long short-term memory. Neural Comput. **9**(8), 1735–1780 (1997)
9. Sundermeyer, M., Ney, H., Schlüter, R.: From feedforward to recurrent LSTM neural networks for language modeling. IEEE/ACM Trans. Audio Speech Lang. Process. (TASLP) **23**(3), 517–529 (2015)
10. Schuster, M., Paliwal, K.K.: Bidirectional recurrent neural networks. IEEE Trans. Signal Process. **45**(11), 2673–2681 (1997)
11. Pennington, J., Socher, R., Manning, C.D.: Glove: global vectors for word representation. In: EMNLP 2014, pp. 1532–1543 (2014)

Playlist Recommendation Based on Reinforcement Learning

Binbin Hu(✉), Chuan Shi, and Jian Liu

Beijing Key Lab of Intelligent Telecommunications Software and Multimedia, Beijing University of Posts and Telecommunications, Beijing 100876, China
{hubinbin,shichuan}@bupt.edu.cn, fullback@yeah.net

Abstract. Recently, there is a surge of recommender system to alleviate the Internet information overload. A number of recommendation techniques have been proposed for many applications, among which music recommendation is a kind of popular Internet services. Unlike other recommendation services, music recommendation needs to consider the interaction and content information, as well as the inherent correlation and feedback among music playlist. Thus, in this paper, we model music recommendation as a Markov Decision Process, and consider the music recommendation as a playlist recommendation task. Along this line, we propose a novel reinforcement learning based model, called RLWRec, to exploit the optimal strategy of playlist. Two novel strategies are designed to solve the curse of state space and efficient music recommendation. Experiments on real dataset validate the effectiveness of our proposed method.

Keywords: Music recommendation · Reinforcement learning · Markov decision process

1 Introduction

With the rapid development of information technology and Internet, human society has gradually entered the era of information overload, where users have more and more difficulties in accessing information in Interest. Thus recommender systems [7] arise at the historic moment, which utilize user-item interaction and/or content information associated with users and items [11] to help users to predict preference and make reasonable recommendation.

Recommender systems have been widely used in various applications [2,5,10]. Among them, the music recommendation is a very popular web service, which recommends songs from huge corpus by matching songs with user preference [15]. Existing recommendation techniques, such as collaborative filtering, matrix factorization and so on [3,9,12,15,17], typically utilize ratings of users on items (may also include some additional information, e.g., social relations and geography) to infer user preference and make proper recommendation. In these methods, there are weak ties among adjacent items. Moreover, there are rare feedbacks

Z. Shi et al. (Eds.): ICIS 2017, IFIP AICT 510, pp. 172–182, 2017.
DOI: 10.1007/978-3-319-68121-4_18

of users on items, and these feedbacks have little swift effects on subsequent items. We think these methods may not be suitable for music recommendation, since there are some specific characteristics in the process of listening to music. We know that people listen to music that reflect their feelings. And the music playlist has the inherent and consistent feelings. In addition, there are many meaningful feedbacks of users on songs, such as "skip", "listen", "download" and "collect". These feedbacks reflect different preferences of users on songs.

In order to consider the song correlation, advanced methods introduce Markov decision process [6] and reinforcement learning [13], and regard the recommendation task as decision problem. Chi et al. [1] models the automatic playlist generation problem as Markov decision process and learns the user preference with reinforcement learning. Furthermore, considering the influence between songs, Liebman et al. [4] relate learning individual preferences with holistic playlist generation and propose a novel reinforcement-learning framework DJ-MC to recommend song sequences. Moreover, Wand et al. present a reinforcement learning framework based on Bayesian model to balance the needs to explore user preferences and to exploit this information for recommendation [15].

In this paper, similarly, we model the recommender system as Markov decision process, and employ reinforcement learning to solve it. However, our research differs from state-of-the-art models in several ways: (1) We integrate the user feedback into model as well as considering the influence between songs. (2) We propose a state compression method to capture enormous state space of MDP. Meanwhile, we design a recommendation strategy to make a trade-off between accuracy and coverage of recommendation. With these two designs, we propose the RLWRec model based on Q-learning with ϵ-greedy strategy. The experiments on real music datasets show the effectiveness of our model. Meanwhile, we analyze the influence of user listening frequency and the window size on our model.

2 Preliminary

In this section, we describe notations used in the paper and present some preliminary knowledge.

Markov decision process is a expansion of Markov chain, whose difference is the actions and rewards to join. In general, MDP is defined as a quintuple form $M = <S, A, T, R, \gamma>$, where S is the set of states, A is the set of actions, $T : S \times S \times A \rightarrow \mathbb{R}$ depicts the function of state transition, representing that state s will transfer to state s' in probability $Pr(s'|s, a)$ when performing action a, R: $S \times S \times A \rightarrow \mathbb{R}$ is the reward function, representing that the rewards of performing action a in state s and reaching state s' is $R(s', s, a)$. $\gamma(0 \leq \gamma \leq 1)$ depicts the attenuation of rewards, the smaller γ means the more importance of immediate rewards. The key of MDP is to find a optimal strategy $\pi : S \rightarrow A$, representing a mapping of states to actions, i.e. the optimal action $a = \pi(s)$ in each state.

Reinforcement learning is the name of a set of methods and algorithms for controlling agents to automatically improve their performance by tying to maximize the rewards received from environment. After modeling the real problem as MDP, we can apply reinforcement learning technology to estimate optimal strategy. Q-learning [16] and SARSA [8] are the most widely employed in actual problems, which are both primarily concerned with estimating the value of performing any action in each state and dynamically update the learned strategy.

3 Reinforcement Learning Based Playlist Recommendation Model

In this section, we describe our recommendation problem and model the recommendation task as MDP, and then propose the *R*einforcement *L*earning with *W*indow for *Rec*ommendation (called RLWRec).

3.1 Problem Description

Listening music on the music platform is actually an interactive process: the music platform recommends a song for us, and we can choose a series of actions as feedback for the music platform, such as skip, listen, download, collect and so on. As a consequence, we can obtain the sequence of interaction between music platform and users, which can be considered as a MDP. Therefore, the recommendation task can be summarized as follows: based on a music corpus and the interactive records, we train the recommender system to recommend songs which will follow user preferences.

3.2 Recommendation Framework Based on Reinforcement Learning

In this section, we model the recommendation task as MDP. Concretely, we will model states, actions and reward function.

Modeling States and Actions. In order to learn user preferences from interactive process, we have to model user's states to reflect his or her listening history. Inspired by the N-Gram model, which predicts the next word while knowing last N words, we model user's states as sliding window of size K, which preserves the K songs recently listened. And we approximatively regard the state as user's listening history on the music platform. Thus, we concretely define the music corpus $M = \{m_1, m_2, \ldots, m_n\}$, and then the state space of our model can be represented as $S = \{(m_i, m_{i+1}, \ldots, m_{i+K-1}) | \forall j \in [i, i+K-1], m_j \in M\}$. With the definition of states, we regard each song as an action. Therefore, each time our model executes an action a (recommending a song) once, user's state will transfer, the sliding window of state s will slide a step backward, and take the song recommended into window, thus reach a new state s'. For an example, we assume user's current state $s = (m_1, m_2, \ldots, m_K)$, the recommender system recommends a song m_{K+1} for the user according to the past strategy, then the user's state will transfer to $s' = (m_2, m_3, \ldots, m_{K+1})$. Thus, the action space of our model can be represented as $A = M$.

Modeling Reward Function. The basis of reinforcement learning lies in the rewards the agent receives, and how it updates state and action values [14]. As for our recommender system, users will express their feelings via a series of implicit or explicit feedback when a song is recommended. Hence, we need to construct a reward function to integrate user's various feedback and measure the achievement to the recommendation goal. In this paper, in consideration of real dataset, we only extract three feedback information: listen, collect and download. Thus, we design a reward function as follows:

$$R(f) = \sum_{i=1}^{3} w_i * I(f == F_i), \tag{1}$$
$$F = \{listen, collect, download\},$$

where F is a feedback set, I is boolean function, w_i is weight, mapping user's feedback to discrete numerical space.

With the definition of the elements of MDP, we can summarize our model as Fig. 1.

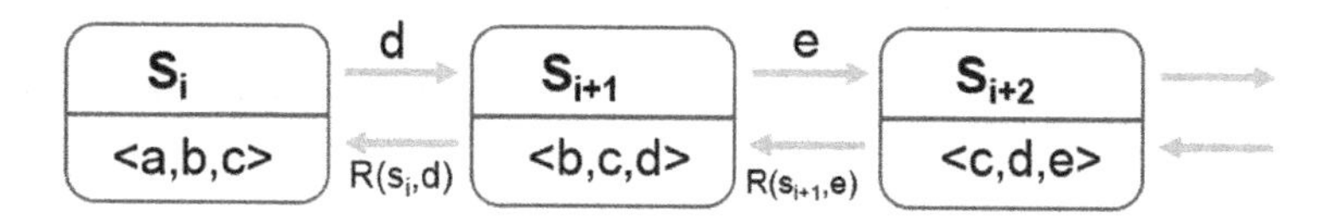

Fig. 1. Model recommender system as MDP.

3.3 Model Challenges

Although we have modeled the recommendation task as MDP, how to solve is not a trivial problem. We will still face the following challenges.

State Space Challenge. Our model preserve user's latest K songs with sliding window of size K, which approximates to user's listening history. However, the model will face a serious problem – the curse of state space. Each state of MDP is composed of K songs, so that our model's state space is $|M|^K$ ($|M|$ represents the size of song set). In terms of time and space complexity, the exponential state space cannot be accepted. Hence, we present a state compression algorithm to transform individual songs to song clusters for further model learning.

Recommendation Challenge. We apply reinforcement learning algorithm in MDP problem, which will estimate a series of strategy for recommender system. Thus, it is necessary to select a proper strategy for a recommendation. In addition, picking a song from a song cluster is another important issue we have to consider. Therefore, we design a novel recommendation strategy based on tree structure to improve the recommendation performance.

4 Model Learning

4.1 State Compression Based on Collaborative Filter

In order to reduce the dimensionality of state space for efficient training and predicting, we propose a state compression algorithm based on collaborative filter. The basic idea is that it clusters songs according to the similar user's performance, and then replaces songs to song clusters in the model learning process.

User Clustering. Our model is mainly used for personalized recommendation of single user. Hence, we extract a user's listening logs from dataset for model training, we name this user as su. We construct each user's feature vector, consisting of user's feedback (listen, download and collect) for all songs in dataset, which reflects user's music preference. Then, we measure similarity between su and each user in dataset. After similarity measure, we extract top k similar users, whose feedback will be feature for song clustering later.

$$sim(u) = \frac{su \cdot u}{|su| * |u|}, \tag{2}$$

where u is arbitrary user except su, $sim(u)$ depicts music preference similarity between u and su.

Song Clustering. In the song clustering process, we transform songs to feature vector of $k + k'$ dimensions consisting of user's feedback feature and song's own feature. User's feedback feature has k dimensions, which are top k similar users' performance (0: never listened, 1: listened, 2: collected, 3: downloaded) for the song. Song's own feature has k' dimensions, including singer, release time, popularity, language, gender and so on. Consequently, we use k-means for clustering, resulting in N song clusters.

RLWRec adopts state compression based on collaborative filter, consisting of user clustering and song clustering, transforms $|M|$ songs to N song clusters ($N << |M|$). Thus, song clusters can replace songs to construct state space in modeling and training, which tend to immensely reduce the size of state space.

4.2 Learning Algorithm

We choose Q-learning as our learning algorithm. This method, combining dynamic programming and Monte Carlo's idea, is an efficient model-free reinforcement learning algorithm. Moreover, we bring ϵ-greedy strategy in learning process, which can make a trade-off between exploration and exploitation.

***Q*-learning.** Q-learning is primarily concerned with estimating an evaluation of performing specific actions in each state, known as Q-values, which not only consider action's immediate rewards but also takes accumulative rewards into account. We denote the Q-value of performing specific action a in state s as $Q(s, a)$. And we design the Q-value update rule as follows:

$$Q_n(s, a) = (1 - \alpha_n)Q_{n-1}(s, a) + \alpha_n[R(s, a) + \gamma \max_{a'} Q_{n-1}(s', a')] \tag{3}$$

with

$$\alpha_n = \frac{1}{1 + VisitCount(s, a)}, \tag{4}$$

where $R(s, a)$ represents the immediate reward of performing action a in state s, s', a' are next state and action of s, γ controls the attenuation of accumulative rewards. Moreover, we bring in α_n, related to the count of performing same action in same state. This rule can partly reveal user's music preference and accelerate the coverage with the decreasing value of α_n.

ϵ-greedy Strategy. The training of reinforcement learning is a process of trying. So when we choose an action in a state, we should consider two aspects: (1) knowing the rewards of each action. (2) choosing the action of most reward recently. Thus, reinforcement learning has two strategies: exploration strategy (trying as many actions as possible) and exploitation strategy (choosing the best action given current information). Obviously, the two strategies are conflicting. Hence, we take advantage of ϵ-greedy strategy, which is an eclectic collection of above two strategies. ϵ-greedy strategy guarantees the trade-off between exploration and exploitation based on a probability ϵ: exploring in the probability of ϵ and exploiting in the probability of $1 - \epsilon$.

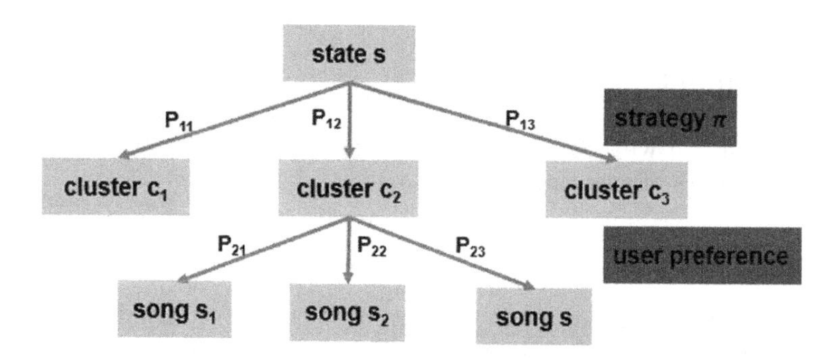

Fig. 2. Recommendation strategy based on tree structure.

4.3 Recommendation Strategy

As mentioned above, Q-learning can evaluate the true rewards of performing each action in each state, knowing as Q-value. Thus, we can get the list of Q-value of actions in each state. Thus, we need design a recommendation strategy to choose a song cluster and then recommend a song for the specific user. First of all, for the specific user su, we let him set a score for each song in the dataset, representing the preference of song of su, denoted as $preference(m)$. We give the calculation rule as follows:

$$preference(m) = \sum_{f \in F} count_f * w_f, \quad (5)$$
$$F = \{listen, collect, download\},$$

where $count_f$ represents the count of user performing feedback f for song m, w_f is the preference weight of user performing feedback f for song m. Meanwhile, in the process of recommendation, we also can update user's preference with the timely feedback, as follows:

$$preference_n(m) = preference_{n-1}(m) + w_f. \quad (6)$$

Based on user's preference score, our model can recommend for user by serval strategies. One naive idea is recommending by the maximal Q-value and preference score. However this strategy will seriously bring down the coverage of recommendation. Moreover, we can combine this strategy with ϵ-greedy strategy to improve the coverage of recommendation. In order to get better performance further, we design a recommendation strategy based on tree structure: recommending song cluster according to the probability of Q-value and recommending song according to the probability of preference score, as shown in Fig. 2. For example, assume in the state s, the recommender system has the probability of P_{12} to choose the song cluster c_2 and further recommends the song s_2 with the probability of P_{22}. Recommending based on probability can not only guarantee accuracy of recommendation but also improve the coverage of recommendation.

5 Experiment

In this section, we will validate the effectiveness and traits of our model by conducting a series of experiments.

5.1 Dataset

In this paper, we use a real music dataset for experiment, including song data and user interaction record of six months. This music dataset consists of 349949 users, 10842 songs and 5652232 feedback (listen, collect, download). Meanwhile, this music dataset includes song's essential features, such as artist, publish time, initial popularity and so on. Based on the raw data, we cut the dataset into three smaller dataset according to user's listening frequency. These three datasets are as follows: low frequency dataset (the frequency of songs listened by users is smaller than 300), medium frequency dataset (between 400 and 700), and high frequency dataset (larger than 800). And in these datasets, we filter out some special users (listen too many or too few) and guarantee the number of users at around 600 and the number of songs at around 10000.

5.2 Comparison Methods and Metrics

Because there are few methods integrating user feedback for music recommendation, and advanced methods based on reinforcement learning [1,4,10] don't use quantitative indicators to measure the recommendation performance, we design the following methods as baselines.

- **RandRec.** Randomly choose songs from user's listening history and recommend for user.
- **CFRec.** Recommend songs for user according to other users with similar music preference.
- **PopRec.** Calculate song's popularity and randomly recommend the top$k$ popular songs for user.

In the experiment, we set the sliding window size $k = 3$, the attenuation factor and the exploration probability in Q-learning $\gamma = 0.8$ and $\epsilon = 0.7$. Meanwhile, optimal parameters are set for other algorithms.

In this paper, we calculate the accuracy and coverage of recommendation. Similar with the traditional F_1 score, we combine the accuracy and coverage and use the *Score* to evaluate the performance of list prediction.

$$Accuracy = \frac{|\{s \in L_p \quad and \quad s \in L_t\}|}{|L_p|}, \tag{7}$$

$$Coverage = \frac{set(L_p) \cap set(L_t)}{|L_t|}, \tag{8}$$

$$Score_{list} = \frac{2 * Accuracy * Coverage}{Accuracy + Coverage}, \tag{9}$$

where L_p is the song list predicted by models, and L_t is the actual listening list. A lager *Score* means a better performance.

5.3 Effectiveness Experiments

In order to validate the effectiveness of our model, we will compare RLWRec to other baselines. For each dataset, we use 90% of data as training and the rest of the dataset for testing. Moreover, the random selection is carried out 10 times independently and the average results are illustrated in Fig. 3.

It is clear that our RLWRec model achieves significant performance improvements compared to other algorithms. It indicates the benefits of recommendation with reinforcement learning. It also shows that RLWRec is more stable with the growth of song list length. Moreover, as the increment of song list length, the performance of RLWRec will gradually converge and tend to be steady, which reveals that the stabilization of users performance of music in a period of time.

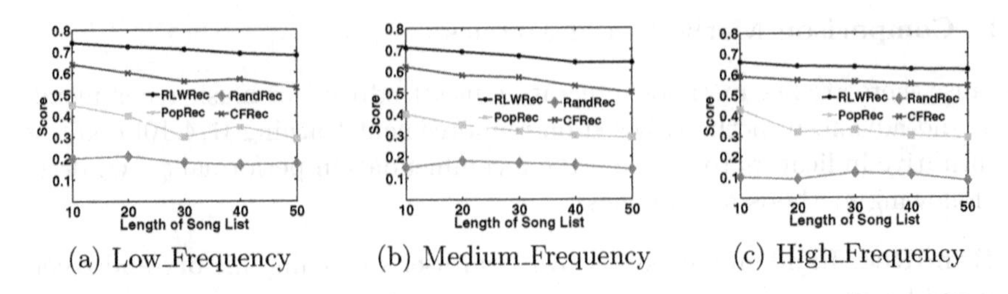

(a) Low_Frequency (b) Medium_Frequency (c) High_Frequency

Fig. 3. Performance comparisons on three datasets

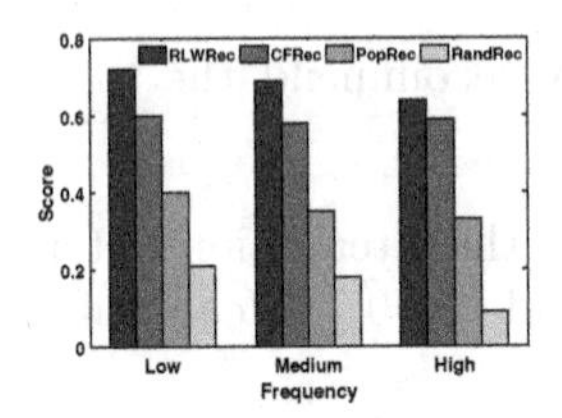

Fig. 4. Performance on different frequency

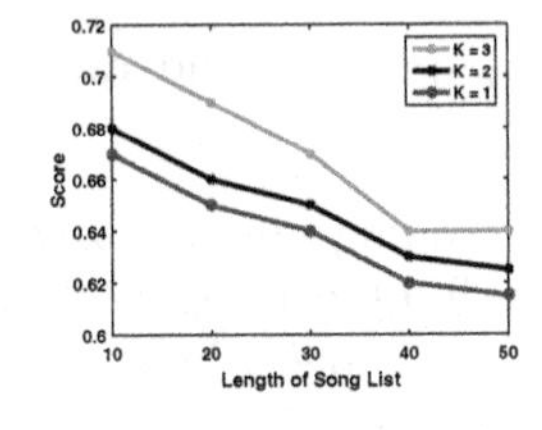

Fig. 5. Performance with different window sizes

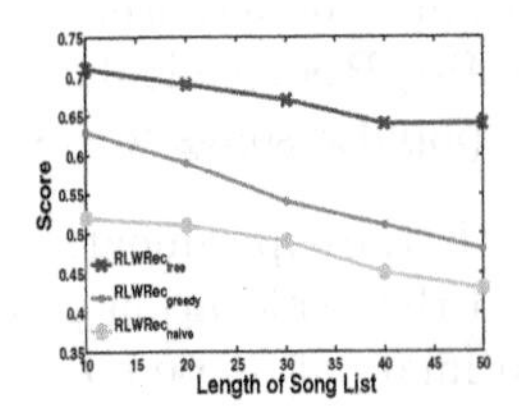

Fig. 6. Performance with different recommendation strategies

5.4 Influence of User's Listening Frequency

In the above experiments, comparing different frequent datasets, we can find that most methods are affected by the listening frequency. In this section, we further study the influence of user's listening frequency to the prediction performance. We do the same experiments, and compare the performances of predicting the song list of length 20.

As illustrated in Fig. 4, it is clear that most of models will be influenced by user's listening frequency. The performance of RLWRec will slightly decrease as the growth of user's listening frequency. It can be inferred that our model's training state space will increase as the growth of user's listening frequency, which leads to the difficulty of choosing the right action in a specific state. As for the recommendation methods base on other users' behaviors, such as CFRec and PopRec, the influence is more slight.

5.5 Influence of the Window Size

RLWRec, based on reinforcement learning, models the recommendation task as a MDP, where we use the sliding window of size K to preserve the K songs user recently listened as user's current listening history. In this section we design a experiment to analyze the influence of the window size K. Due to the restriction of computing ability and memory space, we set K as 1, 2, 3 respectively for experiments and other parameters remain unchanged. The result is shown in Fig. 5.

The result shows that the siding window size can affect our model to some extent. It is obvious that the best result is achieved when the size of window is 3. The reason lies in the window of smaller size (i.e. less listening history) will not hold enough information in memory to make better recommendation. Meanwhile, a large window tend to provide our model a longer memory and achieve a significant performance. However, it will cause a larger state space and need more computing time.

5.6 Influence of Different Recommendation Strategies

In the Sect. 5.3, we show three recommendation strategies in our model: native recommendation strategy, recommendation with ϵ-greedy strategy and our recommendation strategy based on tree structure. In this section, we do experiments in the medium frequency dataset to show the influence of the three recommendation strategies. The result is shown in the Fig. 6.

It is clear that our recommendation strategy based on tree structure well outperforms other recommendation strategies. It indicates that our recommendation strategy can balance the accuracy and coverage of recommendation and improve the performance of the recommender system. Moreover, it also shows that the performance of our recommendation is more stable with the growth of the length of song list. Because the native recommendation strategy will restrict the coverage of the recommender system and the ϵ-greedy strategy will bring in the uncertainty and decrease the accuracy of the recommender system.

6 Conclusion

In this paper, we integrate the user feedback and influence between songs in recommender system and model it as Markov decision process. Then we design a novel reinforcement learning framework RLWRec to generate music playlist. Moreover, we propose the state compression method based on collaborative filter for the dimensionality reduction and design a recommendation strategy based on tree structure to improve the accuracy and coverage of recommendation. Experiments on real datasets verifies the effectiveness of our model.

References

1. Chi, C.Y., Tsai, R.T.H., Lai, J.Y., Hsu, J.Y.J.: A reinforcement learning approach to emotion-based automatic playlist generation. In: 2010 International Conference on Technologies and Applications of Artificial Intelligence, pp. 60–65. IEEE (2010)
2. Konstan, J.A., Riedl, J., Borchers, A., Herlocker, J.L.: Recommender systems: a grouplens perspective. In: Recommender Systems: Papers from the 1998 Workshop (AAAI Technical Report WS-98-08), pp. 60–64 (1998)
3. Koren, Y., Bell, R., Volinsky, C., et al.: Matrix factorization techniques for recommender systems. Computer **42**(8), 30–37 (2009)

4. Liebman, E., Saar-Tsechansky, M., Stone, P.: DJ-MC: a reinforcement-learning agent for music playlist recommendation. In: Proceedings of the 2015 International Conference on Autonomous Agents and Multiagent Systems, pp. 591–599. International Foundation for Autonomous Agents and Multiagent Systems (2015)
5. Mobasher, B., Dai, H., Luo, T., Sun, Y., Zhu, J.: Integrating web usage and content mining for more effective personalization. In: Bauknecht, K., Madria, S.K., Pernul, G. (eds.) EC-Web 2000. LNCS, vol. 1875, pp. 165–176. Springer, Heidelberg (2000). doi:10.1007/3-540-44463-7_15
6. Puterman, M.L.: Markov Decision Processes: Discrete Stochastic Dynamic Programming. Wiley, Hoboken (2014)
7. Resnick, P., Varian, H.R.: Recommender systems. Commun. ACM **40**(3), 56–58 (1997)
8. Rummery, G.A., Niranjan, M.: On-line Q-learning using connectionist systems. Department of Engineering, University of Cambridge (1994)
9. Schedl, M., Schnitzer, D.: Location-aware music artist recommendation. In: Gurrin, C., Hopfgartner, F., Hurst, W., Johansen, H., Lee, H., O'Connor, N. (eds.) MMM 2014. LNCS, vol. 8326, pp. 205–213. Springer, Cham (2014). doi:10.1007/978-3-319-04117-9_19
10. Shani, G., Heckerman, D., Brafman, R.I.: An MDP-based recommender system. J. Mach. Learn. Res. **6**(Sep), 1265–1295 (2005)
11. Shi, C., Liu, J., Zhuang, F., Philip, S.Y., Wu, B.: Integrating heterogeneous information via flexible regularization framework for recommendation. Knowl. Inf. Syst. 1–25 (2016)
12. Song, Y., Dixon, S., Pearce, M.: A survey of music recommendation systems and future perspectives. In: 9th International Symposium on Computer Music Modeling and Retrieval (2012)
13. Sutton, R.S., Barto, A.G.: Reinforcement Learning: An Introduction, vol. 1. MIT Press, Cambridge (1998)
14. Taghipour, N., Kardan, A., Ghidary, S.S.: Usage-based web recommendations: a reinforcement learning approach. In: Proceedings of the 2007 ACM conference on Recommender systems, pp. 113–120. ACM (2007)
15. Wang, X., Wang, Y., Hsu, D., Wang, Y.: Exploration in interactive personalized music recommendation: a reinforcement learning approach. ACM Trans. Multimed. Comput. Commun. Appl. (TOMM) **11**(1), 7 (2014)
16. Watkins, C.J.C.H.: Learning from delayed rewards. Ph.D. thesis, University of Cambridge England (1989)
17. Yoshii, K., Goto, M., Komatani, K., Ogata, T., Okuno, H.G.: Hybrid collaborative and content-based music recommendation using probabilistic model with latent user preferences. In: ISMIR, vol. 6, p. 7th (2006)

Transfer Learning for Music Genre Classification

Guangxiao Song, Zhijie Wang(✉), Fang Han(✉), and Shenyi Ding

College of Information Science and Technology, Donghua University, Shanghai 201620, China
wangzj@dhu.edu.cn, yadiahan@163.com

Abstract. Modern music information retrieval system provides high-level features (genre, instrument, mood and so on) for searching and recommending conveniently. Among these music tags, genre is the most widely used in practice. Machine learning technique has the ability of cataloguing different genres from raw music. A disadvantage of it is that the final performance heavily depends on the used features. As a powerful learning algorithm, deep neural network can extract useful features automatically and effectively instead of time-consuming feature engineering. But deeper architecture means larger data are needed to train the neural network. In many cases, we may not have enough data to train a deep network. Transfer learning solves the problem by pre-training the network in a similar task which has enough data, then fine-tuning the parameters of the pre-trained network using the target dataset. Magnatagatune dataset is used for pre-training the proposed five-layer Recurrent Neural Network (RNN) with Gated Recurrent Unit (GRU). And in order to reduce the input of the network, scattering transform is used in this paper. Then GTZAN dataset is used as the target dataset of genre classification. Experimental results show the transfer learning way can achieve a higher average classification accuracy (95.8%) than the same deep RNN which initials the parameters randomly (93.5%). In addition, the deep RNN using transfer learning converges to the final accuracy faster than using random initialization.

Keywords: Music genre classification · Transfer learning · Deep learning

1 Introduction

Music genre is important to many applications, such as music recommender system and information retrieval. Automatic genre classification system has been developed using machine learning technique recent years. Most of these systems have the ability of cataloguing different music genres from raw music contents [1–3].

Mel-frequency cepstral coefficient (MFCC) and Mel-spectrogram are widely used in genre classification task. Because they can extract variant features from raw data for the learning process. But the performance of genre classification

Z. Shi et al. (Eds.): ICIS 2017, IFIP AICT 510, pp. 183–190, 2017.
DOI: 10.1007/978-3-319-68121-4_19

benefits from features over long-time scale (>500 ms) while MFCC is efficient around time scale of 25 ms, and enlarging the time scale leads the information loss when using mel-spectrogram [4,5]. Differently, scattering transform can recover the information loss by wavelet decompositions, meanwhile, extract long-time scale features by lowpass filters [6,7].

Deep learning makes massive of success in different areas, for instance, computer vision [8–10], speech recognition [11,12], and natural language processing [13,14]. These algorithms can extract high-level features automatically layer by layer, different from traditional machine learning classifiers, such as Support Vector Machine (SVM), Nearest Neighbors, and Decision Trees, which are heavily dependent on the result of feature extraction. Among its several typical models, Recurrent Neural Network (RNN) is widely used for sequential data. And RNN is good at learning the relationship through time [15]. But in purpose of achieving good performance, deep neural network needs large amount of data. In condition of the target dataset need to be classified is not enough, we can use a large data, which is the same or similar to the target dataset, to pre-train the deep neural network, then replace the connections to classifier according to the target classification number and fine-tune the parameters of the pre-trained network. This process is called transfer learning [16]. In this paper, we use Magnatagatune dataset [17] and GTZAN dataset [18] as the large and the target dataset respectively. 5-layer RNN using Gated Recurrent Unit (GRU) [19] and softmax classifier are used. Additionally, for reducing the input of deep RNN, we use scattering transform as its preprocessing.

The results of the experiment show that the proposed 5-layer RNN reaches a high accuracy when using transfer learning, and the same architecture using random initialization converges more slowly to a lower accuracy.

2 Transfer Learning Process

The architecture of the proposed method is shown in Fig. 1. The overall process consists of two parts. One part is deep RNN training on a large musical dataset

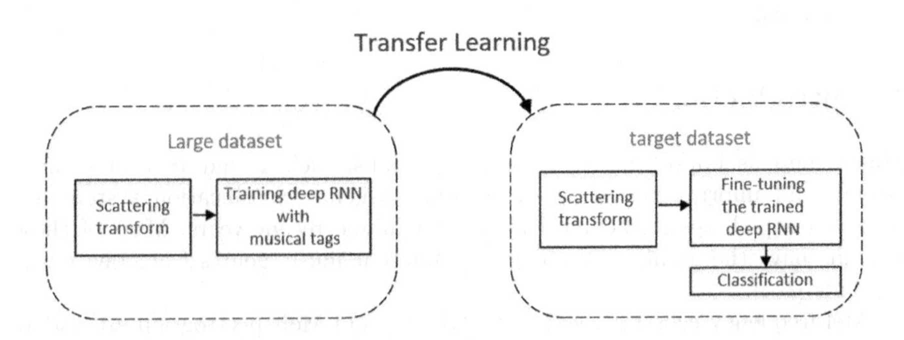

Fig. 1. The architecture of the proposed transfer learning process

(Magnatagatune dataset is used in this paper). The other part is genre classification process after fine-tuning the previous trained deep RNN by target dataset (GTZAN dataset is used in this paper). Specifically, scattering transform is applied at the beginning of each part, in order to reduce the raw music data and to extract features preliminarily for the next process of neural network training. 5-layer RNN with GRU and softmax classifier are trained with tagged music clips as the deep RNN we mentioned. At last, we use the target genre classification dataset (GTZAN) to fine-tune the trained parameters of RNN.

2.1 Scattering Transform

In genre classification task, large time scale (>500 ms) invariant signal representation is important. As widely used methods in audio processing, mel-spectrogram can enlarge the time scale but remove information which is crucial to genre classification. And MFCC is efficient at time scales up to 25 ms. Unlike the previous methods. Scattering transform can provide invariants over large time scales without too much information loss.

For an audio signal x, scattering transform defined as $S_n x$, where n represent the order. $S_0 x = x \star \phi(t)$ has locally invariant property because of the time averaging operation, but it leads to high frequency information loss which can be retrieved by the wavelet modulus coefficients $|x \star \psi_{\lambda_1}(t)|$. To make the wavelet modulus coefficients invariant to translation, a time averaging unit is applied. The first layer of scattering transform defined as:

$$S_1 x(t, \lambda_1) = |x \star \psi_{\lambda_1}| \star \phi(t) \tag{1}$$

Andén [7] indicates that if wavelets filter-bank ψ_{λ_1} have the same frequency resolution as the mel-windows, then $S_1 x$ coefficients can be approximate to the mel-filter-banks coefficients. The difference is that applying a bank of higher frequency wavelet filters ψ_{λ_2} with a modulus to the wavelet modulus coefficients can recover the lost information. The same as previous operation, adding a low-pass filter $\phi(t)$ make the coefficients translation invariant. Then the second layer of scattering transform defined as:

$$S_2 x(t, \lambda_1, \lambda_2) = ||x \star \psi_{\lambda_1}| \star \psi_{\lambda_2}| \star \phi(t) \tag{2}$$

2.2 Deep Recurrent Neural Network

RNNs have an aptitude for handling sequential information, such as speech recognition and NLP. RNN structure can be described as transitions from previous to current states. For classical RNN, this transition is formulized as:

$$h_t = f(W_h[x_t, h_{t-1}] + b_h) \tag{3}$$

In order to solve the problem of vanishing gradients of RNN. Gated structure named LSTM introduced by Hochreiter [15]. The LSTM unit allows that information of more timesteps can be memorized. And the memories are stored by memory cells. Then the LSTM can decide to forget, output, or change the saved memories. As a popular variant of LSTM, GRU is simpler and effective as well. It uses gate Zt and gate Rt to update the hidden state. Theses gates are given by:

$$\begin{aligned} \begin{pmatrix} z_t \\ r_t \end{pmatrix} &= \begin{pmatrix} \sigma(W_z[x_t, h_{t-1}] + b_z) \\ \sigma(W_r[x_t, h_{t-1}] + b_r) \end{pmatrix} \\ g_t &= f(W_g[x_t, r_t * h_{t-1}] + b_g) \\ h_t &= (1 - z_t) * h_{t-1} + z_t * g_t \end{aligned} \tag{4}$$

We use 5-layer GRU neural network which is constructed by stacking each hidden layer on the top of previous layer, in order to improve the ability of representation of our architecture in this paper. Additionally, generalization of the proposed deep RNN is improved by applying dropout between each layer [20].

3 Datasets and Experiment Setup

Magnatagatune and GTZAN dataset are used as the large and target dataset respectively. All the clips are transformed to mono and sampled by 16 kHZ. Magnatagatune has 25863 clips and each clip is annotated with 188 different musical tags such as genre, mood, and instrument. We use the last 2105 clips (distributed in folder 'f') for validation, others for training. We use 512 hidden states in each layer. Dropout is set as 0.7. Learning rate is 0.00001. And we use AUC-ROC score [21] to evaluate the performance of our model to avoid imbalance of the dataset. When the AUC-ROC score is stable, we stop the training and save the model. GTZAN dataset has 1000 clips of 10 genres and each genre contains 100 clips evenly. As the target dataset, it is randomly shuffled and the mean accuracy of 10 times of 10-fold cross validation is used for the final test accuracy. Among the 10 folds in total, we use 1 fold for testing, and the others for training. Each time of 10-fold cross validation, we change the output number of the softmax classifier to 10 (the genre number of GTZAN dataset), then fine-tune the parameters of pre-trained model from Magnatagatune dataset.

4 Experiment Results and Analysis

As shown in Fig. 3, both random initialization and transfer learning models (pre-training process is shown in Fig. 2) of 5-layer RNN with GRU using scattering transform preprocessing converge to quite high accuracy in training. And the models using transfer learning need about 100 epochs to be stable. But the random initialed models need more. This phenomenon not only appears in the three random picked training processes, but also in the unpicked to be shown. It indicates that the transfer learning initials the model better, and improves the speed of convergence.

Comparing with other works of recent years in Table 1, our approach shows a competitive accuracy (95.8%) in genre classification task on GTZAN dataset. Even the model using random initialization can also reach a high accuracy (93.5%) relatively. The combination of scattering transform and deep RNN has been evaluated, and by using this architecture, it performs well in music genre classification.

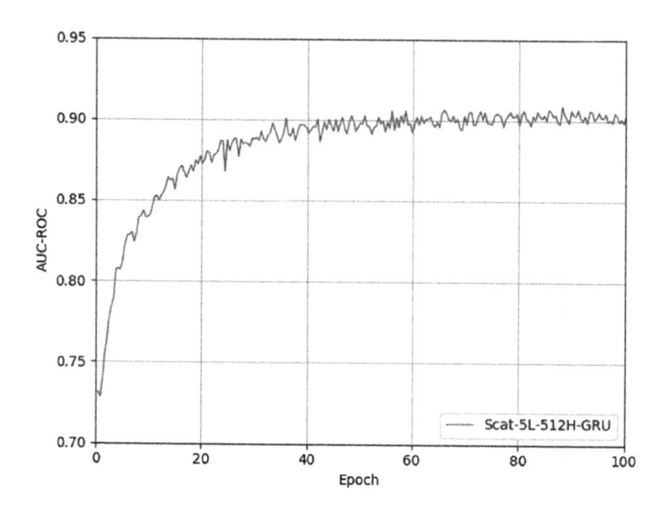

Fig. 2. Validation AUC-ROC score of 5-layer GRU neural network using scattering transformed input

Table 1. Average test accuracy of different models on GTZAN dataset

Model	Average test accuracy
Panagakis and Kotropoulos [22]	92.4%
Andén and Mallat [7]	91.6%
Lee et al. [5]	90.6%
Dai and Liu [23]	93.4%
Random initialization	93.5%
Transfer learning	**95.8%**

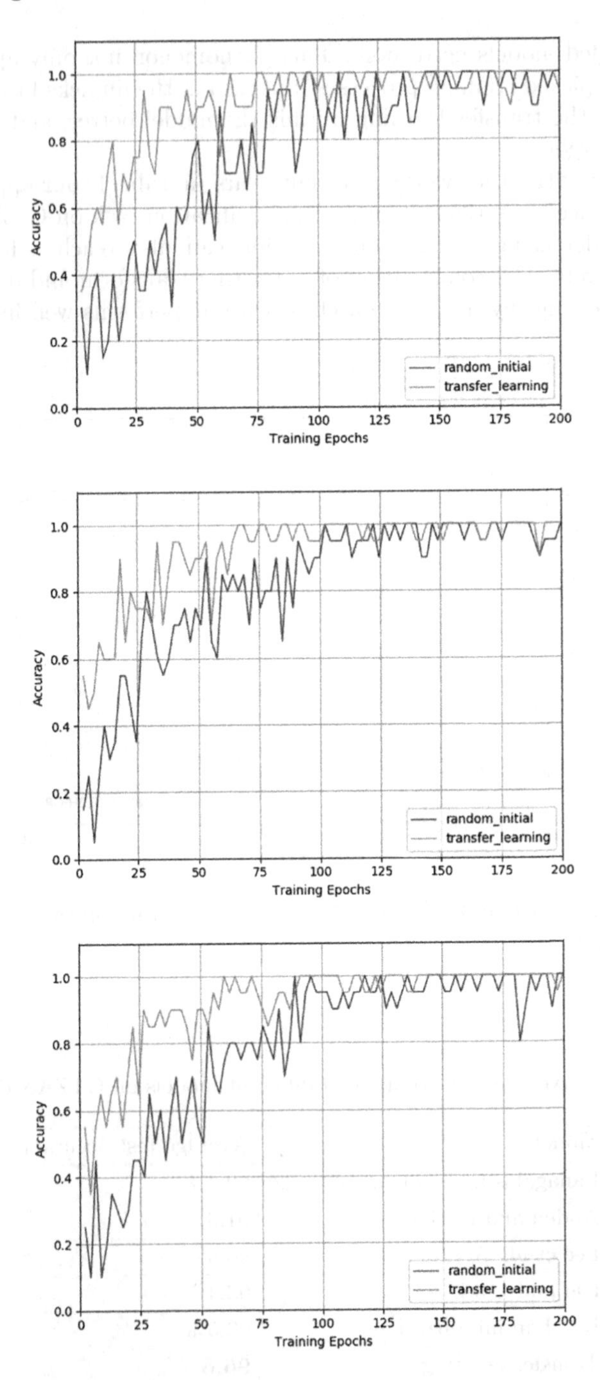

Fig. 3. Three random picked training processes of 10-cross validation, Blue lines represent the RNN using random initialization, and the orange lines represent the RNN using transfer learning. And the accuracy is tested by a random batch of training data (Color figure online)

5 Conclusion

In this paper, we use transfer learning in music genre classification by using 5-layer RNN with GRU and scattering coefficients as its input. When applying the transfer learning from a large music dataset (Magnatagatune is used in this paper), our model shows a faster convergence and higher average accuracy than the same model of random initialization on the target dataset (GTZAN is used in this paper). And the accuracy of transfer learning approach is competitive comparing with the state-of-the-art models as well. The effectiveness of deep RNN combined with scattering transform and transfer learning has been verified in music genre classification task.

Acknowledgements. This work was supported by the National Natural Science Foundation of China (Grants nos. 11572084, 11472061, 71371046), the Fundamental Research Funds for the Central Universities and DHU Distinguished Young Professor Program (No. 16D210404).

References

1. Song, Y., Zhang, C.: Content-based information fusion for semi-supervised music genre classification. IEEE Trans. Multimedia **10**(1), 145–152 (2008). doi:10.1109/tmm.2007.911305
2. Meng, A., Ahrendt, P., Larsen, J., Hansen, L.K.: Temporal feature integration for music genre classification. IEEE Trans. Audio Speech Lang. Process. **15**(5), 1654–1664 (2007). doi:10.1109/tasl.2007.899293
3. Lampropoulos, A.S., Lampropoulou, P.S., Tsihrintzis, G.A.: Music genre classification based on ensemble of signals produced by source separation methods. Intell. Decis. Technol. **4**(3), 229–237 (2010). doi:10.3233/idt-2010-0083
4. McDermott, J.H., Simoncelli, E.P.: Sound texture perception via statistics of the auditory periphery: evidence from sound synthesis. Neuron **71**(5), 926–940 (2011). doi:10.1016/j.neuron.2011.06.032
5. Lee, C.H., Shih, J.L., Yu, K.M., Lin, H.S.: Automatic music genre classification based on modulation spectral analysis of spectral and cepstral features. IEEE Trans. Multimedia **11**(4), 670–682 (2009)
6. Mallat, S.: Group invariant scattering. Commun. Pure Appl. Math. **65**(10), 1331–1398 (2012). doi:10.1002/cpa.21413
7. Andén, J., Mallat, S.: Deep scattering spectrum. IEEE Trans. Signal Process. **62**(16), 4114–4128 (2014). doi:10.1109/tsp.2014.2326991
8. Krizhevsky, A., Sutskever, I., Hinton, G.E.: Imagenet classification with deep convolutional neural networks. In: Advances in Neural Information Processing Systems, pp. 1097–1105 (2012). doi:10.1145/3065386
9. Simonyan, K., Zisserman, A.: Very deep convolutional networks for large-scale image recognition. arXiv preprint arXiv:1409.1556 (2014)
10. He, K., Zhang, X., Ren, S., Sun, J.: Deep residual learning for image recognition. In: Proceedings of the IEEE Conference on Computer Vision and Pattern Recognition, pp. 770–778. doi:10.1109/cvpr.2016.90 (2016)
11. Mohamed, A., Dahl, G.E., Hinton, G.: Acoustic modeling using deep belief networks. IEEE Trans. Audio Speech Lang. Process. **20**(1), 14–22 (2012). doi:10.1109/tasl.2011.2109382

12. Feng, X., Zhang, Y., Glass, J.: Speech feature denoising and dereverberation via deep autoencoders for noisy reverberant speech recognition. In: 2014 IEEE International Conference on Acoustics, Speech and Signal Processing (ICASSP), pp. 1759–1763. IEEE (2014). doi:10.1109/icassp.2014.6853900
13. Sutskever, I., Martens, J., Hinton, G.E.: Generating text with recurrent neural networks. In: Proceedings of the 28th International Conference on Machine Learning (ICML 2011), pp. 1017–1024 (2011)
14. Mikolov, T., Sutskever, I., Chen, K., Corrado, G.S., Dean, J.: Distributed representations of words and phrases and their compositionality. In: Advances in Neural Information Processing Systems, pp. 3111–3119 (2013)
15. Hochreiter, S., Schmidhuber, J.: Long short-term memory. Neural Comput. **9**(8), 1735–1780 (1997). doi:10.1162/neco.1997.9.8.1735
16. Pan, S.J., Yang, Q.: A survey on transfer learning. IEEE Trans. Knowl. Data Eng. **22**(10), 1345–1359 (2010). doi:10.1109/tkde.2009.191
17. Berenzweig, A., Logan, B., Ellis, D.P., Whitman, B.: A large-scale evaluation of acoustic and subjective music-similarity measures. Comput. Music J. **28**(2), 63–76 (2004). doi:10.1162/014892604323112257
18. Tzanetakis, G., Cook, P.: Musical genre classification of audio signals. IEEE Trans. Speech Audio Process. **10**(5), 293–302 (2002). doi:10.1109/tsa.2002.800560
19. Chung, J., Gulcehre, C., Cho, K., Bengio, Y.: Empirical evaluation of gated recurrent neural networks on sequence modeling. arXiv preprint arXiv:1412.3555 (2014)
20. Zaremba, W., Sutskever, I., Vinyals, O.: Recurrent neural network regularization. arXiv preprint arXiv:1409.2329 (2014)
21. Davis, J., Goadrich, M.: The relationship between precision-recall and ROC curves. In: Proceedings of the 23rd International Conference on Machine Learning, pp. 233–240. ACM (2006). doi:10.1145/1143844.1143874
22. Panagakis, Y., Kotropoulos, C., Arce, G.R.: Music genre classification using locality preserving non-negative tensor factorization and sparse representations. In: ISMIR, pp. 249–254 (2009)
23. Dai, J., Liu, W., Ni, C., Dong, L., Yang, H.: "Multilingual" deep neural network for music genre classification. In: Sixteenth Annual Conference of the International Speech Communication Association (2015)

A Functional Model of AIS Data Fusion

Yongming Wang[1(✉)] and Lin Wu[2]

[1] Dalian Maritime University, Dalian 116018, China
trancomm@163.com
[2] Institute of Computing Technology, Chinese Academy of Science, Beijing 100190, China

Abstract. In recent years, maritime situational awareness based on the fusion of AIS (Automatic Identification System) data has attracted more and more researchers. However, the diversity of terms and methodologies hinders the understanding and communication among them. Besides, AIS data was used mainly for traffic pattern discovery and abnormal vessel detection, more values await discovery. To overcome these two problems, this paper proposes a functional model of AIS data fusion. This model may provide a common frame of reference for discussions of AIS data fusion as well as a checklist for functions a maritime situational awareness system should provide. Based on this model, this paper introduces our works.

Keywords: Maritime situation awareness · Automatic Identification System · AIS · Data fusion · Functional model · JDL model

1 Introduction

The AIS [1] is a self-reporting system installed on ships used to exchange kinematic (dynamic) and identity (static) information etc. with other nearby ships, base stations and satellites. The initial purpose of AIS is collision avoidance. It has interfaces for position (GNSS), heading (compass) and rate of turn (gyrocompass or ROT Indicator) sensors. Other information like navigation status, Maritime Mobile Service Identity (MMSI) number, name of ship, type of ship, destination, Estimated Time of Arrival (ETA), etc. is required to input manually. Ships of 300 gross tons and upwards on international voyages, 500 tons and upwards for cargos not in international waters and passenger vessels are required to fit an AIS transceiver [2]. The time interval of receiving AIS messages from a ship is supposed to be no more than 3 min when within the range of shore-based stations (typically less than 60 miles) or on the order of hours otherwise. Global AIS data collected from shore-based stations and satellites is now available on the Internet, through service providers (e.g., China Transport Telecommunications & Information Center (CTTIC)). However, as we will show in this paper, AIS data is rather dirty, full of noises and errors.

As a main source of vessels' near real time information, AIS data has been studied by growing number of researchers for traffic pattern discovery [3, 5, 6]. In their research, vessel objects, waypoint objects and route objects are discovered and updated by fusing AIS data. Based on discovered patterns, abnormal vessels were detected by

Z. Shi et al. (Eds.): ICIS 2017, IFIP AICT 510, pp. 191–199, 2017.
DOI: 10.1007/978-3-319-68121-4_20

comparing vessels' behaviors with normal patterns [9]. For a more comprehensive maritime situational awareness, detailed types of entities should be identified, such as fishing areas [4], ports, anchorages, etc.

Diverse terms and methodologies about AIS data fusion exist in literature, making the communication and reuse among the community difficult. For example, different terms were used to represent the assessment of traffic routes, such as "analysis of motion patterns" [5], "learning for vessel trajectories" [6], "learning of maritime traffic patterns" [7], "extraction of knowledge" [8], "traffic route extraction" [3], "vessel pattern knowledge discovery" [9], "traffic knowledge discovery" [10], "vessel track information mining" [11], etc. These terms stem from the field of machine learning and data mining, which aim at solving the problem of knowledge or pattern discovery. On the other hand, data fusion is about the assessment of entities of interest, including their states, relationships among them and impacts. Patterns discovered by machine learning or data mining is a source of information in data fusion, which is often referred to as "models [12]". We study the problem of maritime situational awareness in the framework of data fusion rather than machine learning or data mining in this paper, and adopt those widely-used terms in fusion community.

To provide a common frame of reference as well as to explore more potential applications of AIS data, we propose a functional model of AIS data fusion in this paper. It describes what analysis functions or processes need to be performed [13] in a maritime situational awareness system. Another category of model is process model (e.g., Boyd's Observe-Orient-Decide-Act model), which describes how analysis is accomplished [13]. A functional model is useful in system engineering by "providing visualization of a framework for partitioning and relating functions and serving as a checklist for functions a system should provide" [12].

The JDL (Joint Directors of Laboratories) data fusion model [15] is a well-accepted functional model in fusion community. Our model is an instantiation of its revised version [14], and it focuses on Level 0 to Level 3. Based on this model, this paper introduces our work.

The remainder of the paper is organized as follows. Section 2 provides a detailed description of the proposed functional model. Section 3 summaries our existing works under the functional model. Section 4 draws the conclusion and future work.

2 Functional Model of AIS Data Fusion

In this section, we describe the functional model from level 0 to level 3, see Fig. 1. This model is based on the revised version [14] of the JDL fusion model [15]. The original JDL model centers on military scenes and "suffered from an unclear partitioning scheme [14]". The revised version is more general and "provides a clear and useful partitioning while adhering as much as possible to current usage across the data fusion community [14]".

Our model is entity-oriented and entities can be vessels, traffic route segments, traffic stops, fishing areas, anchorages, etc. These entities' aggregates or relations can also be regarded as an entity. The "products of processes at each level are estimates of some existing or predicted aspects of reality [15]", and they are stored in database.

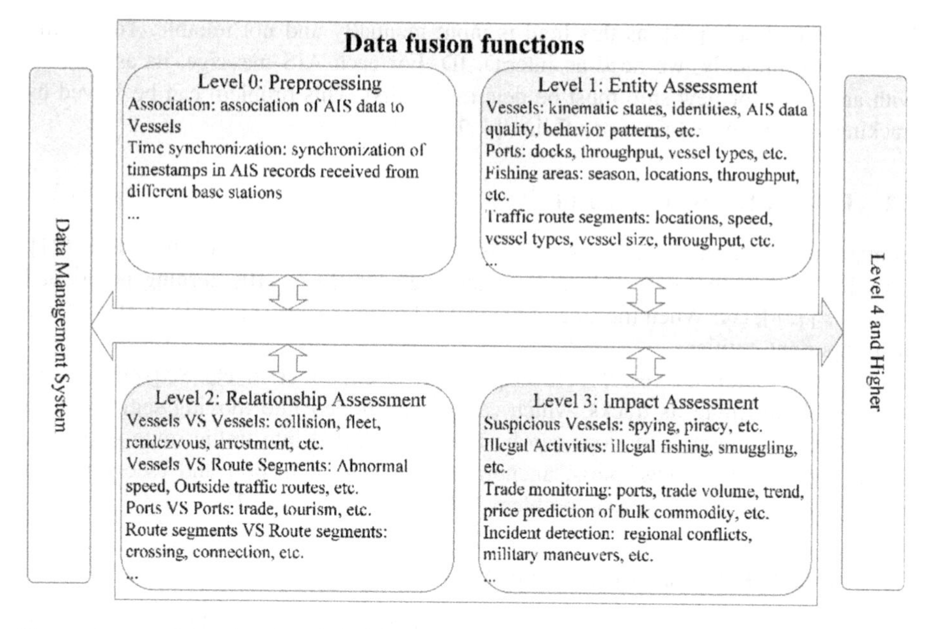

Fig. 1. Proposed functional model of AIS data fusion, which focuses on Level 0 to Level 3. AIS data as well as products of fusion is managed by the data management system on the left of this figure. AIS data

To highlight the data fusion functions in this figure, we put data sources, support databases and fusion databases into the data management system together. According to CTTIC [16], it is gathering about 10 billion records (around 1 TB) each year. So the data management system should be able to handle AIS data on the order of TB.

As level 4 (process assessment) and higher processes (user refinement) are not specifically fusion functions, they are not discussed in this paper.

2.1 Level 0: Preprocessing

AIS data is prone to be highly erroneous and noisy due to the deficiencies of the system and improper use of it. Although this problem was revealed [17, 18], limited work [19, 20] has been done on the preprocessing of AIS data.

As AIS was designed for collision avoidance, many issues may arise when applied for real-time surveillance. For example, an AIS message only contains the seconds part (6 bits) of the time that it was broadcast and it is up to receivers to affix the full time stamp. A comparison was made between the seconds part of the original time stamp and affixed one, and significant differences were found [17]. Besides, as clocks are not synchronized, a ship will jump on the map when its AIS messages are received from different receivers [20].

In addition to flaws in the design of AIS, improper use of it makes the situation even worse. For instance, the MMSI of each ship is supposed to be unique, and it's commonly used [9] to identify and track vessels. But it's not rare that a MMSI is shared by different vessels [19], as this field is input manually and not reliable. To identify each vessel uniquely, we need an internal ID. For each AIS message, its association with an existed or new ship must be determined first. This problem can be solved by tracking vessels in real-time or off-line [4, 10, 19].

2.2 Level 1: Entity Assessment

In maritime situational awareness systems, entities of interest are mainly vessels [21], ports [5, 27], fishing areas [4], traffic route segments [7, 8, 10], turning points and junctions [7, 8], etc. When the area of interest is bounded, entry and exit points [10] are also important entities.

The state vector of a vessel includes position, speed, direction, etc. Sequential states of a vessel constitute its tracks, which can be partitioned into moving segments and stops. These segments and stops can be further classified into traffic route segments, traffic stops, fishing areas, ports, anchorages, etc. The detection and assessment of traffic route segments is a basic problem in Level 1 [3, 9, 10, 22].

2.3 Level 2: Relationship Assessment

Level 2 process is usually referred to as "situation refinement [15]" or "situation assessment [14]". It has been pointed out that situations can be represented as sets of relations [14], so we refer to Level 2 process as relationship assessment in this paper for clarity.

Relations of interest for maritime situational awareness include but are not limited to:

- vessels VS vessels

Maritime authorities are deeply concerned about the relationship among vessels [26], because illegal activities and incidents can be detected by relation assessment. For example, we can discover smuggles by identifying vessels' rendezvous at sea. And encounter of vessels may indicate critical incidents like collision, piracy or arrestment, etc.

- vessels VS route segments

Original AIS data is made up of point sets. After traffic route segments have been extracted in level 1 process, we can represent the track of each vessel as series of segments. This can remove redundancy in raw AIS data and compress it by about two orders. It was modeled as a classification problem in previous works [9], and solved by maximizing the posterior probability that a vessel is sailing over a certain route.

As discovered routes are characterized by spatial, temporal and other attribute-related features (e.g., type and size) of vessels sailing on them [3, 9], and features of vessels include their historical route patterns, we can detect anomalies when a vessel's attributes are not compatible with the route segment, or when it is not compatible with its historical patterns.

- ports VS ports

There is limited work on analyzing the relationship between ports. An example was identifying patterns of ships' transition among ports by AIS data mining using the software R [23]. Based on discovered patterns, they estimated destination ports.

- route segments VS route segments

Extracted route segments are separate lines [22]. We need to know their relations of connection [8], in order to form the global route map.

2.4 Level 3: Impact Assessment

Level 3 process combines products of level 1 and level 2 to assess impacts. Functions provided at this level include predicting vessels' positions, assessing the intention and threat of vessels and detecting anomalies.

As time intervals between messages of the same ship varies from seconds to hours [17] due to limited coverage and bandwidth of receivers, it is necessary to predict positions for situational awareness [5, 9, 17, 19].

Anomaly detection and risk assessment are usually based on the assessment of relationship among entities. Most existing literature only focused on detecting anomalous ships [8, 9, 11, 24–27].

3 Our Existing Works

3.1 Level 0: Preprocessing

We have performed association on global AIS data of 34 months (from August 2012 to May 2015) based on spatial and temporal proximity [28]. When two records with the same MMSI were more than 100 km away and the speed reckoned was more than 120 knots, then they were regarded as from different ships [28]. The results of pre-processing showed that, among 1,998,200 distinct MMSIs recorded, 1,607,664 were suspicious: the average number of AIS messages broadcast by each of them was only 11.3. The remaining 390,536 MMSIs were used by 491,346 vessels, 43,034 messages had been received from one vessel on average.

3.2 Level 1: Entity Assessment

We proposed a vessel trajectory partitioning method based on hierarchical fusion of position data reported by AIS, aiming at improving preciseness and processing speed [29]. Our method consists of two steps: position point fusion and sub-trajectory fusion. The key idea of our method is to describe trajectories by hierarchical concepts. In the first step, trajectories are partitioned by grouping positions in raw AIS records into sub-trajectories represented as straight line segments. In the second step, we aggregate successive sub-trajectories into more abstract concepts: route segments and stops. We applied our method on a data set containing 473963 AIS records from 10 vehicle carriers along the coast of China. Algorithms were implemented by python 2.7.3 on a notebook with Intel Core2 Duo CPU T6600 @2.20 GHz, 2 GB RAM. The average execution time of our two level partitioning on each vessel is around 2.30 s and 0.22 s respectively. After partitioning, we got 250 route segments and 16 stops. Experimental results showed that our method split routes and identified stops precisely at the computational complexity of O(n).

To describe shipping density of each area, we defined vessel and traffic density [28]. The definition of vessel density in a region was taken as the expected number of vessels per unit area at any time, and traffic density as the average number of vessels crossing this region per unit area per unit time. We calculated vessel and traffic density using a grid-based method. The traffic density in 2014 is shown in Fig. 2.

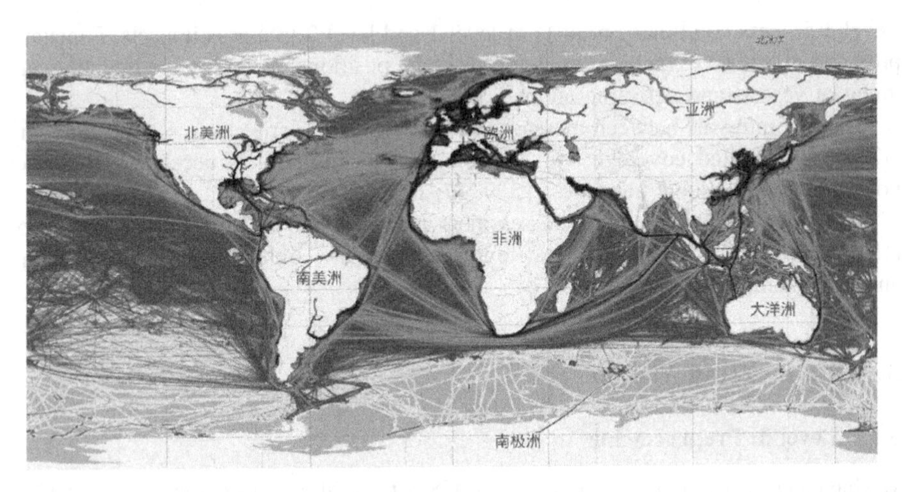

Fig. 2. Global traffic density in 2014, at the spatial resolution of 10 min longitude by 10 min latitude

3.3 Level 2: Relationship Assessment

We are developing an online system detecting vessels' rendezvous. In the system, we divide the earth into grids of the size 1° latitude by 1° longitude, and put ships into them according to their locations. Rendezvous detection is performed every 10 min by calculating distance among ships inside the same grid and surrounding grids.

We have mined ships' mooring positions base on AIS data, including position, speed and status. Each position was represented as a grid of 0.6 s longitude by 0.6 s latitude, and the membership that each grid belonged to a berth was calculated separately according to tracks located in this grid using fuzzy inference. After that, we clustered mooring positions into berths using DBSCAN, taking multiple attributes of positions into consideration in the function of distance. An example of a clustered berth is demonstrated in Fig. 3.

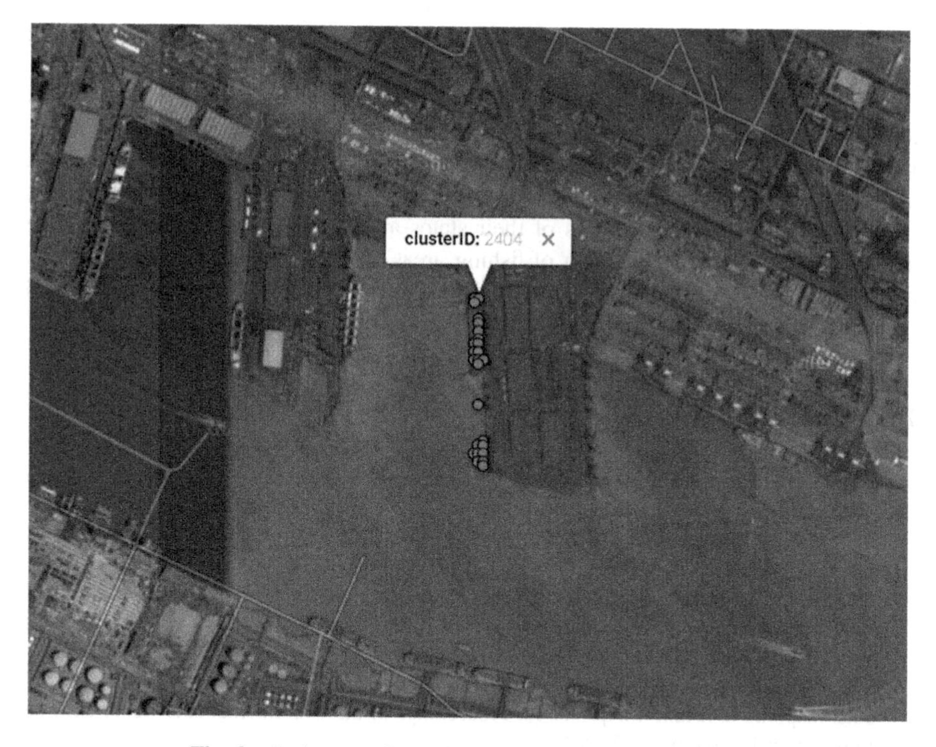

Fig. 3. A clusters of mooring points, belonging to a berth

3.4 Level 3: Impact Assessment

Based on vessel and traffic density maps described in Sect. 3.1, we are developing an online system detecting two types of anomalies: areas with vessel number deviating from expected and ships sailing in unusual routes. The former anomaly is detected by comparing current number of vessels with patterns in vessel density maps. The latter anomaly is detected by comparing each vessel's route with traffic density maps.

4 Conclusion and Future Work

We proposed a functional model of AIS data fusion in this paper. We hope that this model can provide a common frame of reference for maritime situational awareness based on AIS data fusion. This model is an instantiation of the well-accepted JDL model and serves as a checklist for functions a maritime situational awareness system should provide. Based on this model, this paper introduced our works.

Our future work would be to achieve more comprehensive global maritime situational awareness based on AIS data fusion. In the entity assessment step, we will detect and estimate more kinds of entities. Previous work mainly focused on stops and routes. To achieve global maritime situational awareness, we would consider more types of entities: vessels, traffic route segments, traffic stops, fishing areas, anchorages and ports, etc. With those entities, we can assess more interesting kinds of relations in level 2 process, such as fleet identification, which has not been studied yet as far as we know. Although anomaly detection is a hot topic in level 3 process, most efforts were made on identifying ships whose tracks are not compatible with traffic patterns. In our future work, those ships deviating from their historical patterns will also be marked as anomaly. Taking more types of entities into consideration, we plan to detect more kinds of anomaly based on the analysis of their states and relations among them. Possible anomalies include birth and death of fishing areas and anchorages, changes of traffic route segments, etc. Apart from anomaly detection, we will also perform other level 3 processes such as route planning, trade monitoring, piracy detection, etc.

References

1. Technical characteristics for an automatic identification system using time-division multiple access in the VHF maritime mobile band. International Telecommunications Union (ITU) Recommendation ITU-RM, pp. 1371–1374 (2010)
2. Safety of life at sea (solas) convention chapter V, regulation 19
3. Pallotta, G., Vespe, M., Bryant, K.: Traffic route extraction and anomaly detection from AIS data. In: Proceedings of the International COST MOVE Workshop on Moving Objects at Sea, Brest, France (2013)
4. Mazzarella, F., Vespe, M., Damalas, D., et al.: Discovering vessel activities at sea using AIS data: mapping of fishing footprints. In: Proceedings of the 16th International Conference on Information Fusion, pp. 1–7 (2014)
5. Ristic, B., Scala, B.L., Morelande, M., et al.: Statistical analysis of motion patterns in AIS data: Anomaly detection and motion prediction. In: Proceedings of the 11th International Conference on Information Fusion, pp. 1–7 (2008)
6. Vries, G.K.D., Someren, M.: Machine learning for vessel trajectories using compression, alignments and domain knowledge. Expert Syst. Appl. **39**(18), 13426–13439 (2012)
7. Vespe, M., Visentini, I., Bryan, K., et al.: Unsupervised learning of maritime traffic patterns for anomaly detection. In: Proceedings of the 9th IET Data Fusion & Target Tracking Conference: Algorithms & Applications, pp. 1–5 (2012)
8. Guillarme, N.L., Lerouvreurt, X.: Unsupervised extraction of knowledge from S-AIS data for maritime situational awareness. In: Proceedings of the 16th International Conference on Information Fusion, pp. 2025–2032 (2013)
9. Pallotta, G., Vespe, M., Bryant, K.: Vessel pattern knowledge discovery from AIS data: a framework for anomaly detection and route prediction. Entropy **15**(6), 2218–2245 (2013)
10. Pallotta, G., Vespe, M., Bryant, K.: Traffic knowledge discovery from AIS data. In: Proceedings of the 16th International Conference on Information Fusion, pp. 1996–2003 (2013)
11. Deng, F., Guo, S., Deng, Y., et al.: Vessel track information mining using AIS data. In: Proceedings of the International Conference on Multisensor Fusion and Information Integration for Intelligent Systems (MFI), pp. 1–6 (2014)
12. Kesseler, O., White, F.: Data Fusion Perspectives and Its Role in Information Processing. In: Handbook of Multisensor Data Fusion, pp. 15–43 (2009)
13. Antony, R.: Data fusion automation: a top-down perspective. In: Handbook of Multisensor Data Fusion, pp. 137–163 (2009)

14. Steinberg, A.N., Bowman, C.L.: Revisions to the JDL data fusion model. In: Handbook of Multisensor Data Fusion, pp. 45–67 (2009)
15. Hall, D.L., Llinas, J.: An introduction to multisensor data fusion. Proc. IEEE **85**(1), 6–23 (1997)
16. Website of AIS data services provided by CTTIC. http://www.myships.com/ShipLBS/index_en.html
17. Greidanus, H., Alvarez, M., Eriksen, T., et al.: Basin - wide maritime awareness from multi - source ship reporting data. Trans. Nav. Int. J. Mar. Navig. Saf. Sea Transp. **7**(2), 35–42 (2013)
18. Harati-mokhtari, A., Wall, A., Brooks, P., et al.: Automatic Identification System (AIS): A Human Factors Approach. http://217.26.101.136/middenlimburg/downloads/documenten/pdf/ais_human_factors.pdf
19. Mazzarella, F., Alessandrini, A., Greidanus, H., et al.: Data fusion for wide-area maritime surveillance. In: Proceeding of COST MOVE Workshop on Moving Objects at Sea, Brest (2013)
20. Pallotta, G., Horn, S., Braca, P., et al.: Context-enhanced vessel prediction based on Ornstein-Uhlenbeck processes using historical AIS traffic patterns: real-world experimental results. In: Proceedings of 17th International Conference on Information Fusion, pp. 1–7 (2014)
21. Falcon, R., Abielmona, R., Blasch, E.: Behavioral learning of vessel types with fuzzy-rough decision trees. In: Proceedings of 17th International Conference on Information Fusion, pp. 1–8 (2104)
22. Lee, J.G., Han, J., Whang, K.Y.: Trajectory clustering: a partition-and-group framework. In: Proceedings of ACM SIGMOD International Conference on Management of data, pp. 593–604 (2007)
23. Hadzagic, M., St-hilaire, M.O., Webb, S., et al.: Maritime traffic data mining using R. In: Proceedings of 16th International Conference on Information Fusion, pp. 2041–2048 (2013)
24. Balmat, J.F., Lafont, F., Maifret, R., et al.: A decision-making system to maritime risk assessment. Ocean Eng. **38**(1), 171–176 (2011)
25. Mascaro, S., Nicholso, A.E., Korb, K.B.: Anomaly detection in vessel tracks using Bayesian networks. Int. J. Approx. Reas. **55**(1), 84–98 (2014)
26. Lane, R.O., Nevell, D.A., Hayward, S.D., et al.: Maritime anomaly detection and threat assessment. In: Proceedings of 13th International Conference on Information Fusion, pp. 1–8 (2010)
27. Castaldo, F., Palmieri, F.A., Bastani, V., et al.: Abnormal vessel behavior detection in port areas based on Dynamic Bayesian Networks. In: Proceedings of 17th International Conference on Information Fusion, pp. 1–7 (2014)
28. Wu, L., Xu, X.J., Wang, Q., et al.: Mapping global shipping density from AIS Data. J. Nav. **70**, 67–81 (2016)
29. Wu, X.B., Wu, L., Xu, Y., et al.: Vessel trajectory partitioning based on hierarchical fusion of position data. In: Proceedings of 18th International Conference on Information Fusion, pp. 1230–1237 (2015)
30. Natale, F., Gibin, M., Alessandrini, A., et al.: Mapping fishing effort through AIS data. PloS one **10**(6), 1–16 (2015)

Entropy-Based Support Matrix Machine

Changming Zhu(✉)

College of Information Engineering, Shanghai Maritime University,
Shanghai 201306, People's Republic of China
cmzhu@shmtu.edu.cn

Abstract. Support Vector Machine (SVM) cannot process imbalanced problem and matrix patterns. Thus, Fuzzy SVM (FSVM) is proposed to process imbalanced problem while Support Matrix Machine (SMM) is proposed to process matrix patterns. FSVM applies a fuzzy membership to each training pattern such that different patterns can make different contributions to the learning machine. However, how to evaluate fuzzy membership becomes the key point to FSVM. Although SMM can process matrix patterns, it still has no ability to process imbalanced problem. This paper adopts SMM as the basic and proposes an entropy-based support matrix machine for imbalanced data sets, i.e., ESMM. The contributions of ESMM are: (1) proposing an entropy-based fuzzy membership evaluation approach which enhances importance of certainty patterns, (2) guaranteeing importance of positive patterns and getting a more flexible decision surface. Experiments on real-world imbalanced data sets and matrix patterns validate the effectiveness of ESMM.

Keywords: Support matrix machine · Entropy · Fuzzy membership · Imbalanced data set · Pattern recognition

1 Introduction

Support Vector Machine (SVM) constructs a hyperplane or set of hyperplanes in a high- or infinite-dimensional space, which can be used for classification, regression, or other tasks. Intuitively, a good separation is achieved by the hyperplane that has the largest distance to the nearest training-data point of any class (so-called functional margin), since in general the larger the margin the lower the generalization error of the classifier [1]. Conventional SVM can be used in many tasks including text and hypertext categorization, classification of images, classifying proteins in medical science, and recognizing hand-written characters. Although SVM has been validated effect on these applications, it still has two disadvantages. One is that SVM cannot process matrix patterns and another is that it cannot process imbalanced problem.

• Matrix patterns which dimensions are $m \times n$ (where m and n are both larger than 1) are the basic of matrix learning. For example, video and images are both matrix patterns. In order to process matrix patterns, matrix-pattern-oriented

Z. Shi et al. (Eds.): ICIS 2017, IFIP AICT 510, pp. 200–211, 2017.
DOI: 10.1007/978-3-319-68121-4_21

learning machine (MatC), i.e., matrix learning machine, has been developed. Classical learning machines include matrix-pattern-oriented Ho-Kashyap learning machine with regularization learning (MatMHKS) [2], new least squares support vector classification based on matrix patterns (MatLSSVC) [3], and one-class support vector machines based on matrix patterns (OCSVM) [4]. Besides those matrix learning machines, Xie et al. have proposed a Support Matrix Machine (SMM) [5] so as to replace SVM. SMM can leverage the structure of the data matrices and has the grouping effect property. But all of these matrix learning machines cannot process imbalanced problem.

• As is known to all, in many real-world classification problems, such as e-mail foldering [6], fault diagnosis [7], detection of oil spills [8], and medical diagnosis [9], we can always divide a data set into two classes, one is positive class and the other is negative class. When the size of positive class is much smaller than that of negative class, imbalanced problem occurs. Since most standard classification learning machines including Support Vector Machine (SVM) and Neural Network (NN) are proposed with the assumption on the balanced class distributions or equal misclassification costs [10], so they fail to properly represent the distributive characteristics of patterns and result in the unfavorable performance when they are adopted to process imbalanced problem. In order to overcome such a disadvantage, Fuzzy SVM (FSVM) [11] and Bilateral-weighted FSVM (B-FSVM) [12] are proposed. FSVM applies a fuzzy membership to each input pattern and reformulates SVM such that different input patterns can make different contributions to the learning of decision surface. B-FSVM treats every pattern as both positive and negative classes, but with different memberships due to we can not say one pattern belongs to one class absolutely. But for both of them, how to determine the fuzzy membership function is the key point. Furthermore, both of them cannot process matrix patterns.

In this paper, we try to propose a learning machine which can process matrix patterns and imbalanced problem. First, in order to process matrix patterns, we adopt SMM as a basic. Then in order to process imbalanced problem, we adopt the notion of FSVM, namely, applies a fuzzy membership to each input pattern. Furthermore, for the fuzzy membership, we propose a new fuzzy membership evaluation approach which assigns the fuzzy membership of each pattern based on its class certainty. In this paper, class certainty demonstrates the certainty of pattern labeled to a certain class. Due to the entropy is an effective measure of certainty, we adopt the entropy to evaluate the class certainty of each pattern. In doing so, the entropy-based fuzzy membership evaluation approach is proposed. This approach determines the fuzzy membership of training patterns based on their corresponding entropies. By adopting the entropy-based fuzzy membership evaluation and SMM, the Entropy-based Support Matrix Machine (ESMM) is proposed to process the imbalanced data sets. In practice, as the importance of positive class is higher than that of negative class in imbalanced data sets, i.e., the learning machine should pay more attention to positive patterns than negative ones. Thus, positive patterns are assigned to relatively large fuzzy memberships to guarantee the importance of positive class here. While, the fuzzy memberships of negative patterns are determined by the entropy-based fuzzy membership evaluation approach, i.e., patterns with lower class certainty are assigned to small fuzzy memberships based on the criterion that patterns with lower class certainty are more insensitive to noise, and easily mislead the decision surface, thus their importance should be weakened on imbalanced data sets. After evaluating the fuzzy membership of all training patterns, ESMM is adopted to classify imbalanced data sets.

The contributions of this paper can be highlighted as follows:

(1) A new entropy-based fuzzy membership evaluation approach is proposed. This approach adopts entropy to evaluate class certainty of a pattern and determines the corresponding fuzzy membership based on class certainty. In doing so, the learning machine can pay more attention to the patterns with higher class certainty to result in more robust decision surface.
(2) To guarantee the importance of positive class, the positive patterns are assigned to the relatively large fuzzy memberships, which results in the decision surface paying more attention to the positive class so as to increase generalization of learning machine.

The rest of this paper is given below. Section 2 introduces the proposed entropy-based fuzzy membership evaluation approach, then give the details of ESMM. In Sect. 2.1, several experiments on real-world imbalanced data sets and matrix patterns including images are conducted to validate the effectiveness of ESMM. Following that, conclusions are given in Sect. 4.

2 Entropy-Based Matrix Learning Machine (ESMM)

2.1 Entropy-Based Fuzzy Membership

When we process imbalanced data sets, positive class is always more important than negative class. Thus, in the proposed entropy-based fuzzy membership evaluation approach, we assign positive patterns to a relatively high fuzzy membership, e.g., 1.0, to guarantee the importance of positive class. As to the negative class, we fuzzify negative patterns based on their class certainties.

In information theory, entropy is always used to characterize the certainty of source of information. If entropy is smaller, then information is more certain [13]. By employing this character of entropy, we can evaluate the class certainty of a pattern in the training class. After getting the class certainty of each pattern, we assign fuzzy membership of each training pattern based on class certainty. In practice, the patterns with higher class certainty, i.e., lower entropy, are assigned to higher fuzzy memberships to enhance their contributions to the decision surface, and vice versa.

Suppose that there are N training patterns, $\{x_i, y_i\}$ where $i = 1, 2, \ldots, N$ and $y_i \in \{+1, -1\}$ is the class label. When $y_i = 1$, pattern x_i belongs to the positive class, otherwise, it belongs to the negative class. The probabilities of x_i belonging to positive and negative class are p_{+i} and p_{-i} respectively. The entropy of x_i is $H_i = -p_{+i}ln(p_{+i}) - p_{-i}ln(p_{-i})$ where ln represents the natural logarithm operator. Due to neighbors of a pattern can determine local information of it, thus the probability evaluation is based on its k nearest neighbors. For a pattern x_i, we select its k nearest neighbors $\{x_{i1}, x_{i2}, \ldots, x_{ik}\}$ at first. Then we count the number of both positive and negative class in these k selected patterns and denote the numbers of patterns belonging to positive and negative class are num_{+i} and num_{-i} respectively. Finally, the probabilities of x_i belonging to positive and negative class are calculated with $p_{+i} = \frac{num_{+i}}{k}$ and $p_{-i} = \frac{num_{-i}}{k}$. After evaluating the class probabilities of x_i, we can calculate its entropy.

By adopting the above entropy evaluation approach, the entropy of the negative patterns are $H = \{H_{-1}, H_{-2}, \ldots, H_{-n_-}\}$, where n_- is the number of the negative patterns. H_{min} and H_{max} are the minimum and maximum entropy of H. The entropy-based fuzzy memberships for negative patterns are evaluated as follows.

Firstly, separate the negative patterns into m subsets based on their entropy as described in Table 1, i.e., $\{Sub_k\}$ where $k = 1, 2, \ldots, m$.

Table 1. Algorithm of negative class separation.

```
For k=1:m
  Up=H_max - (k-1)/m (H_max - H_min)
  Low=H_max - k/m (H_max - H_min)
  For i=1:n_
    if Low<H_-i ≤Up
      negative pattern x_i is distributed into the subset Sub_k.
  End
End
```

Then, fuzzy memberships of patterns in each subset are set as:

$$FM_k = 1.0 - \alpha \times (k - 1), \quad k = 1, 2, 3, \ldots, m \tag{1}$$

where FM_k is the fuzzy membership for patterns distributed in subset Sub_k, the fuzzy membership parameter $\alpha \in (0, \frac{1}{m-1})$ since FM_k is positive and not larger than 1.0. It should be declared that patterns in the same subset are set to same fuzzy membership so that these patterns selected in the same subsets have same importance to the decision surface. Finally, the fuzzy membership s_i for a training pattern x_i is assigned as: if $y_i = +1$, then $s_i = 1.0$, else if $y_i = -1$ and $x_i \in Sub_k$, then $s_i = FM_k$. So far, the entropy-based fuzzy membership for the training patterns are evaluated.

2.2 Entropy-Based Support Matrix Machine

By adopting the evaluated entropy-based fuzzy membership which is given before, we propose the entropy-based support matrix machine (ESMM). The detailed description on ESMM is given below.

Suppose that there is a binary-class classification problem with N matrix patterns (A_i, y_i, s_i), $i = 1, 2, \ldots, N$. Here $A_i \in R^{m \times n}$ is the matrix representation of x_i and its class label is $y_i \in \{+1, -1\}$. If $y_i = +1$, x_i or A_i belongs to class $+1$ or positive class, and then if $y_i = -1$, the pattern belongs to class -1 or negative class. s_i is the entropy-based fuzzy membership.

The corresponding criterion function of ESMM is defined below.

$$minL(W,b) = \frac{1}{2}tr(W^TW) + \theta\,||W||_\star + C\sum_{i=1}^{N}(s_i(1 - y_i[tr(W^TA_i) + b_i])) \tag{2}$$

where $W \in R^{m\times n}$ is the matrix of regression coefficients and its nuclear norm is $||W||_\star$. θ is the coefficient. tr is the trace of matrix. b_i is a loose variable for pattern A_i. C ($C \in R, C \geq 0$) is the regularization parameter that adjusts the trade-off between model complexity and training error. Here, $b = [b_1, \ldots, b_i, \ldots, b_N]^T$ and b_i is started as $b_i \geq 0$. The iteration for b is given in Eq. (3).

$$b_i(k+1) = b_i(k) + \rho(e_i(k) + |e_i(k)|) \tag{3}$$

where the error vector e at k-th iteration of A_i, i.e., $e_i(k)$ should be $W(k)^TA_i - 1 - b_i(k)$ and $W(k)$ and $b_i(k)$ are k-th component of W and b_i. Then we adopt the similar method of SMM to get the optimal W and b here. After that, we can get the discriminant function of ESMM as below.

$$g(A_i) = W^TA_i \tag{4}$$

If $g(A_i) > 0$, then we label A_i as a positive pattern, then if $g(A_i) < 0$, we label A_i as a negative pattern.

3 Experiments

In this section, we adopt 25 real-world imbalanced data sets and 5 image data sets for examples and the compared learning machines are SVM, FSVM, MatMHKS, B-FSVM, SMM. The used 25 real-world imbalanced data sets are selected from the KEEL imbalanced benchmark ones [14,15]. Information of these data sets are given in Tables 2 and 3.

3.1 Experimental Settings

In terms of the compared learning machines and ESMM, the experimental settings are given here. For the SVM-based learning machines, the used kernel is Radial Basis Function (RBF) kernel $ker(x_i, x_j) = exp(-\frac{||x_i-x_j||_2^2}{\sigma^2})$ where σ is selected from the set $\{10^{-3}, 10^{-2}, \ldots, 100, 1000\}$. For MatMHKS, the experimental setting can be found in [2]. For SMM, the experimental setting can be found in [5]. For ESMM, its setting is similar with SMM. Moreover, for ESMM, the number of the separated subsets $m = 10$ and the fuzzy membership parameter $\alpha = 0.05$ which results in the fuzzy membership $0.5 \leq s_i \leq 1.0$. The reason of restricting $s_i \in [0.5, 1.0]$ is that the class label of x_i should not be neglected when determining the fuzzy membership, i.e., x_i is more likely to be classified to the class which is indicated by its class label. As to negative class patterns,

Table 2. Information of real-world imbalanced data sets (IR represents the imbalanced ratio of the corresponding data set).

Order	Data set	Attributes	Training	Testing	IR
1	Wisconsin	9	546	137	1.86
2	Haberman	3	245	61	2.78
3	Newthyroid2	5	172	43	5.14
4	Ecoli3	7	269	67	8.6
5	Ecoli046vs5	6	162	41	9.15
6	Yeast2vs4	8	411	103	9.08
7	Ecoli067vs5	6	176	44	10
8	Led7digit02456789vs1	7	354	89	10.97
9	Glass2	9	171	43	11.59
10	Yeast1vs7	7	367	92	14.3
11	Pageblocks13vs4	10	378	94	15.86
12	Glass016vs5	9	147	37	19.44
13	Yeast2vs8	8	386	96	23.1
14	Yeast1289vs7	8	758	189	30.57
15	Yeast6	8	1187	297	41.4
16	Pima	8	614	154	1.87
17	Ecoli1	7	269	67	3.36
18	Segment0	19	1846	462	6.02
19	Ecoli034vs5	7	160	40	9
20	Ecoli01vs235	7	195	49	9.17
21	Yeast05679vs4	8	422	106	9.35
22	Glass016vs2	9	154	38	10.29
23	Glass0146vs2	9	164	41	11.06
24	Cleveland0vs4	13	138	35	12.31
25	Ecoli4	7	269	67	15.8

Table 3. Information of image data sets. Imbalanced ratio of each image data set is 9.0.

Order	Data set	Attributes	Training	Testing
26	COIL-20 [16]	32×32	1296	144
27	Letter [17]	24×18	450	50
28	ORL [18]	32×20	360	40
29	PubFig [19]	74×74	52917	5880
30	Gaze [20]	5184×3456	5292	588

we set $s_i > 0.5$ to indicate that a negative pattern x_i is more likely to belong to the negative class. Moreover, the number of nearest neighbors k for calculating the class probability is selected from $\{1, 2, 3, \ldots, 18, 19, 20\}$. In order to measure the performance of compared learning machines on imbalanced data sets, the values of Area Under the ROC Curve (AUC) [25] is given. Besides those, $maxIter = 500$ represents the maximal size of the iteration. One-against-one classification strategy is used for multi-class problems here [21–24]. The 10-fold cross validation approach [26] is adopted for the parameter selection. The computations are performed on Intel Core 2 processors with 2.66 GHz, 8G RAM, Microsoft Windows 7, and Matlab environment.

3.2 Experiments on Real-World Imbalanced Data Sets

For experiments, the AUC values are presented in Table 4. The average AUC on all used data sets are presented. Moreover, the average ranks of the learning machines on the used data sets are listed.

Table 4. AUC values (%) of the compared learning machines on real-world imbalanced data sets and image data sets. The best k for each data sets of ESMM is presented. (Note that the average AUC values and the average ranks of the compared learning machines are listed in the last two rows. Best result for each data set is in bold.)

Data set	k	ESMM	SMM	MatMHKS	B-FSVM	FSVM	SVM
1	9	94.05 ± 2.37	94.02 ± 2.30	89.98 ± 3.33	91.30 ± 3.03	**96.10 ± 2.41**	93.64 ± 3.67
2	13	64.42 ± 1.96	64.47 ± 1.98	**67.06 ± 1.40**	61.58 ± 1.44	58.00 ± 2.05	54.08 ± 1.65
3	11	93.28 ± 1.87	91.47 ± 1.89	89.78 ± 2.57	92.44 ± 2.32	93.43 ± 2.33	**97.12 ± 2.90**
4	17	87.66 ± 2.41	87.37 ± 2.41	93.67 ± 2.70	**94.55 ± 2.54**	76.83 ± 2.36	65.74 ± 1.80
5	14	**98.61 ± 2.73**	97.15 ± 2.74	92.59 ± 2.91	89.07 ± 3.15	86.61 ± 1.06	82.43 ± 1.91
6	7	93.84 ± 2.95	**94.33 ± 2.94**	91.79 ± 1.47	87.37 ± 2.11	87.32 ± 1.85	83.77 ± 2.14
7	5	97.12 ± 2.94	**98.67 ± 2.87**	90.79 ± 1.71	94.39 ± 1.97	78.90 ± 2.28	74.15 ± 2.52
8	17	92.80 ± 2.62	90.53 ± 2.58	88.30 ± 2.29	**93.43 ± 2.35**	87.66 ± 2.14	81.64 ± 2.87
9	18	**81.33 ± 2.22**	80.07 ± 2.20	68.12 ± 1.66	66.99 ± 1.74	69.21 ± 2.09	51.76 ± 1.51
10	18	**78.41 ± 2.53**	76.58 ± 2.50	58.24 ± 1.95	68.95 ± 1.85	61.56 ± 2.14	67.10 ± 2.20
11	15	79.86 ± 2.36	79.29 ± 2.29	77.65 ± 1.93	80.99 ± 2.37	**92.11 ± 2.67**	91.22 ± 3.32
12	14	**87.24 ± 2.27**	85.75 ± 2.21	73.53 ± 2.07	80.66 ± 1.91	74.85 ± 2.18	75.31 ± 1.74
13	15	**71.30 ± 1.67**	69.28 ± 1.69	67.12 ± 1.55	56.86 ± 1.20	57.62 ± 1.87	62.45 ± 1.52
14	10	91.00 ± 2.24	91.62 ± 2.22	**92.81 ± 3.17**	87.97 ± 2.80	71.63 ± 1.54	71.64 ± 2.13
15	5	72.58 ± 1.31	**72.84 ± 1.32**	71.15 ± 2.13	69.81 ± 2.38	68.65 ± 1.53	71.50 ± 1.75
16	3	90.36 ± 1.72	89.80 ± 1.72	81.80 ± 2.18	**93.74 ± 2.44**	85.40 ± 2.59	78.11 ± 2.36
17	10	80.62 ± 2.19	78.88 ± 2.20	82.32 ± 2.82	83.40 ± 2.01	98.24 ± 2.03	**99.72 ± 2.25**
18	6	**93.52 ± 3.00**	91.50 ± 2.89	87.46 ± 1.91	83.81 ± 1.88	89.61 ± 2.98	77.26 ± 2.20
19	18	92.35 ± 2.69	**93.72 ± 2.73**	92.22 ± 3.10	91.20 ± 2.19	74.69 ± 2.30	82.43 ± 2.18
20	19	**83.26 ± 1.96**	81.00 ± 1.99	78.59 ± 2.28	79.42 ± 2.16	64.48 ± 2.02	66.34 ± 1.74
21	5	**75.38 ± 2.08**	75.26 ± 2.06	69.36 ± 2.09	55.51 ± 1.57	61.42 ± 1.84	53.73 ± 1.52
22	16	91.84 ± 2.19	91.08 ± 2.18	85.21 ± 2.01	**95.55 ± 3.40**	87.28 ± 2.53	94.66 ± 2.85
23	4	**81.49 ± 2.00**	80.37 ± 2.00	59.32 ± 1.60	71.62 ± 2.15	66.51 ± 2.05	55.52 ± 1.35
24	1	74.72 ± 2.07	74.44 ± 2.01	63.24 ± 1.72	60.41 ± 1.21	68.05 ± 1.76	**77.90 ± 1.86**
25	12	95.59 ± 2.79	**96.65 ± 2.81**	95.98 ± 2.51	92.53 ± 2.80	84.58 ± 1.72	77.94 ± 2.37
26	5	65.40 ± 1.78	64.90 ± 1.74	59.33 ± 1.97	**71.77 ± 1.76**	71.07 ± 1.76	69.50 ± 2.37
27	18	**66.08 ± 2.15**	65.88 ± 2.15	57.58 ± 1.01	60.62 ± 1.50	49.37 ± 1.45	48.94 ± 1.32
28	20	**90.19 ± 2.06**	88.54 ± 2.01	86.21 ± 3.07	79.27 ± 2.02	79.82 ± 1.71	82.30 ± 2.10
29	8	92.89 ± 2.01	93.59 ± 1.99	88.12 ± 2.36	**95.14 ± 3.19**	89.48 ± 2.26	81.30 ± 2.45
30	18	**94.78 ± 1.70**	92.15 ± 1.68	90.61 ± 2.56	87.60 ± 2.02	66.13 ± 1.23	85.69 ± 2.21
AVG	11.72	**85.06 ± 2.23**	84.37 ± 2.21	79.66 ± 2.20	80.60 ± 2.18	76.55 ± 2.02	75.16 ± 2.16
AVG.rank		2.07	2.53	4.03	3.50	4.30	4.57

From the experimental results, it is found that: (1) ESMM results in the best classification performance on 12 of 30 imbalanced data sets which indicates that the our proposal outperforms the compared learning machines; (2) ESMM outperforms the traditional MatMHKS on 26 of 30 imbalanced data sets; (3) The average AUC of ESMM is respectively greater than B-FSVM, FSVM, and SVM on about 5%, 9%, and 10%, which demonstrates that ESMM is of significate advantage in processing imbalanced data sets compared to SVM, FSVM, and B-FSVM; (4) It is found that for some data sets, for example, COIL-20, the three matrix-pattern-oriented approaches perform worse than SVM. We think it is the coincidence due to the training part is gotten in random. But according to the average result, we still find that our proposed ESMM outperforms others in average; (5) For the used 25 vector data sets, ESMM performs best on the 9 data sets of them. For others, ESMM performs better in average and it does not perform worst on any vector data set. This phenomenon can explain the superiority of the ESMM on the vector data sets; (6) As we said before, ESMM and FSVM both adopt fuzzy membership to each input pattern. Now from the experiments, it is found that compared with FSVM, the better performance of ESMM validates that the proposed entropy-based fuzzy membership evaluation approach boosts the performance of a learning machine.

3.3 Influence of Parameter k on the Performance of ESMM

In ESMM, the entropy-based fuzzy membership is evaluated based on the class probability of each training pattern. Thus, the number of nearest neighbors k might have some influence on the class probability. To further investigate the effectiveness of ESMM, here, we study the influence of k on the classification performance. All data sets given in Table 2 are used and related experimental results are given in Fig. 1. The figure shows AUC on the testing sets of the adopted real-world imbalanced data sets and image data sets with respect to k. It is found that: (1) the number of the selected nearest neighbors for calculating the class probability, i.e., k, has some influence on the classification performance since AUC curves fluctuate with respect to k on most data sets; (2) on some data sets, the classification performances are sensitive to the variation of k, since AUC curves on these data sets fluctuate greatly while on some data sets, are not; (3) in generally, with k from 8 to 12, ESMM always gets best performance. Such a result can give us a guidance that how to determine an appropriate k in practical use.

3.4 Comparison Between ESMM and Entropy-Based MatMHKS (EMatMHKS)

Here, we give the comparison between ESMM and our previous proposed learning machine, entropy-based MatMHKS (EMatMHKS) [27] which is also an entropy-based matrix learning machine. Table 5 shows the comparison between ESMM and EMatMHKS. $\uparrow$ ($\downarrow$) represents that ESMM performance better (worse) than

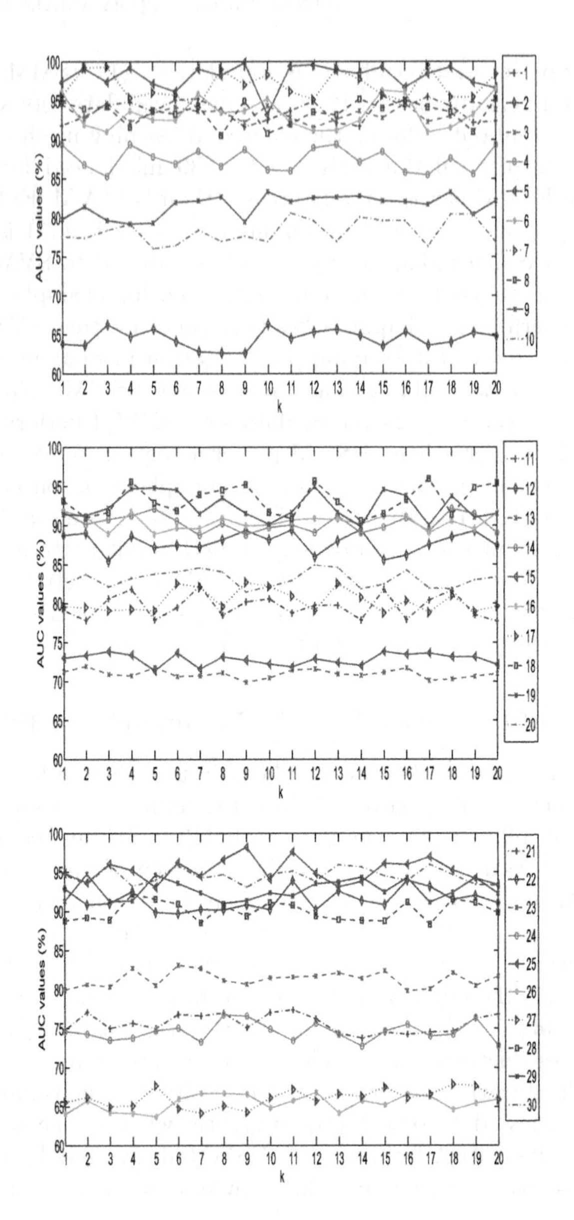

Fig. 1. Variation of AUC values of ESMM with respect to k on the adopted imbalanced data sets and image data sets. Here, in the legend, each number denotes one data set which is given in Tables 2 or 3.

EMatMHKS and ($\star$) represents that the AUC value of ESMM is $\star$ larger or smaller than the one of EMatMHKS. From this table, we find that ESMM performs better than EMatMHKS in average.

Table 5. Comparison between ESMM and EMatMHKS on AUC values (%).

Data set		Data set		Data set		Data set		Data set		Data set	
1	↑(1.02)	2	↑(0.03)	3	↑(2.11)	4	↑(0.02)	5	↑(1.01)	6	↑(0.78)
7	↓(0.23)	8	↑(1.02)	9	↑(2.98)	10	↑(6.18)	11	↓(0.03)	12	↑(3.91)
13	↑(0.82)	14	↑(0.56)	15	↑(0.81)	16	↑(2.87)	17	↓(0.12)	18	↑(1.98)
19	↑(0.04)	20	↑(1.23)	21	↑(3.12)	22	↑(2.91)	23	↑(8.19)	24	↑(2.12)
25	↑(0.01)	26	↑(2.01)	27	↑(2.10)	28	↑(1.81)	29	↑(1.07)	30	↑(0.91)

4 Conclusion

There are two hot spots of present research, one is imbalanced problem and the other is matrix learning. Imbalanced problem occurs when the size of positive class is more smaller than that of negative class. However, most standard classification learning machines result in unfavorable performance on imbalanced data sets since they are originally designed for processing balanced problems. Although SVM can process imbalanced data sets in some extent, it assigns the same importance to each training pattern. This results in the decision surfaces biasing toward the negative class. In order to overcome the disadvantage of SVM, some researchers propose FSVM and B-FSVM by applying fuzzy memberships to the training patterns to reflect different importance of them. Since the key point in FSVM and B-FSVM is how to determine the fuzzy membership, so this paper presents an entropy-based fuzzy membership evaluation approach for imbalanced data sets. Moreover, matrix patterns cannot be solved by those traditional learning machines including SVM well, so some scholars have developed MatMHKS and SMM. This paper adopts the entropy-based fuzzy membership and SMM, and then proposes ESMM. ESMM can not only guarantee importance of the positive class, but also pay more attentions to the patterns with higher class certainties. Thus, ESMM can results in more flexible decision surfaces than both conventional SVM, FSVM, B-FSVM, and MatMHKS on the imbalanced data sets. To validate the effectiveness of ESMM, we adopt 25 real-world imbalanced data sets and 5 image data sets for experiments. Experimental results demonstrates that ESMM outperforms the compared learning machines on real-world imbalanced data sets and the images. Moreover, in the process of ESMM, k has some influence on the classification performance.

Acknowledgment. This work is supported by Natural Science Foundation of Shanghai under grant number 16ZR1414500 and National Natural Science Foundation of China under grant number 61602296 and 51575336, and the authors would like to thank their supports.

References

1. Support Vector Machine. https://en.wikipedia.org/wiki/Support_vector_machine#Applications
2. Chen, S.C., Wang, Z., Tian, Y.J.: Matrix-pattern-oriented Ho-Kashyap classifier with regularization learning. Pattern Recogn. **40**(5), 533–1543 (2007)

3. Wang, Z., Chen, S.C.: New least squares support vector machines based on matrix patterns. Neural Process. Lett. **26**(1), 41–56 (2007)
4. Yan, Y., Wang, Q., Ni, G., Pan, Z., Kong, R.: One-class support vector machines based on matrix patterns. In: Jiang, L. (ed.) Proceedings of the 2011, International Conference on Informatics, Cybernetics, and Computer Engineering (ICCE2011). AINSC, pp. 223–231. Springer, Heidelberg (2011). doi:10.1007/978-3-642-25188-7_27
5. Xie, Y.B., Zhang, Z.H., Li, W.J.: Support Matrix Machines. In: The 32nd International Conference on Machine Learning (ICML 2015)
6. Dai, H.: Class Imbalance Learn. Fuzzy Total Margin Based Support Vector Mach. Appl. Soft Comput. **31**, 172–184 (2015)
7. Deng, X., Tian, X.: Nonlinear process fault pattern recognition using statistics kernel PCA similarity factor. Neurocomputing. **121**, 298–308 (2013)
8. Guo, Y., Zhang, H.Z.: Oil spill detection using synthetic aperture radar images and feature selection in shape space. Int. J. Appl. Earth Obs. Geoinf. **30**, 146–157 (2014)
9. Ozcift, A., Gulten, A.: Classifier ensemble construction with rotation forest to improve medical diagnosis performance of machine learning algorithms. Comput. Methods Programs Biomed. **104**(3), 443–451 (2011)
10. Brown, I., Mues, C.: An experimental comparison of classification algorithms for imbalanced credit scoring data sets. Expert Syst. Appl. **39**(3), 3446–3453 (2012)
11. Lin, C., Wang, S.: Fuzzy support vector machines. IEEE Trans. Neural Netw. **13**(2), 464–471 (2002)
12. Wang, Y., Wang, S., Lai, K.K.: A new fuzzy support vector machine to evaluate credit risk. IEEE Trans. Fuzzy Syst. **13**(6), 820–831 (2005)
13. Shannon, C.E.: A mathematical theory of communication. SIGMOBILE Mob. Comput. Commun. Rev. **5**(1), 3–55 (2001)
14. Alcala-Fdez, J., Fernandez, A., Luengo, J., Derrac, J., Garcia, S., Sanchez, L., Herrera, F.: A software tool to assess evolutionary algorithms for data mining problems. J. Multiple-Valued Logic Soft Comput. **17**, 2–3 (2011)
15. Alcala-Fdez, J., Sanchez, L., Garcia, S., Jesus, M.J., Ventura, S., Garrell, J.M., Otero, J., Romero, C., Bacardit, J., Rivas, V.M.: A software tool to assess evolutionary algorithms for data mining problems. Soft. Comput. **13**(3), 307–318 (2009)
16. Nene, S.A., Nayar, S.K., Murase, H.: Columbia object image library (COIL-20). Technical report CUCS-005-96. Columbia University (1996)
17. Cun, B.B.L., Denker, J.S., Henderson, D., Howard, R.E., Hubbard, W., Jackel, L.D.: Handwritten digit recognition with a back-propagation network. In: Advances in neural information processing systems (1990)
18. Bennett, F., Richardson, T., Harter, A.: Teleporting-making applications mobile. In: Mobile Computing Systems and Applications, pp. 82–84 (1994)
19. Kumar, N., Berg, A.C., Belhumeur, P.N., Nayar, S.K.: Attribute and simile classifiers for face verification. In: International Conference on Computer Vision, pp. 365–372 (2009)
20. Smith, B.A., Yin, Q., Feiner, S.K., Nayar, S.K.: Gaze locking, passive eye contact detection for human-object interaction. In: ACM Symposium on User Interface Software and Technology (UIST), pp. 271–280 (2013)
21. Milgram, J., Cheriet, M., Sabourin, R.: "One Against One" or "One Against All", Which One is Better for Handwriting Recognition with SVMs? In: Tenth International Workshop on Frontiers in Handwriting Recognition (2013)

22. Debnath, R., Takahide, N., Takahashi, H.: A decision based one-against-one method for multi-class support vector machine. Pattern Anal. Appl. **7**(2), 164–175 (2004)
23. Hsu, C.W., Lin, C.J.: A comparison of methods for multiclass support vector machines. IEEE Trans. Neural Netw. **13**(2), 415–425 (2002)
24. Cortes, C., Vapnik, V.: Support vector machine. Mach. Learn. **20**(3), 273–297 (1995)
25. Huang, J., Ling, C.X.: Using AUC and accuracy in evaluating learning algorithms. IEEE Trans. Knowl. Data Eng. **17**(3), 299–310 (2005)
26. Braga-Neto, U.M., Dougherty, E.R.: Is cross-validation valid for small-sample microarray classification? Bioinformatics **20**(3), 374–380 (2004)
27. Zhu, C.M., Wang, Z.: Entropy-based matrix learning machine for imbalanced data sets. Pattern Recogn. Lett. **88**(1), 72–80 (2017)

Using Convolutional Neural Network with Asymmetrical Kernels to Predict Speed of Elevated Highway

Di Zang[1,2(✉)], Jiawei Ling[1,2], Jiujun Cheng[1,2], Keshuang Tang[3], and Xin Li[4]

[1] Department of Computer Science and Technology, Tongji University, Shanghai, China
zangdi@tongji.edu.cn
[2] The Key Laboratory of Embedded System and Service Computing, Ministry of Education, Tongji University, Shanghai, China
[3] Department of Transportation Information and Control Engineering, Tongji University, Shanghai, China
[4] Shanghai Lujie Electronic Technology Co., Ltd., Pudong, Shanghai, China

Abstract. In this paper, we present a deep learning based approach to performing the whole-day prediction of the traffic speed for the elevated highway. In order to learn the temporal features of traffic speed data in a hierarchical way, an improved convolutional neural network (CNN) with asymmetric kernels is proposed. Speed data are collected from loop detectors of Yan'an elevated highway of Shanghai. To test the performance of the presented method, we compare it with some conventional approaches of traffic speed estimation. Experimental results demonstrate that our method outperforms all of them.

Keywords: Convolutional neural network · Kernels · Intelligent transportation system · Speed prediction

1 Introduction

It is an attractive topic for human beings to have the ability to foresee the future, and it is the same in transportation management. It is of great importance for traffic management department to learn the traffic evolution to provide a guide of tomorrow's traffic for people to select an unobstructed route. It is also of value for traffic management department to adjust the traffic strategy in advance [1, 2].

However, it is challenging to define a high-performance prediction model, because the utilization of spatiotemporal relationship was not high, we did not have the ability to form a more efficient prediction model to deal with the spatiotemporal correlation of traffic flow in roads expanding on a two-dimensional field, we were not able to forecast long-term future. Conventional traffic data prediction models usually treat the traffic data as sequential data, so these models usually cannot have a good performance, because of the limitations in implementation, assumptions and hypotheses, noisy or missing data, ineptness to deal with outliers and incapability to determine dimensions [3].

Z. Shi et al. (Eds.): ICIS 2017, IFIP AICT 510, pp. 212–221, 2017.
DOI: 10.1007/978-3-319-68121-4_22

In the existing models, there are two main research methods which dominate the study in traffic forecasting: methods based on statistic and methods based on neural networks [3].

In traffic prediction, statistical methods are widely used. The classic method is autoregressive integrated moving average (ARIMA) model. It is a time-series prediction model which considers the correlations in successive time sequences of traffic variables. Seasonal ARIMA model [4], KARIMA model [5] and ARIMAX model [6] which are the extensions of basic ARIMA model are widely researched and applied. In [7], k-nearest neighbors (KNN) has been used to forecast traffic flow. In [8], support vector machines (SVM) were employed in traffic prediction. Online-SVM and Seasonal SVM were used in [9] and [10] to improve the prediction accuracy. Methods based on statistics have been widely applied in traffic prediction because of their easy implementation and promising results. However, these models did not consider the significant spatiotemporal feature of traffic data, so these models cannot achieve a higher accuracy than models based on neural networks. Besides, some statistical methods are powerless because the model takes a very long time and consumes copious computer memory when it deals with big data.

Neural network based methods, such as artificial neural network (ANN) are usually applied to solve traffic prediction problems. ANN is able to deal with multi-dimensional traffic data. Because of its easy and flexible implementation, strong generalization ability and high performance in traffic prediction, ANN model is favored in recent research in traffic prediction. In [11], ANN was used to predict traffic speed with consideration of weather conditions. In [12], a real-time traffic speed prediction algorithm based on ANN was proposed by Park et al. A model based on ANN combined with conventional Bayes theorem to predict short-term freeway traffic flow was proposed in [13]. Moretti et al. [14] used statistical and ANN bagging ensemble model to predict city traffic flow.

ANN can make use of large amounts of data, but it cannot take advantages of spatiotemporal correlations from large amounts of traffic data. ANN are not able to achieve a better performance than methods based on deep learning. Recently, more and more deep learning models are applied to predict traffic flow because deep learning models are able to learn the deeper level features from the given data. Nowadays, Deep Belief Networks (DBNs) are widely used in traffic volume prediction. The model [15] used the method of heterogeneous multitask learning and K-means clustering to improve the prediction accuracy. Ridha et al. [16] combined DBN with weather condition to predict traffic flow using streams of data. Ma et al. [17] proposed a new model combined Restricted Boltzmann Machine (RBM) and Recurrent Neural Network (RNN) forming a RBM-RNN model, it achieves the advantages of both RNN and RBM. In [18], a Stacked Autoencoder based model was proposed to forecast traffic flow. Based on [18], Duan et al. [19] further improved the SAE model by choosing different appropriate hyperparameters at different times. Tian and Pan [20] first introduce Long Short-term Memory (LSTM) into traffic prediction. The LSTM model outperforms other neural networks in both stability and accuracy. In [21], a deep spatio-temporal residual network was applied in crowd flows prediction. Ma et al. [22] proposed a Convolutional Neural Network (CNN) based model which learns traffic data as image, the model achieves a good result in Beijing road network speed prediction. However, the CNN model in [22] treats the traffic dynamic in time series and space equally, and cannot work with the whole-day traffic data.

To solve the problems in [22], this paper introduces asymmetrical kernels to CNN, which can treat the spatial features and temporal features of traffic data differently. Because of different treatment between spatial features and temporal features, our model gets lower mean squared error (MSE) and mean related error (MRE) than common CNN. In addition, the improved model is applied to predict the whole-day traffic speed of the next day with the help of whole-day traffic speed data of the previous day.

2 Proposed Approach

In this section, we will introduce the method of transforming the loop detectors' data to matrix and the basic theory of our CNN model.

2.1 Loop Detector Data Transformation

The traffic speed of elevated highway can be provided by the loop detectors deployed on the highway. In the time dimension, the loop detectors' data range from 0:00 am to 12:00 pm. The time intervals are usually 5 min. On the elevated highway, each two loop detectors are deployed between 400 m. The loop detector data of elevated highway also can be converted to matrices by a similar method. We let x-axes represents time, and y-axes represents space. We arrange the loop detector data of elevated highway in the order of loop detectors' position and time series to form a 2D matrix. Each row in the matrix denotes speed data in different time periods recorded from a same loop detector in the elevated highway. Each column in the matrix denotes speed data from different loop detectors at a same time period. The time-space traffic speed matrix can be represented as follow:

$$S = \begin{bmatrix} s_{11} & s_{12} & \cdots & s_{1n} \\ s_{21} & s_{22} & & s_{2n} \\ & \vdots & \ddots & \vdots \\ s_{m1} & s_{m2} & \cdots & s_{mn} \end{bmatrix} \quad (1)$$

In the matrix, m denotes the amount of loop detectors, n denotes the length of time intervals and s_{ij} denotes the average speed on the loop detector i at time period j. We also can represent the matrix as a heat map. Figure 1 is the illustration of a heat map transformed from the matrix.

2.2 The Architecture of the Improved CNN

CNN has been widely used in the research of image understanding, because of its strong ability in extracting critical features from images. In the field of image classification, CNN performs better than other deep learning models, even surpasses human beings. As shown in Fig. 2, the Input of our model are matrices of spatiotemporal traffic data. In our model, there are 3 convolution layers and 2 pooling layers. The input matrix goes through two convolution layers, one pooling layer, one convolution layers, one pooling layer and one fully-connected layer in turn. The output of the model is a vector which can be reshaped to a matrix with the same size of the input matrix.

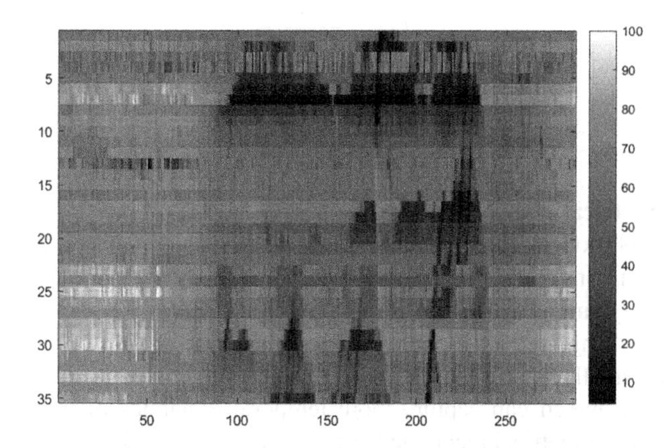

Fig. 1. The visualization of whole-day traffic speed of Shanghai Yan'an elevated highway.

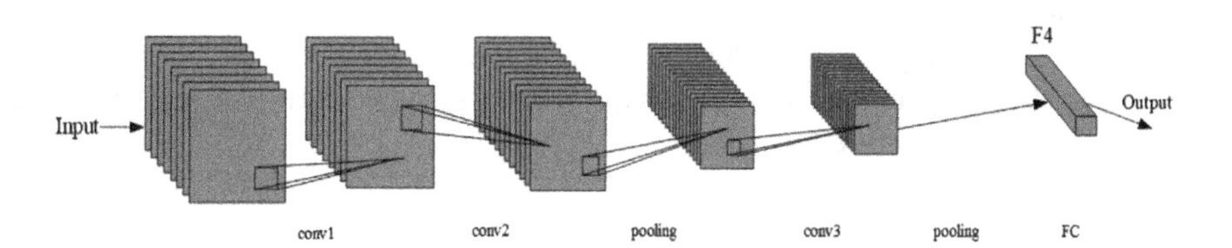

Fig. 2. The architecture of our improved CNN model.

2.2.1 The Model's Input and Output

Like common CNNs, our improved CNN accepts matrices (images) as input. However, in this paper, instead of being used to solve classification problems, we use CNN to finish regressive task. Thus, the output of our model is a vector that can be reshaped to a matrix just the same size as the input matrix which is the prediction of the next day.

2.2.2 Convolution Layers

The previous layer's feature maps or the input matrices are convolved with trainable kernels and then the feature maps are put through the activation function to make up the output feature maps. Each output feature map combines convolutions with multiple input feature maps. In common, the relationship between input and output maps of convolution layers is as follow:

$$x_j^l = f\left(\sum x_i^{l-1} * k_{ij}^l + b_j^l\right) \tag{2}$$

where x^{l-1} denotes the input of the convolution layer, k^l means the convolution layer's kernels, b denotes an additive bias, f is the activation function. We usually use sigmoid (3) function or ReLu (4) function as the activation function.

$$f(x) = \frac{1}{1 + e^{-x}} \tag{3}$$

$$f(x) = \begin{cases} 0, & x \leq 0 \\ x, & x > 0 \end{cases} \tag{4}$$

2.2.3 Asymmetrical Kernels

In common CNN models, the convolution layers' kernels are square, however, in our model, kernels in convolution layers are asymmetrical rectangle matrices, because we consider spatial dynamics and temporal dynamics differently. A traffic congestion event can impact for a long time, sometimes for several hours, while the time intervals in matrices are too small. To solve the problem demonstrated before, we use asymmetrical rectangle kernels, which can capture more temporal dynamics of traffic data. In the models, we use asymmetrical rectangle kernels with 3 × 13 size, 3 × 5 size and 3 × 5 size in different convolution layers.

2.2.4 Pooling Layers

The output of a pooling layer are down-sampled versions of input maps. If there are N input maps, then the number of output maps will be exactly N, but the size of output maps will be smaller. More formally,

$$x_j^l = f\left(\sum \beta_j^l down\left(x_i^{l-1}\right) + b_j^l\right) \tag{5}$$

where the function *down()* denotes a sub-sampling (pooling) function, usually we use max pooling function or average pooling function. Generally, the pooling function will transform each distinct n-by-n block into one pixel of the output map. Then the output maps will be multiplied by a multiplicative bias β and add a bias b.

Although the pooling layer can reduce the amount of model's trainable parameters, it also brings some information loss. In order to reduce the information loss, in our model, we cancel the pooling layer after the first convolution layer. This operation can achieve a bit performance improvement.

2.2.5 Fully Connected Layer

The fully connected layer is similar to the artificial neural network, if we use x to denote the input of fully connected layer, y to represent the output of the layer, the corresponding relation between x and y is as follow:

$$y_j^l = f\left(\sum w_j^l x_i^{l-1}) + b_j^l\right) \tag{6}$$

In the formula (6), w denotes the trainable weights between the input and output, f represents the activation function described before.

2.2.6 Model Optimization

To get an optimized model, we use stochastic gradient decent method to minimize the model's mean squared error (presented in formula 7) with the batch size is one.

$$MSE = \frac{1}{n}\sum_{i=1}^{n}(y_i - \hat{y}_i)^2 \tag{7}$$

In formula (7), y denotes the model output, and $\hat{y}$ represents the model's expected value.

3 Experimental Results

The model is evaluated using loop detectors data of Yan'an elevated highway for the year of 2011. In the experiments, we use a whole day's speed data, to predict the next day's whole-day speed data. The first 320 days of data are selected for training and the left days of data are used for testing. We use the matrix of previous day as the input of the model, the reshaped output vector is applied as the predicted value.

3.1 Handle the Data

We handle the data in the way which is similar to the method used in [22]. There are 35 loop detectors deployed on the Shanghai Yan'an elevated highway. In addition, the observed data is recorded every 5 min. Because of some limitation of loop detectors, we did some work in data cleaning. First, we reset the abnormal data, for example, some elements in the speed matrix are larger than 200 km/h, we set these data to be 100, because there are few cars can run at this speed which is also against Chinese law. According to Chinese law, the max speed on the elevated highway is 80 km/s, we take the slight overspeed into account, thus, we choose 100 km/s as the max speed in the matrix. Second, sometimes, the loop detectors employed on the Yan'an elevated highway do not work during 0:00 am to 4:00 am, in addition, we think that most people do not travel from 0:00 am to 6:00 am so we take no account of the data from 0:00 am to 6:00 am. Third, there are 3 loop detectors often cannot work, thus, 3 rows in the matrix are deleted. Last, in order to reduce the impact of abnormal elements in the matrix, we aggregate the data in time-dimension to obtain a 20-min interval. Eventually, the size of matrix is 32 × 54.

3.2 Experimental Settings

There are three convolution layers and two pooling layers. There 16 kernels with the size of 3 × 13 in the first convolution layer, 512 kernels with 3 × 11 size in the second convolution layer and 1024 kernels with 3 × 5 size in the last convolution layer. In the pooling layers, we do max pooling on the input. The experiments are conducted on a sever with i7-5820 K CPU, 48 GB memory and NVIDIA GeForce GTX1080 GPU. We implement these models on TensorFlow framework of deep learning. The configurations of our CNN model are listed as follow (Table 1):

Table 1. The configurations of our CNN model

Layer	Name	Description
Input	–	A matrix with 32 × 54 size
Layerl	Convolution	16 kernels with 3 × 13 size
Layer2	Convolution	512 kernels with 3 × 11 size
Layer3	Pooling	2 × 2 max pooling
Layer4	Convolution	1024 kernels with 3 × 5 size
Layer5	Pooling	2 × 2 max pooling
Layer6	Fully-connected	
Output	–	A vector with 1728 elements

3.3 Evaluation Metrics

The accuracy of traffic speed prediction is mainly assessed by two performance metrics which are Mean Relative error (MRE) and Mean Squared Error (MSE). MSE evaluate the model's absolute error while MRE shows the relative error of the model. MSE is demonstrated before and MRE is presented as the following:

$$MRE = \frac{1}{n}\sum_{i=1}^{n}\frac{|y_i - \hat{y}_i|}{y_i} \tag{8}$$

where y denotes model's predicted value using the data of the previous day as the model's input, $\hat{y}$ denotes the observed traffic speed value of the next day and n denotes the number of samples.

3.4 Experiment Result

As shown in Figs. 3, 4 and 5, we visualize some kernels in different convolution layers, different feature maps and the output matrix during the experiment. In Fig. 3, The three images in the first row are kernels we choose from the first convolution layers, then the images in the next row are kernels selected from the second convolution layers, the last images are kernels chosen from the third convolution layers. In Fig. 4, The images in

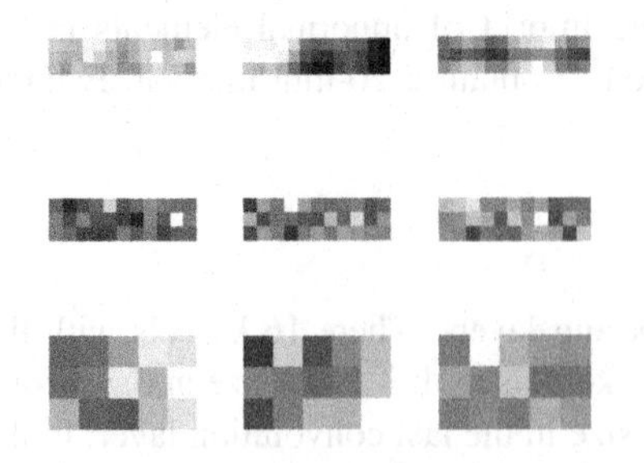

Fig. 3. Kernels' visualization from different convolution layers. The images in the first line are asymmetrical kernels of the first convolution layer and so forth.

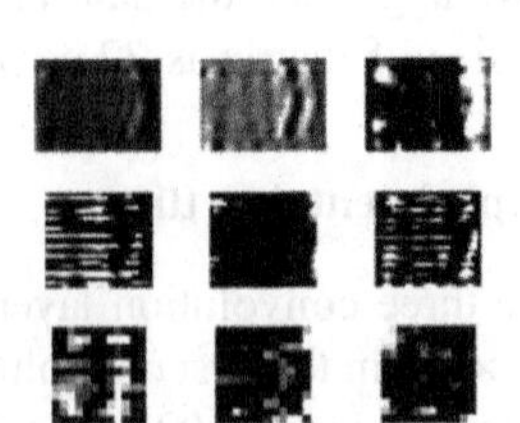

Fig. 4. Feature maps extracted from different convolution layers. The images in the first line are feature maps of the first convolution layer and so forth.

the first row are feature maps extracted from the first convolution layers, then second and third. The left image in Fig. 5 is the model's reshaped output which is used as prediction and the right is the visualized speed data in reality. In Fig. 6 the real data of the next day and the prediction of our model are represented as polylines.

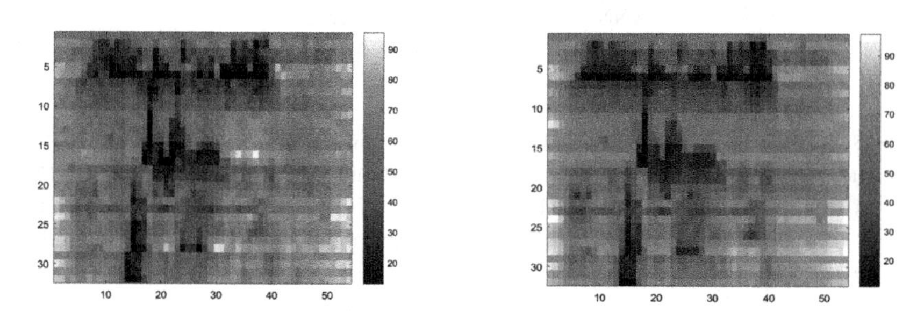

Fig. 5. The left image is the output of the model, which is transformed to heat map and the right image is the visualized real traffic speed of the next day. We can see that the output of our model is very similar to the real data.

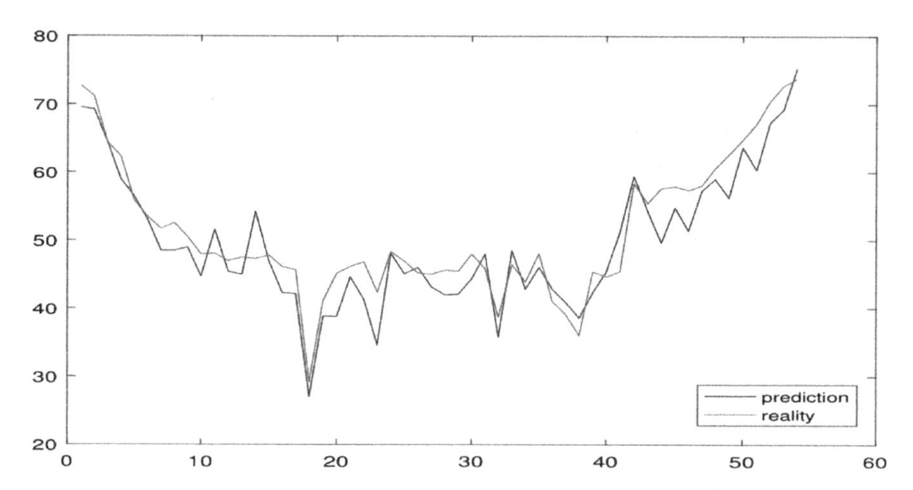

Fig. 6. The blue polyline is the model's prediction, and the red polyline represents the reality of the next day. The blue polyline can reflect the trend of the reality, and fit the reality well. (Color figure online)

We compare our CNN model with the most widely used methods of traffic flow prediction, such as ARIMA, KNN, ANN and common CNN. The performance of these models mentioned before is listed below (Table 2):

Table 2. The performance of different model mentioned before.

Model	MSE	MRE
Our Model	58.7607	0.0998
Common CNN	94.0019	0.1346
KNN	169.1917	0.2105
ANN	126.9821	0.1592
ARIMA	331.6745	0.2562

As is shown in the list, neural network based models get lower MSE and MRE than KNN and ARIMA who are not based on neural networks. In addition, our model achieves the lowest MSE and MRE. The MRE of our model is less than the rest models' over 3%. And for MSE, the MRE of our model is less than the second-best model about 30%.

4 Conclusion

In this paper, we proposed an CNN based deep learning model to predict whole-day traffic speed of elevated highway. In our model, we use asymmetrical kernels in the convolution layer. Our model focuses more on temporal dynamics which solve the problem that common methods cannot treat the special features and the temporal features differently. The experimental result proved that our model can achieve a good performance when comparing with other conventional method.

Acknowledgments. This work has been supported by the Fundamental Research Funds for the Central Universities of China and by National Natural Science Foundation of China under grant 61472284.

References

1. Zhang, J., Wang, F.-Y., Wang, K., Lin, W.-H., Xu, X., Chen, C.: Data-driven intelligent transportation systems: a survey. IEEE Trans. Intell. Transp. Syst. **12**(4), 1624–1639 (2011)
2. Park, J., Li, D., Murphey, Y.L., Kristinsson, J., McGee, R., Kuang, M., Phillips, T.: Real time vehicle speed prediction using a neural network traffic model, pp. 2991–2996. IEEE (2011)
3. Karlaftis, M.G., Vlahogianni, E.I.: Statistical methods versus neural networks in transportation research: differences, similarities and some insights. Transp. Res. Part C Emerg. Technol. **19**(3), 387–399 (2011)
4. Tran, Q.T., Ma, Z., Li, H., Hao, L., Trinh, Q.K.: A multiplicative seasonal ARIMA/GARCH model in EVN traffic prediction. Int. J. Commun. Netw. Syst. Sci. **08**(4), 43–49 (2015)
5. Voort, M.V.D., Dougherty, M., Watson, S.: Combining kohonen maps with ARIMA time series models to forecast traffic flow. Transp. Res. Part C Emerg. Technol. **4**(5), 307–318 (1996)
6. Williams, B.: Multivariate vehicular traffic flow prediction: evaluation of ARIMAX modeling. Transp. Res. Rec. J. Transp. Res. Board **1776**(1), 194–200 (2001)

7. Davis, G.A., Nihan, N.L.: Nonparametric regression and short-term freeway traffic forecasting. J. Transp. Eng. **117**(2), 178–188 (1991)
8. Wu, C.H., Ho, J.M., Lee, D.T.: Travel-time prediction with support vector regression. IEEE Trans. Intell. Transp. Syst. **5**(4), 276–281 (2004)
9. Castro-Neto, M., Jeong, Y.S., Jeong, M.K., Han, L.D.: Online-SVR for short-term traffic flow prediction under typical and atypical traffic conditions. Expert Syst. Appl. **36**(3), 6164–6173 (2009)
10. Hong, W.C.: Traffic flow forecasting by seasonal SVR with chaotic simulated annealing algorithm. Neurocomputing **74**(12–13), 2096–2107 (2011)
11. Asif, M.T., Dauwels, J., Goh, C.Y., Oran, A., Fathi, E., Xu, M., Dhanya, M.M., Mitrovic, N., Jaillet, P.: Spatiotemporal patterns in large-scale traffic speed prediction. IEEE Trans. Intell. Transp. Syst. **15**(2), 794–804 (2014)
12. Huang, S.H., Ran, B.: An application of neural network on traffic speed prediction under adverse weather condition. In: Transportation Research Board Annual Meeting (2003)
13. Zheng, W., Lee, D.H., Zheng, W., Lee, D.H.: Short-term freeway traffic flow prediction: Bayesian combined neural network approach. J. Transp. Eng. **132**(2), 114–121 (2006)
14. Moretti, F., Pizzuti, S., Panzieri, S., Annunziato, M.: Urban traffic flow forecasting through statistical and neural network bagging ensemble hybrid modeling. Neurocomputing **167**(C), 3–7 (2015)
15. Huang, W., Song, G., Hong, H., et al.: Deep architecture for traffic flow prediction: deep belief networks with multitask learning. IEEE Trans. Intell. Transp. Syst. **15**(5), 2191–2201 (2014)
16. Soua, R., Koesdwiady, A., Karray, F.: Big-data-generated traffic flow prediction using deep learning and dempster-shafer theory. In: 2016 International Joint Conference on Neural Networks (IJCNN), pp. 3195–3202. IEEE (2016)
17. Ma, X., Yu, H., Wang, Y., Wang, Y.: Large-scale transportation network congestion evolution prediction using deep learning theory. PLoS ONE **10**(3), e0119044 (2015)
18. Lv, Y., Duan, Y., Kang, W., et al.: Traffic flow prediction with big data: a deep learning approach. IEEE Trans. Intell. Transp. Syst. **16**(2), 865–873 (2015)
19. Duan, Y., Lv, Y., Kang, W., Zhao, Y.: A deep learning based approach for traffic data imputation, pp. 912–917 (2014)
20. Tian Y., Pan L.: Predicting short-term traffic flow by long short-term memory recurrent neural network. In: IEEE International Conference on Smart City, pp. 153–158. IEEE (2015)
21. Zhang, J., Zheng, Y., Qi, D.: Deep spatio-temporal residual networks for citywide crowd flows prediction. In: AAAI, pp. 1655–1661
22. Ma, X., Dai, Z., He, Z., et al.: Learning traffic as images: a deep convolutional neural network for large-scale transportation network speed prediction. Sensors **17**(4), 818 (2017)

Enlightening the Relationship Between Distribution and Regression Fitting

Hang Yu, Qian Yin(✉), and Ping Guo

Image Processing and Pattern Recognition Laboratory, Beijing Normal University, Beijing 100875, China
yuhang@mail.bnu.edu.cn, yinqian@bnu.edu.cn, pguo@ieee.org

Abstract. Statistical distribution fitting and regression fitting are both classic methods to model data. There are slight connections and differences between them, as a result they outperform each other in different cases. A analysis model for processing natural data, say astronomical pulsar data in this paper, is proposed to improve data fitting method performance. Then the insight behind the comprehensive fitting model is given and discussed.

Keywords: Statistical distribution · Regression · Model selection · Goodness of fit

1 Introduction

Statistics is a subject aiming to model the world by rather collecting and analyzing data than to inference and prove precisely. Lots of natural data have been fitted into statistical distributions with significant performances. And with these statistical distributions scientists can utilize their off-the-shelf distribution properties. While unfortunately the diversity of our world is far more complex than the level those statistical distributions can totally model. Alternatively, polynomial regression is also an approach to automatically find empirical laws from data [1]. It is proposed to fit various patterns of data, and has already shown its powerful fitting ability. Researchers who use polynomial regression are conveniently not required to preliminarily estimate data distribution [2]. Regrettably, by implementing polynomial regression we eventually get a polynomial expression with less prior properties than traditional statistical distributions have.

Given that features of natural scientific data are relatively fixed, and they will still be studied in the future, it is significant to determine the best fitting method for certain scientific data feature. And we creatively name it fitting label. As a solution to the dilemma, we propose a discriminative fitting model to help scientists automatically fitting scientific data into the optimal expression. Once scientists get the fitting label of a certain feature of natural scientific data, those later scientists are able to directly update the fitting expression parameters on new data's arrival. We operated experiments on an astronomical data set, ANTF

Z. Shi et al. (Eds.): ICIS 2017, IFIP AICT 510, pp. 222–227, 2017.
DOI: 10.1007/978-3-319-68121-4_23

Pulsar Catalogue [3]. Besides, given the reliability of P value is still controversial [4], and mean squared error and KL divergence are trustworthy measures for both fields of statistics and machine learning, we complementally took them in consideration.

The structure of this paper is shown as follows: the comprehensive fitting model is proposed in Sect. 2; and with it its corresponding experiments are showed and discussed in Sects. 3 and 4; the section of conclusion comes as the end.

2 Methodology

2.1 Comprehensive Fitting Model

As is shown in Algorithm 1, the selection model consists of mainly 3 steps. Step 1 is to determine whether the data is subject to a certain natural distribution or not through significance testing. Step 2, fitting data into significant distributions and operating regression, then calculating their respective similarities. Step 3 is to classify the data's fitting label with distance-based classifier according to the similarities. Consequently, scientists are able to learn as more properties as possible of the data.

Algorithm 1. Comprehensive model

Input: Scientific Data $\mathcal{D}$;
Candidate Statistical Distribution Families $\boldsymbol{\Pi} = \{\pi_1, ..., \pi_i, ..., \pi_m\}$;
Selected Significance Testing Methods $\boldsymbol{\Gamma} = \{\Gamma_1, ..., \Gamma_j, ...\Gamma_n\}$;
Selected Similarity Measures $\boldsymbol{\Sigma} = \{\sigma_1, ..., \sigma_j, ...\sigma_n\}$;
Polynomial Regression Order e;
Cross-Validation Times t and Fold k.
Output: Fitting Labels π^*.
1: $\hat{\boldsymbol{\Pi}} \leftarrow Algorithm2\,(\mathcal{D}, \boldsymbol{\Pi}, \boldsymbol{\Gamma}, \boldsymbol{\Omega})$
2: $\hat{\boldsymbol{\Sigma}} \leftarrow Algorithm3\left(\mathcal{D}, \boldsymbol{\Sigma}, \hat{\boldsymbol{\Pi}}\right)$
3: Fitting Labels $\pi^* \leftarrow kNN\left(\hat{\boldsymbol{\Sigma}}\right)$
return π^*

2.2 Statistical Significance Testing

As is shown in Algorithm 2, firstly we need human experts, astronomers for example, to give us candidate statistical distribution families for each feature. These distribution families with empirical parameters are then be tested one by one for statistical significance under selected statistical significance measures, usually P value. Concretely with specified natural science feature data, we perform statistical significance test to screen out distribution families that have statistically large significance levels for the feature. Candidate statistical families and selected significance measures are given into the testing process, If the output P value reaches the threshold, a significance level of 0.05 by convention, the distribution will be recorded as a significant distribution.

Algorithm 2. Statistical significance testing

Input: Scientific Data $\mathcal{D}$;
Candidate Statistical Distribution Families $\boldsymbol{\Pi} = \{\pi_1, ..., \pi_m\}$;
Selected Significance Testing Methods $\boldsymbol{\Gamma} = \{\Gamma_1, ..., \Gamma_n\}$;
Testing Methods Weights $\boldsymbol{\Omega} = \{\omega_1, ..., \omega_n\}$;
Significance Level τ.
Output: Significant Statistical Distribution Families $\hat{\boldsymbol{\Pi}} = \{\hat{\pi}_1, ..., \hat{\pi}_l\}$.
1: **for** each $\pi_i \in \boldsymbol{\Pi}$ **do**
2: **for** each $\Gamma_j \in \boldsymbol{\Gamma}$ **do**
3: $\hat{\psi}_j^i \leftarrow \Gamma_j(\mathcal{D}, \pi_i)$;
4: $\hat{\boldsymbol{\Psi}}^i \leftarrow \hat{\boldsymbol{\Psi}}^i \cup \hat{\psi}_j^i$;
5: **end for**
6: $\psi^i \leftarrow \boldsymbol{\Omega}^{\mathsf{T}} \times \hat{\boldsymbol{\Psi}}^i$;
7: **if** $\psi^i \geqslant \tau$ **then**
8: $\hat{\boldsymbol{\Pi}} \leftarrow \hat{\boldsymbol{\Pi}} \cup \pi_i$;
9: **end if**
10: **end for**
return $\hat{\boldsymbol{\Pi}}$

2.3 Similarity Evaluation

For this step, we use similarity measures to detect how well the expression fits with the data, which is shown in Algorithm 3. Significant statistical distributions

Algorithm 3. Similarity evaluation

Input: Pulsar Feature Data $\mathcal{D}$;
Significant Statistical Distribution Families $\hat{\boldsymbol{\Pi}} = \{\hat{\pi}_1, ..., \hat{\pi}_l\}$;
Selected Similarity Measures $\boldsymbol{\Sigma} = \{\sigma_1, ..., \sigma_n\}$;
Similarity Measure Weights $\boldsymbol{\Theta} = \{\theta_1, ..., \theta_n\}$;
Polynomial Regression Highest Order e;
Cross-Validation Times t and Fold k.
Output: Similarity/Goodness of Fit $\tilde{\boldsymbol{\Sigma}} = \{\tilde{\sigma}^1, ..., \tilde{\sigma}^{l+1}\}$.
1: **for** each $\hat{\pi}_i \in \hat{\boldsymbol{\Pi}}$ **do**
2: **for** each $\sigma_j \in \boldsymbol{\Sigma}$ **do**
3: $\hat{\sigma}_j^i \leftarrow CV(t, k, \mathcal{D}, \hat{\pi}_i, \sigma_j)$;
4: $\hat{\boldsymbol{\sigma}}^i \leftarrow \hat{\boldsymbol{\sigma}}^i \cup \hat{\sigma}_j^i$;
5: **end for**
6: $\tilde{\sigma}^i \leftarrow \Theta^{\mathsf{T}} \times \hat{\boldsymbol{\sigma}}^i$;
7: $\tilde{\boldsymbol{\Sigma}} \leftarrow \tilde{\boldsymbol{\Sigma}} \cup \tilde{\sigma}^i$
8: **end for**
9: **for** each $\sigma_j \in \boldsymbol{\Sigma}$ **do**
10: $\hat{\sigma}_j^0 \leftarrow CV(t, k, \mathcal{D}, Polyfit(e), \sigma_j)$;
11: $\hat{\boldsymbol{\sigma}}^0 \leftarrow \hat{\boldsymbol{\sigma}}^0 \cup \hat{\sigma}_j^0$;
12: **end for**
13: $\tilde{\sigma}^0 \leftarrow \Theta^{\mathsf{T}} \times \hat{\boldsymbol{\sigma}}^0$;
14: $\tilde{\boldsymbol{\Sigma}} \leftarrow \tilde{\boldsymbol{\Sigma}} \cup \tilde{\sigma}^0$
return $\hat{\boldsymbol{\Sigma}}$

obtained from last step are introduced here, and with it selected similarity measures. We perform t times k-fold cross-validation test to acquire the average similarity between the distribution and scientific data.

2.4 Distance-Based Classification

It is the final outcome, the fitting label for the feature data, producing step. We can take those similarity measures as distances, and run a distance-based classifier to eventually select the expression with minimal distance as the optimal fitting label. Here we bring a classic classifier k-Nearest Neighbours (abbreviated as kNN) with its parameter k chosen 1.

3 Experiments

A logical starting point is to test a distribution's normality for its popularity in natural world [5]. While for features whose scatter diagram is shown in Fig. 1, they seem obviously subject to lognormal distribution [6,7]. We took the logarithm of that data first, then tested the resulting data's normality, which is shown in Fig. 2. As we know, logarithm of scale parameter and shape parameter of lognormal distribution are respectively expectation and standard deviation of normal distribution.

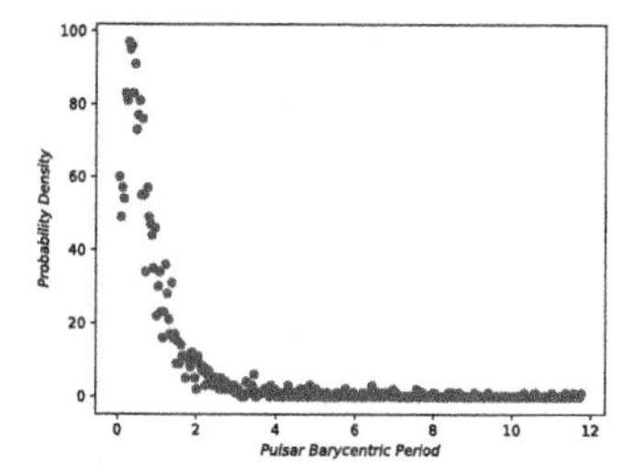

Fig. 1. Original plot

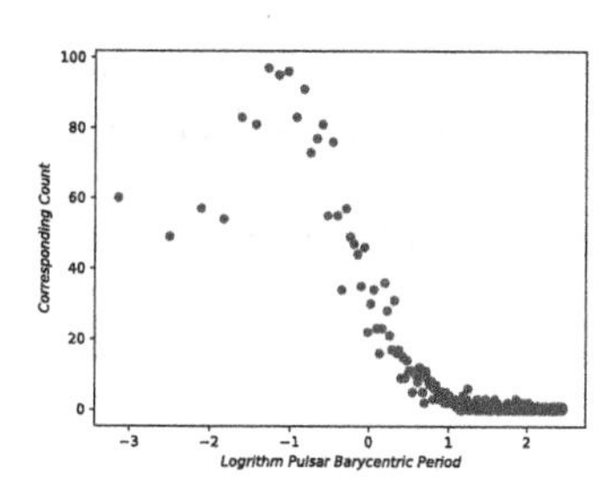

Fig. 2. Logarithm plot

Significance results are shown in Table 1. For those classic statistical testing methods, the values recorded are their P values. Results shows that lognormal has the highest significance among expert-selecting distributions.

Table 1. Significance testing of P0

P0	Normal	Lognormal	Power-law
KS test	4.21349250398968713e−61	0.487092039062517	0.0
Shaprio-Wilk test	0.0	0.155729278922081	–
Kurtosis test	9.7779914011373847e−154	0.10520924193801827	–
Skewness test	0.0	0.97680081829094689	–

Similarity evaluation results are shown in Table 2. The polynomial order is empirically chosen 2. Thus lognormal distribution has the lowest relative entropy and mean squared error. So it is the optimal fitting label for pulsar period feature. The temporary parameters of its corresponding normal distribution are mean −0.53455956344351729 and standard deviation 0.99686283436845236.

Table 2. Similarity Evaluation of P0

P0	Lognormal	Polynomial
KL divergence	0.0188633190407	0.565469835801
MSE	2.49154711009	115.375894222

4 Discussions

As we know, the insight is a thought on inductive bias of Occam's razor and the principle of maximum entropy. Occam's razor has been the default heuristic technique in natural science for hundreds of years. It can be interpreted as stating that the one among competing hypotheses with the fewest assumptions should be selected [8,9]. The preference for simplicity also comes to machine learning's avoiding overfitting, a low-order fitting curve is almost always better than a high-order one. On the contrary, the second law of thermodynamics states that the total entropy of an isolated system can only increase over time. It is seen as an axiom of statistical thermodynamics. Here comes the principle of maximum entropy which states that the probability distribution that best represents the current state of knowledge is the one with largest entropy. And it also emphasizes a natural correspondence between statistical mechanics and information theory [10]. Numbers of statistical distributions widely used by natural scientists to model scientific data can be derived under the principle of maximum entropy. For any certain mean μ and variance σ^2, the maximum entropy distribution $p\left(x|\mu,\sigma^2\right)$ is normal distribution $N\left(\mu,\sigma^2\right)$ [11].

5 Conclusions

For scientists we propose a discriminative fitting model through which diverse scientific data will be fitted into the optimal expression. And with our model as much as possible sample information can be used in future study. For data that is subject to a common statistical distribution, they then will be fitted into that distribution with parameters tuned by numerical optimization methods; comparatively for those who are not we model them by polynomial regression. Among main fields of natural science are still features whose fitting label remains unfixed, so it will be a great work to build a fitting label database with our model for scientists.

Acknowledgments. The research work in this paper was supported by the grants from National Natural Science Foundation of China (61472043, 61375045) and the Joint Research Fund in Astronomy (U1531242) under cooperative agreement between the NSFC and CAS, Beijing Natural Science Foundation (4142030). Prof. Qian Yin is the author to whom all the correspondence should be addressed.

References

1. Edwards, J.R., Parry, M.E.: On the use of polynomial regression equations as an alternative to difference scores in organizational research. Acad. Manag. J. **36**(6), 1577–1613 (1993)
2. Theil, H.: A rank-invariant method of linear and polynomial regression analysis, 3; confidence regions for the parameters of polynomial regression equations. Stichting Mathematisch Centrum. Statistische Afdeling (SP 5a/50/R), 1–16 (1950)
3. Manchester, R.N., Hobbs, G.B., Teoh, A., et al.: The Australia telescope national facility pulsar catalogue. Astron. J. **129**(4), 1993 (2005)
4. Nuzzo, R.: Statistical errors. Nature **506**(7487), 150 (2014)
5. Faucher-Giguere, C.A., Kaspi, V.M.: Birth and evolution of isolated radio pulsars. Astrophys. J. **643**(1), 332 (2006)
6. Lorimer, D.R., Faulkner, A.J., Lyne, A.G., et al.: The Parkes multibeam pulsar SurveyCVI. Discovery and timing of 142 pulsars and a Galactic population analysis. Mon. Not. R. Astron. Soc. **372**(2), 777–800 (2006)
7. Bates, S.D., Lorimer, D.R., Rane, A., Swiggum, J.: PsrPopPy: an open-source package for pulsar population simulations. Monthly Not. Roy. Astrono. Soc. **439**(3), 2893–2902 (2014). https://academic.oup.com/mnras/article/439/3/2893/1108811/PsrPopPy-an-open-source-package-for-pulsar
8. Gauch, H.G.: Scientific Method in Practice. Cambridge University Press, Cambridge (2003)
9. Hoffmann, R., Minkin, V.I., Carpenter, B.K.: Ockham's razor and chemistry. Bulletin de la Socit chimique de France **133**(2), 117–130 (1996)
10. Jaynes, E.T.: Information theory and statistical mechanics. Phys. Rev. **106**(4), 620–630 (1957)
11. Jaynes, E.T.: Probability Theory: The Logic of Science. Cambridge University Press, Cambridge (2003)

Application and Implementation of Batch File Transfer in Redis Storage

Hu Meng(✉), Yongsheng Pan, and Lang Sun

HEFEI City Cloud Data Center Co., Ltd., Hefei 230601, China
menghu@citycloud.org.cn

Abstract. In the face of massive information, batch processing of files is an important way of information transmission and storage, and the application is quite common. With the increasing demand for batch files processing reliability and speed, and the problem of low storage efficiency for current batch file processing, the paper proposes a storage method that combine distributed storage system HDFS file storage advantage and Redis cache technology to form a rapid batch merge files. The files that meet the conditions are merged into the Sequence File and stored in the HDFS. The multiple linear regression analysis method is used to determine the load factor, so that the load balancing is adjusted and the Redis cache hash data is used to ensure the efficiency. Through experiments on the corresponding file platform for file upload, query, delete and memory usage, we analysis batch processing method and non-batch method comparatively. It can be concluded that compared with the non-batch direct upload file to HDFS way, improved batch file processing method can process files more faster and ensure the stability and reliability of the file at the same time.

Keywords: Redis · HDFS · Batch processing · Distributed file system

1 Introduction

File system is an important way to transmit and store information. The use of the scale continues to expand, such as the office system, mail, message system through which information can be shared and distributed quickly. Users in such applications, not only requires high-speed processing speed, but also requires the reliability of storage. Therefore, massive files in cloud storage research has important practical value.

Massive file storage is generally based on HDFS. HDFS is a distributed file system, through the cheap multi-machine support large-scale data sets of large file storage, with strong scalability, and solve the storage problem of space constraints. Meantime, HDFS can provide high-throughput data access. It is ideal solution for large-scale data set applications, and even in the case of error can guarantee the reliability of data storage. It assumes that the calculation elements and storage would fail, so it maintains multiple copies of the work data to ensure that they can be redistributed against the failed nodes. It works in parallel to ensure efficient processing. But the storage efficiency of small files in HDFS is not high. It uses NameNode to maintain the mapping of file path to the data block and the mapping of the data block to DataNode, and also monitor DataNode

Z. Shi et al. (Eds.): ICIS 2017, IFIP AICT 510, pp. 228–233, 2017.
DOI: 10.1007/978-3-319-68121-4_24

heartbeat and maintain the number of data block copies. When a large number of small files stored in HDFS then the NameNode will run out of most of the memory, resulting in low storage efficiency, limiting the file access speed.

Taking into account the above-mentioned problems, we use a separate server with large memory to cache the data to be merged. It would improve the performance of the management node, and avoid the main server bottlenecks. The cache server uses Redis to store data. Redis is a memory-based high-performance Key/Value database. It writes updated data to disk or writes modified operation to additional log files periodically to ensure data persistence. And the Master-Slave synchronization provides a high availability and reliable platform to users. The first upload files cache in Redis, writing to disk operation only need to execute one time after merging the files which would reduce the times of disk I/O. And file uploading processes in memory which can provide a significant reduction of response time of file uploading.

2 Algorithm Summary of Batch Processing

The file storage scheme designed in this paper is to build an intermediate platform between users and HDFS system to handle the upload, query and delete operations of received files. As large files can be stored directly and efficiently in HDFS, the platform only process small files. Processing of files that larger than 32 M would return a processing-received tag directly. Consolidated storage scheme as shown in Fig. 1. Users interact with the platform through socket. Redis is used to cache user files. Caching files merge and store in the Sequence File of HDFS through the HDFS interface. The metadata records cache in Redis.

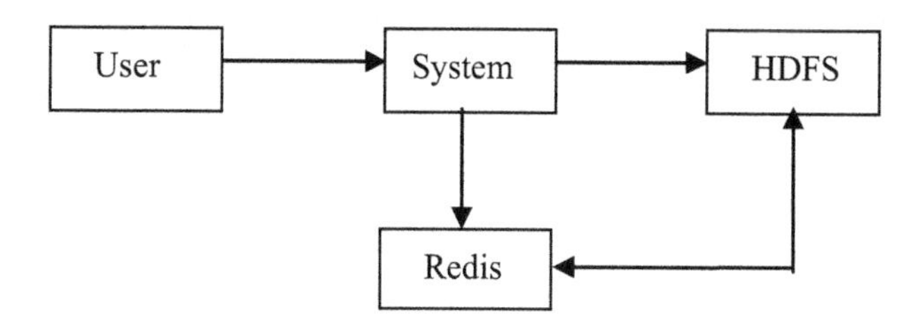

Fig. 1. Files consolidated storage solution

2.1 Storage Structure

Redis as a file cache database, save the cache file content and metadata records. The cache structure design is as follows (Table 1):

The storage structure of HDFS platform is the Sequence File stored and combined under basic directory, named after timestamp. The Sequence File uses the sequential storage structure so that we can quickly locate the contents of the small files through the file location Store Position recorded in metadata corresponding to File Name. And, Sequence File uses Block compression to reduce disk usage and increase transmission speed. Block compression is a series of records, that is, the small files here, organized

Table 1. Redis cache data storage structure

Name	Type	Description
RCF	Hash	Cache the data of the file, including the contents of the file. Key is the file name, value is the contents of the file
RCFL	Long	The length of the file in the cache, that is, the total length of the file data stored in the RCF
MH:DID	Hash	File information that update to metadata record after serialization. For example, MH: 12 store all the metadata structure of the folder identified as 12. Key for the file name, value for the metadata
SDIR:DID	Hash	Folder structure, key for the folder, value for the folder name
DID	Long	Automatic growth of the folder identification

together, unified compressed into a Block. Block information mainly store: the number of records contained in the block, the set of the length of each record Value, and the set of the value of each record Value.

2.2 Storage Platform Implementation

According to the above design proposal, based on the load cost model, the file platform is divided into two parts: basic processing and background processing. Users process basic file operation such as upload, modify and delete through the platform interacted with Redis and Sequence File; The timer combined with the basic operation triggers the event to invoke background processing to ensure the reliability and speed of the system.

When uploading a file, the received file is stored in the file cache RCF in Redis and the RCF Length of the file (RCFL) stored in the RCF is updated. Then, to determine the RCFL, if the length achieved the size for merge, a "merge file" message MF is send to the background processing module. When reading a file, the file will be returned directly if it exists in the upload buffer of Redis. Otherwise, the contents of the file will be read by the cache processing module and returned to the client. When deleting a file, first determine whether it exists in RCF. If true then delete it. Otherwise, the metadata of delete flag will be set to 0 and mark the file would be deleted.

2.3 Load Cost Model

As a complete system, not only to improve the efficiency of file storage, but also take into account the system load conditions. The load cost of existing server resources generally evaluated by the usage of independent CPU or memory. This statistics is not comprehensive. For example, high CPU usage will not affect the operation which only occupy high memory and disk I/O usage, and in actual use, the various types of resources requirements of the thread operation cannot be comprehensively evaluated either.

In order to make up for the shortage of resource statistics, this paper puts forward a load cost model which use the user's response time as an estimate criteria, based on the experimental analysis of statistical data to determine the cost of the formula, so to evaluate the effects of variety factors more reasonable. The process is: while the system is running normally, gather statistics and analysis the various types of resources, such as CPU, memory, disk I/O and other in real-time, record the response time of task processing threads and main customer service thread, to determine various factors. In this way, to avoid the lack of timeliness of traditional statistical estimates, the use of real-time computing can ensure reliable and comprehensive analysis with all kinds of environmental resources.

This paper combines the features of Sequence access and based on the GD-SIZE algorithm, calculate the cost H with the formula (1), and archive the small file cache replacement strategy.

$$H_i = N/S \tag{1}$$

The general GD-SIZE algorithm is: Each document in the buffer has a corresponding cost. When the document is brought into the buffer, the H value of the document is the reciprocal of the document size. When the replacement occurs, document which has the smallest H value H_{min} is swapped out, and the H value of the remaining document becomes the H value before the replacement minus H_{min}. According to the characteristics of Sequence File, reading the file in a single block may need to traverse many times. The value of H that GZ-SIZE algorithm used cannot actually reflect the cost of the document. The cost of the document has positive correlation with the traversed files number N for visiting the file. We can multiply the reciprocal of the file size S by N as the initial value of the GZ-SIZE algorithm, to achieve cache replacement.

3 Experiment

To establish the load cost model and determine the load cost formula, the influence of various factors on the response time of the main thread needs to be quantified. The coefficient of influence of the factor is obtained by the method of multiple linear regression analysis. In this system, CPU usage (C), memory usage (M), and disk I/O (D) have a major impact on performance, so they are used as dependent variables and uploading response time (T) as response variable, the multiple regression equation is expressed as:

$$T = k_1C + k_2M + k_3D \tag{2}$$

The specific operation is as follows: In the running nodes of the platform, execute multiple processes that have great impact to C, M and D to get different resource occupancy results, and gather the file upload time statistics. The results are as follows (Table 2):

Regression analysis of the results can calculate that k1 = 0.257, k2 = 0.332, k3 = 0.103. In order to enable the user get responding within 100 ms, the response time T calculated with C, M, and D should be less than 100 ms as the expected threshold of the load. In the experimental environment, the value of k1, k2, k3 is input

Table 2. Response time of the user request

T/ms	C/%	M/%	D/(Blk_wrtn/s)
49	13	33	17.03
59	15	31	184.00
92	63	21	188.00
55	23	22	0.00
57	12	21	0.00
73	36	32	116.00
51	45	43	8.00
68	89	21	32.00

to the running configuration. When the background message MF is received, the load threshold is calculated by the formula.

In order to eliminate the impact of unstable factors (such as speed), randomly selected 10 small files in the standard HDFS and the use of optimization modules in the file system to upload, and ultimately get the cost time shown below (Fig. 2):

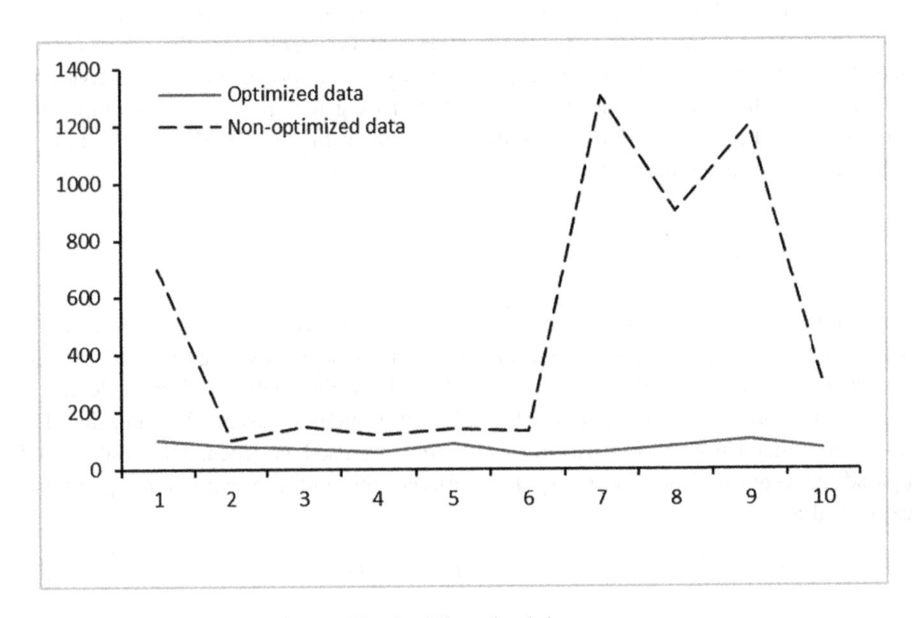

Fig. 2. File upload time

It can be seen that the time for file uploading is significantly reduced by batch merging of files, which is reduced from an average of 453.1 ms with traditional way to an average of 52.3 ms by 88.45%. In the file uploading in a batch file, the implementation changes from receiving files through original HDFS memory and writing the disk later to receiving files by Redis memory directly, no longer need to wait for the slow disk I/O operation of writing. The upload speed is significantly improved.

4 Conclusion

In this paper, we consider the storage method of small files in HDFS, design and implement the small files storage optimization based on reliable HDFS file system. Combining the Redis cache mechanism effectively reduces the memory usage of the NameNode node, the disk I/O is reduced compared to the traditional HDFS files merge method, speeding up the file uploading and acquiring speed in a large number of frequent file reads. It can be seen from the results of the experiment that the Redis-based HDFS file batch merge storage optimization method can improve memory utilization and speed up the file retrieval speed, and not affect the speed of file updating and querying, ensure the fast and reliable file operation and preservation.

References

1. Codd, E.F.: A relational model of data for large shared data banks. Commun. ACM **13**(6), 377–387 (1970)
2. Stonebraker, M.: SQL databases v. NoSQL databases. Commun. ACM **53**(4), 10–11 (2010)
3. Karger, D., Lehman, E., Leighton, T., et al.: Consistent hashing and random trees: distributed caching protocols for relieving hot spots on the World Wide Web. In: ACM Symposium on Theory of Computing, pp. 654–663. ACM, New York (1997)
4. Kubiatowicz, J., Bindel, D., Chen, Y., et al.: OceanStore: an architecture for global-scale persistent storage. In: Proceedings of the Ninth International Conference on Architectural Support for Programming Languages and Operating Systems, pp. 190–201. ASPLOS, Boston, MA (2000)
5. Pester, M.: Multidisciplinary conceptual aircraft design using CEASIOM. Hochschule für Angewandte Wissenschaften Hamburg, Hamburg(2010)
6. Heath, C., Gray, J.: OpenMDAO: framework for flexible multidisciplinary design, analysis and optimization methods. In: AIAA/ASME/ASCE/AHS/ASC Structures, Structural Dynamics and Materials Conference, AIAA/ASME/AHS Adaptive Structures Conference, AIAA (2012)

The Optimization Algorithm of Circle Stock Problem with Standard Usage Leftover

Chen Yan[1,2](✉), Qiqi Xie[2], Qiulin Chen[2], Li Zhang[2], Yaodong Cui[2], and Zuqiang Meng[2]

[1] Department of Business Administration, South China University of Technology, Guangzhou 510640, China
gxcy@foxmail.com

[2] Department of Computer and Electronic Information, Guangxi University, Nanning 530004, China

Abstract. In view of the cutting stock problem of the plate fragments and scrap cannot make full use, this paper introduces the concept of standard usage leftover, uses the recursive algorithm and the sequential heuristic procedure to solve the circle cutting problem, achieving the follow-up orders specification to use the usage leftover of the previous orders. This method can make the cutting process more simple, convenient usage leftovers inventory management, as well as conform to the requirements of the blanking for a long time. Computational experimental results show that the algorithm has a high material utilization, and playing a guiding role to the actual industrial production.

Keywords: Standard usage leftover · Recursive algorithm · Sequential heuristic procedure · Circle stock problem

1 Introduction

The problem of circle stock cutting exists in many industrial processes. It is inevitable to produce leftover, which results of material waste and inventory management difficulties [1]. There are usually three models are used to solve the Cutting Stock Problem with Leftover (CSPL): minimizing bar cost; minimizing leftover and waste; minimizing the number of sheet. The model Cui [2] proposed aims at minimizing bar cost. It discusses One-dimensional Multiple Stock Size Cutting Stock Problem (1DMSSCSP). Residual length of a cutting pattern is taken as a leftover if it is longer than a threshold; as trim loss otherwise. The model Trkman [3] proposed aims at minimizing leftover and waste. It discusses General One-dimensional Cutting Stock Problem (G1D-CSP). According to the order is cyclical or not, Trkman classifies the problem into three types: (1) all items will be produced in one production cycle; (2) all items will be produced in multiple production cycle when order and material is ensure; (3) all items will be produced in multiple production cycle. But only the first production cycle task is ensured. The model Chen [1] proposed aims at minimizing sheet number. She pointed out two aspects: (1) Stock number is taken as the main part while usage

Z. Shi et al. (Eds.): ICIS 2017, IFIP AICT 510, pp. 234–240, 2017.
DOI: 10.1007/978-3-319-68121-4_25

leftover value is neglected; (2) As a part of stock, leftover has been included in production cost. In concrete solution method, Cui [2] used integer programming (IP) and column generation algorithm to solve 1DMSSCSP. Andrade [4] came up with mixed integer programming to solve how to use leftover. Cherri [5] discussed leftover length and defined the concept of waste. Yurij [6] proposed the concept of standard existing algorithms are restricted to 2D objects; Miyazawa [7] presented iterative separation management which could simplified process.

In this paper, a concept of standard leftover is proposed, and Recursive Algorithm (RA) with Sequential Heuristic Procedure (SHP) is used to maximize the sum of items and leftover value.

2 Mathematical Model

The meaning of circle stock problem with standard usage leftover is direct cutting along the optimal line of a vertical plate on the rectangular stock. The left side is the sheet stock used for the current order and the right is the standard leftover for the subsequent order; the material for the current order only uses the sheet on the left. There are three concepts.

(1) Leftover: remained part in the cutting stock process. The leftover sheet stock won't be placed any items in this order, it will be used in the subsequent order.
(2) Standard leftover: the usage leftover generated in current order. It has the same width with stock. The standard leftover can be used in the next order. As Fig. 1 shows, the part in the lower right is standard leftover, and the cross slash area is waste.

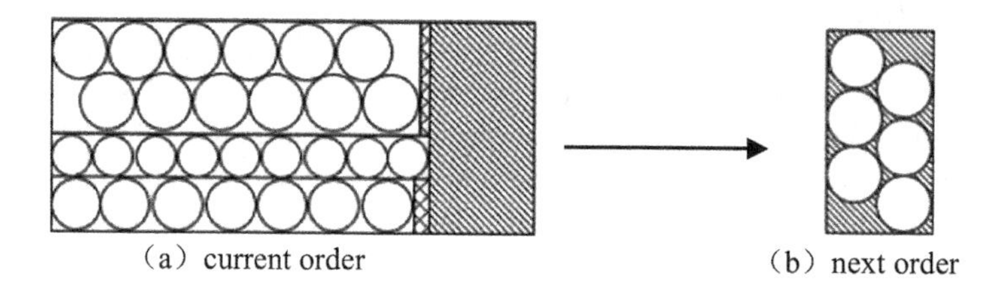

Fig. 1. Standard leftover and the waste

(3) Waste: the remaining part in the cutting stock process. The leftover sheet stock won't be placed any items in this order, it won't be used in the subsequent order.

The problem is characterized by the following data:

L: stock length
W: stock width
K: number of cutting patterns
τ: number of standard leftover types;

l_k: length of leftover appearing in cutting pattern k, $k = 1, \ldots, K$
V_k: value of leftover appearing in cutting pattern k
e_{ik}: number of leftover type i appearing in cutting pattern k
a_{ik}: number of item type i appearing in cutting pattern k
x_k: frequency of cutting pattern k
v_i: value of item type i; $i = 1, \ldots, m$
$N_{\max}$: maximum number of standard leftover type;
B: $B = [b_1, \ldots, b_m]$, b_i is the demand of item type i

The problem can be formulated as the following programming problem:

$$\text{Maximize}: \sum\nolimits_{k=1}^{K} (a_{ik} v_i) x_k + V_k \tag{1}$$

$$\text{Subject to:} \sum\nolimits_{k=1}^{K} a_{ik} x_k \geq b_i \tag{2}$$

$$V_k = \sum\nolimits_{k=1}^{K} (e_{ik} v_i) x_k \tag{3}$$

$$\tau \leq N_{\max}\ 0 \leq l_k \leq l_{\max}\ x_k \in N\ \tau \in N \tag{4}$$

Function (1) is to maximize the sum value of items. Constrains (2) constrain the number of item produced. Constrains (3) is to circulate leftover value.

In fact, the total value of stocks is the sum of items value and usage leftover value because of the introduction of standard leftover concept.

3 Algorithm Description

(1) Generate Cutting Pattern

Standard leftover length is l_k, width is W_k. There will be one or multiple rows in sheets. When $l_k = 0$, no leftover is produced. Traversing all items for this year, making inventory management of leftover to make it usable for future. RA [9] is used to solve the direct cutting problem. It will assemble a horizontal homogeneous sheet along the direction perpendicular to the width of the plate. 2 sheets are considered when calculating sheet value. One sheet length is x and width is y−1 while another sheet length is x−1 and width is y. The sheet with bigger value will be chosen.

$F(x, y)$: value of stock whose length is x and width is y
$d_{\min}$: length of the smallest item
u_i: value of sheet consists of item i
n_i: number of item i appearing in the sheet whose width is w_i.

The recursive algorithm can be expressed as:

If $y < d_{\min}$ or $(L - l_k) < d_{\min}$, $F(L - l_k, y) = 0$
If $y \geq d_{\min}$ and $(L - l_k) \geq d_{\min}$,

$$F(x,y) = \begin{cases} \max\{F(L-l_k-1,y), F(L-l_k,y-1)\} \\ \max\limits_{1 \le i \le m} [u_i + F(L-l_k, y-w_i)], w_i \le y, n_i > 0 \end{cases} \quad (5)$$

When sheet stock length or width is less than $d_{\min}$, no item can be cut from the sheet. So stock value is equal to leftover value; otherwise, it is equal to the sum value of sheet and leftover.

(2) Generate Cutting Plan

The sequential heuristic algorithm is used to solve the circle stock problem with standard usage leftover. It refers to getting a series of cutting pattern with given constraints by modifying the current remaining demand and item value until the demand is 0. Finally, stocks are cut into sheets and standard leftover. The cutting plan will be completed, and an optimized one which has the highest utilization ratio is obtained.

Input: $L * W$, m, d_i, b_i;

Step1: input initialize $v_i = \pi\left(\frac{d_i}{2}\right)^2$, $P = \Phi$, residual demand of item i $h_i = b_i$;

Step2: use RA;

Step3: circulate the frequency x_k and add the cutting pattern into set P;

Step4: modify item value;

Step5: update h_i, if $h_i \neq 0$, then go to step2; otherwise, end loop.

Output: optimal cutting plan.

4 Experiment

Experiments are on VS2013 platform with C# on windows10, and the computer is 2.60 GHz, 8 GB RAM. Experiments are compared with literature [9]. The range of parameter values is shown in Table 1. Random generating 500 test instances. The upper bound of leftover length and type is 700 mm and 10 respectively.

Table 1. Range of parameter value in experiment

Name	Range	Name	Range
Stock length (mm)	2000–3000	Stock width (mm)	1000–1500
Item type	2–5	Item diameter	100–500
Item demand	500–3000	Maximum rows in sheet	3

Figure 2 is a cutting plan of one of the random instances. Amount of item is 4, and diameters are 188, 278, 232, and 182. Demands are 2770, 1350, 700, 950. The computational experimental results show that stock number is 115, the cutting pattern is 4, and the standard leftover is 3. The cutting patterns are shown in Table 2. The average utilization is 74.69% while the one in paper [9] is 70.81%, which shows the average is enhanced by 3.88%.

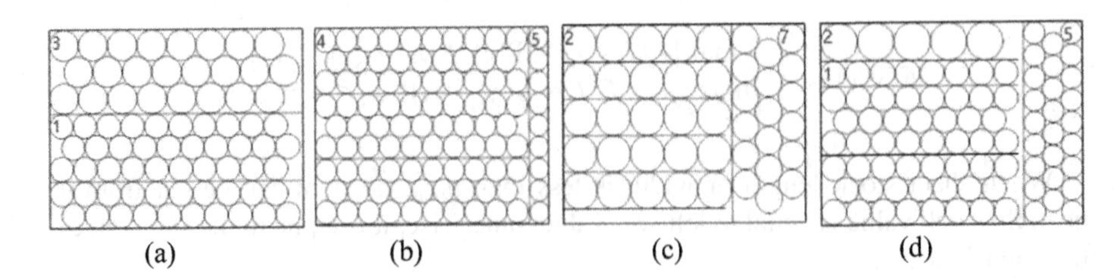

Fig. 2. Random instances

Table 2. Cutting patterns of instance

Cutting pattern	Item type	Item number	Stock number	Leftover (item diameter * rows)	Utilization ratio
(a)	4	87	11	166 * 1	82.16%
(b)	1, 3	50, 24	30	No leftover	80.15%
(c)	1, 2	54, 5	24	166 * 3	79.05%
(d)	2	25	50	220 * 3	73.59%

In experiments, according to whether or not to produce leftover, the effective of producing standard leftover on the utilization ratio improvement of stocks is analyzed. Stock size is 2000 * 1000. Randomly produce 10 circle items. Item diameter: 209, 159, 318, 179, 244, 375, 117, 348, 401, 157; Item required: 2857, 2747, 1514, 1993, 2108, 551, 2794, 2689, 1518, 992.

Table 3 shows the cutting stock plan when leftover is allowed. There are 10 type cutting patterns. When leftover is not allowed, the cutting patterns are shown in Table 4. The same type cutting patterns is 10. But the average utilization and the total stocks are different. The difference shows in Table 5. We can learn that the former utilization is 5.35% higher than the latter, and the former plates are 34 more than the latter.

Table 3. The patterns with leftovers

Cutting pattern	Item type	Item number	Stock number	Leftover size (item diameter mm * rows)	Utilization
(a)	4, 7	22, 100	28	No leftover	81.66%
(b)	1, 10	27, 36	28	No leftover	81.38%
(c)	4	64	22	No leftover	80.74%
(d)	1, 2	18, 48	58	No leftover	78.74%
(e)	3, 5	6, 23	92	No leftover	77.81%
(f)	1	34	32	189 * 3	78.17%
(g)	8	14	193	257 * 1	74.56%
(h)	3, 9	9, 4	107	189 * 2	73.83%
(i)	9	8	137	189 * 2	63.31%
(j)	6	8	69	473 * 1	61.92%

Table 4. The patterns without leftovers

Cutting pattern	Item type	Item number	Stock number	Leftover (item diameter * rows)	Utilization
(a)	4, 7	22, 100	28	No leftover	81.66%
(b)	1, 10	27, 36	28	No leftover	81.38%
(c)	4	64	22	No leftover	80.74%
(d)	1, 2	18, 48	58	No leftover	78.74%
(e)	3, 5	6, 23	92	No leftover	77.81%
(f)	1	45	24	No leftover	77.40%
(g)	3	18	54	No leftover	71.67%
(h)	8	15	180	No leftover	71.53%
(i)	6	10	56	No leftover	55.37%
(j)	9	8	190	No leftover	50.65%

Table 5. The difference between with leftovers and without leftovers

	Used stocks	Average utilization
With leftovers	732	72.22%
Without leftovers	766	66.87%

5 Conclusion

In this paper, the concept of standard leftover generated, so that it can be available for the later orders. The constraints of leftover will make it convenient for inventory management, which is benefit for long period order. Experiments show the combination of sequential heuristic algorithm and recursive algorithm has a higher material utilization. It will play a guiding role in actual production.

Acknowledgments. This research is part of Projects 71371058 and 61363026 supported by National Natural Science Foundation of China.

References

1. Chen, Q.L.: Optimization algorithms for the two-dimensional three-staged guillotine cutting stock problem. Institutes of Technology of South China, Guangzhou (2016)
2. Cui, Y.D., Song, X., Chen, Y.: New model and heuristic solution approach for one-dimensional cutting stock problem with usable leftovers. J. Oper. Res. Soc. 1–12 (2016)
3. Trkman, P., Gradisar, M.: One-dimensional cutting stock optimization in consecutive time periods. Eur. J. Oper. Res. **179**(2), 291–301 (2007)
4. Andrade, R., Birgin, E.G., Morabito, R.: Two-stage two-dimensional guillotine cutting stock problems with usable leftover. Int. Trans. Oper. Res. 1–25 (2014)
5. Cherri, A.C., Arenales, M.N., Yanasse, H.H.: The one-dimensional cutting stock problem with usable leftover-a heuristic approach. Eur. J. Oper. Res. **196**, 897–908 (2009)

6. Yurij, S., Johannes, T., Guntram, S., et al.: Φ-functions for primary 2D-objects. Stud. Inform. Univers. **2**, 1–32 (2002)
7. Miyazawa, F.K., Pedrosa, L.L.C., Schouery, R.C.S., et al.: Polynomial-time approximation schemes for circle and other packing problems. Algorithmica **76**(2), 536–568 (2016)
8. Cui, Y.P., Tang, T.B.: Parallelized sequential value correction procedure for the one-dimensional cutting stock problem with multiple stock lengths. Eng. Optim. **46**(10), 1352–1368 (2014)
9. Cui, Y.D., Xu, D.Y.: Strips minimization in two-dimensional cutting stock of circular items. Comput. Oper. Res. **37**(1), 621–629 (2010)
10. Cui, Y.D., Yang, Y.L.: A heuristic for the one-dimensional cutting stock problem with usable leftover. Eur. J. Oper. Res. **204**(2), 245–250 (2010)
11. Chen, Y., Xie, Q.Q., Liu, Y.: The cutting stock problem of circular items based on sequential grouping heuristic algorithm. Eur. J. Oper. Res. **38**(1), 5–9 (2017)

Weighting Features Before Applying Machine Learning Methods to Pulsar Search

Dayang Wang, Qian Yin(✉), and Hongfeng Wang

Image Processing and Pattern Recognition Laboratory,
Beijing Normal University, Beijing 100875, China
wdyan9@163.com, yinqian@bnu.edu.cn, dzuwhf@163.com

Abstract. In recent years, different Artificial Intelligence methods have been applied to pulsar search, such as Artificial Neural Network method, PEACE Sorting Algorithm, Real-time Classification method. In this paper, Weighting Feature method before applying machine learning (ML) was proposed. We give weight to each feature according to its ability to distinguish pulsar and non-pulsar candidates. The ability is determined by the separation degree of the distribution of pulsars and non-pulsars on particular feature. And then use the ML methods to classify different types of candidates. The results show that this method is significant. The accuracy of identifying pulsars and modeling time were both improved after weighting.

Keywords: Weighting · Machine learning · Pulsar search · WEKA

1 Introduction

Pulsar is fast rotated neutron star, which periodically sends pulse signal whose period is short and very stable. Pulsar plays an important part in physics, astronomy and many other fields. In recent years, AI methods like image pattern recognition [2], artificial neural network method and scheduling algorithm are used in pulsar search. Lee et al. [6] proposed the PEACE sorting algorithm to search pulsar, which had obtained good results. Lyon et al. [5] used the GH-VFDT (Gauss-Hellinger Very Fast Decision Tree) to distinguish the candidate, with recognition rate of pulsars over 90% [3].

While GH-VFDT obtained a high recognition rate of pulsars, the difference between the abilities of different features to distinguish the pulse and non-pulsar are not reflected. Thus, in this paper, we add different weights to the eight features before the machine learning process according to their separation degree. Results show that weighting improves both the accuracy rate of classification and modeling time.

The structure of this paper is shown as follows: the related work is mentioned in Sect. 2; the Feature Weighting method is proposed in Sect. 3; and with its corresponding experiments are showed and analyzed in Sects. 4 and 5; the section of conclusion comes as the end.

Z. Shi et al. (Eds.): ICIS 2017, IFIP AICT 510, pp. 241–247, 2017.
DOI: 10.1007/978-3-319-68121-4_26

2 Related Work

2.1 Feature

In the process of searching for pulsar signals with radio telescope, the most basic data are obtained. These data are subjected to Removing signal interference, de dispersion, FFT [4] and periodic search. Then a pulsar candidate is generated which has some basic Features. Lyon et al. used eight new features to describe the pulsar candidate. The eight features are Mean of the integrated profile $Prof_{\mu}$, Standard deviation of the integrated profile $Prof_{\sigma}$, Excess kurtosis of the integrated profile $Prof_{k}$, Skewness of the integrated profile $Prof_{s}$, Mean of the DM-SNR curve DM_{μ}, Standard deviation of the DM-SNR curve DM_{σ}, Excess kurtosis of the DM-SNR curve DM_{k}, Skewness of the DM-SNR curve DM_{s} [5]. Pulsar Feature Lab and Presto [6] are used to process the primitive data into these eight features.

2.2 Dataset

Three separate datasets were used to the measure the performance of ML methods on pulsar search. The small scale dataset is LOTAAS which was obtained during the LOTAAS survey and is currently private. The medium scale dataset HTRU2 was obtained during an analysis of HTRU Medium Latitude data by Thornton (2013). The large scale dataset HTRU1 is produced by Morello et al. The detailed information of the three datasets is summarized in the Table 1.

Table 1. Three pulsar candidate datasets

Dataset	Creator	Time	Volume	Pulsar	Non-pulsar
LOTAAS	Morello et al.	2012	5053	66	4987
HTRU2	Thornton	2013	17898	1639	16259
HTRU1	Lofar	2013	91191	1196	89995

3 Methodology-Feature Weighting

Analyzing the statistic distribution of the eight features from the sample data of pulsars and non-pulsars, feature data was extracted from 90,000 labelled pulsar candidates produced by Morello et al. [8], via Pulsar Feature Lab. As it is showed in Fig. 1, the data were scaled to the interval of [0, 1]. For each feature, there are two box plots. The orange red box shows the feature distribution of known pulsars, while the box in light blue describes the RFI/noise distribution.

It is obvious that when we are classifying a pulsar candidate via its feature, the feature that has a high degree of separation between pulsars and non-pulsars weighs more than other features. Therefore, this paper naturally adds different weights to the eight features according to their separation degree between different types of candidates. As a specific feature, this paper defines the separation degree as follows:

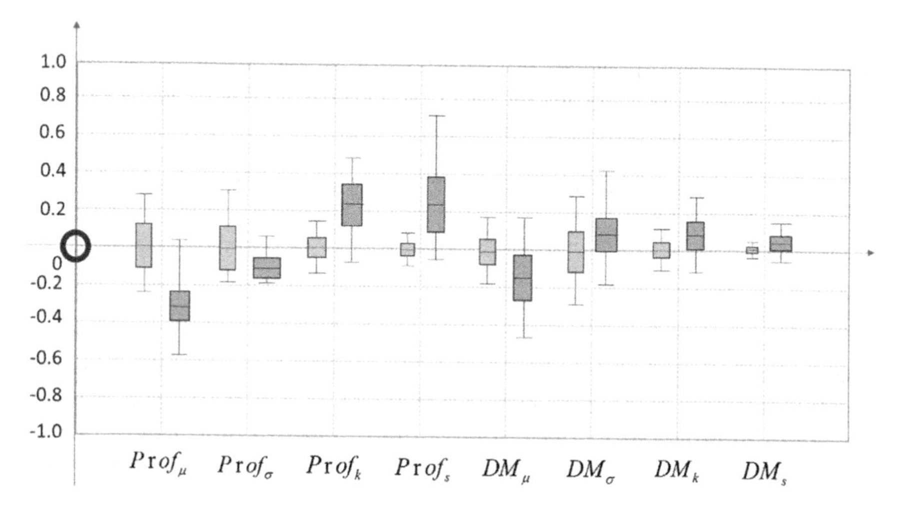

Fig. 1. Feature distribution of pulsars and non-pulsars (Color figure online)

$$Ab = -\frac{l}{R_P} - \frac{l}{R_I} + 2 \tag{1}$$

In this formula, as can be seen from Fig. 2, for a particular feature, Ab denotes the separation degree, l means the coincident area of pulsar and non-pulsar, Rp denotes the width of the distribution of the pulsars on the feature, while R_I means the distribution width of non-pulsars.

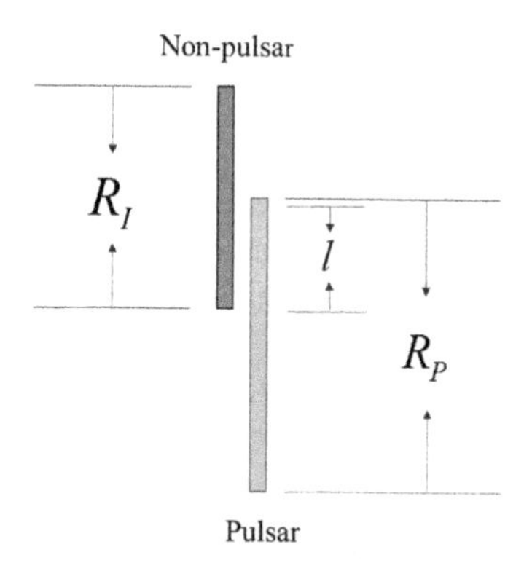

Fig. 2. Distribution of pulsar and non-pulsars and their coincident area (Color figure online)

The distribution of features between candidates can be considered as natural distribution. According 3σ principle, features of almost all candidates will be within the range of feature box. By analyzing the data from LOTAAS, HTRU2 and HTRU1, this paper get the weight of each feature W_i (i = 1–8).

$$W = (1.64, 0.85, 1.91, 1.38, 1.29, 1.50, 2.03, 1.18) \quad (2)$$

4 Experiments

In this part, weighting each feature before utilizing ML methods on the datasets are proposed. Classification accuracy and modelling time are both taken to be criterion to judge the performance of the methods. The paper supposes weighting is useful if methods improves the accuracy or improves the modeling time. What's more, accuracy goes before modeling time.

For small scale dataset LOTAAS, the experimental results are shown in Table 2, after weighting, accuracy rate of SMO, IBK and JRIP are improved. Modeling time of J48 and RandomForest are improved.

Table 2. Accuracy and modeling time before and after weighting for LOTAAS.

Methods	Accuracy		Modeling time	
	Before	After	Before	After
SMO	99.7625%	**99.8813%**	0.05	0.02
IBK	99.8417%	**99.8615%**	0	0
JRIP	99.8417%	**99.8615%**	0.08	0.11
J48	99.8615%	99.8615%	0.05	**0.03**
RandomForest	99.8615%	99.8615%	0.7	**0.47**

For medium scale dataset HTRU2, the experimental results are shown in Table 3, after weighting, accuracy rate of SMO is improved. Modeling time of JRIP, J48 and Random-Forest are improved, while IBK remains the same.

Table 3. Accuracy and modeling time before and after weighting for HTRU2.

Methods	Accuracy		Modeling time	
	Before	After	Before	After
SMO	97.5640%	**97.5696%**	0.22	0.52
IBK	97.1449%	97.1449%	0	0
JRIP	97.8154%	97.8154%	1.84	**1.44**
J48	97.8433%	97.8433%	0.28	**0.2**
RandomForest	97.9942%	97.9942%	9.94	**7.16**

For large scale dataset HTRU1, as is shown in Table 4, after weighting, accuracy rate of JRIP, J48 and RandomForest are improved. Modeling time of IBK is improved, while SMO remains the same.

Table 4. Accuracy and modeling time before and after weighting for HTRU1.

Methods	Accuracy		Modeling time	
	Before	After	Before	After
SMO	99.5866%	99.5866%	0.36	0.36
IBK	99.5340%	99.5340%	0.06	**0.01**
JRIP	99.6129%	**99.6228%**	4.78	18.0
J48	99.6074%	**99.6085%**	3.16	3.23
RandomForest	99.6721%	**99.6798%**	77.72	83.8

In conclusion, for the five ML methods SMO, IBK, JRIP, J48 and RandomForest, weighting either improves the accuracy or modeling time, or in the worst cases, weighting will at least be the same as not weighting.

5 Discussions

This part explains why SMO, IBK, JRIP, J48 and RandomForest are selected to test the effects of weighting instead of other ML methods. In this paper, we actually experimented various ML methods using WEKA.

In Table 5, the purple number means the corresponding methods performs better than others. As is shown, with the scale of datasets becomes larger, SMO, IBK, JRIP, J48 and RandomForest have better performance over other algorithms.

Table 5. Accuracy rate of pulsar recognition of various ML methods before weighting

Types	Methods	LOTAAS	HTRU2	HTRU1
Bayes	NaiveBayes	99.5448%	94.4966%	98.9155%
Functions	LibSVM	98.7928%	91.1443%	98.8585%
	SMO	99.7625%	97.5640%	**99.5866%**
Lazy	IBK	99.8417%	97.1449%	**99.5340%**
	LWL	**99.9010%**	97.7539%	99.5175%
Meta	AdaBoostML	**99.8615%**	97.6534%	99.5175%
	Stacking	98.6938%	90.8426%	98.6885%
Misc	InputMappedClassifier	98.6938%	90.8426%	98.6885%
Rules	JRIP	99.8417%	**97.8154%**	**99.6129%**
Trees	HoeffdingTree	99.8417%	97.4411%	99.5822%
	J48	**99.8615%**	**97.8433%**	**99.6074%**
	RandomForest	**99.8615%**	**97.9942%**	**99.6721%**
	RandomTree	99.8021%	**96.8432%**	99.5241%

6 Conclusion

Due to its stable cycle, Pulsar plays a very important part in physics, astronomy and many other fields. Traditional ways of pulsar search are manual. In recent years, Artificial intelligence is widely used in various fields and achieves great success. Therefore, AI methods are gradually applied to pulsar search. This paper is based on the principled real-time classification approach. Eight features are used to describe a pulsar candidate. Before applying ML methods on datasets, this paper weights each feature according to their separation degree, and then find out that either the accuracy or modeling time is improved after weighting.

Acknowledgments. The research work in this paper was supported by the grants from National Natural Science Foundation of China (61472043, 61375045) and the Joint Research Fund in Astronomy (U1531242) under cooperative agreement between the NSFC and CAS, Beijing Natural Science Foundation (4142030). Prof. Qian Yin is the author to whom all the correspondence should be addressed.

References

1. Pan, Z., Qian, L., Yue, Y.: Pulsar search technology and FAST telescope pulsar search foreground. National Astronomical Observatories Chinese Academy of Sciences, Beijing (2016)
2. Zhu, W.W., Berndsen, A., Madsen, E.C., Tan, M., Stairs, I.H., Brazier, A., Lazarus, P., Lynch, R., Scholz, P., Stovall, K., Ransom, S.M., Banaszak, S., Biwer, C.M., Cohen, S., Dartez, L.P., Flanigan, J., Lunsford, G., Martinez, J.G., Mata, A., Rohr, M., Walker, A., Allen, B., Bhat, N.D.R., Bogdanov, S., Camilo, F., Chatterjee, S., Cordes, J.M., Crawford, F., Deneva, J.S., Desvignes, G., Ferdman, R.D., Freire, P.C.C., Hessels, J.W.T., Jenet, F.A., Kaplan, D.L., Kaspi, V.M., Knispel, B., Lee, K.J., van Leeuwen, J., Lyne, A.G., McLaughlin, M.A., Siemens, X., Spitler, L.G., Venkataraman, A.: Searching for pulsars using image pattern recognition. University of British Columbia (2014)
3. Xu, Y., Fan, C.: Application of artificial intelligence in pulsar screening. National Astronomical Observatories, Chinese Academy of Sciences, Beijing (2006)
4. Zhou, Q., Ji, J., Ren, H.: X-ray pulsar cycle fast search algorithm for non-equal interval time data. National Key Laboratory of Geographic Information Engineering, Xi'an (2012)
5. Lyon, R.J., Stappers, B.W., Cooper, S., Brooke, J.M., Knowles, J.D.: Fifty years of pulsar candidate selection: from simple filters to a new principled real-time classification approach. The University of Manchester (2016)
6. Lee, K.J., Stovall, K., Jenet, F.A., Martinez, J., Dartez, L.P., Mata, A., Lunsford, G., Cohen, S., Biwer, C.M., Rohr, M., Flanigan, J., Walker, A., Banaszak, S., Allen, B., Barr, E.D., Bhat, N.D.R., Bogdanov, S., Brazier, A., Camilo, D., Champion, J., Chatterjee, S., Cordes, J., Crawford, F., Deneva, J., Desvignes, G., Ferdman, R.D., Freire, P., Hessels, J.W.T., Karuppusamy, R., Kaspi, V.M., Knispel, B., Kramer, M., Lazarus, P., Lynch, R., Lyne, A., McLaughlin, M., Ransom, S., Scholz, P., Siemens, X., Spitler, L., Stairs, I., Tan, M., van Leeuwen, J., Zhu, W.W.: PEACE: pulsar evaluation algorithm for candidate extraction – a software package for post-analysis processing of pulsar survey candidates. Royal Astronomy Society, 27 May (2013)

7. Yuan, M.: Application of Data Mining and Machine Learning Using WEKA, 2nd edn, pp. 498–526. Tsinghua University Press,Beijing (2016)
8. Morello, V., Barr, E.D., Bailes, M., Flynn, C.M., Keane, E.F., van Straten, W.: SPINN: a straightforward machine learning solution to the pulsar candidate selection problem. Swinburne University of Technology (2014)

Machine Perception

Patch Image Based LSMR Method for Moving Point Target Detection

Weina Zhou[1,2(✉)], Xinwei Lin[1], Zhijing Xu[1], and Xiangyang Xue[2]

[1] Fudan University, Shanghai 201203, China
{wnzhou, xwlin, zjxu}@shmtu.edu.cn
[2] Shanghai Maritime University, Shanghai 201306, China
xyxue@fudan.edu.cn

Abstract. The quick and high accuracy detection of point targets is a difficult but important technique in infrared surveillance. In this work, we proposed a patch image based low-rank and sparse matrices recovery (LSMR) method to detect moving point target in infrared marine surveillance. After analyzing the relationship between detection speed and image size, we wisely use frame difference and local threshold to efficiently exclude majority non-target patches before LSMR, which greatly accelerated the detection speed. Then integrated with the powerful classification ability of LSMR, the method proposed in this paper finally gains a high accuracy in point target detection in a marine environment. The experiment results show that the proposed method could effectively enhance the system's signal-to-clutter ratio gain and background suppression factor while keeping a high detection speed compared to other similar algorithms.

Keywords: Point target detection · Patch image · Low rank and sparse matrix recovery · Frame difference · Local threshold

1 Introduction

Moving point target detection is the most difficult and significant task in infrared target surveillance. Although a lot of methods have been proposed to detect point target during the past few years, their performance is still unsatisfactory due to many reasons. The main reasons resulting in the problems lie in two aspects: infrared point targets always have less texture, color and shape information, and the detection environment is commonly very harsh.

Recently, low-rank and sparse matrices recovery (LSMR) theory was proposed and proved to be much more effective than many traditional methods in a lot of fields [1–11]. And it is also applied in target detection and tracking to further improve the performance. They also can be classified into two kinds. The first kind prefers to detect target in a single image [5–8]. It can solve the problems when imaging backgrounds change quickly due to rapid relative motion between the imaging sensor and the target. And the other kind of methods makes use of information of multi-frames to predict the target position more quickly and accurate [9–11]. However, both of their performance could degrade rapidly when the signal-to-clutter ratio (SCR) is low, or the computation is too time-consuming to be realized in real-time.

Z. Shi et al. (Eds.): ICIS 2017, IFIP AICT 510, pp. 251–259, 2017.
DOI: 10.1007/978-3-319-68121-4_27

In this paper, we fully exploited the advantage of LSMR algorithm, and proposed a patch-image based LSMR method to detect moving point targets in infrared marine surveillance. Different from previously proposed LSMR based algorithms, our method takes full advantage of the information from intra and inter frames, and greatly reduces computation time while obtaining a good performance on signal-to-clutter gain (SCRG) and background suppression factor (BSF).

This paper is organized as follows. In Sect. 2, preliminary work including the LSMR algorithm and its application in small target detection will be introduced briefly. In Sect. 3, the relationship between the computation amount and image size is analyzed, thus a patch image based LSMR method is proposed and discussed in detail. In Sect. 4, experiments are executed to validate the performance of the proposed approach. And in final section, a conclusion and our future work is presented for further research.

2 LRMR Theory and Small Target Detection

2.1 Low-Rank and Sparse Matrices Recovery (LRMR) Algorithm

As a theory stretching from compressive sensing (CS) and sparse represent algorithm, the LRMR supposes an ideal image or video as a low-rank matrix. Take image signal as an instance, its rank is usually much smaller than its size.

$$\gamma \prec\prec \min(m, n) \tag{1}$$

In (1), r is the rank of an ideal image, m, n are image's width and height respectively. In fact, the entries of the matrix are often corrupted by errors or noises, making the image or video not a low-rank matrix any more. Consequently, every image/video can then be regarded as the combination of a low-rank matrix and a noise matrix in LRMR theory, which could be represented as

$$\mathrm{I} = \mathrm{L} + \mathrm{S} \tag{2}$$

I is an usual image/video, L is an ideal image/video which is low-rank, S represents random noise and errors, or other unregular signal caused by various of factors, which can be regarded as sparse matrix. It has been proved that, under surprisingly broad conditions, L and S can be exactly recovered from I via Robust Principal Component Analysis (RPCA) by solving the following convex optimization problem.

$$\min_{\mathrm{L,S}} = \|\mathrm{L}\|_* + \lambda\|\mathrm{S}\|_1, \quad \textit{subject to} \quad \mathrm{I} = \mathrm{L} + \mathrm{S} \tag{3}$$

Here, $\|\bullet\|_*$ represents the nuclear norm of a matrix, $\|\bullet\|_1$ denotes the norm-1, and λ is a positive weighting parameter, which could be utilized to fine tune L and S for optimum results.

2.2 LSMR in Point Target Detection

Generally speaking, there are three components in an infrared image (II): the background (B), the point targets (T), and various kinds of noises (N). And the infrared image can be expressed as follow:

$$\mathrm{II} = \mathrm{B} + \mathrm{T} + \mathrm{N} \tag{4}$$

When detecting point target in a single image, we found out that small targets have no concrete shape and texture, and are commonly not larger than 10 × 10 pixels because of the long imaging distance. They can be considered as "sparse" with respect to the extensive background. However, background patches are always "low-rank" and approximately context correlated with each other, even though the pixel distance between two patches may be large. As for detecting targets in multi-frames, background in each frame is also 'low-rank' due to the consistency between adjacent frames. And target is "sparse" because of its moving characteristics.

Thus the point target detection could be regarded as a typical problem of recovering a sparse component from the infrared image in LSMR theory. This assumption has been proved valid in [8], under the assumption that the random noise is i.i.d. and its Frobenius norm is smaller than some σ ($\sigma > 0$). Positive weighting parameter λ plays a promoting role in enhancing the detection stability and suppressing random noise.

$$\|\mathrm{II} - \mathrm{B} - \mathrm{T}\|_F \leq \sigma \tag{5}$$

Nevertheless, the recovery of low-rank and sparse matrix usually leads to high computational cost due to demanding convex optimization of RPCA especially with multi-frames. And the method of detecting in a single frame always cannot ensure the detection accuracy.

3 Patch Based LSMR Algorithm

How to improve the detection speed while keeping a high accuracy? A figure was drawn up to show the factor that may increase the detection time. In Fig. 1, we can see the relationship between the processing time of five randomly selected infrared marine images and their downscaled versions. Accelerated proximal gradient (APG) algorithm is used to solve the convex optimization problem in detection. From Fig. 1, we can find out that it is hard to meet the real-time requirements of most applications for detecting targets even in a single frame. However, at the same time, we also find out that the processing time is closely related to the image size. As the image shrinks, the processing time falls sharply. In that case, the processing time could be decreased when the image size is downscaled.

In view of the hugeness of the computation amount in detection with all frames in video, we use image difference to extract information between frames and divide difference image into patches to further smaller the candidate region to be processed by LSMR. These methods restrict the processing time to a reasonable range. Figure 2 is the schematic diagram of the proposed method, which can be described in detail as the following steps.

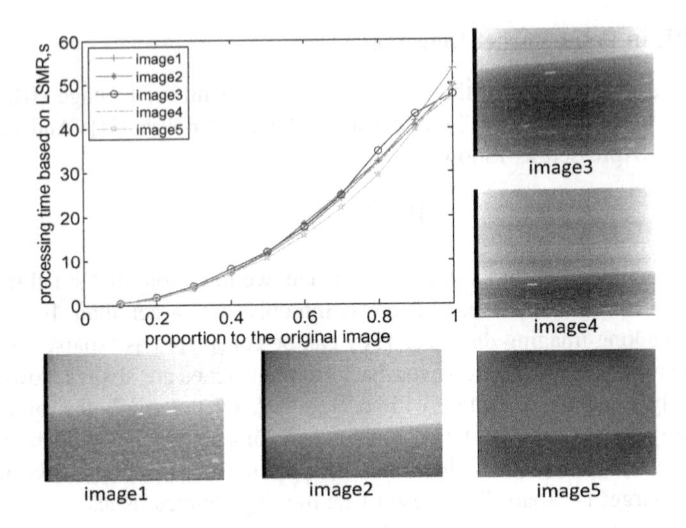

Fig. 1. Processing time of different sized images

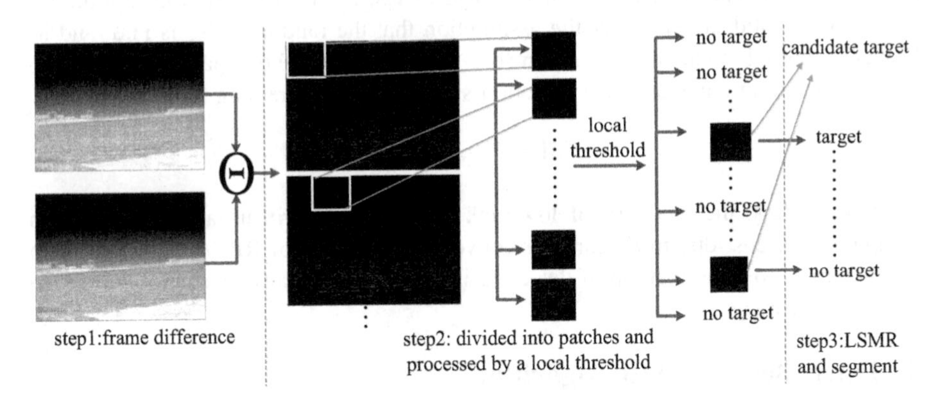

Fig. 2. Schematic diagram of the proposed method framework

Step 1: Frame difference

The frame difference is used to narrow the candidate region with little computation cost by utilizing information between adjacent frames. It could not only extract target position information from adjacent frames but also exclude the stationary disturbances for obtaining the candidate moving target regions which we are most interested in. Suppose C as the candidate area including moving point target, it can be obtained by (6).

$$\mathrm{C} = |\Pi(\mathrm{t}) - \Pi(\mathrm{t}-1)| \tag{6}$$

Here, $\Pi(t)$ is the t *th* frame, $\Pi(t-1)$ is the frame before the I(t). $| * |$ is the absolute operation.

Step 2: Divided into patches and processed by a local threshold

After obtaining the frame difference image C, we divide the image into partially overlapped patches, which further decreases the data to be processed at a time.

Obviously, the size of the patch and the overlapping area will influence the detection capability. Since point target is usually smaller than 10×10 pixels, the vertical and horizontal steps are set as P-10 to ensure no loss of targets with minimum redundant calculation. P is the width of patches, which will be discussed in the following section.

The patch is then processed by a local self-adaptive threshold to further narrow the detection region. The local self-adaptive threshold is described in (7).

$$C(x,y) = \begin{cases} 1 & if(C(x,y) \geq \alpha M_{\max})\ \&\ (M_{\max} - M_{\min} > \beta) \\ 0 & others \end{cases} \quad (7)$$

where M_{max} and M_{min} are respectively the maximum and minimum pixel values of the local region. β is the threshold coefficient and set as 0.6, α is the threshold for judging the existing of the target. For the targets brightness is usually relative high in a local region, no target is expected to exist in the patches whose difference result of M_{max} and M_{min} is smaller than β. In the paper, we choose $\beta = 40$.

Step 3: LSMR and segment

Last but not least, we recover the target region T by LSMR in the left candidate region C and then use a simple segment method to extract targets. APG algorithm is used to solve the convex optimization problem of (3). It can achieve dramatically better performance among all the RPCA algorithms [6]. The algorithm is explained as follows:

Algorithm: RPCA via APG

Input: infrared difference image with target $C \in R^{m \times n}$, λ

1. $B_0 = B_{-1} = 0; T_0 = T_{-1} = 0, t_0 = t_{-1} = 1; \mu_0 \rangle 0; \bar{\mu} = \delta\mu_0; 0 \langle \eta \langle 1.$

While not converged do

2. $Y_k^B = B_k + \frac{t_{k-1} - 1}{t_k}(B_k - B_{k-1}), Y_k^T = T_k + \frac{t_{k-1} - 1}{t_k}(T_k - T_{k-1}).$

3. $G_k^B = Y_k^B - \frac{1}{2}(Y_k^B + Y_k^T - C).$

4. $(U, S, V) = \mathrm{svd}(G_k^B)$, $B_{k+1} = US_{\frac{\mu_k}{2}}[S]V'.$

5. $G_k^T = Y_k^T - \frac{1}{2}(Y_k^B + Y_k^T - C).$

6. $T_{k+1} = S_{\frac{\lambda\mu_k}{2}}[G_k^T].$

7. $t_{k+1} = \frac{1 + \sqrt{4t_k^2 + 1}}{2}; \mu_{k+1} = \max(\eta\mu_k, \bar{\mu}).$

8. k=k+1.

End while

Output: $B = B_k; T = T_k$

4 Experimental Results and Analysis

4.1 The Performance Comparison with Different Weight of Sparse Error Term (λ) and Patch Size

To verify the capability of the proposed method, a real and representative cluttered marine surveillance video sequence with 48 frames is used to execute experiments. The resolution of the frames is 498 × 696, and the moving point targets in the frames occupy about 5 × 5 pixels. All experiments were implemented by Matlab software on a PC with 2-GB RAM and 2.60-GHz Intel-i5 processor. Average detection time (ADT), SCRG, BSF are employed for objective evaluation. SCRG, BSF are same to the metrics presented in [12]. They are defined by (8) and (9). Where, S is the target amplitude and C is the clutter standard deviation within the original frame or the processed frame.

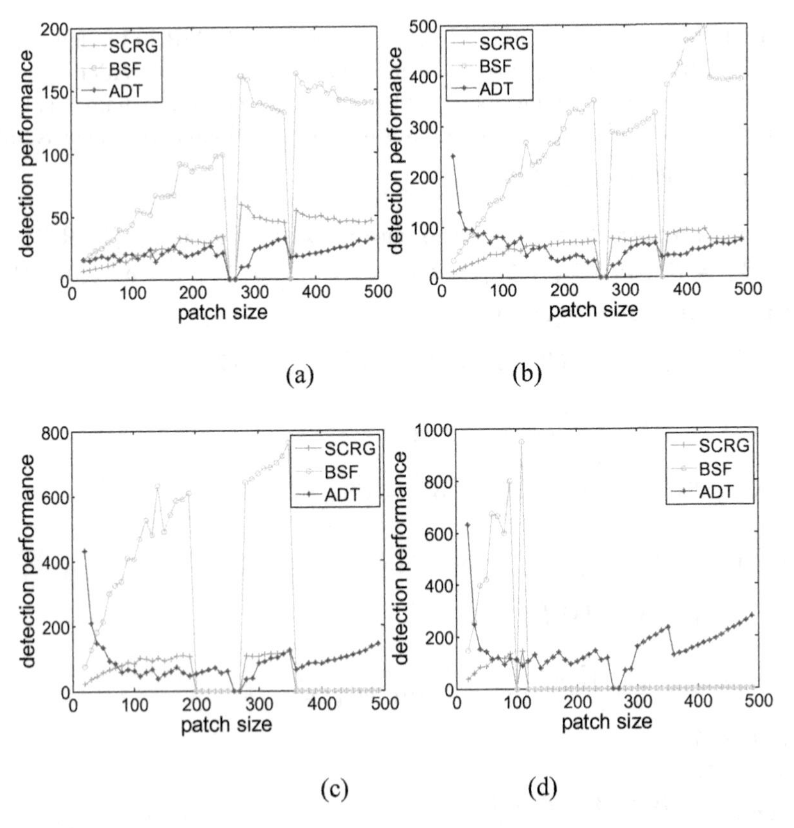

Fig. 3. Proposed performance comparison with different weight of sparse error term (λ). (a) Performance with different patch sizes (λ = 0.1); (b) Performance with different patch sizes (λ = 0.2); (c) Performance with different patch sizes (λ = 0.3); (d) Performance with different patch sizes (λ = 0.4).

$$\mathrm{SCRG} = \frac{(S/C)_{out}}{(S/C)_{in}} \tag{8}$$

$$\mathrm{BSF} = \frac{C_{in}}{C_{out}} \tag{9}$$

The patch sizes range from 20 × 20 pixels to 490 × 490 pixels with an increasing of 10 pixels. And the average results of all frames in the above mentioned video sequence are used for comparison.

Figure 3 shows the performance comparison with different weight of sparse error term (λ), which is an important parameter in APG algorithm. The ADT has been enlarged by 5, 10, 20 times in Fig. 3(b), (c) and (d) respectively for convenient observation. And it means the loss of targets when the SCRG and BSF are zero. In addition, it can be seen from Fig. 3 that, optimum detection time always exists with each determined λ, which is directly related to the patch size. Patches with too small or too large sizes always consume more time to be processed. The least detection time is obtained with patch size of 280 × 280 pixels in our experiment. The SCRG and BSF roughly increase with the patch size under determined λ, and increase with λ under determined patch size. Taking every aspect into consideration, we finally choose the optimum patch size as 280 × 280 pixels with $\lambda = 0.3$ in the proposed method. Its ADT is 3.53 s. And SCRG, BSF reach 107.0 and 641.1 respectively.

4.2 Evaluation Comparison

The detection capability of the proposed method is also compared with the result of the same algorithm with non-partitioned image and IPI model [6].

From Table 1, we can see that, the proposed method has better performance than other two methods under the same circumstance. Compared to IPI, our method reduces the detection time by 97.4% while improving SCRG, BSF by 180.1% and 389.4% respectively. What's more, the detection speed could be further accelerated by parallelism due to independence of patches in our method.

Table 1. Performance comparison of point target detection methods

Detection methods	IPI	Non-partitioned image	Proposed
SCRG	38.2	85.6	107.0
BSF	131.0	141.5	641.1
ADT(s)	136.67	21.29	3.53

By analyzing the results in above figure and table, we can see that: the proposed method could really improve the detection capability. And the reasons are as follows. First, the difference between adjacent frames and local threshold excluded a large amount of non-target patches before LSMR algorithm, and greatly decreased the computation load of LSMR. Since the frame difference and threshold process need much less computation than LSMR, the total computation time could be reduced tremendously. Second, the LSMR is still effective in detection as long as the point targets are "sparse" in each patch. Third, the proposed method integrates methods of frame difference, local threshold, image segment and LSMR, which takes full advantage of multi-methods to exactly distinguish targets from complex background.

5 Conclusion and Future Work

In this paper, we proposed a patch-image based LSMR method for moving point target detection in infrared coastal surveillance. It is the first time the patches of frames are independently processed by LSMR algorithm, and the optimum patch size is also analyzed with metrics of ADT, SCRG and BSF. Experiment results show that the propose method prompts the detection speed while keeping a high detection capability.

However, detection results of the proposed method would be affected by dynamic noises which are also irregular. Thus, more research work should continually focus on suppressing noise and improving detection speed and accuracy in future. A comprehensive method which can both handle the detection velocity and accuracy parameters is essential for actual industrial applications.

Acknowledgments. This research was financially supported in part by the National Natural Science Foundation of China under Grant Nos. 61404083 and 51579143, the China Postdoctoral Science Foundation under Grant No. 2015M581527, the Innovation Program of Shanghai Municipality Education Commission under Grant No. 15ZZ080.

References

1. Filho, K.N., Romano, J.M.T.: Low-rank decomposition based on disjoint component analysis with applications in seismic imaging. IEEE Trans. Comput. Imaging **3**(2), 275–281 (2017). IEEE Press
2. Guo, K., Liu, L., Xu, X., Xu, D., Tao, D.: GoDec+ : fast and robust low-rank matrix decomposition based on maximum correntropy. IEEE Trans. Neural Netw. Learn. Syst. **18**(3), 636–653 (2016)
3. Hao, S., Ma, X., Fu, Z., Wang, Q., Li, H.: Landing cooperative target robust detection via low rank and sparse matrix decomposition. In: 2016 International Symposium on Computer, Consumer and Control, Xi'an, pp. 18–29 (2016)
4. Yang, X., Gao, X., Tao, D., Li, X., Han, B., Li, J.: Shape-constrained sparse and low-rank decomposition for auroral substorm detection. IEEE Trans. Neural Netw. Learn. Syst. **27**(1), 32–46 (2016)
5. Dai, Y., Wu, Y.: Reweighted infrared patch-tensor model with both nonlocal and local priors for single-frame small target detection. IEEE J. Sel. Top. Appl. Earth Obs. Remote Sens. 499–508 (2017)
6. Gao, C.Q., Meng, D.Y., Yang, Y., Wang, Y.T., Zhou, X.F., Hauptmann, A.G.: Infrared patch-image model for small target detection in a single image. IEEE Trans. Image Process. **22**(12), 4996–5009 (2013)
7. Li, L., Li, H., Li, T., Gao, F.: Infrared small target detection in compressive domain. Electron. Lett. **50**(7), 510–512 (2014)

8. Zheng, C., Li, H.: Small infrared target detection based on harmonic and sparse matrix decomposition. Opt. Eng. **52**(6), 066401 (2013)
9. Wen, J.J., Xu, Y., Tang, J.H., Zhan, Y.W., Lai, Z.H., Guo, X.T.: Joint video frame set division and low-rank decomposition for background subtraction. IEEE Trans. Circ. Syst. Video Technol. **24**(12), 2034–2048 (2014)
10. Yang, C.W., Liu, H.P., Liao, S.Y., Wang, S.C.: Structured sparse coding method for infrared small target detection in video sequence. In: International Joint Conference on Neural Networks (IJCNN), Beijing, pp. 1179–1184, July 2014
11. Kim, H., Paik, J.: Low-rank representation-based object tracking using multitask feature learning with joint sparsity. Abstr. Appl. Anal. 1–12 (2014)
12. Hilliard, C.I.: Selection of a clutter rejection algorithm for real-time target detection from an airborne platform. In: Proceedings of the SPIE, pp. 74–84, Orlando, USA (2000)

An Improved Image Transformation Network for Neural Style Transfer

Hong-Xu Qiu and Xiao-Xia Huang(✉)

Department of Computer Sciences, College of Information & Engineering, Shanghai Maritime University, Shanghai 201306, China
qiuhongxu93@163.com, lhuangxiaoxia@163.com

Abstract. By using Convolutional Neural Networks (CNNs), the semantics and styles of images can be separated and recombined to create fascinating images. In this paper, an image transformation network for style transfer is proposed, which consists of convolution layers, deconvolution layers and Fusion modules composed of two 1×1 convolution layers and a residual block. The output of each layer in the network is normalized using batch normalization to speed up the training process. Compared with other networks, our network has fewer parameters and better real-time performance while generating similar quality images.

Keywords: Style transfer · Batch normalization · Residual block · Convolutional neural networks

1 Introduction

For centuries, painting has been a popular form of art, having produced plenty of valuable masterpieces which attract people's attention. But in the past, it would take a long time for a well-trained artist to draw a painting of a particular style.

Recently, Gatys et al. first studied how to use CNN to reproduce famous painting styles in the natural picture. They obtained the representations of the image from the CNN and found that the content of image and the style of image were separable. Based on the above findings, Gatys et al. proposed a Neural Style Transfer algorithm [1] to recombine the contents of a given image and the style of famous artworks. However, the efficiency of his algorithm can't meet the real-time requirements. Johnson et al. introduced a fast method based on the algorithm proposed by Gatys et al. Firstly, they trained an equivalent feed-forward generator network by using the perceptual loss function [2] they proposed for each particular style. The perceptual loss function calculates the loss by using high-level features extracted from the images using the

Fund project: The 48th Project Sponsored by the Scientific Research Foundation for the Returned Overseas Chinese Scholars, State Education Ministry; Shanghai Science and Technology Commission Innovation Project (16DZ1201402).

Z. Shi et al. (Eds.): ICIS 2017, IFIP AICT 510, pp. 260–267, 2017.
DOI: 10.1007/978-3-319-68121-4_28

16-layer VGG network [3] pretrained on ImageNet dataset [4]. When there is a content image to be stylized, only a single forward transfer is required to produce the result.

Due to the amazing stylized results, the study of Neural Style Transfer has led to many successful industrial applications. The mobile application Prisma [5] is one of the first industrial applications that offer the Neural Style Transfer algorithm as a service. Before Prisma appeared, people never thought that one day they could turn their images into artworks in just a few minutes. In order to meet the growing needs of the mobile end, a smaller and faster network is urgently needed.

In this paper, a new module which can be used to construct an image transformation network is proposed. In order to train the network, the pretrained 16-layer VGG network is used to extract the advanced features of the image, then train our network by minimizing the perceptual loss function. At test-time, compared with the network of Johnson et al., our transformation networks reduce network parameters by 62.3% and the running time reduced by 12%.

2 Related Work

2.1 Feed-Forward Image Transformation

In recent years, training a deep convolutional neural network can solve many image transformation problems. Since the purpose of our image transformation network is to convert an image into a stylized image, we referred to the architecture of the Fully Convolutional Networks [6] and the Deconvolution Network [7]. In the architecture of our image transformation network, instead of using the pooling layer, the convolution layer and the deconvolution layer are used to perform the down-sampling and up-sampling operations (Fig. 1).

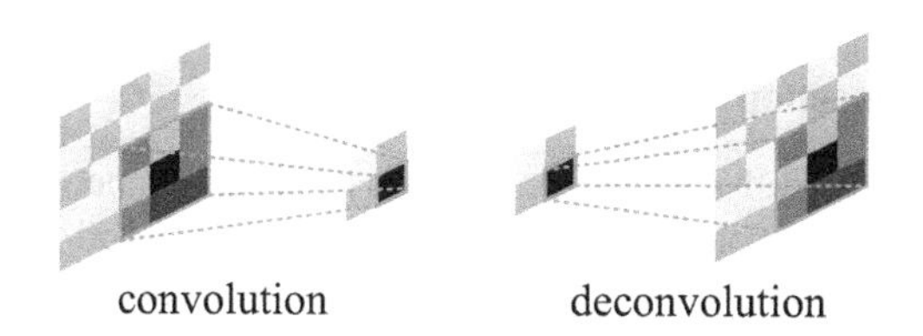

Fig. 1. Convolution and deconvolution operations for downsampling and upsampling.

2.2 Neural Style Transfer

The method proposed by Gatys et al. starts with random noise and updating the stylized image constantly by back propagation. The method of Johnson et al. is to train a feedforward network on a large image dataset for each particular style of image. Gradient descent is used to optimize the network by iteratively updating the network. Those two methods use similar objective function.

The perceptual loss function is improved by Johnson et al. on the basis of Gatys et al. Given the style and content images y_s and y_c, and the layer j and J used in the network ϕ for feature and style reconstruction, the stylized images can be generated by minimizing the total loss.

$$\text{loss} = \lambda_c l_{feat}^{\phi,j}(y, y_c) + \lambda_s l_{style}^{\phi,J}(y, y_s) + \lambda_{TV} l_{TV}(y) \tag{1}$$

λ_c, λ_s and λ_{TV} are scalars representing the weights of feature reconstruction loss, style reconstruction loss and total variation regularizer in the total loss. In this paper, the network ϕ is the pretrained 16-layer VGG network.

2.3 Batch Normalization

Because of change in the parameters of the previous layer, the distribution of each layer's inputs changes during training. It makes the depth neural network difficult to train.

To solve this problem, Ioffe et al. proposed batch normalization [8]. By using batch normalization, the internal covariate shift is reduced and the training process of network is shortened greatly. Batch normalization is implemented by normalizing each feature map so that the mean is zero and the variance is one. For a layer with d-dimensional input $\mathrm{x} = \left(x^{(1)} \ldots x^{(d)}\right)$, each dimensional will be normalized

$$\hat{x}^{(k)} = \frac{x^{(k)} - E\left[x^{(k)}\right]}{\sqrt{Var\left[x^{(k)}\right]}} \tag{2}$$

where the expectation and variance are computed over the training data set.

In this paper, batch normalization is added after the output of each convolutional and deconvolutional layer.

2.4 Residual Connection

He et al. found the use of residual connections [9] in the network which made it possible to train deep convolutional neural networks. They used residual connections on various datasets and proved the effect of residual connections. The residual block is defined as:

$$\mathrm{y} = \mathrm{F}(x, \{W_i\}) + x \tag{3}$$

Here x and y are the input and output vectors of the layers. The function $\mathrm{F}(x, \{W_i\})$ is a residual mapping that needs to be learned. The structure of the residual block is shown in Fig. 2.

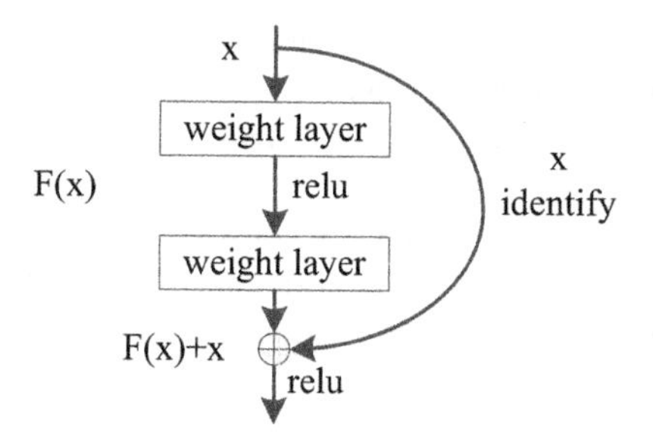

Fig. 2. Residual block.

3 Image Transformation Network

A module we designed with fewer parameters is proposed in this part. First, we introduced design ideas of modules with fewer parameters. Then, we proposed a module called **Fusion** which enabled us to build an image transformation.

3.1 Design Idea

Our objective is to define a CNN module with fewer parameters while keeping the image transformation network capability of generating similar quality images. To achieve it, the following ideas are used in the design of module:

Use 1 × 1 Filters. Given a certain number of filters, the module will use more of the 1 × 1 filters because the parameters of 1 × 1 filters are 9 times less than the parameters of 3 × 3 filters.

Reduce the Number of Input Channels to the 3 × 3 Filters. When the layer contains 3 × 3 filters, the convolution layer will have (number of input channels) * (number of filters) * (3 * 3) parameters. Obviously, the parameters of the convolution layer can be effectively reduced by reducing the number of filters and the number of input channels. It is crucial to reduce the number of input channels and the number of filters to reduce the convolution layer parameters.

These ideas will be applied to getting a novel module and utilizing it on the main body of the image transformation network. The first layers of the network are the convolution layers used to downsampling, and the last layers of the network are the deconvolution layers for upsampling.

3.2 The Fusion Module

The **Fusion** module is defined as follows.

$$y = \text{joint}\left\{W_{1\times1}^{2}\left(W_{1\times1}^{1}x\right), F\left(W_{1\times1}^{1}x\right) + W_{1\times1}^{1}x\right\} \tag{4}$$

Here, x and y are module inputs and outputs, the function F is the residual mapping mentioned above. $W_{1\times1}^{1}$ is the first 1 × 1 convolution layer used to reduce the number of input channel. Then, the output of the first 1 × 1 convolution layer is fed into second 1 × 1 convolution layer $W^2_{1\times1}$ and a residual block respectively. Finally, the output of the second 1 × 1 convolution layer and the output of the residual block are jointed as output of the module.

In the **Fusion** module, the residual block consists of two 3 × 3 convolution layers. The output of each convolution layer is normalized by batch normalization and the activation function is ReLU. The **Fusion** module is illustrated in Fig. 3.

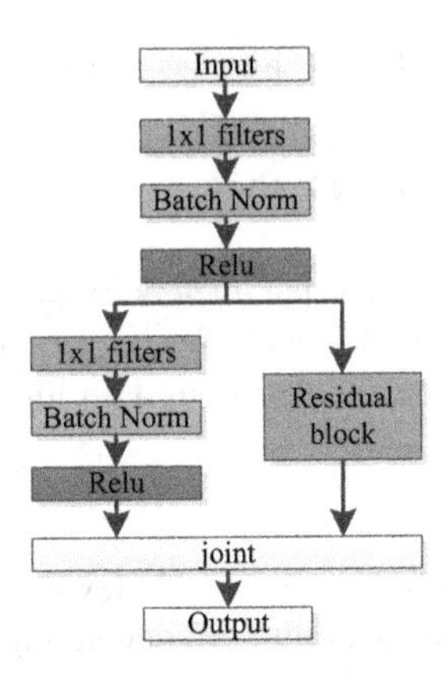

Fig. 3. The architecture of **Fusion** module.

3.3 The Transformation Network Architecture

The transformation network architecture is shown in Fig. 4. The image transformation network starts with three convolution layers used for downsampling, followed by 5 **Fusion** modules, and finally ends with three deconvolution layers used for upsampling to generate the final image.

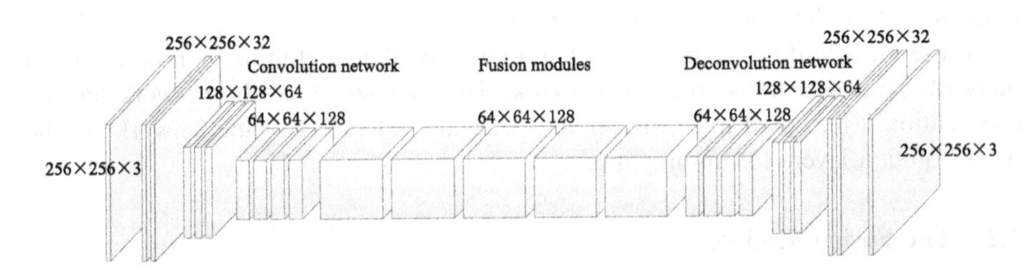

Fig. 4. The architecture of transformation network.

In Fig. 5, it is obvious that parameters in our network are much lower than those in Johnson et al. Compared with their network, our network parameters decreased by 62.3%.

Layer name	output size	filter size/stride	depth	parameter
input image	256×256×3			
conv1	256×256×32	9×9/1	1	7,808
conv2	128×128×64	3×3/2	1	18,496
conv3	64×64×128	3×3/2	1	73,856
resid1	64×64×128		2	295,168
resid2	64×64×128		2	295,168
resid3	64×64×128		2	295,168
resid4	64×64×128		2	295,168
resid5	64×64×128		2	295,168
conv5	128×128×64	3×3/(1/2)	1	73,792
conv6	256×256×32	3×3/(1/2)	1	18,464
conv7	256×256×3	9×9/1	1	7,779
output image	256×256×3			
			16 (total)	1,676,035 (total)

(a)

Layer name	output size	filter size/stride	depth	parameter
input image	256×256×3			
conv1	256×256×32	9×9/1	1	7,808
conv2	128×128×64	3×3/2	1	18,496
conv3	64×64×128	3×3/2	1	73,856
fusion1	64×64×128		3	86,272
fusion2	64×64×128		3	86,272
fusion3	64×64×128		3	86,272
fusion4	64×64×128		3	86,272
fusion5	64×64×128		3	86,272
conv5	128×128×64	3×3/(1/2)	1	73,792
conv6	256×256×32	3×3/(1/2)	1	18,464
conv7	256×256×3	9×9/1	1	7,779
output image	256×256×3			
			21 (total)	631,555 (total)

(b)

Fig. 5. Parameters for each layer of the network. (a) The network of Johnson et al. (b) Our network.

4 Experiments

We use the perceptual loss function as the objective function to train the image transform network. The Microsoft COCO dataset [10] is used to train our image transformation network. Our implementation uses Tensorflow [11] and cuDNN [12]. The whole training process is accomplished on a single GTX 950 M GPU. At the same time, we also have reimplemented the work of Johnson et al. In Fig. 6, our image transformation network has generated images with similar quality to Johnson et al.

(a) (b) (c) (d)

Fig. 6. (a) The content images. (b) The style images. (c) Results of Johnson et al. (d) Our results.

In Table 1, we compared the runtime of two networks on different sizes of images. In different sizes of images, our network has a faster speed. Compare to the network of Johnson et al., our network is 12% faster on average. Experiments show that our network can meet the higher real-time requirements.

Table 1. For different sizes of images, the table lists the speed of our network and the speed of the network of Johnson et al. Our network gets stylized images of similar quality but is faster than the network of Johnson *et al.* Both methods are implemented on GTX 950 M GPU.

Image size	Johnson et al.	Ours	Speedup
256 × 256	0.098 s	0.084 s	14.3%
512 × 512	0.335 s	0.296 s	11.6%
1024 × 1024	1.244 s	1.1 s	11.6%

Since the image transformation network is a fully convolutional neural network, it can be applied to any resolution image as long as the machine memory is enough.

5 Conclusion

In this paper, **Fusion** module is proposed based on some design ideas, and an image transformation network is constructed with the module to obtain a network with fewer parameters and better real-time performance. Experimental results show that, compared with the network of Johnson et al., the number of parameters in our network is greatly reduced and the runtime is shortened when similar quality images are generated.

In the future, we would like to explore a better network architecture in image transformation tasks. We hope that an image transformation network with smaller and higher real-time performance can be applied to smart phones, which enable them to partly break away from the constraints of mobile phone hardware conditions.

References

1. Gatys, L.A., Ecker, A.S., Bethge, M.: A neural algorithm of artistic style. arXiv preprint arXiv:1508.06576 (2015)
2. Johnson, J., Alahi, A., Fei-Fei, L.: Perceptual losses for real-time style transfer and super-resolution. In: Leibe, B., Matas, J., Sebe, N., Welling, M. (eds.) ECCV 2016. LNCS, vol. 9906, pp. 694–711. Springer, Cham (2016). doi:10.1007/978-3-319-46475-6_43
3. Simonyan, K., Zisserman, A.: Very deep convolutional networks for large-scale image recognition. In: ICLR (2015)
4. Russakovsky, O., Deng, J., Su, H., Krause, J., Satheesh, S., Ma, S., Huang, Z., Karpathy, A., Khosla, A., Bernstein, M., Berg, A.C., Fei-Fei, L.: ImageNet large scale visual recognition challenge. Int. J. Comput. Vis. (IJCV) **115**(3), 211–252 (2015)
5. I. Prisma Labs. Prisma: Turn memories into art using artificial intelligence. (2016)
6. Long, J., Shelhamer, E., Darrell, T.: Fully convolutional networks for semantic segmentation. In: CVPR (2015)

7. Noh, H., Hong, S., Han, B.: Learning deconvolution network for semantic segmentation. In: ICCV (2015)
8. Ioffe, S., Szegedy, C.: Batch normalization: accelerating deep network training by reducing internal covariate shift. In: ICML (2015)
9. He, K., Zhang, X., Ren, S., Sun, J.: Deep residual learning for image recognition. In: CVPR (2016)
10. Lin, T.-Y., Maire, M., Belongie, S., Hays, J., Perona, P., Ramanan, D., Dollár, P., Zitnick, C.L.: Microsoft COCO: common objects in context. In: Fleet, D., Pajdla, T., Schiele, B., Tuytelaars, T. (eds.) ECCV 2014. LNCS, vol. 8693, pp. 740–755. Springer, Cham (2014). doi:10.1007/978-3-319-10602-1_48
11. Abadi, M., Barham, P., Chen, J., Chen, Z., Davis, A., Dean, J., Devin, M., Ghemawat, S., Irving, G., Isard, M., Kudlur, M., Levenberg, J., Monga, R., Moore, S., Murray, D., Steiner, B., Tucker, P., Vasudevan, V., Warden, P., Wicke, M., Yu, Y., Zheng, X.: TensorFlow: a system for large-scale machine learning. arXiv preprint arXiv:1605.08695 (2016)
12. Chetlur, S., Woolley, C., Vandermersch, P., Cohen, J., Tran, J., Catanzaro, B., Shelhamer, E.: cuDNN: efficient primitives for deep learning. arXiv preprint arXiv:1410.0759 (2014)

An Improved Algorithm for Redundant Readers Elimination in Dense RFID Networks

Xin Zhang(✉) and Zhiying Yang

College of Information Engineering, Shanghai Maritime University, Shanghai 201306, People's Republic of China
Langjueyun2010@163.com, zzyang@shmtu.edu.cn

Abstract. In a dense RFID sensor network system, in order to prolong the life of the sensor network and reduce the interference between readers and the problem of eliminating redundant readers has always been the concern of the people. Now there are many algorithms and optimization technology to eliminate redundant reader in the RFID system. Inspired by the previous algorithms such as SCBA and RRE, this paper presents an improved algorithm, ICBA (Improved Count Based Algorithm) which based on the RRE and SCBA, using the count of covered tags and the Greedy Algorithm to eliminate redundant readers. The simulation experiments show that the algorithm ICBA can eliminate more redundancy readers than other algorithms. The same case, ICBA algorithm compared with other algorithm is increased by 3.2%–15% of redundant reader eliminating rate.

Keywords: Radio Frequency Identification (RFID) · Redundant Reader Elimination · CBA · RRE · LEO

1 Introduction

Radio Frequency Identifier (RFID) is a kind of commonly used wireless communication technology, which can read and write target's data through the radio signal, and in the identifying process the identify system does not need to establish any contact with the identified targets. Because the RFID system is low cost, high safety, large capacity, nowadays it is widely used in all aspects of people's life, such as automation and identity identification, access control, supply chain inventory tracking. In a densely deployed RFID network, redundant readers can cause a waste of energy and easily lead to mutual interference, because the radio frequencies of two or more neighboring readers may overlap and interfere. Therefore how to use the minimum number of readers effectively cover every tag, which is called redundant readers elimination problem. Redundant readers elimination problem is a basic problem in the RFID system, which has proved to be NP - hard problem [1].

In this paper, an Improved Count Based Algorithm (ICBA) is designed to detect and eliminate more redundant RFID readers in the complicated RFID networks. ICBA is based on the every tag's count of readers which covered this tag and the every

Z. Shi et al. (Eds.): ICIS 2017, IFIP AICT 510, pp. 268–278, 2017.
DOI: 10.1007/978-3-319-68121-4_29

reader's count of tags which are covered by this reader. Simulation results illustrate that ICBA has a better capacity to detect more redundant readers, a maximum of up to 15% more effectively than LEO and RRE in the conditions we set,and increased by 3.2% of redundant reader eliminating rate compared with the CBA which is a good algorithm for redundant readers elimination.

2 Relevant Research

The RFID system mainly is composed by two components, tags and readers. The RFID tag can store information and be detected by readers. The readers can read and write data which stored at tag's memory through radio signal. For reader's convenience, a definition of redundant reader is given as following.

Definition of redundant reader: In a RFID network, a RFID reader covers a group of RFID tags, which are also covered by the other RFID readers simultaneously. This reader is called as the redundant reader [2–4].

As the simulative RFID network showed in Fig. 1, Tag T1 and tag T2 are covered by reader R1 and they also are covered by R2. All tags in R1 are covered by other readers, so the R1 is a redundant reader. Similarly, R2 and R3 are also redundant readers due to the same reason [4]. Although all the readers are redundant readers in this RFID network, we can only eliminate at most two redundant readers without affecting the work of the RFID system.

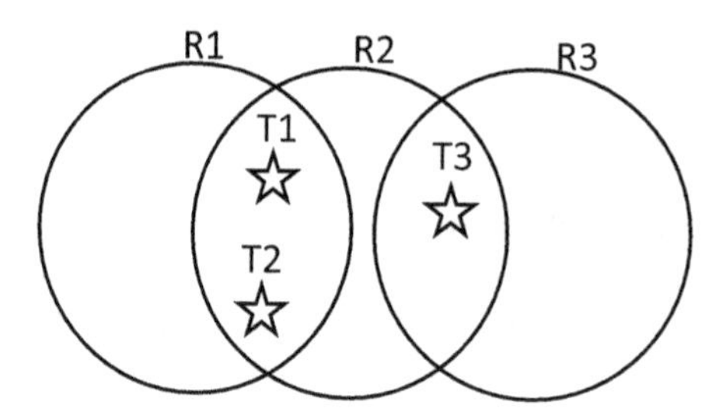

Fig. 1. Redundant reader

Nowadays there are many algorithms and optimization technique to eliminate the redundant reader of the RFID system. Some introduction and analysis of algorithms for eliminating redundant readers are given as following:

2.1 Algorithm RRE

The Redundant Reader Elimination (RRE) is the first distributed algorithm for eliminating redundant readers, which was proposed by Carbunar et al. in 2005 [1]. The main idea of Algorithm RRE is using the greedy algorithm: If a tag T1 is covered by multiple readers, find the reader which covered T1 and covered most of tags, then set T1's holder to the reader's ID. With this strategy, the tags will be greatly covered by the few readers and the goal of eliminating some redundant readers can be reached. Algorithm RRE steps are described below:

① Each reader delivers write commands that write the tagcount (the number of tags which are covered by this readers) to all tags which this reader covers. Each tag stores only the highest value of tagcount that it received and the ID of the corresponding reader.

② The reader queries each of the tags in the inquiry area and reads the tag's holder of each tag in the coverage area. A non-redundant reader should be locked at least to one tag and this reader will remain active. While readers that do not lock any of the tags are marked as redundant readers and can be safely turned off (Fig. 2).

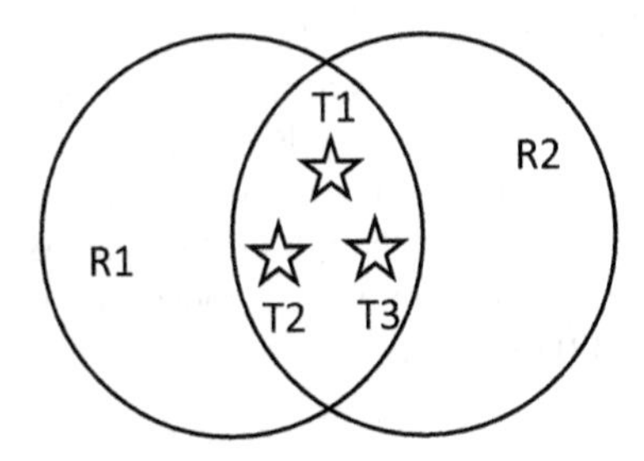

Fig. 2. An example of RFID system network

The RRE algorithm has its flaw that the selection of tag's holder is random and unreliable [4]. Because when a tag is in the overlapping areas of two or more reader and the number of tags covered by the two or more readers is the same, each reader of them has a same chance to become this tag's holder and this tag randomly chooses a reader of them as its holder. In Fig. 1 R1 and R2 have same chance to become the holder of T1, T2 and T3 according to Algorithm RRE, because R1's tagcount is same as the R2's tagcount. R1 and R2 randomly become the holder of these tags. The time-complexity of RRE is O(n logβ logm) m is the number of readers, n is the number of tags and β is the bit length of RFID tag identifiers [1, 7].

2.2 Algorithm LEO

In order to effectively reduce the read and write operations of readers, Ching - Hsien et al. proposed the LEO (Layered Elimination Optimization) algorithm [5]. This algorithm is based on the principle of "first come, first served".

Algorithm RRE steps are described below:

① RFID reader (Ri) delivers a query signal that check the tag's holder (holder) to all tags in its query area and the tags respond the corresponding readers.
② There are two possibilities for receiving a tag response, the tag's holder = "NULL" or the tag's holder = "Rk". If the holder = "NULL", the reader delivers a write command that set the reader's Id to the tag holder. If the holder = "Rk" and Rk $\neq$ Ri, the reader skips the reply information of this tag.
③ When a reader receives all tags' response that their holder $\neq$ Ri, that means all tags have belonged to other readers, the reader is a redundant reader, close the reader.

LEO algorithm can effectively reduce the read and write operation compared with other algorithms [2]. But the LEO algorithm randomly determines the holder of the tag, so the algorithm is unreliable. In Fig. 1 R1 and R2 have same chance to become the holder of T1, T2 and T3 and which is the holder depends on the first query command. The time-complexity of LEO is O(mn), m is the number of readers and n is the number of tags [2].

2.3 Algorithm CBA

Algorithm CBA is based on the number of covered readers and involves the influence of the neighboring readers. In algorithm CBA, the redundant readers will be eliminated according to the sum of the count of the neighboring readers and the count of tags covered [6].
Algorithm RRE steps are described below:

① RFID reader (Ri) delivers a query signal that check the tag's holder (holder) to all tags in its query area and records the number of tags covered by this reader as tagcount. Each tag records how many readers cover it and writes this value as its readercount.
② If the reader queries a tag's readercount = 1 in the inquiry area, change the holder of all tags that are covered by this reader to the reader ID. If there is readercount > 1 for all tags in the reader's query area, mark the reader as a redundant reader and close the reader. Decrement the reader count of all tags in the query area.
③ When all readers in the RFID system have run step 2, the algorithm ends and some redundant readers are eliminated.

Algorithm CBA involves the influence of the neighboring readers and finds the un-redundant readers in the first step while RRE and LEO don't, but the RRE algorithm also has its flaw that the elimination rate is connected with the order of reader selection.

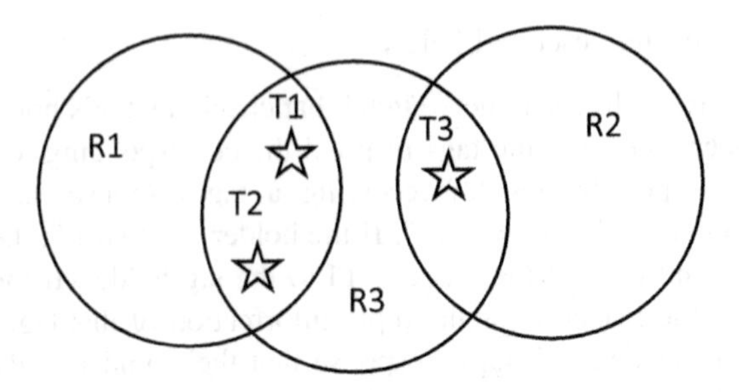

Fig. 3. An example of RFID system network

Such as shown in the Fig. 3, if in the order of R1-> R2-> R3, T1's readercount = 2 and T2's readercount = 2, the R1 and R2 are eliminated according the Algorithm CBA. If in the order of R3-> R1-> R2, the R3 is the only redundant reader which should be eliminated in this wireless network according to Algorithm CBA. So this means that the redundant reader eliminating rate of Algorithm CBA is related with the order of reader selection. The time-complexity of CBA algorithm is $O(mn^2)$, m is the number of readers and n is the number of tags [6].

3 Algorithm ICBA

Algorithm ICBA (Improved Count Based Algorithm) is designed by improving CBA and RRE. Algorithm ICBA concerns about the influence of the neighboring readers while RRE and LEO don't. In algorithm ICBA, the redundant readers will be eliminated according to readercount from small to large order while CBA don't. So Algorithm ICBA can solve the problem that the elimination rate is connected with the order of reader selection in the Algorithm CBA and the problem that the selection of tag's holder is random and unreliable in Algorithm LEO and RRE.

3.1 Algorithm ICBA

Input: The information of readers and tags, such as reader coordinates and tag coordinates (The value of readercount and tagcount will be calculated by the reader coordinates and tag coordinates).

Output: The ID of redundant readers and the number of the redundant readers.

```
For i=1 to readernum
  reader[i].tagcount=scan();
endFor
For j=1 to tagnum
  tag[j].readercount=get();
endFor
tag.sort(readercount);
//Sort all tags according the value of readercount from
//small to large order
for j=1 to tagnum
  If(readercount(j)==1)
    for i=1 to readernum
      If(tag[j] under reader[i])
        tag[j].holdid= reader[i].ID;
//The holder of all the tags in this reader is set as the
//reader ID
          for m=1 to tag[j].readerlist.count
            If (tag[j].readerlist.ID != tag[j].holdid)
              tag[j].readerlist[m].tagcount--;
          endFor
    Endfor
  If(readercount>1)
     tag[j].holdid=max((reader of tag[j] under).tagcount);
//Find a reader that covers this tag and has the maximum
//tagcount value
    For m=1 to tag[j].readerlist.count
      If (tag[j].readerlist.ID != tag[j].holdid)
        tag[j].readerlist[m].tagcount--;
    endFor
       For i=0 to readernum
          If(readers[i].tagcount==0)
             return(readers[i].ID);
       EndFor
End
```

3.2 Process

Algorithm ICBA steps are described below:

① The first step is information gathering phase. Every reader broadcasts a query command to tags which are covered by this reader and the reader records the number of tags it covers as the value of tagcount. Meanwhile each tag records the number of readers cover it as the value of readercount. Then readers deliver a query command again and readers send the data of their tagcount and every tag's readercount to the centralized processing host.

② All tags are sorted by the value of readercount from small to large in the centralized processing host. Meanwhile, the centralized processing host assigns a list of reader to each tag that used to record which readers cover this tag.

③ According to readercount from small to large order, find every tag's holder. If the tag's readercount = 0, it means no reader cover this tag and we set its holder to −1. If the tag's readercount = 1, it means this tag is only covered by a unique reader and this reader is non-redundant reader. All the tags in this non-redundant reader's interrogation zones are held by this reader. The tag that has the holdid value only responds to the reader access that corresponds to the holdid value (for other readers do not respond to). When each tag set the tag' holder, the tagcount of the corresponding readers also update. If a reader's tagcount = 0, this reader can be safely turned off, because no tag is covered by this reader.

④ To those tags whose readercount $\geq$ 2, using the ideas of RRE greedy algorithm, set the tag's holdid to the ID of reader which is maximum tagcount value and covers this tag.

⑤ When all tags have their holders, the algorithm ends and some redundant readers are eliminated.

3.3 Analysis

Algorithm ICBA is based on CBA and RRE. Through the order of tags according the value of readercount, the ICBA can solve the problem of the eliminating rate is related with the order of reader selection. And the second step of ICBA is a kind of improved Greedy Algorithm that tags which has holder only respond the query commands of theirs holders, so Algorithm ICBA can avoid the happening of random selection in the RRE. The algorithm firstly calculates the value of readercount and tagcount. The time-complexity of the first step is O(mn). Then the time-complexity of sorting tags is O(nlogn) (Using the Quicksort). The time-complexity of finding tag's holder and updating the value of other readers is O(mn^2). In summary, the time-complexity of ICBA is O(mn^2), m is the number of readers and n is the number of tags. The algorithm ICBA has the same time complexity as the algorithm CBA. Compared with other algorithms such as RRE(O(n logβ logm)) and LEO(O(nm)), the ICBA(O(mn^2)) needs more time but has a better elimination effect.

4 Simulation Experiments and Analysis

A RFID network simulation environment was designed and then compared with the classic algorithms RRE, LEO and CBA. ICBA algorithm turns out to be highly efficiency.

The following experiment data are all from the average of 3 experiments.

4.1 Experiment 1

The first experiment is designed to illustrate the performance of different algorithms under different number of tags. At the invariable regional size of 1000×1000, randomly deploying 500 readers, and the working radius of readers are 50 m. The numbers of tags increases from 1000 to 5000. The following Fig. 4 shows the comparison of results among RRE, LEO, CBA and ICBA under the condition of the above variables.

As it is shown in Fig. 4, ICBA algorithm can detect and eliminate the most of redundant readers in all situations. As the number of tags increases, the number of redundant readers eliminated by each algorithm has declined. Because more readers might become non-redundant readers, with more tags covered by a certain reader.

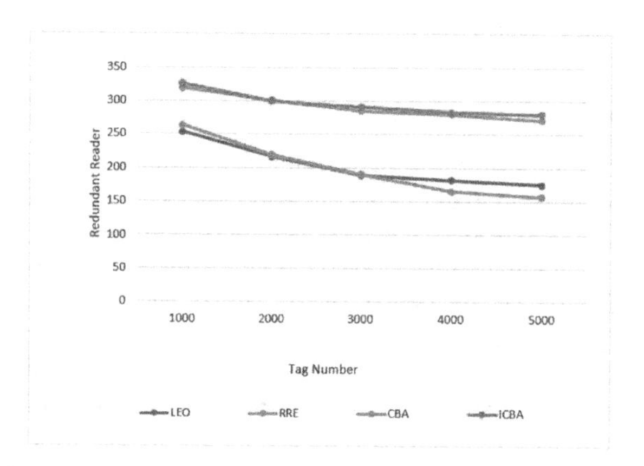

Fig. 4. Performance comparison of each algorithm under different number of tags.

4.2 Experiment 2

In the second experiment is designed to figure out the performance of different algorithms under different number of readers. At the invariable regional size of 1000×1000, we randomly generate 1000 tags and the numbers of readers increases from 200 to 600 and their working distance is 50 m. when the number of randomly deployed RFID. The following Fig. 5 shows the comparison of results among RRE, LEO, CBA and ICBA under the condition of the above variables.

As it is shown in Fig. 5, the numbers of redundant readers eliminated in all four algorithms increase gradually when the number of readers deploying in this area is rising. When more readers come into the system, the network becomes denser and the probability of every reader with more neighbors gets larger. ICBA algorithm involves a condition of neighbor count, therefore an increasing number of redundant readers will be detected.

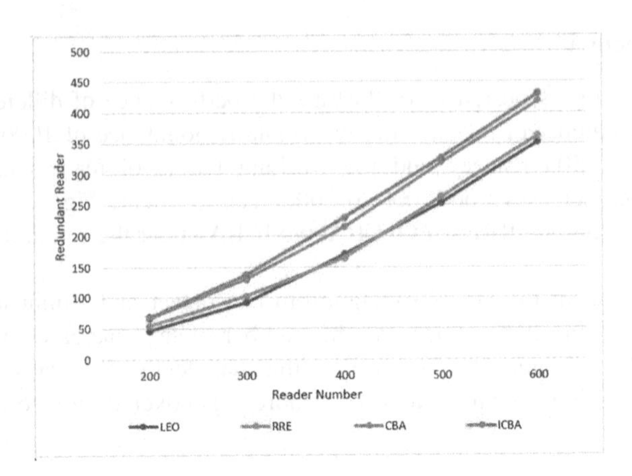

Fig. 5. Performance comparison of each algorithm under different number of readers

4.3 Experiment 3

The third experiment is designed to figure out the performance of different algorithms under different length of working radius of RFID readers. At the invariable regional size of 1000 × 1000, 500 readers and 1000 tags are randomly generated. The length of radius of readers in the experiments are from 20 m to 60 m. The following Fig. 6 shows the comparison of results among RRE, LEO, CBA and ICBA under the condition of the above variables.

As it is shown in Fig. 6, the tendencies of all algorithms are similar in this experiment. At the beginning, with the increase of working radius of readers, the count of tags each reader covering gradually rises and the probability of a reader becoming redundant. Later, when the working radius goes up to 30 m, more multiple readers might cover the same tags simultaneously, which causes more readers become redundant (Fig. 7).

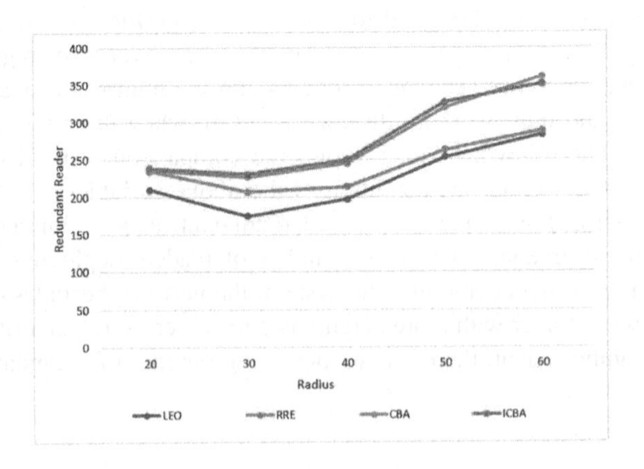

Fig. 6. Performance comparison of each algorithm under different radius of readers

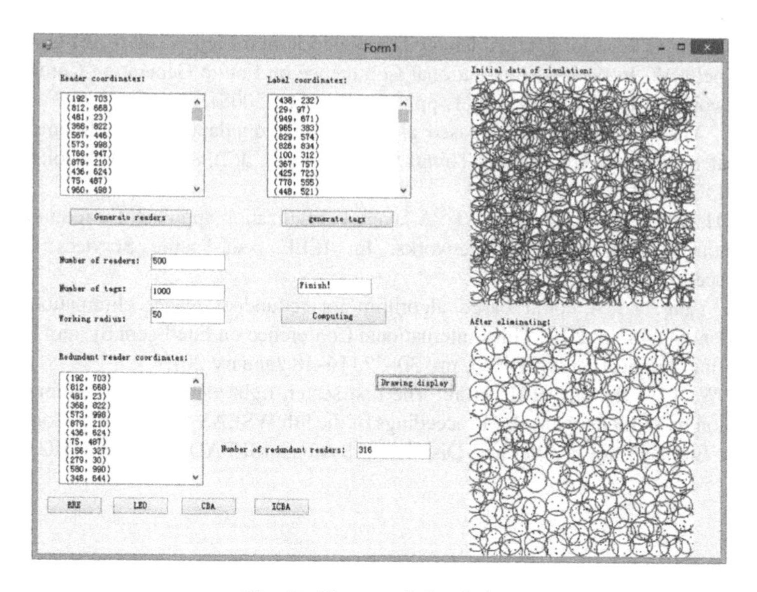

Fig. 7. Picture of simulation

5 Conclusion

In this paper, we have designed a new centralized algorithm named ICBA (Improved Count Based Algorithm) which based on the value of readercount (how many reader covers this tag) and tagcount (how many tags are covered by this reader) to solve the redundant reader elimination problem in wireless RFID network system.

Algorithm ICBA solves the problems of eliminating rate is relate with order in CBA and the randomness in RRE and LEO. Simulations show that compared to other algorithms the performance in redundant reader elimination of ICBA algorithm is more efficient in most of the situations and maximum of up to 15% more effectively than Algorithm LEO.

The data of simulation experiments is based on the ideal experimental environment. Although it is made close to the deployment of RFID system actual applying environment as much as possible, there is still a difference from the actual situations, which needs further testing. Meanwhile it is also necessary to consider the indoor environment and other interferential factors on the signal attenuation.

References

1. Carbunar, B., Ramanathan, M.K., Koyuturk, M., Hoffmann, C., Grama, A.: Redundant-reader elimination in RFID systems. In: 2005 Second Annual IEEE Communications Society Conference on Sensor and Ad Hoc Communications and Networks, pp. 176–184 (2005)
2. Ma, M., Wang, P., Chu, C.H.: A novel distributed algorithm for redundant reader elimination in RFID networks. In: 2013 IEEE International Conference on RFID - Technologies and Applications (RFID-TA), pp. 1–6. IEEE (2013)

3. Yu, K.M., Yu, C.W., Lin, Z.Y.: A density-based algorithm for redundant reader elimination in a RFID network. In: Second International Conference on Future Generation Communication and Networking, FGCN 2008, vol. 1, pp. 89–92. IEEE (2008)
4. Shen, Z., Yang, Z.: Three-count based algorithm for redundant readers elimination in the complicated RFID system. Internet Things Cloud Comput. **3**(2), 8–13 (2015). doi:10.11648/j.iotcc.20150302.11
5. Hsu, C.-H., Chen, Y.-M., Yang, C.-T.: A layered optimization approach for redundant reader elimination in wireless RFID networks. In: IEEE Asia-Pacific Services Computing Conference, pp. 138–145 (2007)
6. Pan, S., Yang, Z.: A count based algorithm for redundant reader elimination in RFID application system. In: 2013 Third International Conference on Intelligent System Design and Engineering Applications (ISDEA), pp. 30–33, 16–18 January 2013
7. Hung, J.W., Li, I.H., Lin, H.H., et al.: The first search right algorithm for redundant reader elimination in RFID network. In: Proceedings of the 9th WSEAS International Conference on Software Engineering, Parallel and Distributed Systems (SEPADS), pp. 177–183 (2010)

A Coding Efficiency Improvement Algorithm for Future Video Coding

Xiantao Jiang[1(✉)], Xiaofeng Wang[1], Yadong Yang[1], Tian Song[2], Wen Shi[2], and Takafumi Katayama[2]

[1] Department of Information Engineering, Shanghai Maritime University, Shanghai, China
xtjiang@shmtu.edu.cn
[2] Department of Electrical and Electronics Engineering, Tokushima University, Tokushima, Japan

Abstract. Motion estimation is critical in motion coding. However, the fixed pattern of search center decision method in HEVC is lack of precision. Taking advantage of the surrounding coding unit information, a universal motion vector prediction framework is presented to improve the coding efficiency in HEVC. The proposed framework is composed of two parts: motion vector prediction (MVP) and search range (SR) selection. Firstly, a novel motion vector prediction (NMVP) method is presented to improve the coding efficiency. Secondly, an adaptive search range selection (ASRS) method is developed to reduce the coding complexity in motion estimation. The simulation results demonstrate that the overall bitrate can be reduced by 5.49% on average, up to 9.18% compared with HEVC reference software.

Keywords: High efficiency video coding · Coding efficiency · Motion vector prediction · Search range

1 Introduction

In HEVC standard, the performance of motion estimation (ME) highly depended on the selection of advanced motion vector prediction (AMVP) technology [1]. The motion vector prediction (MVP) is selected from a motion vector candidate list which consists of one motion vector from neighboring units on the left of the current coding unit, one motion vector from above neighboring units, and the motion vector of the spatially the same position in the previous encoded frame. And the motion vector in the list with minimum cost is selected as the final MVP. AMVP is significantly simplified to provide a good trade-off between coding efficiency and an implementation cost. However, the fixed pattern of the MVP decision process without consideration of the reliability of the surrounding motion vectors makes it has lower estimation accuracy.

Some previous works are proposed to improving the performance of MV coding. The main idea based on the spatial and temporal MVP candidate

Z. Shi et al. (Eds.): ICIS 2017, IFIP AICT 510, pp. 279–287, 2017.
DOI: 10.1007/978-3-319-68121-4_30

schemes for video coding have one assumption in common. The motion of neighboring blocks has to be similar [2–8]. Lin et al. present a new location of the temporal motion vector predictor, a priority-based derivation algorithm of spatial and temporal MVPs [1,2]. Jung et al. propose the motion vector competition (MVC) scheme to select one motion vector predictor among the given motion vector candidates [3]. These methods can increase the coding efficiency of motion vector coding in HEVC. Yang et al. define a predictor candidate set, and the motion vector predictor can be generated by the minimum motion vector difference criterion [4]. Chien et al. design an enhanced AMVP mechanism to get an accurate motion vector predictor [5]. However, the spatial and temporal MVP candidates lack precision, and these approaches improve the performance of motion vector coding limitedly.

Motion estimation is a core part of video coding, and it can improve the coding efficiency significantly. Meanwhile, the coding complexity has significantly increased. Some adaptive search range (SR) methods have been presented to reduce the encoding complexity. Determining a suitable SR, they can be classified two categories: MV-based method and SAD-based method.

The MV-based methods [10–15] persist in that, when the distribution of the MVD is concentrated in zero with a small variance, the SR can be adjusted to the small. Lou et al. present an adaptive motion search range method to reduce the memory access bandwidth [10–12]. In this method, an applicable SR is chosen to contain the optimal MV by using a probability model. In Dai's work [13], an adaptive SR method is proposed by using a Cauchy distribution. These methods can reduce the encoding complexity significantly.

The SAD-based methods [16,17] set a threshold on SAD value to decide whether the video content is motion or not. However, the SAD-based methods are unreliable.

In summary, a universal motion vector prediction framework is proposed to improve the coding efficiency in this work. For the fixed pattern of MV coding in HEVC, the main disadvantages of the previous work are that the precision is not sufficient and the robustness is not high. Firstly, a novel motion vector prediction method is used to generate the optimal MV. For the fixed SR adopting in motion search processing, there are a large number of redundant computation. Thus, an adaptive search range selection method is developed to reduce the encoding complexity of ME owing to the benefit of the accurate motion vector prediction. Different from the state-of-the-art, my approach treats the MVP selection and the SR selection jointly.

2 Proposed Method

2.1 Novel Motion Vector Prediction (NMVP)

Considering the video content with strong spatial correlations, motion vector predictor of the current CU can be generated from the adjacent CUs. The novel motion vector prediction is based on the previous work [9]. Different from the fixed pattern AMVP technology, the spatial neighborhood cluster G is composed

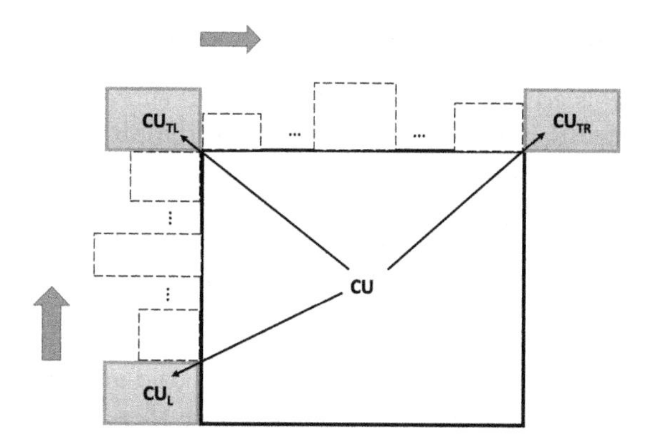

Fig. 1. Spatial correlation neighborhood cluster.

of all spatial neighbor CUs. The cluster G is shown as in Fig. 1. Where CU_L, CU_{TR} and CU_{TL} denote the left, top right and top left CU of the current CU, separately. The cluster G is defined as

$$G = \{CU_L, \ldots CU_{TL}, \ldots CU_{TR}\} \tag{1}$$

The MVs and depths information of G can be used to predict the MVP of the current CU. However, the computation complexity is high by the checking all of the information. Thus, the relatively reliable sub-cluster should be developed for the MVP. Therefore, in order to utilize the spatial correlation, the sub-cluster M is defined as

$$M = \{CU_L, CU_{TL}, CU_{TR}\} \tag{2}$$

The sub-cluster M is contained in the cluster G ($M \subset G$). MV_L, MV_{TR} and MV_{TL} indicate the MV candidates in the left, top right and top left of the current CU.

The basic idea of the proposed MVP method is to prejudge the MVP of the current CU according to the MVs of the spatial adjacent CUs. When the sub-cluster M is available, the information of M is used to predict the MVP of the current CU. In contrast, when the sub-cluster M is unavailable, the information of G is used to predict the MVP of the current CU.

When the MVs of sub-cluster M are available, a simple MV can be selected as the optimized MV for the current CU. On the contrary, when the MVs of sub-cluster M are not available, the reliability of the candidate MVs is the lowest and it is hard to get the accurate MVP using the fixed AMVP mechanism. In this case, the MVP position may tend to be near to the left of CU, and it is possible to tend to be near to the top of CU. Thus, all available MVs of the spatial neighborhood cluster G need to be checked. In order to get the accurate MVP, all surrounding MVs of G can be added to the candidates MVs, and the cost of these MVs are checked to get the optimized MVP.

2.2 Adaptive Search Range Selection (ASRS)

An accurate MVP is critical to SR reduction. In this subsection, the importance of the MVP to the SR reduction is studied. As to the SR, the optimized MV is close to the MVP so that a smaller SR can be used in the ME. In the contrast, a larger SR should be used in the ME if the optimized MV is far away the MVP. The difference between the optimized MV and the MVP is named the MV prediction difference (MVPD). Thus, when the distribution of the MVPD is concentrated near the center with a small variance, a smaller SR can be adopted.

Previous works on adaptive search range selection (ASRS) methods are reported in [10,12]. The MV variance (σ^2) of the spatial and temporal neighboring CUs is used to adjust the search range in ME. In these methods, the Cauchy distribution is proposed to model the distribution of the MVPD. The probability density function of MVPD is used to calculate the probability of the optimized MV within the SR in [10], the probability density function of zero mean Cauchy distribution is can be defined as

$$f_{ME}(x,y) = f_{ME,X}(x) \cdot f_{ME,Y}(y) \tag{3}$$

$$f_{ME,X}(x) = \frac{C_x}{|\frac{x}{\zeta_x}|^{\frac{5}{3}} + 1} \tag{4}$$

$$f_{ME,Y}(y) = \frac{C_y}{|\frac{y}{\zeta_y}|^{\frac{5}{3}} + 1} \tag{5}$$

where C_x, C_y are normalization constants, and ζ_x and ζ_y are parameters of the modified zero-mean Cauchy function, which can be computed by the sample variances of MVPDs (σ^2). Thus, the probability of the optimized MV can be defined as

$$F_{ME}(SR_x, SR_y) = F_{ME,X}(SR_x) \cdot F_{ME,Y}(SR_y) \tag{6}$$

$$F_{ME,X}(SR_x) = \int_{-SR_x-0.5}^{SR_x+0.5} f_{ME,X}(x)dx \tag{7}$$

$$F_{ME,Y}(SR_y) = \int_{-SR_y-0.5}^{SR_y+0.5} f_{ME,Y}(y)dy \tag{8}$$

where SR_x and SR_y are horizontal and vertical component of search range, separately. For the given probability C_{prob} that denotes the search range contains the optimized MV, the dynamic search range can be determined by the closest SR with a probability no less than C_{prob}, and it satisfies

$$F_{ME,X}(SR_x) = F_{ME,Y}(SR_y) = \sqrt{C_{prob}} \tag{9}$$

where the computational dynamic search range can be represented by SRx_dyn and SRy_dyn, respectively. In this work, the default SR is represented by SR_def. When the MVP is accurate enough, the minimal search range (represented by SRx_min and SRy_min) is set to 1/8 of the default SR.

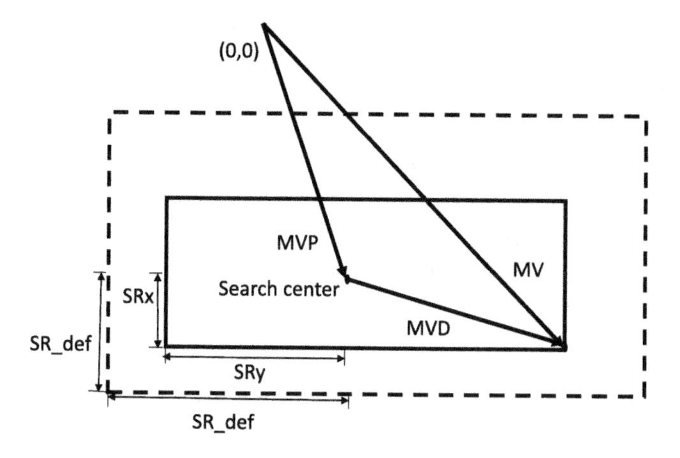

Fig. 2. Search range refinement

Recall that the spatial correlation based MVP decision algorithm is studied in above section, and the reliability of the neighboring candidate MVs is used to get the optimized MVP. To treat the MVP selection and the SR reduction jointly, when a high precision MVP is selected to be close to the optimized MV, the SR will be small. An example of the search range refinement method is shown in Fig. 2. In this case, the SR of modified search range is set to SR_x and SR_y, while the default SR (SR_def) is set to 64×64 in the ME of HEVC reference software.

2.3 Universal Motion Vector Prediction Framework

Jointing the NMVP and ASRS methods together, the universal motion vector prediction algorithm is shown as algorithm 1. Firstly, the optimized MVP is decided by the MVs of the spatial neighbor CUs. Secondly, the search range is adjusted by the probability of the optimized MV within the SR and the reliability of the MV candidates. It is noted that, when the reliability of the neighboring candidate MVs is the highest, the SR can be reduced significantly and it is set to the (SRx_min, SRy_min). Moreover, when the reliability of the neighboring candidate MVs is the lowest, the SR is set to the $\{\min(SR\_def,\ SRx\_min + SRx\_dyn),\ \min(SR\_def,\ SRy\_min + SRy\_dyn)\}$ respectively.

3 Experiment Results

The proposed algorithm is implemented and verified based on HEVC test model HM16.12. The test conditions are set to evaluate the performance of the proposed algorithm at different profiles (RA and LD) [18]. The quantization parameters (QP_i) are set to 22, 27, 32 and 37, respectively. In this work, the search strategy is TZsearch, and the SR is adjusted adaptively.

Algorithm 1. Universal motion vector prediction.

```
1  Start inter prediction for CU
2  if the sub-cluster M exist then
3      if MV_TR = MV_TL = MV_L then
4          MV_L is selected as best MVP;
5          SR_x = SRx_min, SR_y=SRy_min;
6      else if |MV_TR − MV_TL|>|MV_TL − MV_L| then
7          Add MV_TL and MV_L to MV candidate list;
8          SR_x = min(SR_def, SRx_dyn),
9          SR_y=min(SR_def, SRy_min + SRy_dyn);
10     else
11         Add MV_TR and MV_TL to MV candidate list;
12         SR_x = min(SR_def, SRx_min + SRx_dyn),
13         SR_y= min(SR_def, SRy_dyn);
14 else
15     Add all MVs of the cluster G to MV candidate list;
16     Reduce the redundant MV candidates;
17     SR_x = min(SR_def, SRx_min + SRx_dyn), SR_y=min(SR_def,
       SRy_min + SRy_dyn);
18 Process motion search
```

The performance of the proposed algorithm is evaluated Bjontegarrd Delta bitrate (BDBR) [19], and the average time increasing (TI) is defined as

$$TI(\%) = \frac{1}{4}\sum_{i=1}^{i=4} \frac{T_{Pro}(QP_i) - T_{HM}(QP_i)}{T_{HM}(QP_i)} \times 100\% \tag{10}$$

where $T_{HM}(QP_i)$ and $T_{pro}(QP_i)$ are the encoding time by using the HEVC reference software and the proposed method with different QP_i.

Table 1 shows the performance of the universal motion vector prediction algorithm. In RA case, the bitrate can be reduced by 5.49% on average, while the encoding time increasing is 51.13%. In LD case, the bitrate can be reduced by 5.34% on average, while the encoding time increasing is 44.96%. The proposed algorithm can improve the coding efficiency significantly.

It is noted that, for the motion severe sequence, the proposed algorithm can improve the performance, which is the greatest contribution of this paper. The bitrate can be reduced by 7.85% for BasketballDrive sequence in RA case, and the R-D curve is shown as Fig. 3.

It would be specially mentioned that this proposed method causes the encoding complexity increasing with the encoding efficiency raising. However, for the application that does not care about the real-time encoding, and care more about the coding efficiency, and it is an efficient approach for coding efficiency improvement in HEVC.

Table 1. Performance of universal motion vector prediction.

Class	Sequence	RA		LD	
		BDBR	TI (%)	BDBR	TI (%)
1920 × 1080	Kimono	−5.92	51.64	−4.46	45.41
	ParkScene	−4.35	56.79	−4.88	52.13
	Cactus	−4.76	54.68	−4.55	48.29
	BasketballDrive	−7.85	51.14	−6.04	51.03
	BQTerrace	−2.15	54.67	−2.77	50.59
1280 × 720	Vidyo1	−3.61	63.92	−5.30	60.75
	Vidyo3	−4.66	59.53	−4.29	51.20
	Vidyo4	−5.17	59.12	−5.84	52.72
High resolution	Average	**−4.81**	**56.44**	**−4.77**	**51.52**
832 × 480	BasketballDrill	−7.30	48.46	−7.13	42.56
	BQMall	−4.69	53.65	−5.10	47.36
	PartyScene	−4.60	45.68	−4.51	35.83
	RaceHorses	−8.93	41.63	−6.72	33.40
416 × 240	BasketballPass	−6.18	56.13	−6.17	34.97
	BQSquare	−2.79	54.77	−3.84	36.29
	BlowingBubbles	−5.74	47.41	−5.92	36.59
	RaceHorses	−9.18	18.89	−7.95	40.23
Low resolution	Average	**−6.18**	**45.83**	**−5.92**	**38.40**
Average		−5.49	51.13	−5.34	44.96

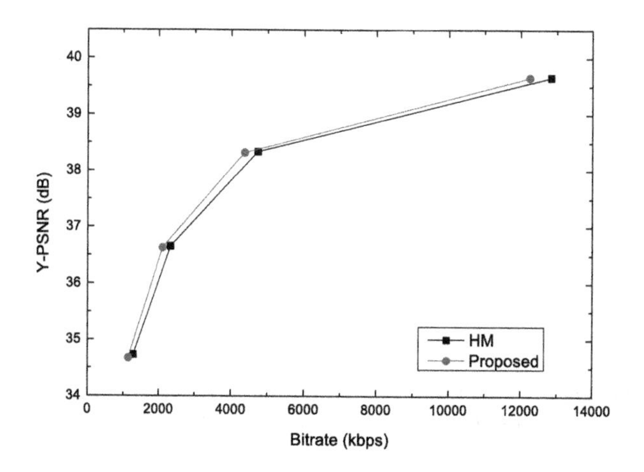

Fig. 3. R-D curve of BasketballDrive (RA).

4 Conclusion

In this work, a universal motion vector prediction framework is presented to improve the performance of HEVC. The simulation results demonstrate that the proposed overall algorithm can improve the encoding efficiency by 5.34–5.49% on average, which the encoding time increasing is about 45–51%.

References

1. Lin, J.L., Chen, Y.W., Huang, Y.W., Lei, S.M.: Motion vector coding in the HEVC standard. IEEE J. Sel. Top. Signal Process. **7**(6), 957–968 (2013)
2. Lin, J.L., Chen, Y.W., Tsai, Y.P., Huang, Y.W., Lei, S.M.: Motion vector coding techniques for HEVC. In: International Workshop on Multimedia Signal Processing (MMSP), pp. 1–6 (2011)
3. Jung, J., Bross, B., Chen, P., Han, W.-J.: Description of core experiment 9 (CE9): MV coding and skip/merge operations. In: Document of Joint Collaborative Team on Video Coding, JCTVC-D609, January 2011
4. Yang, W., Au, O.C., Dai, J., Zou, F., Pang, C., Liu, Y.: Motion vector coding algorithm based on adaptive template matching. In: International Workshop on Multimedia Signal Processing (MMSP), pp. 222–227 (2010)
5. Chien, W.D., Liao, K.Y., Yang, J.F.: Enhanced AMVP mechanism based adaptive motion search range decision algorithm for fast HEVC coding. In: IEEE International Conference on Image Processing (ICIP), pp. 3696–3699, October 2014
6. Tok, M., Glantz, A., Krutz, A., Sikora, T.: Parametric motion vector prediction for hybrid video coding. In: Picture Coding Symposium (PCS), pp. 381–384 (2012)
7. Tok, M., Eiselein, V., Sikora, T.: Motion modeling for motion vector coding in HEVC. In: Picture Coding Symposium (PCS), pp. 154–158 (2015)
8. Springer, D., Simmet, F., Niederkorn, D., Kaup, A.: Robust rotational motion estimation for efficient HEVC compression of 2D and 3D navigation video sequences. In: IEEE International Conference on Acoustics, Speech and Signal Processing, pp. 1379–1383 (2013)
9. Ikegita, J., Jiang, X., Song, T., Leu, J.-S., Shimamoto, T.: Efficient prediction motion vector candidate selection algorithm for HEVC. In: ITC-CSCC: International Technical Conference on Circuits Systems, Computers and Communications, pp. 402–403 (2015)
10. Lou, C.C., Lee, S.W., Kuo, C.C.J.: Motion vector search window prediction in memory-constrained systems. In: SPIE Optical Engineering Applications. International Society for Optics and Photonics, p. 74430E (2009)
11. Lou, C.C., Lee, S.W., Kuo, C.C.J.: Adaptive search range selection in motion estimation. In: IEEE International Conference on Acoustics, Speech and Signal Processing, pp. 918–921 (2010)
12. Lou, C.C., Lee, S.W., Kuo, C.C.J.: Adaptive motion search range prediction for video encoding. IEEE Trans. Circuits Syst. Video Technol. **20**(12), 1903–1908 (2010)
13. Dai, W., Au, O.C., Li, S., Sun, L., Zou, R.: Adaptive search range algorithm based on Cauchy distribution. In: Visual Communications and Image Processing (VCIP), pp. 1–5 (2012)
14. Shi, Z., Fernando, W.A.C., Fernando, A.: Adaptive direction search algorithms based on motion correlation for block motion estimation. IEEE Trans. Consum. Electron. **57**(3), 1354–1361 (2011)

15. Ko, Y.H., Kang, H.S., Lee, S.W.: Adaptive search range motion estimation using neighboring motion vector differences. IEEE Trans. Consum. Electron. **57**(2), 726–730 (2011)
16. Lee, S.W., Park, S.M., Kang, H.S.: Fast motion estimation with adaptive search range adjustment. Opt. Eng. **46**(4), 040504–0405043 (2007)
17. Saponara, S., Casula, M., Fanucci, L., Rovati, F., Alfonso, D.: Dynamic control of motion estimation search parameters for low complex H. 264/AVC video coding. In: IEEE International Conference on Consumer Electronics, pp. 481–482 (2006)
18. Bossen, F.: Common test conditions and software reference configurations. In: JCTVC-L1100, January 2013
19. Bjontegaard, G.: Calculation of average PSNR differences between RD-curves. In: ITU-T SG16 Q.6, VCEG-M33, April 2001

A Two-Step Pedestrian Detection Algorithm Based on RGB-D Data

Qiming Li[1,2](✉), Liqing Hu[2], Yaping Gao[3], Yimin Chen[3], and Lizhuang Ma[1]

[1] Department of Computer Science and Engineering, Shanghai Jiaotong University, Shanghai 200240, China

[2] College of Information Engineering, Shanghai Maritime University, Shanghai 201306, China

qmli@shmtu.edu.cn

[3] Department of Computer Science and Technology, Shanghai University, Shanghai 200444, China

Abstract. A two-step pedestrian detection algorithm based RGB-D data is presented. Firstly, down-sample the depth image and extract the key information with a voxel grid. Second, remove the ground by using the random sample consensus (RANSAC) segmentation algorithm. Third, describe pedestrian characteristics by using Point Feature Histogram (PFH) and estimate the position of pedestrian preliminarily. Finally, calculate the pedestrian characteristics based on color image by Histogram of Oriented Gradient (HOG) descriptor and detect pedestrian using Support Vector Machine (SVM) classifier. The experimental results show that the algorithm can accurately detect the pedestrian not only in the single-pedestrian scene with pose variety but also in the multi-pedestrian scene with partial occlusion between pedestrians.

Keywords: Pedestrian detection · RGB-D · Down-sample · PFH · HOG · SVM

1 Introduction

Pedestrian detection is a hot research topic in computer vision. It plays an important role in vehicle assistant driving, intelligent video surveillance, human abnormal behavior detection, and so on. Pedestrians are both rigid and flexible, and the appearance is susceptible by wearing, occlusion, scale zoom, perspective and other factors, so the pedestrian detection is still a challenging issue.

Pedestrian detection technology based on RGB image has been mature, but the RGB feature is unitary. With the popularity of depth acquisition device, more and more researchers have begun to study pedestrian detection based on depth image. The classical HOG features used in color image is used to depth image by Spinello [1, 2], the Histogram of Oriented Depth (HOD) descriptor is proposed. Based on the algorithm presented by Spinello, Choi et al. [3] proposed a framework of multiple detector fusion, including skin color detection, face detection, trunk detection and so on. Wu et al. [4] proposed the histogram of depth difference feature. Yu et al. [5] proposed simplified

Z. Shi et al. (Eds.): ICIS 2017, IFIP AICT 510, pp. 288–294, 2017.
DOI: 10.1007/978-3-319-68121-4_31

local ternary patterns features based on depth image. Xia et al. [6] detect human targets by using two-dimensional contour model and 3D facial model. Wang et al. [7] presented a pyramid depth self-similar feature extraction algorithm according to the strong similar depth information of the human local body. Ikemura and Fujiyoshi [8] think that the frequency distribution of the depth values of different objects has a great diversity. They divide the depth map into nonoverlapping blocks and carry on histogram statistics of depth values for each block. The relational depth similarity Features of all blocks in series is used as the feature of the image. In recent years, Cheng et al. [9] proposed a semi supervised learning framework. Wang et al. [10] extract the fusion features of RGB and depth information, and reduce the dimension of the feature by using sparse automatic encoding. Gupta et al. [11] reselect exactly after primary select. Liu et al. [12] use synthetic image to detect pedestrian. Linder and Arras [13] detect pedestrian by using the local multi feature.

In summary, most of the existing pedestrian detection techniques are based on RGB images, so the feature is single, while the pedestrian detection based on depth images is prone to drift. Therefore, a pedestrian detection method based on RGB-D image is proposed in this paper. The algorithm is executed in two steps. Based on the depth image detection, the initial detection results are given first, and then the final detection results are obtained based on the RGB image.

2 Two-Step Pedestrian Detection Algorithm

As shown in Fig. 1, firstly, the point cloud expression of the scene is constructed based on depth data, and the down sampling is performed to the point cloud by using the voxel grid method in order to reduce the data size. Then the random sampling consistency segmentation algorithm is adopted to remove the ground plane from point cloud to reduce point cloud data further. Finally the PFH and SVM are used to screen out the candidate pedestrians.

Further detection needs to combine with RGB image. At present, the HOG feature is still the most effective and widely used in the pedestrian detection method based on RGB image, so the HOG feature and the SVM classifier are used in the further accurate detection to get the final pedestrian detection results.

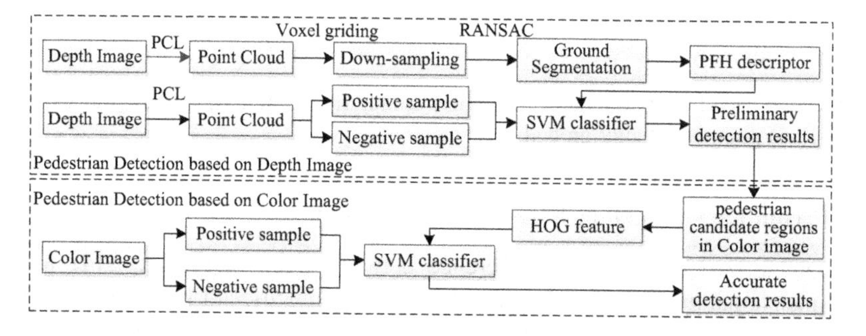

Fig. 1. Algorithm flow

2.1 Down-Sampling to Depth Point Cloud

The point cloud expression of the scene is constructed based on depth image. However, the data size of point cloud is relatively large. In order to improve the detection speed, the point cloud data must be down-sampled on the premise of keeping the shape of the cloud. In each frame, the spatial data is divided into a set of voxels. The coordinates of all points within each voxel are approximated to the coordinates of the voxel's center. A 3D voxel grid is created based on the point cloud data by using the point cloud library (PCL), and the center of gravity is approximately expressed by the coordinates of all points in voxel. Voxel can reduce the order of magnitude of data processing, reduce the calculation of pedestrian characteristic, and achieve the real-time performance. In addition, this operation is also conducive to constant point cloud density, makes the cloud data volume consistent in the unit space, and makes the point cloud density dependent on the distance from the sensor no longer. In this paper, the selected size of voxel is 0.08 m, and the original point cloud and the filtered result are shown in Fig. 2a and b respectively.

2.2 Ground Segmentation

We make a hypothesis that people walk on the ground in this paper. The ground plane is estimated by several points marked in advance on the ground plane. According to the estimated ground plane, we can remove the corresponding voxel point cloud, and further reduce the irrelevant data. The RANSAC segmentation algorithm is adopted to design a model for segmentation judging, and the input data is iteratively extracted by the judgment criterion. First, set thresholds for all points of the point cloud dataset, and randomly select points from the grid point cloud data and calculate the parameters of the given judgment model. If the distance between the point and the determining model is smaller than the threshold, then the point is internal, otherwise it is external out of points. The external points will be removed. As shown in Fig. 2c, the voxels of the ground plane are removed and the volume of the target image is greatly reduced.

Fig. 2. Point cloud grid: (a) the initial point cloud; (b) point cloud after grid and down-sampling; (c) point cloud after ground segmentation

2.3 Preliminary Detection Based on Point Cloud

By parameterized querying the difference between point and point, PFH forms a multidimensional histogram to describe the spatial geometric attributes of the K points in neighborhood. The high-dimensional hyperspace provides measurable information space for features. The six dimensional pose of a surface is invariant and robust to different sampling densities or neighborhood noises. PFH representation is based on the relationship between the point and its K neighborhood and their estimated normal, considers the interaction between the normal directions, captures the best changes of the point cloud surface, and describes the geometric features of data samples. For example, for the query point P, the radius of its affected region is r, and all its K neighborhood elements are completely connected to each other in a connected region. By computing the relation between the two connected points in the connected region, the point feature histogram is the final PFH descriptor.

After removing the ground data from the point cloud, the false positive rate strategy is adopted for preliminary pedestrian detection, and the accurate detection is performed in the next step. The advantage of this approach is that it can allow as many candidate targets as possible pass this initial screening to minimize the omission ratio. First, scan the whole image with a fixed size detection window, then describe the characteristics of pedestrian cloud by using PFH, and select the candidate pedestrians using SVM classifier at last.

Kinect sensor is used to capture data in 4 different indoor scenes, including classroom, restaurant, dormitory and library. In order to make the depth information more accurate, only the data the shooting distance of which is between 3–8 m is selected. After manually tailoring the picture, 1218 pairs of positive samples and 3000 pairs of negative samples are obtained. Among them, the negative samples are randomly extracted from pictures that contain no human body. The sample size is 64×128. Finally, extract the PFH features of positive and negative samples from the depth images and input them into SVM for training.

2.4 Accurate Detection Based on RGB Image

The accurate pedestrian detection algorithm using HOG features is as following. A fixed size detection window is used, and the window is divided into a grid the unit of which is cell. The gradient direction of the pixels in each cell is computed into a one-dimensional histogram. The intuitive formulation is that the local appearance and shape can be well described by the distribution of local gradients without having to know the exact locations of these gradients in cell. A set of cell is aggregated into block for local contrast normalization. The histogram of all the blocks is concatenated to form the descriptor vector of the detection window. This descriptor vector is used to train the linear SVM classifier. When detecting pedestrian, the detection window slides in different scale spaces of the image and the HOG descriptors of each position and scale are calculated. Then a trained SVM classifier is used to classify, and finally the location of the pedestrian is obtained.

Compared with the depth SVM classifier trained in Sect. 2.3, the training process of the RGB SVM classifier is basically the same, but it uses HOG features. Considering the real-time application requirement, the modified HOG [14] is adopted to reduce dimensionality. Similarly, the training set constructed in Sect. 2.3 is also used for training the RGB SVM classifier.

3 Results

Based on the above idea, the related algorithms are implemented. The development environment is the Visual C++ platform, the software that drives the Kinect is OpenNI interface (Prime Sense Company), and the PCL is used for processing the 3D point clouds, deep and color images. The hardware platform is core duo CPU@2.2 GHz, 4 GB memory, GTX 660 graphics card and Kinect for Windows.

Experiments were carried out on 382 pairs of images containing 1201 human bodies, and the detection rate achieves an average performance of 97.75%. The experiment is divided into two types: single scene and multi person scene. The experimental results show that the pedestrian detection algorithm based on RGB-D has good detection results even in the circumstance when the pose of pedestrian is changeable in single scene. The multi person scene is divided into two cases: with occlusion and without occlusion. The algorithm has good performance in both cases. Some experimental results are shown in Fig. 3.

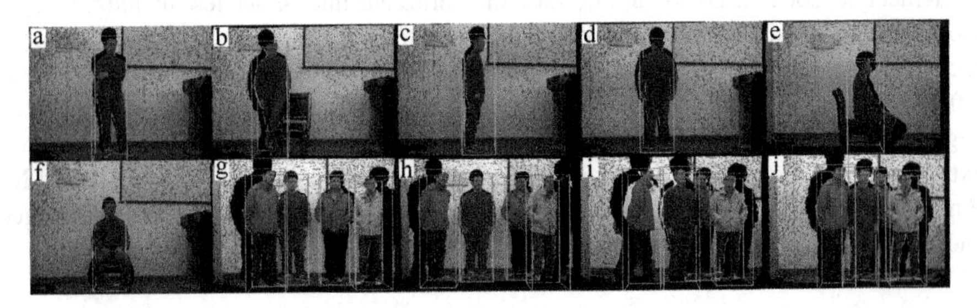

Fig. 3. Pedestrian detection results: (a) single scene (stand, front); (b) single scene (stand, front); (c) single scene (stand, side); (d) single scene (stand, back); (e) single scene (sit down, side); (f) single scene (sit down, front); (g) multi person scene (no occlusion); (h) multi person scene (no occlusion); (i) multi person scene (occlusion); (j) multi person scene (occlusion).

4 Conclusion and Future Work

A two-step pedestrian detection algorithm based RGB-D data is presented in this paper. Firstly, the algorithm captured the depth image and color image by using OpenNI SDK to drive Kinect camera. Then down-sampled the depth image and extracted the key information by voxel gridding method. After that, eliminated the ground by using the RANSAC segmentation algorithm and further screened the point cloud data independent with the target pedestrian. Next, described pedestrian characteristics by using PFH histograms and preliminary estimated the position of pedestrian. Finally, the pedestrian characteristics based on color image were further calculated by HOG descriptor and SVM classifier was used for pedestrian detection. The experimental results show that the algorithm can accurately detect the pedestrian not only in the single scene with pedestrian posture changing (stand, sit down, front, side and back) but also in the multi person scene with occlusion and without occlusion.

Although the algorithm has a high performance in a certain range of indoor applications, there are still some problems to be solved in the wide range scene applications. Limited by Kinect capability, the scope that can obtain the depth information is limited. When the distance between the target and the camera is more than 4 m, the amount of target information will sharply reduce. So, how to detect pedestrian based on RGB-D in a large range application scene is a further research direction.

Acknowledgement. This work is supported by Natural Science Foundation of China (Grant No. 61472245) and Shanghai Municipal Natural Science Foundation (Grant No. 14ZR1419700 and 13ZR1455600).

References

1. Spinello, L., Arras, K.O.: People detection in RGB-D data. IEEE/RSJ Int. Conf. Intell. Robots Syst. (IROS) **38**(2), 3838–3843 (2011)
2. Luber, M., Spinello, L., Arras, K.O.: People Tracking in RGB-D Data with On-line Boosted Target Models. IEEE/RSJ Int. Conf. Intell. Robots Syst. (IROS) **30**(1), 3844–3849 (2011)
3. Choi, W., Pantofaru, C., Savarese, S.: Detecting and tracking people using an RGB-D camera via multiple detector fusion. In: IEEE International Conference on Computer Vision Workshops (ICCV), pp. 1076–1083 (2011)
4. Wu, S., Yu, S., Chen, W.: An attempt to pedestrian detection in depth images. In: Third Chinese Conference on Intelligent Visual Surveillance (IVS), pp. 97–100 (2011)
5. Yu, S., Wu, S., Wang, L.: Sltp: a fast descriptor for people detection in depth images. In: IEEE Ninth International Conference on Advanced Video and Signal-based Surveillance (AVSS), pp. 43–47 (2012)
6. Xia, L., Chen, C.C., Aggarwal, J.K.: Human detection using depth information by kinect. In: IEEE Computer Society Conference on Computer Vision and Pattern Recognition Workshops (CVPR), pp. 15–22 (2011)
7. Wang, N., Gong, X., Liu, J.: A new depth descriptor for pedestrian detection in RGB-D images. In: 21st International Conference on Pattern Recognition (ICPR), pp. 3688–3691 (2012)
8. Ikemura, S., Fujiyoshi, H.: Real-time human detection using relational depth similarity features. In: Kimmel, R., Klette, R., Sugimoto, A. (eds.) ACCV 2010. LNCS, vol. 6495, pp. 25–38. Springer, Heidelberg (2011). doi:10.1007/978-3-642-19282-1_3
9. Cheng, Y., Zhao, X., Huang, K., Tan, T.: Semi-supervised Learning for RGB-D Object Recognition. In: 22nd International Conference on IEEE Pattern Recognition (ICPR), pp. 2377–2382 (2014)
10. Wang, A., Lu, J., Wang, G., Cai, J., Cham, T.J.: Multi-modal unsupervised feature learning for RGB-D scene labeling. In: Fleet, D., Pajdla, T., Schiele, B., Tuytelaars, T. (eds.) ECCV 2014. LNCS, vol. 8693, pp. 453–467. Springer, Cham (2014). doi:10.1007/978-3-319-10602-1_30
11. Gupta, S., Arbeláez, P., Girshick, R., et al.: Aligning 3D models to RGB-D images of cluttered scenes. In: IEEE Conference on Computer Vision and Pattern Recognition (CVPR), pp. 4731–4740 (2015)
12. Liu, J., Liu, Y., Zhang, G., et al.: Detecting and tracking people in real time with RGB-D camera. Pattern Recogn. Lett. **53**, 16–23 (2015)

13. Linder, T., Arras, K.O.: Real-time full-body human attribute classification in RGB-D using a tessellation boosting approach. In: IEEE/RSJ International Conference on Intelligent Robots and Systems (IROS), pp. 1335–1341 (2015)
14. Maji, S., Berg, A., Malik, J.: Classification using intersection kernel support vector machines is efficient. In: IEEE Conference on Computer Vision and Pattern Recognition (CVPR), pp. 1–8 (2008)

Inferring and Analysis Drivers Violation Behavior Through Trajectory

Zouqing Cai(✉), Wei Pei, Yongying Zhu, Mingyu Lu, and Lei Wu

Information Science and Technology College, Dalian Maritime University, Dalian, Liaoning Province, China
Zouqing.Cai@gmail.com, Wei.Pei@gmail.com, Peiweidl@gmail.com, Yongying.Zhu@gmail.com, Mingyu.Lu@gmail.com, Lei.Wu@gmail.com

Abstract. In this paper, we present an algorithm for inferring violation movements and categorizing levels of driving behavior. With this algorithm we extract the speeding and retrograde behavior from the real trajectories datasets of Xinjiang, analyze the changing regulation of six streets in the working day and day off on the overall and explore the driving characteristics which is very dangerous (level 4). The results of this study can not only be used for early warning of drivers violations, but also provide the data support and decision basis for the traffic management department to master the situation of traffic violations and formulate the traffic management counter measures.

Keywords: Trajectory data mining · Violation driving behavior · Violation movements · Driver classification

1 Introduction

The global status report on road safety 2015 indicates that worldwide the total number of road traffic deaths has plateaued at 1.25 million per year [1]. In 2015, the number of all kinds of production safety accident deaths has plateaued at 66182 in our country and 54% of the deaths are caused by traffic accidents. Drunk driving, speeding, retrograde, running the red light and other illegal operations account for more than 60% of the proportion of traffic accidents [2]. Due to a rising trend in vicious traffic accidents, the study of violation driving behavior has been a hot spot in recent years.

Several works have been proposed for driving behavior analysis in simulation systems [4–8]. For instance, sensors and simulators are used to recognize the fatigue driving state [6].

With the popularity of mobile devices such as GPS and cell phones, large amounts of real driving traces are available to analyze the behavior of drivers [9–14].

By inferring violation behavior and sum up the law of the illegal driving behavior, the traffic management departments can grasp the situation of traffic violations and take effective measures to reduce or even avoid major traffic accidents. At the same time, it has a warning effect on the drivers who often break the rules. In this paper, we focus on real trajectories of drivers, propose an efficient algorithm to identify the violation behaviors of drivers based on individual trajectories and classify the drivers according to

Z. Shi et al. (Eds.): ICIS 2017, IFIP AICT 510, pp. 295–304, 2017.
DOI: 10.1007/978-3-319-68121-4_32

the risk weights. According to the different function area, six roads including four one-way streets are picked up to demonstrate the algorithm. The first two steps of the algorithm extract the speeding and retrograde behavior, analyze the change regulation of illegal driving behaviors either workday or rest day and discuss the place where the driver often has retrograde behaviors. Furthermore the violation behavior of drivers is evaluated with the last two steps of the algorithm and the very dangerous drivers' (level 4) personality characteristics is analyzed.

The rest of the paper is organized as follows. Section 2 introduces the related work in illegal driving behavior analysis. Section 3 presents the main definitions used in this paper, experiments with real data and proposes a method to find speeding and retrograde behavior. Section 4 shows and discusses the experiments results and Sect. 5 concludes the paper.

2 Related Work

Several works have been done to study the driver behaviors in simulation systems. Kedar-Dongarkar and Das presented a new method of driver classification for optimizing energy usage [5]. Vehicle acceleration, braking, speed, and throttle pedal were used to classify the drivers into three categories: aggressive, moderate and conservative. Hu presented a new identification method of fatigue driving state, which is obtained from driving behavior data analysis both about normal driving state and the fatigues one [6]. Quantitative evaluation of driving styles by normalizing driving behavior was studied in [7].

Some works which analyze driving behavior using real GPS trajectories will be introduced as follows. For example, Carboni was interested in abnormal behaviors of individual trajectories of drivers, and presented an algorithm for finding anomalous movements and categorizing levels of driving behavior [10]. Li proposed a method to realize the safety analysis of vehicle driving behavior by using GPS vehicle trajectory data [11]. They studied the proportion of the bus over-speed time in entire driving time. Dueholm presented a method of latent semantic information mining for trajectory data [12]. He proposed WhozDriving to solve the issues of detecting abnormal driving trajectory in driver's history trajectories [13]. Chen and Zhang focused on abnormal trajectories of taxis that deviated the standard route from origin and destination, where the standard route represented the path followed by the majority of taxis [14].

Although the previous detailed works analyze several illegal characteristics of driving, most of them have not been developed with real trajectories. Some works used the real trajectory analysis of dangerous driving behavior, but they concentrated on how to judge the abnormal behavior such as rapid change, sharp turn speeding etc. They rarely conduct a comprehensive analysis of illegal behavior and summarize the rules of violation driving. In this paper we focus on real trajectories of drivers, propose an algorithm to identify violation behavior and classify drivers into different levels of danger. The effectiveness of the algorithm is greatly proved by experiments.

3 Discover Traffic Violation

In this section we firstly present some basic definitions and formulas (Sect. 3.1). Secondly, filter trajectories will be introduced in Sect. 3.2. Finally, an algorithm will be proposed to find and evaluate violation driving behavior in Sect. 3.3.

3.1 Main Definitions

The basic definitions for trajectories are described as follows.

Definition 1. Point. A point is defined as p = (id, x, y, t, s, h), where id is a vehicle ID represents a driver, x and y are the latitude and longitude that represent space, t is the *timestamp* in which the point has been collected and h is a direction angle that represents the driving direction.

Definition 2. Trajectory. A trajectory is expressed as $\mathrm{T} = \{r_1, r_2 \ldots r_n\}$, where $r_i = \{p_1, p_2 \ldots p_n\}$, $p_i = (\mathrm{id}, x_i, y_i, t_i, s_i, d_i)$, $t_1 < t_2 < \ldots < t_n$. A trajectory includes one or several roads, and r represents a road.

Definition 3. Subtrajectory. A subjectory s of T is a series of points $<p_k, p_{k+1}, \ldots, <p_j$, where $p_k \subset T$ and $1 \leq \mathrm{k} \leq \mathrm{j}$ and $\mathrm{j} \leq \mathrm{n}$. (In this paper, subtrajectory division is based on road. A road of trajectory is corresponds to a subtrajectory.)

Definition 4. Trajectories. A trajectories is $T_s = \{T_1, T_2 \ldots T_n\}$, where $T_i = \{p_1, p_2 \ldots p_n\}$, $\mathrm{pi} = (\mathrm{id}, x_i, y_i, t_i, s_i, d_i)$, $t_1 < t_2 < \ldots < t_n$.

Definition 5. Violation movement. A trajectory has violation movements when it has at least one subtrajectory with violation s such as the driving speed exceeds the road speed limit, the driving direction is contrary to the normal driving direction.

The case which meets the Eq. (1) is judged as the over speed and the one meeting the Eq. (2) is retrograde. v is the speed of the point, and h stands for the direction of the point. r represents a road, and r = 1 presents a one-way street. v' represents the maximum value of road speed limit, and h' is the direction of road restrictions.

$$c = \begin{cases} 1, & v > v' \\ 0, & v \leq v' \end{cases} \tag{1}$$

$$d = \begin{cases} 1, r = 1 \cap |h - h'| \leq 45 \\ \quad 0, \; other \end{cases} \tag{2}$$

After defining the violation movement, the movements are analyzed deeply. Some characteristics are found to evaluate driving behavior. In this analysis we mainly consider three features:

F1: The driver only has a violation movement in a road.
F2: Continuous illegal behaviors in a road.
F3: Continuous illegal behaviors in different roads.

Based on these three features, we defined four levels of drivers:

Level 1 (Careful driver): A careful driver is without any violation behavior. Although someone may complain that it makes no sense to discover careful drivers, it is very useful for the company to give a reward or compliment to the good drivers.

Level 2 (Distract driver): A distract driver is with feature F1. The driver is distracted while driving with the performance of many secondary tasks, including texting and dialing cell phones and so on.

Level 3 (Dangerous driver): A dangerous driver is someone with feature F2. Such as a driver continues to over-speed on the road.

Level 4 (Very dangerous driver): A very dangerous driver is someone with features F3. For example, the driver has a continuous speeding or retrograde driving behavior in different roads.

3.2 Filter Trajectories

We get real trajectories of taxis collected at February 2016 in the city of Xinjiang, China. The dataset has about 300 million data with points collected at intervals of 30 s, including vehicleid, gettime, storetime, speed, direction, latitude and longitude. The taxi dataset is for a city scale but we just focus on several specific routes, so we implement filter trajectories which has the same origin and the same destination on a road [13]. Using this method, trajectories are filtered in six routes where close to different function area.

3.3 Discover Speeding or Retrograde

In this paper an algorithm is proposed to discover violation driving behaviors. Firstly it identifies violation movements based on over-speed and retrograde driving behavior. Secondly it evaluates the driving behavior according to the formulas which we define.

Because the violations driving behavior is very transitory, the subtrajectories with violations behavior are normally only a few points. If considering the violation movements of each point, noise can be introduced. If considering too many points (three or more) the violation movement may not be captured. So after some analysis and experiments on real trajectory data, it consider that at least two consecutive points should have violation movements for a subtrajectory to characterise the violation behavior. Violation behavior can be captured well for trajectories with frequently sampled points, like 1 or 2 s. A dataset with sampling rate as 30 s, for instance, can also reveal violation behavior.mm.

The algorithm of findViolation is as follows.

findViolation:

$Ts=\{t_1, t_2 \ldots t_n\}$. For each trajectory t_i, using the length of the road to overlap violations, $t_i=\{r_1, r_2 \ldots r_n\}$, $vio(t_i)=vio(r_1)+vio(r_2)+\ldots+vio(r_n)$.

$r_i=\{p_1, p_2 \ldots p_n\}$. For each point, c_i and d_i can be computed according to equation (1)and(2).

The count of subtrajectories violations movments can be computed according to the road ..

$$violation(r_i)=\sum(c_i*c_{i+1}+d_i*d_{i+1}) \quad (3)$$

The violation behavior of drivers is evaluated using formulal(4)

$$level = \begin{cases} level1, vio(t_i) = 0 \\ level2, vio(r_i) = 1 \text{ and } vio(r_i) + \cdots + vio(r_{i-1}) + vio(r_{i+1}) + \cdots + vio(r_n) = \\ level3, 4 > vio(r_i) > 1 \text{ and } vio(r_1) + \cdots + vio(r_{i-1}) + vio(r_{i+1}) + \cdots + vio(\\ level4, vio(r_i) > 1 \text{ and } vio(r_1) + \cdots + vio(r_{i-1}) + vio(r_{i+1}) + \cdots + vio(r_n) > \\ \quad \text{or } vio(r_i) > 4 \text{ and } vio(r_1) + \cdots + vio(r_{i-1}) + vio(r_{i+1}) + \cdots + vio(r_n) = \end{cases} \quad (4)$$

In findViolation, the first two steps are used to discover the violation movements, which named fv-1. The last two steps of the algorithm are used to divide the drivers into different levels of violation, which named fv-2.

4 Experiental Results and Dicussion

In this section our algorithm is demonstrated and the experimental results are analyzed with the real-world GPS trajectory dataset. The real dataset is processed in Sect. 4.1. We analyzed the different change laws of different roads in working days and day off in Sect. 4.2. The violation behavior of drivers is evaluated and the characteristics of the very dangerous driver are analyzed in Sect. 4.3.

4.1 Data Processing

The dataset that mentioned in Sect. 4.2 is used in the experiment. Firstly, the error points are removed. Such as the latitude or the longitude is null, or the value of speed is less than zero. Secondly, a day is divided into 24 time periods. Thirdly, all the tracks are adapted to the spatial division and established the index to speed up the search. Finally, the trajectory points are matched with the map using AMAP API.

4.2 Part One

The superscript numeral used to refer to a footnote appears in the text either directly after the word to be discussed or – in relation to a phrase or a sentence – following the punctuation mark (comma, semicolon, or period). Footnotes should appear at the bottom of the normal text area, with a line of about 5 cm set immediately above them.

In this section we use February 1st, 2016 (Monday) as the example of working day and February 27th, 2016 (Saturday) as the example of day off to explore the temporal and spatial evolution of speeding and retrograde on the same roads. We pick up a region which has a large number of taxis. Figure 1 shows the heatmap of the area on Monday. Based on the color change, it is clear to show where the number of operation taxis is the largest. According to the heatmap, we choose six busy roads and use the method which is mentioned in Sect. 3.2 to filter trajectories. The relevant information of the roads is shown in Table 1. As we all known that roads can cross over different functional areas. In order to expediently discuss the relationship between the distribution of the taxi and the different functional areas, the roads are divided into four groups according the main utilization of land along the roads. For example, although there are two top grade residences nearby Zhongshan Rd, it is defined as a commercial road because there are several supermarkets and shopping malls on both sides of the road. Table 2 shows the results of the division.

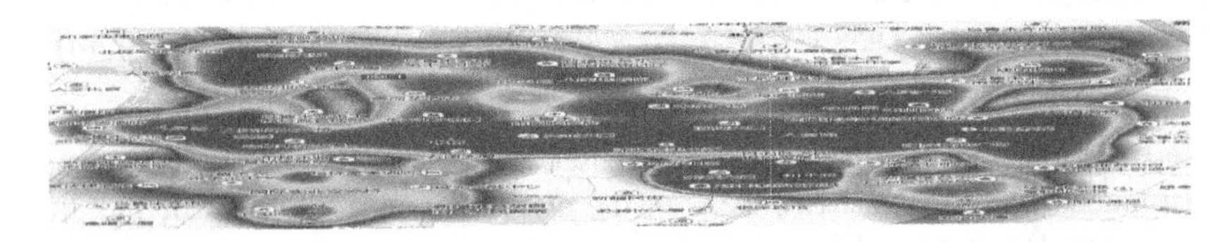

Fig. 1. The thermodynamic chart of the area. (Color figure online)

Table 1. The information of the six streets

Detail	Renmin Rd	Democracy Rd	Liberation Rd	Zhongshan Rd	Xingfu Rd	Jianshe Rd
Rd level	Main rd	Main rd	Main rd	Main rd	Main rd	Secondary rd
Speed	60 km/h	50 km/h	50 km/h	50 km/h	50 km/h	50 km/h
Direction	Null	West-east	Null	East-west	Northwest-Southeast	Southeast-Northwest

Table 2.

Character	Streets
Commercial	Zhongshan Rd, Democracy Rd
Health and education	Liberation Rd, Xingfu Rd
Administrative office	Renmin Rd
Residence community	Jianshe Rd

Figures 2 and 3 show the changing laws of the operation taxis in February 1st, 2016 (workday) and February 27st, 2016 (rest day). The changing rules of operating taxis during the two days have similar characteristics. For example, the number of the operation taxis gradually decreases from 00:00:00 to 07:00:00 and there are lots of taxis from 10:00:00 to 17:00:00. The distribution of operating vehicles between weekdays and rest days are different. For example, there are more operation taxis on workday than the rest day, the rest day only has a low peak and the proportion of vehicles in the night are bigger. From the micro perspective, Zhongshan Rd has a large amount of vehicles at 10:00:00−12:00:00 and 13:00:00−17:00:00 on Saturday which is related to the commercial character of this road.The character of Jianshe Rd is a residence community. The variation of the number of vehicles on Monday is quite similar to the one of Saturday, which has more vehicles on Monday at 10:00:00 −12:00:00 and 16:00:00−18:00:00. This period is the peak of commuting. By comparing the trends of the two graphs, we find a strange phenomenon that the Renmin Rd on Monday at 12:00:00−14:00:00 has a peak valley but on Saturday is a trough. The reason for it is that there are many office buildings near the Renmin Rd, and more staff go out to eat or go home at the lunch time of 12:00:00−14:00:00 on Monday (working day) etc.

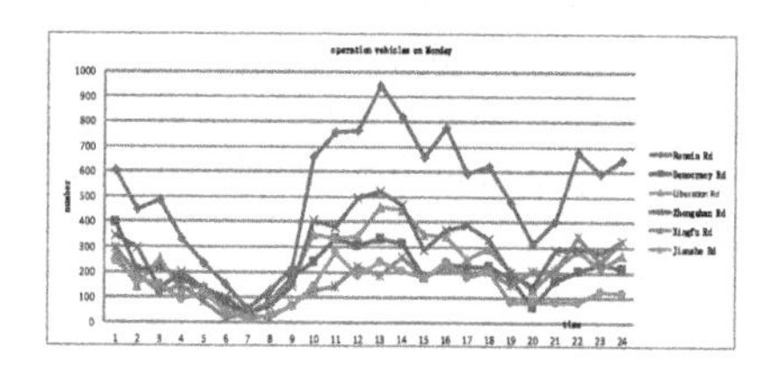

Fig. 2. The regulation of taxis on Monday.

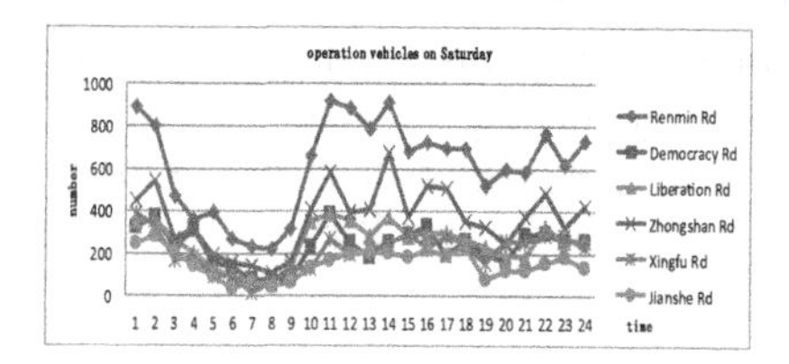

Fig. 3. The regulation of taxis on Saturday.

The retrograde behavior is extracted from the real dataset by fv-1. Figures 4 and 5 show the changing law of the retrograde vehicles. In Fig. 4 we find that there are many vehicles with retrograde behavior from 10:00:00 to 19:00:00. The number of the retrograde taxis gradually decreases from 00:00:00 to 07:00:00 because of the reducing total number of vehicles. Comparing the trend of each road in Figs. 2 and 4 we find that the general trend of illegal vehicles and operating vehicles is similar. Of course, it also has individual differences. For example, On Monday, the overall trend of operating vehicles in Zhongshan Rd from 12:00:00 to 14:00:00 is a peak valley and reaches its maximum at 13:00:00, but the trend of retrograde vehicles is a trough and drops to the lowest at 13:00:00. Owing to the trips of day off are random, the retrograde vehicles are also distributed randomly in Fig. 5.

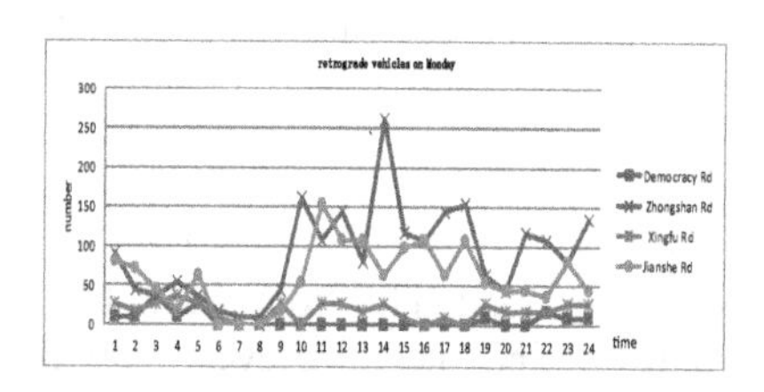

Fig. 4. The retrograde taxis on Monday.

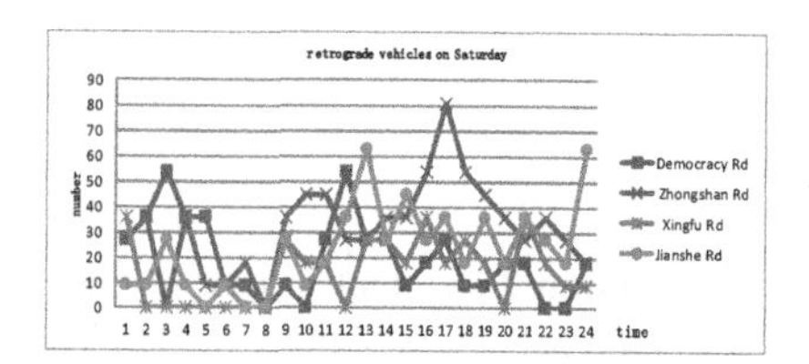

Fig. 5. The retrograde taxis on Saturday.

The statistics show that in addition to Xignfu Rd, the numbers of retrograde vehicles of the other three roads in the working days are more than the rest day. On Xingfu Rd there are 234 drivers with retrograde behavior on working day and there are 342 on rest day. The Xingfu Rd is used for Health and Education. There are a primary school, a secondary school and a vocational college near the road. The reason for the number of violation vehicles in the working day less than the rest day is that the students and their parents can comply with the traffic rules. Figure 6 shows the heatmap of retrograde vehicles. We find that retrograde behavior occurs at the crossroads with a higher frequency. So it suggest that the traffic management department should increase the supervision on the intersection.

The speeding behavior is extracted from the real dataset by the fv-1. Through analyzing Figs. 7 and 8 we find that the number of over-speed vehicles at night is higher than the one at the daytime. The number of serious over-speed vehicles in workday is more than that in the rest day. At 9:00:00−10:00:00 the number of speeding vehicles is very small. According to Figs. 2 and 3 we find that the traffic flow during this period is very large, so we speculate that in the absence of speeding vehicles during the period, the road may be in a congested state. The low-level over-speed behavior that no more than 20% is the top of the over-speed ranking. Severe speeding more than 50% is the least.

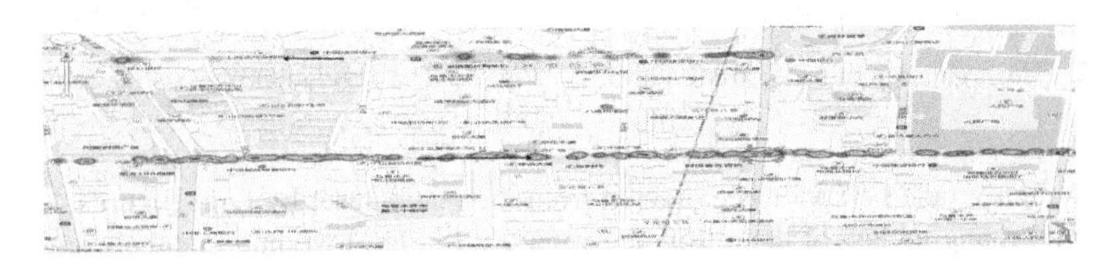

Fig. 6. The heat map of retrograde taxis on Monday.

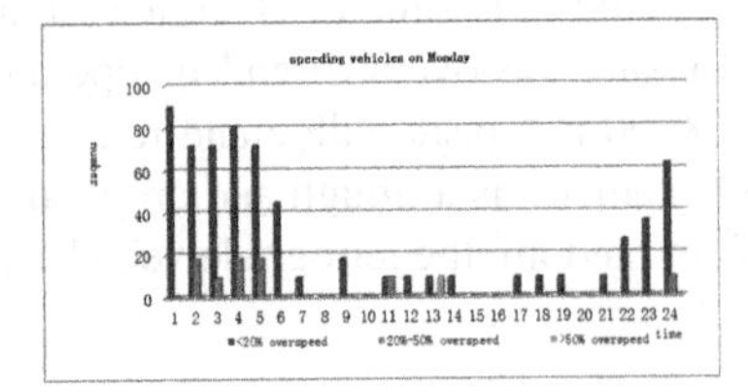

Fig. 7. The level of speeding taxis on Monday.

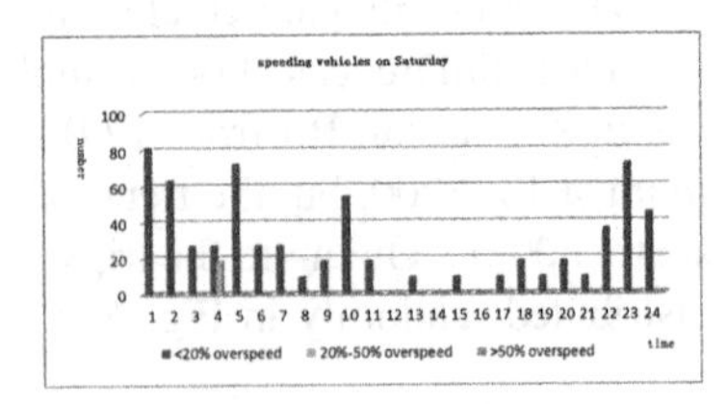

Fig. 8. The level of speeding taxis on Saturday.

4.3 Part Two

In this section 20 vehicles trajectories are picked up randomly from Sect. 4.2 to findViolation behaviors of individual trajectories of drivers and categorize levels of driving behavior with the algorithm. The results are shown in Table 3. In order to illustrate the violations trajectories in detail, we show a trajectory of a very dangerous driver.

Table 3. The results of part two

Level	Results
Precise driver	15
Imprecise driver	2
Dangerous driver	0
Very dangerous driver	1

Figure 9 shows the part of a very dangerous driver's trajectory, where the red line represents the trajectories with violation subtrajectories, the blue line represents the normal trajectory and the red point represents the driver with violation behavior in this position. As mentioned in Sect. 3.3, at least two consecutive points with the violation movement can be characterized with violation behavior for a subtrajectory. So we draw red lines to match the consecutive points with the same violation behavior. It is easy to find that the driver has violation movements on different roads.

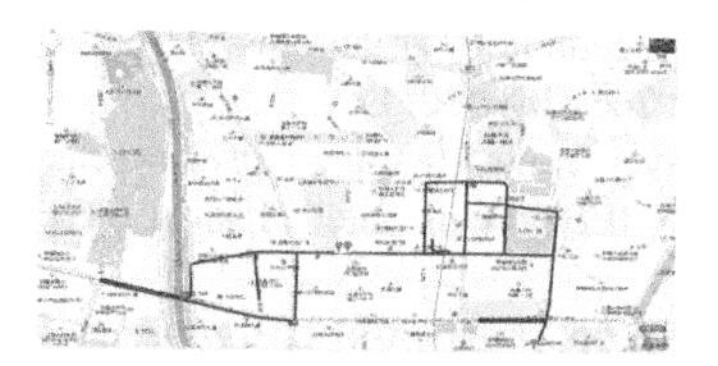

Fig. 9. The very dangerous driver. (Color figure online)

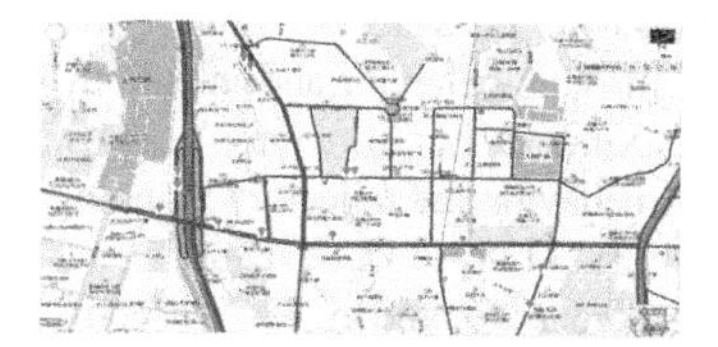

Fig. 10. The very dangerous driver for a whole day. (Color figure online)

The driver's trajectories for a whole day are shown in Fig. 10. It can find that many discontinuous trajectory points have violation behavior and some of points are located at the intersection. Due to the dataset with sampling rate as 30 s and having few taxis in the area at the same time, it can infer that the drivers may have illegal behavior such as running a red light. It is not only very dangerous to passengers and themselves, but also to other cars on the road.

5 Conclusions and Future Works

Trajectory behavior analysis is becoming very useful in our life. In this paper we presented an algorithm to measure the violation behavior of drivers. Firstly, the algorithm finds violation movements such as speeding and retrograde. Secondly, the driver is divided into different danger levels according to the characteristics values related to the violation movements. Experiments show that this algorithm can correctly detect the violation movement and mark the violation trajectory. Through the experimental analysis, we suggest that the traffic management departments increase supervision at intersections. At the same time, we remind that the drivers do not have violation driving behavior to guarantee the safety of yourself and others.

Tens millions of taxis information are generated every day and it is difficult for us to deal with large amounts of data so we just analysis two days' information. In the future, we will intent to perform more analysis with more real data, to explore speeding, retrograde and other illegal driving behavior such as running a red light.

Acknowledgment. The work was supported by the National Natural Science Foundation of Chinas (Nos. 61001158, 61272369 and 61370070), Liaoning Provincial Natural Science Foundation of China (Grant No. 2014025003), Scientific Research Fund of Liaoning Provincial Education Department (Grant No. L2012270), Specialized Research Fund for the Doctoral Program of Higher Education (SRFDP No 200801511027), and Fundamental Research Funds for the Central Universities (Grant No. 3132015084).

References

1. WHO - World Health Organization, 2015 Global status report on road safety (2015). http://www.who.int/violence_injury_prevention/road_safety_status/2015/zh/
2. http://auto.people.com.cn/n1/2016/0702/c1005–28517746.html
3. Zhu, Y., Xiao, X., Zhu, J.: Research on active safety technology of automobile based on driving behavior and intention. Mach. Des. Manuf. **1**, 266–268 (2011)
4. Xu, G., Liu, L., Ou, Y.: Dynamic modeling of driver control strategy of lane-change behavior and trajectory planning for collision prediction. IEEE Trans. Intell. Transp. Syst. **13** (3), 1138–1155 (2012)
5. Kedar-Dongarkar, G., Das, M.: Driver classification for optimization of energy usage in a vehicle. Procedia Comput. Sci. **8**, 388–393 (2012)
6. Hu, D.I., Zhao, X.H., Mu, Z.C.: Distinguish method of fatigue state based on driving behavior wavelet analysis. In: Control Conference, pp. 3590–3596. IEEE (2013)
7. Shi, B., Xu, L., Hu, J.: Evaluating driving styles by normalizing driving behavior based on personalized driver modeling. IEEE Trans. Syst. Man Cybern. Syst. **45**(12), 1 (2015)
8. Xu, L., Hu, J., Jiang, H.: Establishing style-oriented driver models by imitating human driving behaviors. IEEE Trans. Intell. Transp. Syst. **16**(5), 2522–2530 (2015)
9. Guo, H., Wang, Z., Yu, B.: TripVista: triple perspective visual trajectory analytics and its application on microscopic traffic data at a road intersection. In: 2011 IEEE Pacific Visualization Symposium, PACIFICVIS. pp. 163–170 (2011)
10. Carboni, E.M., Bogorny, V.: Inferring drivers behavior through trajectory analysis. In: Angelov, P., Atanassov, K.T., Doukovska, L., Hadjiski, M., Jotsov, V., Kacprzyk, J., Kasabov, N., Sotirov, S., Szmidt, E., Zadrożny, S. (eds.) Intelligent Systems 2014. AISC, vol. 322, pp. 837–848. Springer, Cham (2015). doi:10.1007/978-3-319-11313-5_73
11. Xu, T., Li, X.: Analysis on the Safety of bus driving using vehicle GPS track data. In: GEOMATICS and Information Science of Wuhan Univers, pp. 739–744 (2014)
12. Dueholm, J.V., Kristoffersen, M.S., Satzoda, R.K.: Trajectories and maneuvers of surrounding vehicles with panoramic camera arrays, p. 1 (2016)
13. He, M., Guo, B., Chen, H., Chin, A., Tian, J., Yu, Z.: WhozDriving: abnormal driving trajectory detection by studying multi-faceted driving behavior features. In: Wang, Y., Yu, G., Zhang, Y., Han, Z., Wang, G. (eds.) BigCom 2016. LNCS, vol. 9784, pp. 135–144. Springer, Cham (2016). doi:10.1007/978-3-319-42553-5_12
14. Chen, C., Zhang, D., Castro, P.S.: iBOAT: isolation-based online anomalous trajectory detection. IEEE Trans. Intell. Transp. Syst. **14**(2), 806–818 (2013)
15. Krüger, R., Thom, D., Wörner, M.: TrajectoryLenses – a set-based filtering and exploration technique for long-term trajectory data. In: Computer Graphics Forum, pp. 451–460 (2013)

Improved CNN Based on Super-Pixel Segmentation

Yadong Yang(✉) and Xiaofeng Wang

College of Information Engineering, Shanghai Maritime University, Shanghai, China
yangyadong03@stu.shmtu.edu.cn, xfwang@shmtu.edu.cn

Abstract. Convolutional neural network has unique superiority in images processing, it can effectively extract features and reduce data dimensions by convolution and pooling. But it takes the "violent segmentation" method in the process of pooling. This method cannot guarantee the final selected pixel value can be a good representative of the partial features, neither average pooling nor max pooling. Therefore, three pooling methods based on super-pixel segmentation are proposed, called "super-pixel average pooling", "super-pixel max pooling" and "super-pixel smooth pooling". Firstly, the super-pixel segmentation is performed on the feature images, and then the value of the point which has the smoothest gradient in each super pixel is selected to represent the feature of the local area. Compared to the violent segmentation pooling operation, this method exhibits more stable characterization ability, it can retain image features perfectly while reduce the data dimensions. Experiments show that the improved convolutional neural network achieved better results than normal algorithm in the standard data sets.

Keywords: Convolutional neural network · Super-pixel segmentation · Super-pixel average pooling · Super-pixel max pooling · Super-pixel smooth pooling

1 Introduction

Convolutional neural network (CNN) is a deep neural network model with convolution structure. It can effectively reduce the number of weights and lower the complexity of the network. This model has some invariance for scaling and other forms of deformation. Each neuron only needs to perceive the local image area, The network obtain global information by combining these neurons that perceive different local regions at higher levels. In 1989, LeCun et al. proposed a CNN model "LeNet-5 [1] " for character recognition. LeNet-5 consists of convolutional layers, sub-sampling layers and fully connected layers. The system achieves good results in small-scale handwritten numeral recognition. In the ImageNet contest, "AlexNet [2] ", a new network architecture of CNN, designed by Krizhevsky et al. won the 2012 championship. AlexNet and OverFeat [3] are quite powerful CNN models trained on large natural image datasets. By improving the network performance, Girshick et al. proposed R-CNN [4] (Regions with CNN) can complete the target detection task

Z. Shi et al. (Eds.): ICIS 2017, IFIP AICT 510, pp. 305–310, 2017.
DOI: 10.1007/978-3-319-68121-4_33

effectively. He utilize spatial pyramid pooling (SPP) to deal with different size and aspect ratio of the input images, this new network called SPP-Net [6]. In recent years, more and more popular method to optimize CNN models is to develop deeper and more complex network structures, and then to train them with massive training data. VGG [5] (visual geometry group), a 19-layer depth network, mainly explores the importance of depth for the network. GoogLeNet [6] is the particular incarnation for ILSVRC14 submit by Szegedy et al., a 22 layers deep network, the quality of which is evaluated in classification and detection. In order to improve the performance of CNN, the researchers not only discuss the structure of CNN and its applications, but also improve the design of the network layers, the loss function, activation function, regular items and many other aspects of the existing network. They achieved a series of results, for example, Inception network [5], stochastic pooling, ReLU [2], Leakly ReLU [7], Cross entropy loss, Batch normalization [8], etc.

2 Related Work

2.1 Convolutional Neural Network

In a typical CNN model, the beginning layers are usually alternating between the convolution layer and the sub-sampling layer. The last few layers of the network near the output layer are usually fully connection networks. The four basic elements of CNN are convolutional layers, sub-sampling layers, all connected layers and back propagation (BP) algorithm.

The convolutional layers receive the input images or the feature maps from the previous layer, they carry out the convolution operation through the N convolution cores and activate operation using the activation function, then N new feature maps are generated and directed into the next sub-sampling layer. The function of convolution is defined by

$$x_j^l = f(\sum_{i \in M_j} x_i^{l-1} * k_{ij}^l + b_j^l) \tag{1}$$

where l denotes the convolution layer, j is the j-th channel, x_j^l is the output value from the j-th channel of convolution layer l, f(·) is the activation function, M_j represents the subset of feature maps, k_{ij}^l is the convolution kernel matrix, * is the convolution operation, b_j^l is offset of l.

The sub-sampling layers perform sub-sampling operations on the input feature maps, which can effectively reduce the computational complexity and extract more representative local features. The function of sub-sampling is defined by

$$x_j^l = f(\beta_j^l down(x_j^{l-1}) + b_j^l) \tag{2}$$

where β_j^l denotes is the sub-sampling weight coefficient, down(·) is sub-sampling function.

The fully connected layers consist of an input layer, several hidden layers and an output layer. The upper layer's feature maps is spliced into one-dimensional vector as input data, and the final outputs is obtained by weighting and activating.

$$x^l = f(\varpi^l x^{l-1} + b^l) \tag{3}$$

where ϖ^l is the neural network weight.

The BP algorithm is a common neural network method in supervised learning. For the convolutional neural network, the main job is to optimization the convolution kernel parameters, the sub-sampling layers weights, the network weights of the full connection layer and the bias parameters of all layers. The essence of the BP algorithm is to allow us to calculate each network layer's effective errors and to derive the learning rules of network parameters, and then drive the actual network outputs are closer to the target values.

2.2 Super-Pixel Segmentation

Super pixels are divided a pixel-level image into district-level image or sets of pixels by image segmentation. The goal of super pixel segmentation is to change or simplify the representation of an image into something that is more convenient and significant to analyze. The super pixel segmentation method has been under intensive study, now there are many super pixel segmentation algorithms.

Super pixels are obtained by using NCut [9] and SLIC [10] algorithm has a high compactness. Using SLIC and Watershed algorithm for ultra-pixel segmentation is very productive. But if you put more emphasis on edge accuracy and regional merging, you can choose Marker-based Watershed and Meanshift algorithm.

3 Improved CNN

CNN have been substantiated to provide a powerful approach for image processing. In this work, we focus on the pooling part of the CNN.

The common CNN pooling methods include average pooling, max pooling and multi-scale mixing pooling. These methods have achieved some good results, they calculate a value representing the local region feature in the pre-divided local area of feature maps. However, those violent segmentation methods have their own unreasonable place, including ignore some local features or neutralize some salient features. This paper presents three pooling methods based on super pixel segmentation. Firstly, take super-pixel segmentation action on the feature maps, and the pixel feature values in each local region (i.e., super pixels) we got are similar. Then three kinds of pooling methods are come up with super pixel segmentation.

(1) Calculate the average value of pixels within each super pixel called super pixel average pooling.
(2) Get the max value of pixels within each super pixel, that is, super pixel max pooling.
(3) And super pixel smooth pooling trying to take the value of the point which has the smoothest gradient in each super pixel.

This paper only considers the super pixel level, does not deal with regional merger after image segmentation. Based on SLIC has the low time complexity and high compactness, we choose to use SLIC algorithm for super pixel segmentation (Fig. 1).

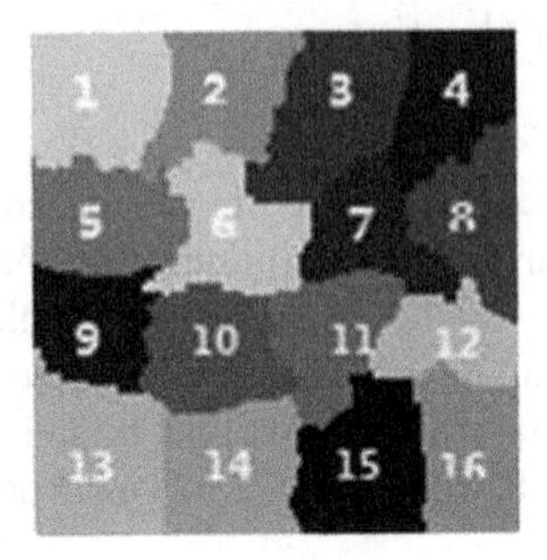

1	2	3	4
5	6	7	8
9	10	11	12
13	14	15	16

Fig. 1. The super-pixel pooling schematic diagram

These segmentation methods can not only extract the necessary features, but also the local stability of the images has no destruction at all.

4 Results

The structure of CNN we used in experiment for MNIST has 7 layers. The network is composed of 2 convolutional layers, 2 pooling layers and 3 fully connected layers. The experiment for CIFAR-10 use AlexNet. While the convolutional layers are followed by rectified linear operator (ReLU) layers and the pooling layers take the value of local regions with two-pixel strides, the dropout layers having a 0.5 dropout ratio. The last fully connected layer employ softmax function as a multi-class activation function.

Because of the MNIST data have relatively simple characteristics. Here divide the MNIST test set into six parts randomly, each part is transformed into different scales or angles, and then normalize the images in 28*28 pixels to get a new test set. The training set remains the same. Test on data sets MNIST, New MNIST (Fig. 2).

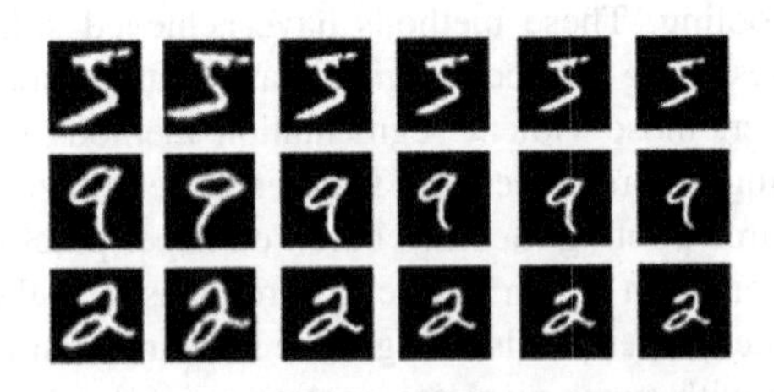

Fig. 2. Part of the new MNIST test images

In the following tables: “max”, “avg”, “sps”, “sp-max”, “sp-avg” each representing max pooling CNN method, average pooling CNN method, super-pixel smooth pooling CNN method, super-pixel max pooling CNN method and super-pixel average pooling CNN method. The results are shown as follows (Tables 1, 2, and 3).

Table 1. Test results of the five methods on the MNIST dataset.

Methods	Training error (%)	Test error (%)
max	0.01	0.8
avg	0.06	0.9
sps	0.04	0.7
sp+max	0.05	0.8
sp+avg	0.06	0.8

Table 2. Test results of the five methods on the new MNIST dataset

Methods	Training error (%)	Test error (%)
max	0.04	10.7
avg	0.1	9.8
sps	0.06	4.7
sp+max	0.06	8.9
sp+avg	0.08	6.6

Table 3. Test results of the five methods on the CIFAR-10 dataset

Methods	Training error (%)	Test error (%)
max	2.67	29.5
avg	12.50	28.8
sps	5.65	24.2
sp+max	6.54	26.4
sp+avg	8.26	25.1

Based on the above results it can be seen that: the average pooling method is better than the max pooling methods, because of the local regions average values are more representative. Super-pixel pooling method is better than the standard pooling method, which is due to better image segmentation. The super pixel smooth pooling method gets the best results, because this method can find the most representative values.

5 Conclusions

A neoteric and universal approach is proposed for improving CNN performance through using the super-pixel pooling to train the network. This approach makes the models have more stable characterization and better generalization, and it can be used for different CNN network structures. Extensive experiments on several standard data sets for the image classification prove that using the super-pixel pooling in the training process can significantly enhance performance of CNN models, in comparison with the same model trained without employing this method.

The super-pixel segmentation technique may be further used for the convolution operation. By introducing the fuzzy segmentation method, adjust the number of extra pixels and the number of pixels per super-pixel contains after segmentation. This method is suitable for convolution of large-size images, and convolution of image sets containing different sizes of images. Moreover, we can use the improved CNN to resolve many practical problems such as pedestrian detection, ship classification, text recognition or medical images processing.

References

1. LeCun, Y., Bottou, L., Bengio, Y., Haffner, P.: Gradient-based learning applied to document recognition. Proc. IEEE **86**(11), 2278–2324 (1998)
2. Krizhevsky, A., Sutskever, I., Hinton, G.E.: Imagenet classification with deep convolutional neural networks. In: Proceedings of Advances in Neural Information Processing Systems, vol. 25, pp. 1097–105. Curran Associates, Inc., Lake Tahoe (2012)
3. Girshick, R., Donahue, J., Darrell, T., Malik, J.: Rich feature hierarchies for accurate object detection and semantic segmentation. In: Proceedings of the 2014 IEEE Conference on Computer Vision and Pattern Recognition, Columbus, USA, pp. 580–587. IEEE (2014)
4. He, K.M., Zhang, X.Y., Ren, S.Q., Sun, J.: Spatial pyramid pooling in deep convolutional networks for visual recognition. IEEE Trans. Pattern Anal. Mach. Intell. **37**(9), 1904–1916 (2015)
5. Szegedy, C., Liu, W., Jia, Y.Q., Sermanet, P., Reed, S., Anguelov, D., Erhan, D., Vanhoucke, V., Rabinovich, A.: Going deeper with convolutions. In: Proceedings of the 2015 IEEE Conference on Computer Vision and Pattern Recognition, Boston, MA, pp. 1–9. IEEE (2015)
6. Simonyan, K., Zisserman, A.: Very deep convolutional networks for large-scale image recognition. http://arxiv.org/abs/1409.1556. Accessed 16 May 2016
7. Maas, A.L., Hannun, A.Y., Ng, A.Y.: Rectier nonlinearities improve neural network acoustic models. In: Proceedings of ICML Workshop on Deep Learning for Audio, Speech, and Language Processing, Atlanta, USA. IMLS (2013)
8. Ioffe, S., Szegedy, C.: Batch normalization: accelerating deep network training by reducing internal covariate shift. In: Proceedings of the 32nd International Conference on Machine Learning, Lille, France, pp. 448–456. IMLS (2015)
9. Shi, J., Malik, J.: Motion segmentation and tracking using normalized cuts. In: Sixth International Conference on Computer Vision, pp. 1154–1160 (1998)
10. Jung, E.S., Ranka, S., Sahni, S.: Bandwidth allocation for iterative data-dependent e-science applications. In: EEE/ACM International Conference on Cluster, Cloud and Grid Computing, pp. 233–242 (2010)

Speaker Verification Channel Compensation Based on DAE-RBM-PLDA

Shuangyan Shan(✉) and Zhijing Xu

College of Information Engineering, Shanghai Maritime University, Shanghai, China
993480720@qq.com

Abstract. In the speaker recognition system, a model combining the Deep Neural Network (DNN), Identity Vector (I-Vector) and Probabilistic Linear Discriminant Analysis (PLDA) proved to be very effective. In order to further improve the performance of PLDA recognition model, the Denoising Autoencoder (DAE) and Restricted Boltzmann Machine (RBM) and the combination of them (DAE-RBM) are applied to the channel compensation on PLDA model, the aim is to minimize the effect of the speaker i-vector space channel information. The results of our experiment indicate that the Equal Error Rate (EER) and the minimum Detection Cost Function (minDCF) of DAE-PLDA and RBM-PLDA are significantly reduced compared with the standard PLDA system. The DAE-RBM-PLDA which combined the advantages of them enables system identification performance to be further improved.

Keywords: Speaker recognition · I-vector · Denoising Autoencoders · Restricted Boltzmann Machine

1 Introduction

Speaker recognition is a kind of biometric technology, which is a technique for extracting effective feature information from speaker voice for speaker recognition. The more popular speaker recognition model is based on the Gaussian Mixture Model-Universal Background Model (GMM-UBM) [1]. Then, Patrick proposed Joint Factor Analysis (JFA) [2], Najim proposed modeling methods such as Identity-Vector (i-vector) [3]. The current i-vector has become the most effective technology which is the text-independent speaker recognition, this framework can be divided into three steps: First, using the GMM-UBM to express the speech acoustics feature sequence as a sufficient statistic, and then converted to low-dimensional features vector i-vector, after the i-vector is extracted, the Probabilistic Linear Discriminant Analysis (PLDA) model is used for channel compensation and the vertexes are obtained by comparing the i-vector generation verification scores of the different speech segments.

In recent years, the deep neural network (DNN) has been successfully applied to the field of speech recognition [4]. In the field of speaker recognition, Lei [5] used DNN to classify phonetic features into different phoneme spaces based on phoneme features, and then extract the acoustic features of different utterances in each space, and propose

Z. Shi et al. (Eds.): ICIS 2017, IFIP AICT 510, pp. 311–318, 2017.
DOI: 10.1007/978-3-319-68121-4_34

i-vector based on DNN. This model uses the output of the DNN output layer softmax in the UBM to calculate the various posterior probability, which results in significant performance improvement for the speaker confirmation.

Denoising Autoencoders (DAE) can reconstruct the raw data from the corrupted data by training. The feature of the speaker indicate that i-vector is affected by the influence of the speaker channel information as damaged data. Therefore, channel compensation can be achieved by DAE reconstruction method to obtain a more robust effect, resulting in noise immunity, thereby reducing the channel diversity of the speaker. In [6], RBM-PLDA-based channel compensation technology is proved to be superior to standard PLDA. RBM reconstructs i-vector by separating the speaker information and channel information, and then applies the factor containing the speaker information to the PLDA side for comparison. Based on the advantages of DAE and RBM, this paper proposes a channel compensation method based on DAE-RBM-PLDA to further reduce the influence of speaker channel diversity.

2 I-Vector-Based Speaker Recognition System

2.1 GMM I-Vector Technology

The i-vector is a compact representation of a GMM supervector, containing both the speaker and channel characteristics of a given speech utterance. The model is based on the mean supervector represented by GMM-UBM. The mean super-vector of a speaker's speech can be decomposed into the following equation:

$$M = m + T\omega \tag{1}$$

where m is the Universal Background Model (UBM), a GMM mean supervector, T is a low-rank matrix defines the total variability space, ω is a speaker-and channel-dependent latent variable with standard normal distribution, and its posterior mean is i-vector.

In the process of i-vector extraction, we need to use the EM algorithm to estimate the global difference space matrix T, and extract the Baum-Welch statistic. The zeroth-order statistics and the first-order statistics of the speech segment h of the speaker s in the j-th GMM mixed component are as follows:

$$N_{j,h}(s) = \sum P(j|x_t) \tag{2}$$

$$F_{j,h}(s) = \sum P(j|x_t)(x_t - m_j) \tag{3}$$

where $P(j|x_t)$ represents the posterior probability of generating the x_t in the Gaussian mixture component j in the UBM model:

$$P(j|x_t) = \frac{w_j P(j|x_t)}{\sum_{j-1}^{M} w_j P(j|x_t)} \tag{4}$$

And then the following calculation can be obtained corresponding i-vector.

$$\omega_h = E[W_h] = I^{-1} T^T \sum^{-1} F_h \tag{5}$$

2.2 DNN I-Vector Technology

GMM has a strong ability to fit, but its shortcoming is that it cannot effectively model nonlinear or near nonlinear data. Therefore, DNN is applied to acoustic modeling, DNN's multi-layer nonlinear structure makes it a powerful characterization capability, it uses unsupervised generation algorithm for pre-training, and then use the back propagation algorithm for parameter fine tuning.

DNN consists of input layer, multiple hidden layer and Softmax output layer. The Softmax layer gives the posterior probability $P(j|x_t)$ of the bound three-factor state class on the speech frame, which is used as the corresponding Gaussian occupancy rate, substituting the formulas (2) and (3) to estimate the DNN i-vector zero-order statistics and first-order statistics, and then extract i-vcetor according to formula (5). DNN-based i-vector extraction process and the identification process are as shown in Fig. 1:

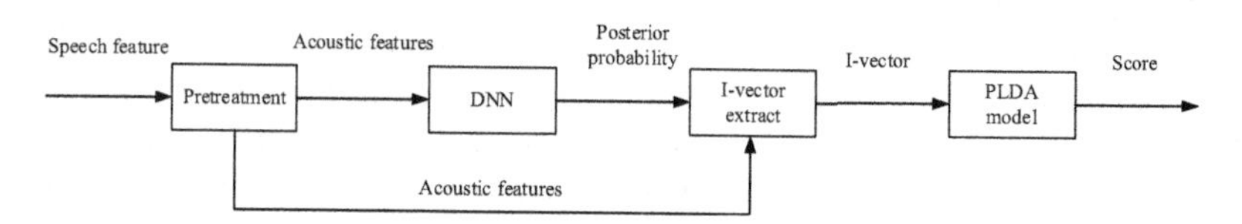

Fig. 1. DNN-based speaker identification system flow chart

3 Analysis of Back - End PLDA Technology

3.1 PLDA Model

PLDA is an i-vector-based channel compensation algorithm, i-vector feature contains speaker information and channel information. We only need to extract the speaker information, so channel compensation is needed to remove channel interference. The simplified PLDA proved to be an effective method of channel compensation [7]. The simplified PLDA model is as follows:

$$\omega_{sh} = \mu + Vy_s + Z_{sh} \tag{6}$$

where ω_{sh} is the i-vector representing h-th session of s-th speaker, μ is the mean of all training data, matrix V describes the subspace of the speaker, characterizes the differences between human beings, y_s is the hidden speaker factor, z_{sh} is the residual noise. The above parameters satisfy the following distribution:

$$y_s \sim N(0,1) \tag{7}$$

$$z_{sh} \sim N(0,D) \tag{8}$$

The purpose of the PLDA training phase is to estimate the parameter $\theta = \{\mu, V, D\}$ required by the model using the EM algorithm based on the speaker's speech data set for a given sample. Identify the score after the model is trained, the i-vector which is given the same speaker registration and testing are ω_e and ω_s respectively, the formula for calculating the likelihood ratio score is as follows:

$$\text{Score} = \ln \frac{P(\omega_e, \omega_s | H_0)}{P(\omega_e, \omega_s | H_1)} \tag{9}$$

where H_0 indicates that ω_e and ω_s are from the same speaker, and H_1 is from different speakers. Calculate the likelihood ratio of the two Gaussian functions as the final decision for the score.

3.2 PLDA Based on DAE and RBM

Denoising Autoencoders (DAE) is a self-coding device by special training. Accept the damaged data as input in the input, and training to predict the original undamaged data as an output of the automatic encoder, to produce anti-noise ability, resulting in a more robust data reconstruction effect. DAE training process is shown in Fig. 2. Introducing a damage process $C(y|x)$, which represents the probability that the given data x will produce a corrupted sample y. The automatic encoder assumes that x is the original input, and the noise reduction automatic encoder uses $C(y|x)$ to introduce the damaged sample y. And then taking y as the damage input with noise, taking x as an output, and self-coding for learning and training. The application of DAE to the speaker recognition system back-end model was first proposed in [8], and this paper will continue to explore further improve system performance. The i-vector can be regarded as a damaged data which is influenced by the speaker's channel information in this system. The training can be simplified as follows.

In the experiment, the DAE training starts from generative supervised training of the denoising RBM as shown in Fig. 3. This RBM has binary hidden layer and Gaussian visible layer, taking a concatenation of two-valued vectors as an input. The first vector $i(s)$ is the average over all sessions of this speaker, the second vector $i(s,h)$

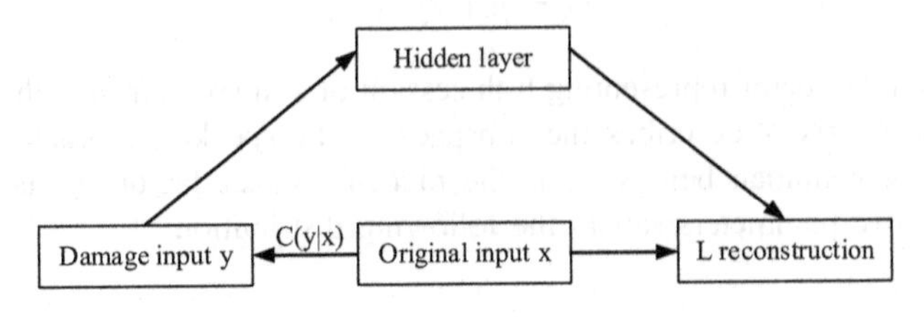

Fig. 2. DAE structure schematic diagram

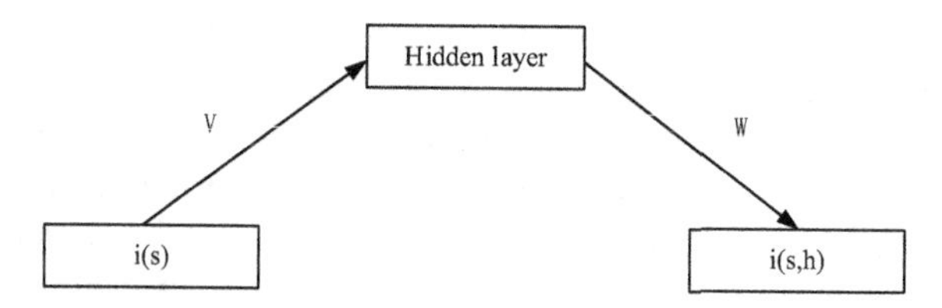

Fig. 3. RBM pre-training

is an i-vector extracted from the h-th session of s-th speaker. RBM is trained by CD algorithm [9], weight matrix parameters *V*, *W* are used to initialize the DAE model.

After the pre-training model is expanded as shown in Fig. 4, this model can be seen as a standard DAE model to rebuild i-vector. The output uses the speaker's average i-vector to reduce the difference in speaker channel information. Then the back propagation algorithm is used to tune the network parameters. The output of the DAE is whitened and length normalizated and it can be validated directly as a standard PLDA model input (DAE-PLDA), the judgment is based on a pre-set threshold.

RBM is an undirected model consisting of a random layer of visible neurons and a layer of hidden neuron. It can act on the PLDA channel compensation side, the hidden layer is decomposed by the speaker information factor and the channel information factor, as shown in Fig. 5. We use the similar algorithm [6] to carry out training, the

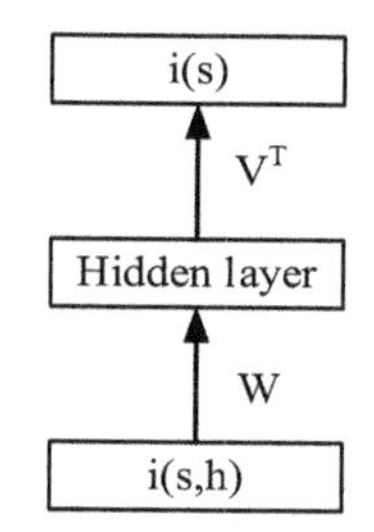

Fig. 4. DAE

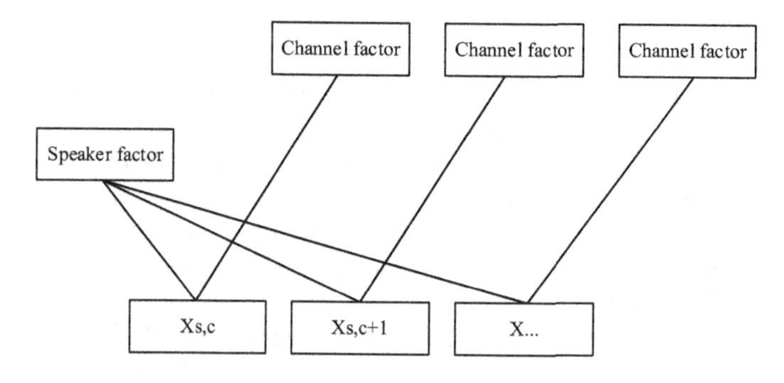

Fig. 5. RBM-PLDA

difference is that it is consistent with the hidden layer values for the previous DAE pre-training, where the hidden layer uses binary values and subjects to the Gaussian Bernoulli distribution. In the recognition phase, the visible layer inputs the i-vector of the speaker and the speaker's speech containing the speaker information at the output as input of the PLDA model (RBM-PLDA) is used to score the comparison.

From the above analysis we can see the effective feature extraction principle based on DAE and RBM. Use DAE and RBM mixed method, the first layer is the DAE, after the whitening and length normalization as RBM input, RBM and standard PLDA is combined to form a discriminant model, recorded as DAE-RBM-PLDA. The block diagram for the system is shown in Fig. 6.

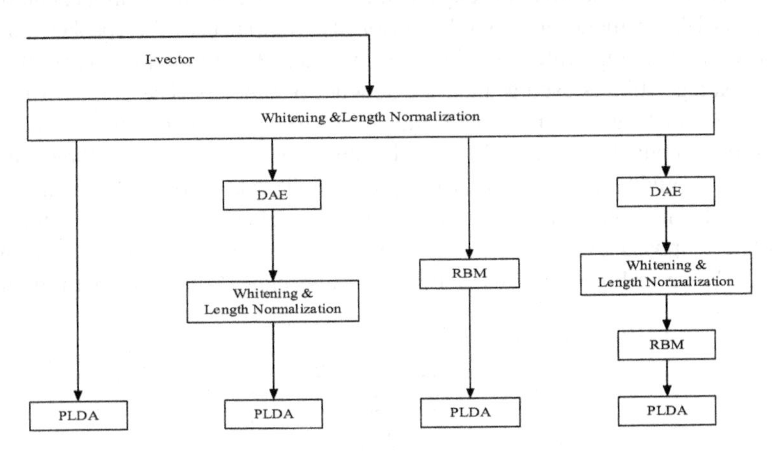

Fig. 6. PLDA, DAE-PLDA, RBM-PLDA, DAE-RBM-PLDA process

4 Experiments and Results

In this paper, DAC13 is used as experimental phonetic database, and the Equal Error Rate (EER) and minimum Detection Cost Function (DCF) are used as performance evaluation indexes.

In the UBM i-vector system, MFCC and the one-dimensional energy and its first and second order differences, which are total 39-dimensional MFCC features. Voice frame length is 25 ms, frame shift is 10 ms. In the DNN i-vector system, DNN speaker features are 40-dimensional Filter Bank features and its first and second order differences, which are total 120-dimensional. DNN has 5 hidden layers, each layer includes 2048 nodes. We first compared the performance of the standard PLDA model in the UBM i-vector and DNN i-vector systems. Experiments show that the recognition performance of DNN system is significantly improved by comparing with GMM-UBM system. DNN i-vector PLDA is the baseline system, the performance comparison is shown in Fig. 7 and Table 1.

From the experimental results of Table 1, it can be seen that the equal error rate and the minimum detection cost function of DAE-PLDA and RBM-PLDA back-end

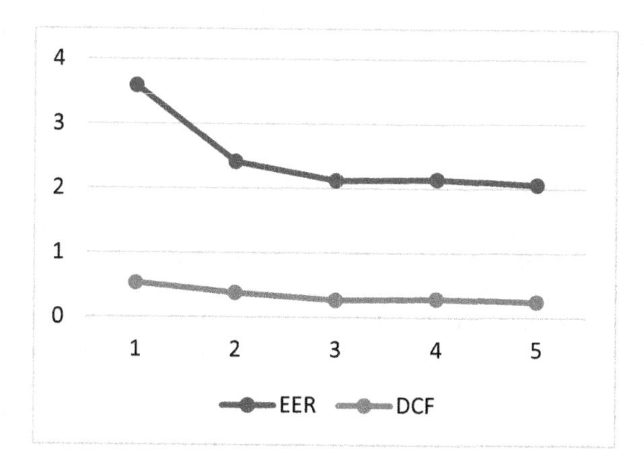

Fig. 7. Model performance line chart

Table 1. PLDA, DAE-PLDA, RBM-PLDA, DAE-RBM-PLDA performance comparison

System	Back-end model	EER, %	minDCF
1 UBM	PLDA	4.49	1.432
2 DNN	PLDA	3.32	1.275
3 DNN	DAE-PLDA	2.79	0.863
4 DNN	RBM-PLDA	2.91	0.954
5 DNN	DAE-RBM-PLDA	2.82	0.913

channel compensation model with deep learning model are significantly lower than those of standard PLDA model system. The performance improvement of the DAE-RBM-PLDA model is more obvious than the baseline system, which is 15.1% higher than the baseline system, which verified the effectiveness of the channel compensation method.

5 Conclusions

In this paper, we proposed the speaker verification channel compensation method based on DAE-RBM-PLDA, the method is combining the advantages of DAE and RBM. This method first performs RBM pre-training and initializes the DAE model with the i-vector processed by whitening and length normalization. The output of DAE is the average i-vector of over all sessions of the speaker, so it reduced the influence of the speaker channel information. And then combined with the RBM, the output i-vector of DAE is used as the input of RBM, hidden layer reconstruction separates speaker information and speaker channel information, select the speaker information needed for the experiment to perform the final likelihood ratio score of the back-end PLDA, further reducing the channel diversity of the speaker. The speaker confirmation experiment on the DAC13 dataset shows that the DAE-RBM-PLDA model, which combines the advantages of both DAE and RBM, can effectively improve the recognition rate.

Acknowledgment. This study was funded by the National Natural Science Foundation of China (Grant No.61404083).

References

1. Reynolds, D.A., Quatieri, T.F., Dunn, R.B.: Speaker verification using adapted Gaussian mixture models. Digit. Signal Process. **10**(1–3), 19–41 (2000)
2. Kenny, P., Ouellet, P., Dehak, N., et al.: A study of interspeaker variability in speaker verification. IEEE Trans. Audio Speech Lang. Process. **16**(5), 980–988 (2008)
3. Dehak, N., Kenny, P., Dehak, R., et al.: Front-end factor analysis for speaker verification. IEEE Trans. Audio Speech Lang. Process. **19**(4), 788–798 (2011)
4. Hinton, G., Deng, L., Yu, D., et al.: Deep neural networks for acoustic modeling in speech recognition: the shared views of four research groups. IEEE Signal Process. Mag. **29**(6), 82–97 (2012)
5. Variani, E., Lei, X., McDermott, E., et al.: Deep neural networks for small footprint text-dependent speaker verification. In: 2014 IEEE International Conference on Acoustics, Speech and Signal Processing (ICASSP), pp. 4052–4056. IEEE (2014)
6. Stafylakis, T., Kenny, P., Senoussaoui, M., et al.: PLDA using Gaussian restricted Boltzmann machines with application to speaker verification. In: INTERSPEECH, pp. 1692–1695 (2012)
7. Garcia-Romero, D., Espy-Wilson, C.Y.: Analysis of i-vector length normalization in speaker recognition systems. In: INTERSPEECH, pp. 249–252 (2011)
8. Novoselov, S., Pekhovsky, T., Kudashev, O., et al.: Non-linear PLDA for i-vector speaker verification. In: INTERSPEECH, pp. 214–218 (2015)
9. Hinton, G.E.: A practical guide to training restricted boltzmann machines. In: Montavon, G., Orr, G.B., Müller, K.R. (eds.) Neural Networks: Tricks of the Trade. LNCS, vol. 7700, pp. 599–619. Springer, Heidelberg (2012). doi:10.1007/978-3-642-35289-8_32

Intelligent Information Processing

Channels' Matching Algorithm for Mixture Models

Chenguang Lu(✉)

College of Intelligence Engineering and Mathematics,
Liaoning Engineering and Technology University,
Fuxin 123000, Liaoning, China
lcguang@foxmail.com

Abstract. To solve the Maximum Mutual Information (MMI) and Maximum Likelihood (ML) for tests, estimations, and mixture models, it is found that we can obtain a new iterative algorithm by the Semantic Mutual Information (SMI) and $R(G)$ function proposed by Chenguang Lu (1993) (where $R(G)$ function is an extension of information rate distortion function $R(D)$, G is the lower limit of the SMI, and $R(G)$ represents the minimum R for given G). This paper focus on mixture models. The SMI is defined by the average log normalized likelihood. The likelihood function is produced from the truth function and the prior by the semantic Bayesian inference. A group of truth functions constitute a semantic channel. Letting the semantic channel and Shannon channel mutually match and iterate, we can obtain the Shannon channel that maximizes the MMI and the average log likelihood. Therefore, this iterative algorithm is called Channels' Matching algorithm or the CM algorithm. It is proved that the relative entropy between the sampling distribution and predicted distribution may be equal to $R - G$. Hence, solving the maximum likelihood mixture model only needs minimizing $R - G$, without needing Jensen's inequality. The convergence can be intuitively explained and proved by the $R(G)$ function. Two iterative examples of mixture models (which are demonstrated in an excel file) show that the computation for the CM algorithm is simple. In most cases, the number of iterations for convergence (as the relative entropy <0.001 bit) is about 5. The CM algorithm is similar to the EM algorithm; however, the CM algorithm has better convergence and more potential applications.

Keywords: Shannon channel · Semantic channel · Semantic information · Likelihood · Mixture models · EM algorithm · Machine learning · Statistical inference

1 Introduction

To obtain maximum likelihood mixture models, The EM algorithm [1] and the Newton method [2] are often used. There have been many papers on applying or improving the EM algorithm. Lu proposed the semantic information measure (SIM) and the $R(G)$ function in 1993 [3–5]. The $R(G)$ function is an extension of Shannon's information rate distortion function $R(D)$ [6, 7]. The $R(G)$ means the minimum R for given SIM G. It is found that using SIM and $R(G)$ function, we can obtain a new iterative

Z. Shi et al. (Eds.): ICIS 2017, IFIP AICT 510, pp. 321–332, 2017.
DOI: 10.1007/978-3-319-68121-4_35

algorithm, i.e, Channels' Matching algorithm (or the CM algorithm). Compared with the EM algorithm, the CM algorithm proposed by this paper is seemly similar yet essentially different[1].

In this study, we use the sampling distribution instead of the sampling sequence. Assume the sampling distribution is $P(X)$ and the predicted distribution by the mixture model is $Q(X)$. The goal is to minimize the relative entropy or Kullback-Leibler (KL) divergence $H(Q||P)$ [8, 9]. With the semantic information method, we may prove $H(Q||P) = R(G) - G$. Then, maximizing G and modifying R alternatively, we can minimize $H(Q||P)$.

We first introduce the semantic channel, semantic information measure, and $R(G)$ function in a way that is as compatible with the likelihood method as possible. Then we discuss how the CM algorithm is applied to mixture models. Finally, we compare the CM algorithm with the EM algorithm to show the advantages of the CM algorithm.

2 Semantic Channel, Semantic Information Measure, and the *R*(*G*) Function

2.1 From the Shannon Channel to the Semantic Channel

First, we introduce the Shannon channel.

Let X be a discrete random variable representing a fact with alphabet $A = \{x_1, x_2, \ldots, x_m\}$, and let Y be a discrete random variable representing a message with alphabet $B = \{y_1, y_2, \ldots, y_n\}$. A Shannon channel is composed of a group of transition probability functions [6]: $P(y_j|X)$, $j = 1, 2, \ldots, n$.

In terms of hypothesis-testing, X is a sample point and Y is a hypothesis or a model label. We need a sample sequence or sampling distribution $P(X|.)$ to test a hypothesis to see how accurate it is.

Let Θ be a random variable for a predictive model, and let θ_j be a value taken by Θ when $Y = y_j$. The semantic meaning of a predicate $y_j(X)$ is defined by θ_j or its (fuzzy) truth function $T(\theta_j|X) \in [0,1]$. Because $T(\theta_j|X)$ is constructed with some parameters, we may also treat θ_j as a set of model parameters. We can also state that $T(\theta_j|X)$ is defined by a normalized likelihood, i.e., $T(\theta_j|X) = k\, P(\theta_j|X)/P(\theta_j) = k\, P(X|\theta_j)/P(X)$, where k is a coefficient that makes the maximum of $T(\theta_j|X)$ be 1. The θ_j can also be regarded as a fuzzy set, and $T(\theta_j|X)$ can be considered as a membership function of a fuzzy set proposed by Zadeh [10].

In contrast to the popular likelihood method, the above method uses sub-models $\theta_1, \theta_2, \ldots, \theta_n$ instead of one model θ or Θ. The $P(X|\theta_j)$ is equivalent to $P(X|y_j, \theta)$ in the popular likelihood method. A sample used to test y_j is also a sub-sample or a conditional sample. These changes will make the new method more flexible and more compatible with the Shannon information theory.

---

[1] Excel files demonstrating iterative process for tests, estimations, and mixture models can be download from http://survivor99.com/lcg/CM-iteration.zip.

A semantic channel is composed of a group of truth value functions or membership functions: $T(\theta_j|X)$, $j = 1, 2, \ldots, n$.

Similar to $P(y_j|X)$, $T(\theta_j|X)$ can also be used for Bayesian prediction to produce likelihood function [4]:

$$P(X|\theta_j) = P(X)T(\theta_j|X)/T(\theta_j), \quad T(\theta_j) = \sum_i P(x_i)T(\theta_j|x_i) \tag{1}$$

where $T(\theta_j)$ is called the logical probability of y_j. The author now know that this formula was proposed by Thomas as early as 1981 [11]. We call this prediction the semantic Bayesian prediction. If $T(\theta_j|X) \propto P(y_j|X)$, then the semantic Bayesian prediction is equivalent to the Bayesian prediction.

2.2 Semantic Information Measure and the Optimization of the Semantic Channel

The semantic information conveyed by y_j about x_i is defined by normalized likelihood as [3]:

$$I(x_i; \theta_j) = \log\frac{P(x_i|\theta_j)}{P(x_i)} = \log\frac{T(\theta_j|x_i)}{T(\theta_j)} \tag{2}$$

where the semantic Bayesian inference is used; it is assumed that prior likelihood function $P(X|\Theta)$ is equal to prior probability distribution $P(X)$.

After averaging $I(x_i;\theta_j)$, we obtain semantic (or generalized) KL information:

$$I(X; \theta_j) = \sum_i P(x_i|y_j)\log\frac{P(x_i|\theta_j)}{P(x_i)} = \sum_i P(x_i|y_j)\log\frac{T(\theta_j|x_i)}{T(\theta_j)} \tag{3}$$

The statistical probability $P(x_i|y_j)$, $i = 1, 2, \ldots$, on the left of "log" above, represents a sampling distribution to test the hypothesis y_j or model θ_j. Assume we choose y_j according to observed condition $Z \epsilon C$. If $y_j = f(Z|Z \epsilon C_j)$, where C_j is a cub-set of C, then $P(X|y_j) = P(X|C_j)$.

After averaging $I(X;\theta_j)$, we obtain semantic (or generalized) mutual information:

$$\begin{aligned} I(X; \Theta) &= \sum_j P(y_j)\sum_i P(x_i|y_j)\log\frac{P(x_i|\theta_j)}{P(x_i)} \\ &= \sum_j\sum_i P(x_i)P(y_j|x_i)\log\frac{T(\theta_j|x_i)}{T(\theta_j)} = H(X) - H(X|\Theta) \\ H(X|\Theta) &= -\sum_j\sum_i P(x_i, y_j)\log P(x_i|\theta_j) \end{aligned} \tag{4}$$

where $H(X)$ is the Shannon entropy of X, $H(X|\Theta)$ is the generalized posterior entropy of X. Each of them has coding meaning [4, 5].

Optimizing a semantic Channel is equivalent to optimizing a predictive model Θ. For given $y_j = f(Z|Z\epsilon C_j)$, optimizing θ_j is equivalent to optimizing $T(\theta_j|X)$ by

$$T^*(\theta_j|X) = \underset{T(\theta_j|X)}{\arg\ \max} I(X;\theta_j) \tag{5}$$

It is easy to prove that when $P(X|\theta_j) = P(X|y_j)$, or

$$\frac{T(\theta_j|X)}{T(\theta_j)} = \frac{P(y_j|X)}{P(y_j)}, \quad \text{or } T(h_j|E) \propto P(h_j|E) \tag{6}$$

$I(X;\ \theta_j)$ reaches the maximum. Set the maximum of $T(\theta_j|X)$ to 1. Then we can obtain

$$T^*(\theta_j|X) = P(y_j|X)/P(y_j|x_j^*) = [P(X|y_j)/P(X)]/[P\left(x_j^*|y_j\right)/P\left(x_j^*\right)] \tag{7}$$

In this equation, x_j^* makes $P(x_j^*|y_j)/P(x_j^*)$ be the maximum of $P(X|y_j)/P(X)$.

2.3 Relationship Between Semantic Mutual Information and Likelihood

Assume that the size of the sample used to test y_j is N_j; the sample points come from independent and identically distributed random variables. Among these points, the number of x_i is N_{ij}. Assume that N_j is infinite, $P(X|y_j) = N_{ij}/N_j$. Hence, there is log normalized likelihood:

$$\log \prod_i \left[\frac{P(x_i|\theta_j)}{P(x_i)}\right]^{N_{ji}} = N_j \sum_i P(x_i|y_j) \log \frac{P(x_i|\theta_j)}{P(x_i)} = N_j I(X;\theta_j) \tag{8}$$

After averaging the above likelihood for different y_j, $j = 1, 2, \ldots, n$, we have the average log normalized likelihood:

$$\begin{aligned} \sum_j \frac{N_j}{N} \log \prod_i \left[\frac{P(x_i|\theta_j)}{P(x_i)}\right]^{N_{ji}} &= \sum_j P(y_j) \sum_i P(x_i|y_j) \log \frac{P(x_i|\theta_j)}{P(x_i)} \\ &= I(X;\Theta) = H(X) - H(X|\Theta) \end{aligned} \tag{9}$$

where $N = N_1 + N_2 + \cdots + N_n$. It shows that the ML criterion is equivalent to the minimum generalized posterior entropy criterion and the Maximum Semantic Information (MSI) criterion. When $P(X|\theta_j) = P(X|y_j)$ (for all j), the semantic mutual information $I(X;\ \Theta)$ is equal to the Shannon mutual information $I(X;Y)$, which is the special case of $I(X;\ \Theta)$.

2.4 The Matching Function $R(G)$ of R and G

The $R(G)$ function is an extension of the rate distortion function $R(D)$ [7]. In the $R(D)$ function, R is the information rate, D is the upper limit of the distortion. The $R(D)$ function means that for given D, $R = R(D)$ is the minimum of the Shannon mutual information $I(X;Y)$.

Let distortion function d_{ij} be replaced with $I_{ij} = I(x_i; y_j) = \log[T(\theta_j|x_i)/T(\theta_j)] = \log[P(x_i|\theta_j)/P(x_i)]$, and let G be the lower limit of the semantic mutual information $I(X; \Theta)$. The information rate for given G and $P(X)$ is defined as

$$R(G) = \min_{P(Y|X):I(E;\Theta) \geq G} I(X;Y) \tag{10}$$

Following the derivation of $R(D)$ [12], we can obtain [3]

$$\begin{aligned} G(s) &= \sum_i \sum_j I_{ij} P(x_i) P(y_j) 2^{sI_{ij}}) / \lambda_i = \sum_i \sum_j I_{ij} P(x_i) P(y_j) m_{ij}^s / \lambda_i \\ R(s) &= sG(s) - \sum_i P(x_i) \log \lambda_i \end{aligned} \tag{11}$$

where $m_{ij} = T(\theta_j|x_i)/T(\theta_j) = P(x_i|\theta_j)/P(x_i)$ is the normalized likelihood; $\lambda_i = \sum_j P(y_j) m_{ij}^s$. We may also use $m_{ij} = P(x_i|\ \theta_j)$, which results in the same m_{ij}^s/λ_i. The shape of an $R(G)$ function is a bowl-like curve as shown in Fig. 1.

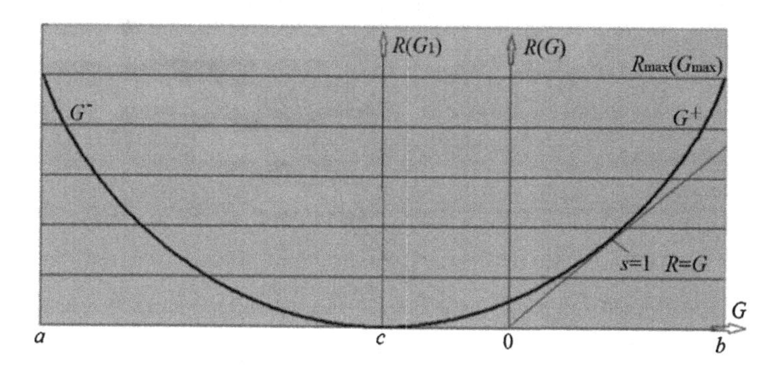

Fig. 1. The $R(G)$ function of a binary source. When the slope $s = 1$, $G = R$, and information efficiency G/R reaches its maximum 1.

The R(G) function is different from the $R(D)$ function. For a given R, we have the maximum value G^+ and the minimum value G^-, which is negative and means that to bring a certain information loss $|G|$ to enemies, we also need certain objective information R.

In the rate distortion theory, $dR/dD = s$ ($s \leq 0$). It is easy to prove that there is also $dR/dG = s$, where s may be less or greater than 0. The increase of s will raise the model's prediction precision. If s changes from positive s_1 to $-s_1$, then $R(-s_1) = R(s_1)$ and G changes from G^+ to G^- (see Fig. 1).

When $s = 1$, $\lambda_i = 1$, and $R = G$, which means that the semantic channel matches the Shannon channel and the semantic mutual information is equal to the Shannon mutual information. When $s = 0$, $R = 0$ and $G < 0$. In Fig. 1, $c = G(s = 0)$.

3 The CM Algorithm for Mixture Models

3.1 Explaining the Iterative Process by the $R(G)$ Function

Assume a sampling distribution $P(X)$ is produced by the conditional probability $P^*(X|Y)$ being some function such as Gaussian distribution. We only know that the number of the mixture components is n, without knowing $P(Y)$. We need to solve $P(Y)$ and model (or parameters) Θ, so that the predicted probability distribution of X, denoted by $Q(X)$, is as close to the sampling distribution $P(X)$ as possible, i.e. the relative entropy or Kullback-Leibler divergence $H(Q||P)$ is as small as possible. The Fig. 2 shows the convergent processes of two examples.

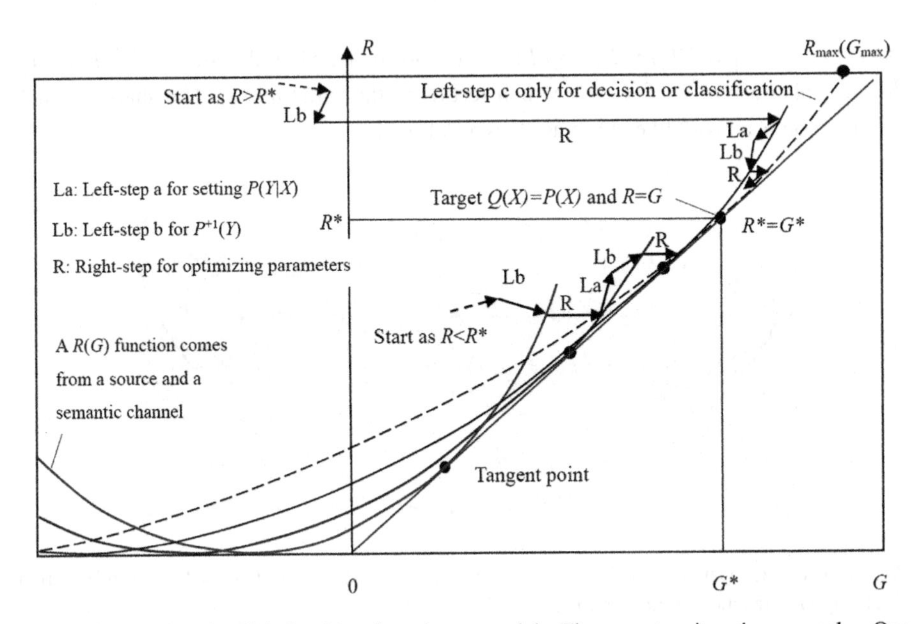

Fig. 2. Illustrating the CM algorithm for mixture models. There are two iterative examples. One is for $R > R^*$ and another is for $R < R^*$. The Left-step a and Left-step b make R close to R^*; whereas the Right-step increases G so that (G, R) approaches line $R = G$.

We use $P^*(Y)$ and $P^*(X|Y)$ to denote the $P(Y)$ and $P(X|Y)$ that are used to produce the sampling distribution $P(X)$, and use $P^*(Y|X)$ and $R^* = I^*(X;Y)$ to denote the corresponding Shannon channel and Shannon mutual information. When $Q(X) = P(X)$, there should be $P(X|\Theta) = P^*(X|Y)$, and $G^* = R^*$.

For mixture models, when we let the Shannon channel match the semantic channel (in Left-steps), we do not maximize $I(X;\Theta)$, but seek a $P(X|\Theta)$ that accords with $P^*(X|Y)$ as possible (Left-step a in Fig. 2 is for this purpose), and a $P(Y)$ that accords with $P^*(Y)$ as possible (Left-step b in Fig. 2 is for this purpose). That means we seek a R that is as close to R^* as possible. Meanwhile, $I(X;\Theta)$ may decrease. However, in popular

EM algorithms, the objective function, such as $P(X^N, Y|\Theta)$, is required to keep increasing without decreasing in both steps.

With CM algorithm, only after the optimal model is obtained, if we need to choose Y according to X (for decision or classification), we may seek the Shannon channel $P(Y|X)$ that conveys the MMI $R_{\max}(G_{\max})$ (see Left-step c in Fig. 2).

Assume that $P(X)$ is produced by $P^*(X|Y)$ with the Gaussian distribution. Then the likelihood functions are

$$P(X|\theta_j) = k_j \exp\left[-(X - c_j)^2/(2d_j)^2\right], j = 1, 2, \ldots, n$$

If $n = 2$, then parameters are c_1, c_2, d_1, d_2. In the beginning of the iteration, we may set $P(Y) = 1/\text{n}$. We begin iterating from Left-step a.

Left-step a: Construct Shannon channel by

$$\begin{array}{ll} P(y_j|X) = P(y_j)P(X|\theta_j)/Q(X) & \\ Q(X) = \sum\limits_j P(y_j)|(X|\theta_j) & ,j = 1, 2, \ldots, n \end{array} \tag{12}$$

This formula has already been used in the EM algorithm [1]. It was also used in the derivation process of the $R(D)$ function [12]. Hence the semantic mutual information is

$$G = I(X;\Theta) = \sum_i \sum_j P(x_i) \frac{P(x_i|\theta_j)}{Q(x_i)} P(y_j) \log \frac{P(x_i|\theta_j)}{P(x_i)} \tag{13}$$

Left-step b: Use the following equation to obtain a new $P(Y)$ repeatedly until the iteration converges.

$$P(y_j) \Leftarrow \sum_i P(x_i)P(y_j|x_i) = \sum_i P(x_i) \frac{P(x_i|\theta_j)}{\sum\limits_k P(y_k)P(x_i|\theta_k)} P(y_j), j = 1, 2, \ldots, n \tag{14}$$

The convergent $P(Y)$ is denoted by $P^{+1}(Y)$. This is because $P(Y|X)$ from Eq. (12) is an incompetent Shannon channel so that $\sum_i P(x_i)P(y_j|x_i) \neq P(y_j)$. The above iteration makes $P^{+1}(Y)$ match $P(X)$ and $P(X|\Theta)$ better. This iteration has been used by some authors, such as in [13].

When $n = 2$, we should avoid choosing c_1 and c_2 so that both are larger or less than the mean of X; otherwise $P(y_1)$ or $P(y_2)$ will be 0, and cannot be larger than 0 later.

If $H(Q||P)$ is less than a small number, such as 0.001 bit, then end the iteration; otherwise go to Right-step.

Right-step: Optimize the parameters in the likelihood function $P(X|\Theta)$ on the right of the log in Eq. (13) to maximize $I(X;\Theta)$. Then go to Left-step a.

3.2 Using Two Examples to Show the Iterative Processes

3.2.1 Example 1 for $R < R^*$

In Table 1, there are real parameters that produce the sample distribution $P(X)$ and guessed parameters that are used to produce $Q(X)$. The convergence process from the starting (G, R) to (G^*, R^*) is shown by the iterative locus as $R < R^*$ in Fig. 2. The convergence speed and changes of R and G are shown in Fig. 3. The iterative results are shown in Table 1.

Table 1. Real and guessed model parameters and iterative results of Example 1 ($R < R^*$)

Y	Real parameters			Start parameters $H(Q\|\|P) = 0.410$ bit			Parameters after 5 iterations $H(Q\|\|P) = 0.00088$ bit		
	c	d	$P^*(Y)$	c	d	$P(Y)$	c	d	$P(Y)$
y_1	35	8	0.7	30	15	0.5	35.4	8.3	0.72
y_2	65	12	0.3	70	15	0.5	65.2	11.4	0.28

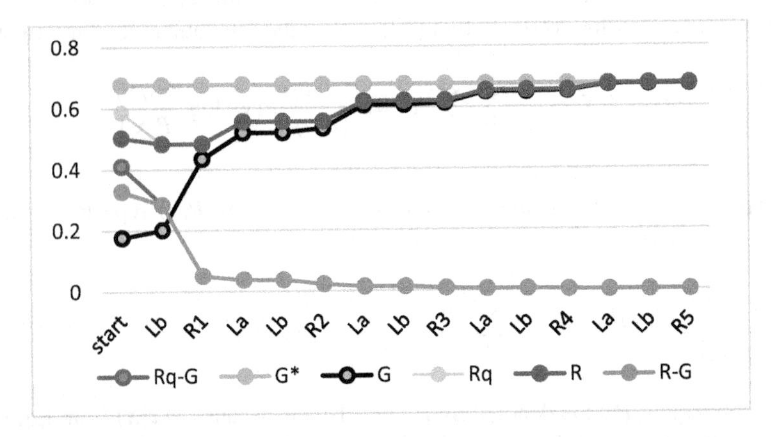

Fig. 3. The iterative process as $R < R^*$. Rq is R_Q in Eq. (15). $H(Q||P) = R_Q - G$ decreases in all steps. G is monotonically increasing. R is also monotonically increasing except in the first Left-step b. G and R gradually approach $G^* = R^*$ so that $H(Q||P) = R_Q - G$ is close to 0.

Analyses: In this iterative process, there are always $R < R^*$ and $G < G^*$. After each step, R and G increase a little bit so that G approaches G^* gradually. This process seams to tell us that each of Right-step, Left-step a, and Left-step b can increase G; and hence maximizing G can minimize $H(Q||P)$, which is our goal. Yet, it is wrong. The Left a and Left b do not necessarily increase G. There are many counterexamples. Fortunately, iterations for these counterexamples can still converge. Let us see Example 2 as a counterexample.

3.2.2 Example 2 for $R > R^*$

Table 2 shows the parameters and iterative results for $R > R^*$. The iterative process is shown in Fig. 4.

Table 2. Real and guessed model parameters and iterative results for Example 2 ($R > R^*$)

Y	Real parameters			Start parameters $H(Q\|\|P) = 0.680$ bit			Parameters after 5 iterations $H(Q\|\|P) = 0.00092$ bit		
	c	d	$P^*(Y)$	c	d	$P(Y)$	c	d	$P(Y)$
y_1	35	8	0.1	30	8	0.5	38	9.3	0.134
y_2	65	12	0.9	70	8	0.5	65.8	11.5	0.886

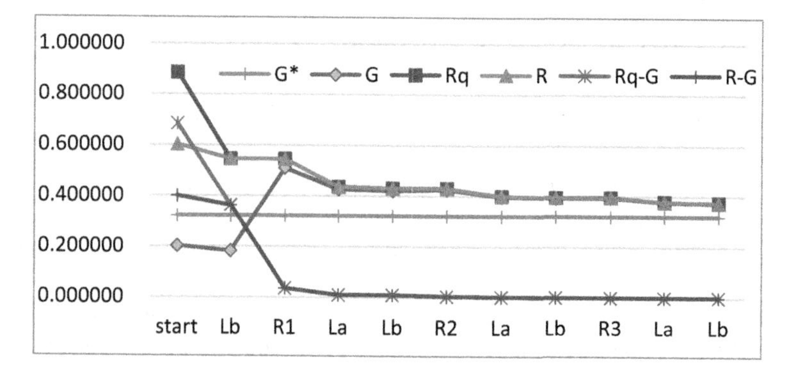

Fig. 4. The iterative process as $R > R^*$. Rq is R_Q in Eq. (15). $H(Q||P) = R_Q - G$ decreases in all steps. R is monotonically decreasing. G increases more or less in all Right-steps and decreases in all Left-steps. G and R gradually approach $G^* = R^*$ so that $H(Q||P) = R_Q - G$ is close to 0.

Analyses: G is not monotonically increasing nor monotonically decreasing. It increases in all Right steps and decreases in all Left steps. This example is a challenge to all authors who prove that the standard EM algorithm or a variant EM algorithm converges. If G is not monotonically increasing, it must be difficult or impossible to prove that $\log P(X^N, Y|\Theta)$ or other likelihood is monotonically increasing or no-decreasing in all steps. For example, in Example 2, $Q^* = -NH^*(X, Y) = -6.031\ N$. After the first optimization of parameters, $Q = -6.011\ N > Q^*$. If we continuously maximize Q, Q cannot approach less Q^*.

We also use some other true models $P^*(X|Y)$ and $P^*(Y)$ to test the CM algorithm. In most cases, the number of iterations is close to 5. In rare cases where R and G are much bigger than G^*, such as $R \approx G > 2G^*$, the iterative convergence is slow. In these cases where $\log P(X^N, Y|\Theta)$ is also much bigger than $\log P^*(X^N, Y)$, the EM algorithm confronts similar problem. Because of these cases, the convergence proof of the EM algorithm is challenged.

3.3 The Convergence Proof of the CM Algorithm

Proof. To prove the CM algorithm converges, we need to prove that $H(Q||P)$ is decreasing or no-increasing in every step.

Consider the Right-step. Assume that the Shannon mutual information conveyed by Y about $Q(X)$ is R_Q, and that about $P(X)$ is R. Then we have

$$R_Q = I_Q(X;Y) = \sum_i \sum_j P(x_i) \frac{P(x_i|\theta_j)}{Q(x_i)} P(y_j) \log \frac{P(x_i|\theta_j)}{Q(x_i)} \tag{15}$$

$$\begin{aligned} R = I(X;Y) &= \sum_i \sum_j P(x_i) \frac{P(x_i|\theta_j)}{Q(x_i)} P(y_j) \log \frac{P(y_j|x_i)}{P^{+1}(y_j)} \\ &= R_Q - H(Y||Y^{+1}) \\ H(Y||Y^{+1}) &= \sum_j P^{+1}(y_j) \log[P^{+1}(y_j)/P(y_j)] \end{aligned} \tag{16}$$

According to Eqs. (13) and (15), we have

$$H(Q||P) = R_Q - G = R + H(Y||Y^{+1}) - G \tag{17}$$

Because of this equation, we do not need Jensen's inequality that the EM algorithm needs.

In Right-steps, the Shannon channel and R_Q does not change, G is maximized. Therefore $H(Q||P)$ is decreasing and its decrement is equal to the increment of G.

Consider Left-step a. After this step, $Q(X)$ becomes $Q^{+1}(X) = \sum_j P(y_j)P^{+1}(X|\theta_j)$. Since $Q^{+1}(X)$ is produced by a better likelihood function and the same $P(Y)$, $Q^{+1}(X)$ should be closer to $P(X)$ than $Q(X)$, i.e. $H(Q^{+1}||P) < H(Q||P)$ (More strict mathematical proof for this conclusion is needed).

Consider Left-step b. The iteration for $P^{+1}(Y)$ moves (G, R) to the $R(G)$ function cure ascertained by $P(X)$ and $P(X|\theta_j)$ (for all j) that form a semantic channel. This conclusion can be obtained from the derivation processes of $R(D)$ function [12] and $R(G)$ function [3]. A similar iteration is used for $P(Y|X)$ and $P(Y)$ in deriving the $R(D)$ function. Because $R(G)$ is the minimum R for given G, $H(Q||P) = R_Q - G = R - G$ becomes less.

Because $H(Q||P)$ becomes less after every step, the iteration converges. **Q.E.D.**

3.4 The Decision Function with the ML Criterion

After we obtain optimized $P(X|\Theta)$, we need to select Y (to make decision or classification) according to X. The parameter s in $R(G)$ function (see Eq. (11)) reminds us that we may use the following Shannon channel

$$\begin{aligned} P(y_j|X) &= P(y_j)[P(X|\theta_j)]^s/Q(X) \\ Q(X) &= \sum_j P(y_j)[P(X|\theta_j)]^s \end{aligned} \quad ,j = 1, 2, \ldots, n \tag{18}$$

which are fuzzy decision functions. When $s \rightarrow +\infty$, the fuzzy decision will become crisp decision. Different from Maximum A prior (MAP) estimation, the above decision function still persists in the ML criterion or MSI criterion. The Left-step c in Fig. 2 shows that (G, R) moves to $(G_{\max}, R_{\max})$ with s increasing.

3.5 Comparing the CM Algorithm and the EM Algorithm

In the EM algorithm [1, 14], the likelihood of a mixture model is expressed as $\log P(X^N|\Theta) > L = Q - H$. If we move $P(Y)$ or $P(Y|\Theta)$ from Q into H, then Q will become $-NH(X|\Theta)$ and H becomes $-NR_Q$. If we add $NH(X)$ to both sides of the inequality, we will have $H(Q||P) \leq R_Q - G$, which is similar to Eq. (17). It is easy to prove

$$Q = NG - NP(X) - NH(Y) \tag{19}$$

where $H(Y) = -\sum_j P^{+1}(y_j)\log P(y_j)$ is a generalized entropy. We may think the M-step merges the Left-step b and the Right-step of the CM algorithm into one step. In brief,

$$\begin{gathered} \text{The E-step of EM} = \text{the Left-step a of CM} \\ \text{The M-step of EM} \approx \text{the Left-step b} + \text{the Right-step of CM} \end{gathered}$$

In the EM algorithm, if we first optimize $P(Y)$ (not for maximum Q) and then optimize $P(X|Y, \Theta)$, then the M-step will be equivalent to the CM algorithm.

There are also other improved EM algorithms [13, 15–17] with some advantages. However, no one of these algorithms facilitates that R converges to R^*, and $R - G$ converges to 0 as the CM algorithm.

The convergence reason of the CM algorithm is seemly clearer than the EM algorithm (see the analyses in Example 2 for $R > R^*$). According to [7, 15–17], the CM algorithm is faster at least in most cases than the various EM algorithms.

The CM algorithm can also be used to achieve maximum mutual information and maximum likelihood of tests and estimations. There are more detailed discussions about the CM algorithm[2].

4 Conclusions

Lu's semantic information measure can combine the Shannon information theory and likelihood method so that the semantic mutual information is the average log normalized likelihood. By letting the semantic channel and Shannon channel mutually match and iterate, we can achieve the mixture model with minimum relative entropy. The iterative convergence can be intuitively explained and proved by the $R(G)$ function.

[2] https://arxiv.org/abs/1706.07918.

Two iterative examples and mathematical analyses show that the CM algorithm has higher efficiency at least in most cases and clearer convergence reasons than the popular EM algorithm.

Acknowledgment. The author thanks Professor Peizhuang Wang for his long term supports and encouragements.

References

1. Dempster, A.P., Laird, N.M., Rubin, D.B.: Maximum likelihood from incomplete data via the EM algorithm. J. Roy. Stat. Soc. B **39**, 1–38 (1977)
2. Kok, M., Dahlin, J.B., Schon, T.B., Wills, A: Newton-based maximum likelihood estimation in nonlinear state space models. In: IFAC-PapersOnLine 48, pp. 398–403 (2015)
3. Lu, C.: A Generalized Information Theory. China Science and Technology University Press, Hefei (1993). (in Chinese)
4. Lu, C.: Coding meanings of generalized entropy and generalized mutual information. J. China Inst. Commun. **15**, 37–44 (1994). (in Chinese)
5. Lu, C.: A generalization of Shannon's information theory. Int. J. Gen. Syst. **28**, 453–490 (1999)
6. Shannon, C.E.: A mathematical theory of communication. Bell Syst. Tech. J. **27**, 379–429, 623–656 (1948)
7. Shannon, C.E.: Coding theorems for a discrete source with a fidelity criterion. IRE Nat. Conv. Rec. **4**, 142–163 (1959)
8. Kullback, S., Leibler, R.: On information and sufficiency. Ann. Math. Stat. **22**, 79–86 (1951)
9. Akaike, H.: A new look at the statistical model identification. IEEE Trans. Autom. Control **19**, 716–723 (1974)
10. Zadeh, L.A.: Fuzzy sets. Inf. Control **8**, 338–353 (1965)
11. Thomas, S.F.: Possibilistic uncertainty and statistical inference. ORSA/TIMS Meeting, Houston, Texas (1981)
12. Zhou, J.: Fundamentals of Information Theory. People's Posts & Telecom Press, Beijing (1983). (in Chinese)
13. Byrne, C.L.: The EM algorithm theory applications and related methods. https://www.researchgate.net/profile/Charles_Byrne
14. Wu, C.F.J.: On the convergence properties of the EM algorithm. Ann. Stat. **11**, 95–103 (1983)
15. Neal, R., Hinton, G.: A view of the EM algorithm that justifies incremental, sparse, and other variants. In: Jordan, M.I. (ed.) Learning in Graphical Models, pp. 355–368. MIT Press, Cambridge (1999)
16. Huang, W.H., Chen, Y.G.: The multiset EM algorithm. Stat. Probab. Lett. **126**, 41–48 (2017)
17. Springer, T., Urban, K.: Comparison of the EM algorithm and alternatives. Numer. Algorithms **67**, 335–364 (2014)

Understanding: How to Resolve Ambiguity

Shunpeng Zou[1] and Xiaohui Zou[1,2](✉)

[1] China University of Geosciences (Beijing), 29 Xueyuan Road, Beijing 100083, China
949309225@qq.com, zouxiaohui@pku.org.cn

[2] SINO-US Searle Research Center, 10 Ronghua Middle Road, Beijing 100176, China

Abstract. This article aims to explore the question: understanding, interpreting, translating, how to resolve ambiguity? Or: how does man-machine combination resolve ambiguity? In order to focus on the essence of the problem, the method is that: target analysis butterfly model and its use cases, macroscopic analysis ambiguity model and its use cases, microscopic analysis matrix model or search model within a series of bi-list and its use cases. The result is through the three examples, from manual translation to machine translation and translation memory on view, pointed out that the fundamental way to resolve ambiguity. Its significance is that the method can be advanced to the generalized translation and corresponding interpretation and final practical understanding, the specific performance is that through man-machine collaboration, and its verifiable results with this method, we can work to resolve various ambiguities better, to ensure accurate understand, prevent and eliminate all kinds of misunderstandings.

Keywords: Linguistic cognition · Mind philosophy · Brain-machine integration · Attribute theory method

1 Introduction

This article aims to explore the question: understanding, interpreting, translating, how to resolve ambiguity? Or: how does human-machine better resolve ambiguity?

Nature language contains a variety of ambiguities. It is difficult to bypass problem of understanding, interpretation and translation. That is a great challenge for human and computer. We know that "ambiguity in language is an essential part of language, it is often an obstacle to be ignored or a problem to be solved for people to understand each other" [1]. James Allen introduces the concepts required to build a NL system without losing you in the psycholinguistics, psychology and philosophy of language [2]. "Interpretation is a complex practice that requires the interpreter to fully understand, analyze, and process spoken or signed messages. The interpreter, after processing this information, renders the message into another language" [3]. "The Language of food: a linguist reads the menu" [4].

Not only the difficulty of natural language understanding is ambiguity everywhere, but also generalized language or generalized text (including size character, formula, graphic, table, sound, image, 3D, living) passing information, imparting experience, expressing knowledge (expert knowledge acquisition and its formal representation and

Z. Shi et al. (Eds.): ICIS 2017, IFIP AICT 510, pp. 333–343, 2017.
DOI: 10.1007/978-3-319-68121-4_36

even repetitive reuse) is bound to be due to the various ambiguities (including disparity and disagreement) for both human and machine and their combination to make judgments or decisions, such problems are a major challenge to both human intelligence and artificial intelligence. "We saw earlier how to use ctrl + = to enter natural language input. Now we're going to talk about how to set up functions that understand natural language. Interpreter is the key to much of this. You tell Interpreter what type of thing you want to get, and it will take any string you provide, and try to interpret it that way" [5]. "While most of the neural network techniques are easy to apply, sometimes as almost drop-in replacements of the old linear classifiers, there is in many cases a strong barrier of entry". [6] "provides a more coherent and integrated framework for performing both bottom-up and top-down knowledge acquisition" [7]. "Turing himself provides a hint of the answer, noting that his deceptively simple question requires a robust definition of "think"; without one, the question itself is meaningless." "Indeed, Turing's original question" "Brilliant as the Turing test is, its popularity has had one pernicious effect: It has reinforced for many people the comfortable illusion that human intelligence is a meaningful measure of intelligence in general. This is understandable: If you can't define think or intelligence, refer back to human intelligence as the gold standard" [8]. "Artificial intelligence meets human intelligence". "Why did it take so long for neural networks to recognize speech and objects in images at human levels? [9]. "Linguistically, Harmony maximization corresponds to minimization of markedness or structural ill-formedness" [10].

2 Method

In order to focus on the essence of the problem, the following examples, from manual translation to machine translation and translation memory of practical point of view, and then advance to general translation and its corresponding interpretation and the final understanding of the basic point of ambiguity resolution (involving targeting analytic butterfly model, macro analytic triangular model and microscopic analytic enumeration matrix model or search model within a series of bi-list query).

2.1 Butterfly Model and its Use Cases

From the "pyramid model" (Fig. 1a) through the reverse thinking and model reconstruction - that is, the construction of "butterfly model" (Fig. 1b, it is for us to "understand, explain, translate, how to resolve ambiguity?" What is the role, value and significance of these two models?). We find a paradox existing in the translation pyramid model (Fig. 1a) [11]. From artificial and machine translation "pyramid model" (Fig. 1a) to bilingual "butterfly model" (Fig. 1b), there are differences and links between the two basic assumptions involved in the conversion principle. The former is the father of machine translation Weaver (interlanguage hypothesis) and computational linguistics of the international master B. Vauquois (translation pyramid) combination, the latter is the younger generation Zou Xiaohui's unintentional discovery between the two sides both in the actual process of natural language understanding and in establishment of bilingual coprocessing system ManMachine interaction. To this end, published a corresponding articles on the model was introduced, the focus of which also led to the academic director Professor Feng Zhiwei's concern and the occurrence of the corresponding academic debate and exchange of ideas, and now, not only this academic idea exchange has been effectively promoted, but also the (the narrow sense, the generalized, the alternative) three kinds of bilingual co - processing system, and further discovered its implied the three basic laws (principles or rules) in logic, mathematics and linguistics [12].

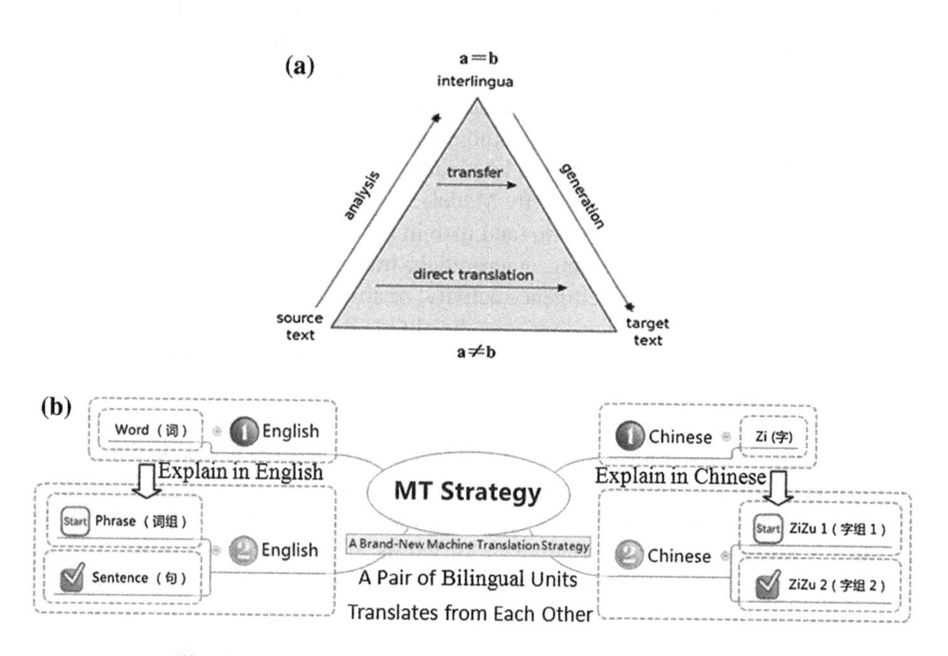

Fig. 1. From the pyramid model (a) to the butterfly model (b)

Figure 1 Butterfly Model (pyramid model to remove the spire after the reverse re-construction of the reverse thinking results).

You can see that from artificial and machine translation "pyramid model" (Fig. 1a) to bilingual "butterfly model" (Fig. 1b), there are differences and links between the two basic assumptions involved in the conversion principle.

Targeted analysis of butterfly model is characterized by first explained each kind of language both in bilingual model internally, and then translated from each other.

The specific performance of wisdom is for value-taken and confidence-building.

Through using basic methods of value-taking and confidence-building of language by the author's fundamental research in general linguistics taking Chinese language as an example, particularly in exploring how to solve the sharp conflict between two schools of scholars among workers, who adhere to Chinese character based linguist theory and word based linguist theory respectively. Specifically, the purpose of this paper is to solve the conflict between both Chinese character based view and word based view by using the basic unit of Chinese theory [13].

Identification, distinguishing and selection of lexical ambiguity are achieved by the selection of term or value-taken. Structural ambiguity distinguishes the function of the phrase and the sentence in their context. English morphemes and Chinese characters and their internal radicals, English words and Chinese words, English phrases and sentences and Chinese corresponding structural units, are almost all difficult to recommend a corresponding function of the relationship. Its roots lies in characters, not only the existence of the word information (its internal radicals), and the existence of inter-word information and even outside the word information. These three aspects of the value-taken and confidence-building on the two aspects, although the system can be used to do a series of processing, but a series of values, a series of confidence, how these two aspects of specific match? Human, computer, bi-brain, between the three, their operation and algorithm and the combination of the model there are differences. So, natural language understanding and how does resolve ambiguity? Be sure to follow the basic program given by the Butterfly Model, i.e., (in the respective native language system, respectively) to explain them, (and then in both the phrase and the sentence) to translate them. Thus, understanding, interpreting, translating, three intellectual activities, incompetence is mental intelligence activity, or artificial intelligence activities, and even man-machine bi-brain wisdom, can be divided into two basic activities: the value-taken (select each match) and the confidence-building (sure that the common choice of the selected item is matched).

The following is illustrated and demonstrated in connection with the embodiments:

In Chinese, the relationship between characters and their strings as partial word is understood as formal structure between the object language and the explanatory or interpretation language or meta - language. At the same time, in English, the formal structure is understood by the relationship of words, phrases and sentences.

Figure 1b covers: lexical ambiguity (1) and structural ambiguity (2) which divided into two levels: phrase (ZiZu1) and sentence (ZiZu2). As a result "first explanation" and "re-translation" of the principles, therefore, the above two types of ambiguity and confidence-building process, resolved in the mother tongue groups. In other words, the translation is based on the bilingual expression of knowledge ontology. That is, un-ambiguous text conversion between source language and target language (which means that bilingual formal expressions refer to a specific object or concept).

For example, the performances of physics, meaning, grammar in English are three words, but three phrases (ZiZu1) in Chinese. Therefore, from the level of Chinese characters to the level of English words, there is no way one by one correspondence. In other words, English words and Chinese words are not entirely at one level.

English and Chinese phrases (ZiZu1) and sentences (ZiZu2) these two levels can be in the same level [14].

How does this ambiguity resolve? We use butterfly model to try the list explained as follows: Table 1 with large string (of character) formula to interpret Chinese characters and their relationship between characters, thus, resolve ambiguity.

Table 1. Resolve ambiguity with string of character formula.

String (Character) formula	Example of word	String of character
物 + 理=物理 (Material + Law = Physical)	Physical 物理	= Material + Law 物质(运动的)基本规律
意 + 义=意义 (Idea + Principle = Meaning)	Meaning 意义	= Idea + Principle 意念(确立的)原则含义
文 + 法=文法 (Text + Rule = Grammar)	Grammar 文法	= Text + Rule 文本(组合的)基本法则
x + y = z	z	$= x + y$

The equation of string (character) formula 1.

$$x + y = z \tag{1}$$

2.2 Triangular Pyramid (Tetrahedron) Model and its Use Cases

From the "semantic triangle" and "physics, meaning, grammar" (that covering "grammar, semantics, pragmatics") through the combination of Chinese and Western create three-dimensional Triangular pyramid (tetrahedron) model (the most ambitious knowledge system frame and the most fundamental knowledge classification).

What is the use, value, and significance of that? Understanding, explaining, translating: How better to resolve ambiguity by using the models?

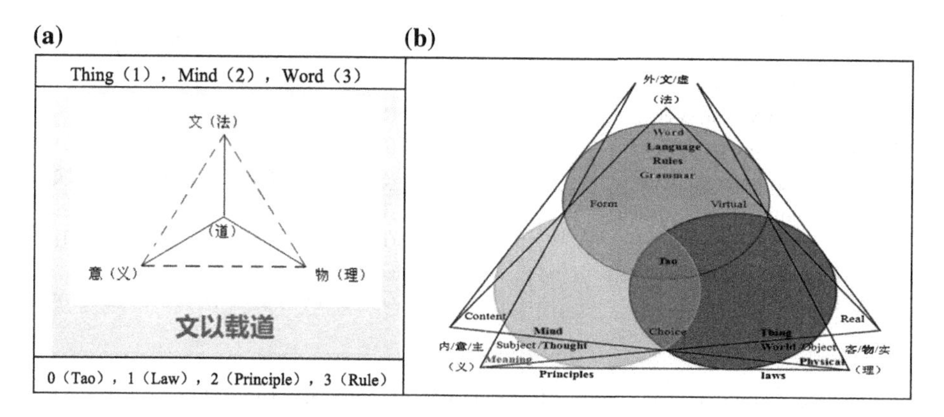

Fig. 2. Tetrahedral model (a) and its interpretation (b)

Figure 2 Triangular Pyramid (Tetrahedron in Fig. 2a) Model is Macro model. It is a broad text (文), Tao is basic true information (covering all phenomenon information).

Combining with Table 1, you can understand the four fundamental categories and the six basic concepts both in Fig. 2a and b. The four fundamental categories: physical world (real object), meaning thought (subject content), grammar language (virtual form),

Tao (laws or principles or rules); the six basic concepts: object and subject, content and form, virtual and real.

First of all, the three composite concept of decomposition, in the Triangle model will be a character set of Chinese visualization of the intuitive advantages, specifically expressed as three sets of character string formula (See: Table 2).

Table 2. Resolve ambiguity by using the model in Fig. 2a and b.

Phenomenon	Essential nature	Resolve ambiguity
(物理的)物:世界 (Physical) World	Law (物之)理	Law of material movement 物质运动的规律
(意义的)意:思想 (Meaningful) Thought	Principle (意之)义	Principle of idea/intention 意念生成的原则
(文法的)文:语言 (Grammatical) Language	Rule (文之)法	Rules of text construction 文本建构的法则
	道Tao	

In the fundamental categories (See: Fig. 2 and Table 3), introduction of three pairs of concepts: object and subject, content and form, virtual and real, highlight the scope of the phenomenon (world/thing, thought/mind, language/word).

Table 3. Characters = Yan, Chinese = Yu, word and sentence have attributes of both Yan and Yu.

Yan: Monosyllabic character (Zi) (Yan + Yu = Word or Sentence)	Example	Yu: Two-syllable and multi-syllable phrase or sentence (ZiZu1or2)
(音)字(Phonological) Yan		(Linguistic) Yu语 (言)
(形)字(Morphological) Yan		(Semantic + Lexical) Yu语 (辞)
(象)字(Object Language) Yan		(Grammatical) Yu语 (链)
(释)字(Meta-Language) Yan		(Semantic + Grammatical) Yu语 (块)
(实)字(Semantic) Yan (虚)字(Grammatical) Yan (用)字(Pragmatic) Yan (解)字(Lexical) Yan		(Pragmatic + Classical) Yu语 (读) (Pragmatic + Vernacular) Yu语 (句) (Pragmatic + Rhetoric) Yu语 (段) (Pragmatic + Article) Yu语 (篇)

Chinese category Tao is fourth fundamental category (See: Fig. 2), the basic concept different from these Western concepts "physics, meaning, grammar" and the two sides behind the outlook. The theory of "holism" and "reduction theory" are not clear (the biggest ambiguity problem is highlighted). Therefore, the categories of thing, mind, word and the essence (law, principle and rule) can be further Explored.

The fundamental category contains more basic categories, which is composed of two groups of three pairs of concepts, they are: object and subject, content and form, virtual and real. The following is illustrated in connection with the embodiments:

2.3 Query Model and its Use Case

Figure 3 Query Model (Three kinds of views in linguistics and three bilingual division methods for that from translation to query, laid the basis for both theoretical analysis and practical operation) [15].

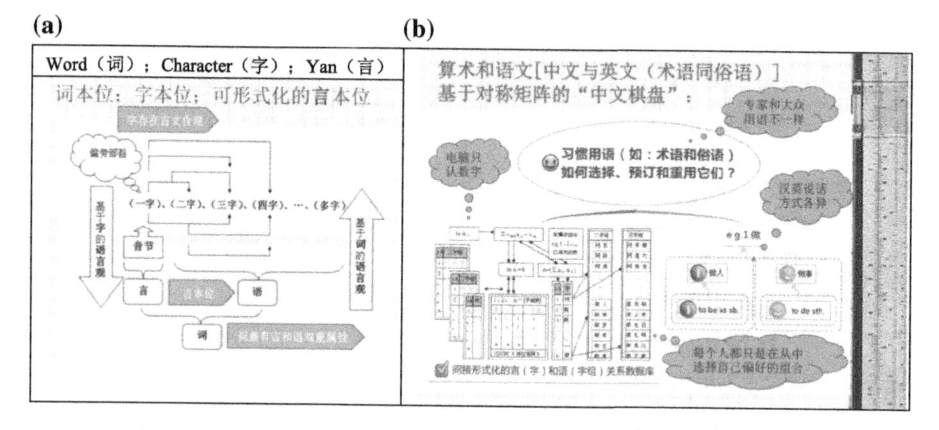

Fig. 3. Comparison of views (a) and actual use cases (b)

As it can be seen from Fig. 3a, from English word and Chinese character, we extract bilingual Yan and Yu, for all kinds of object language and mete-language that is, out of semantic language, laid a solid foundation for formal understanding [18].

It can be seen from Fig. 3b that the relational data structure within generalized bilingual object language and mete-language, man-machine cooperation to complete both Chinese information processing (find the best algorithm) and natural language understanding (to resolve ambiguity), laid solid basis by using two categories formal strategy. It is a actually global (language)positioning system (GLPS).

Three types of bilingual co-processing with multiple interfaces and can reflect a series of bilingual conversion, not only to answer “understanding, interpretation, translation, how to resolve ambiguity?” for us, but also the problem has a practical effect, value and significance, moreover, it can also be seen as a test “how to better resolve ambiguity with man-machine combination?” Bi-brain and bi-intelligence collaboration system is intuitive prototype [16, 17].

From “monolingual model” and “multilingual model”, we abstract “bilingual model” (especially the “three kinds of bilingual co-processing” and its implied logical sequence and position, function linkage, translation promotion or explanation expansion. Resolve ambiguity and understand language.

Human-Machine Twin System as Collaboration Translation.

Based on the three basic laws of human-computer cooperation (three basic rules of logical sequence-position, function linkage, translation promotion) and constructed a series of bi-list, using computer-aided, not only can achieve the joint operation between arithmetic and language, but also to the joint operation combined fully automatic, semi-automatic, big collaboration, by the narrow language and its plain text of the string, through the promotion of characters such as the intermediary model, to further promote to generalized language or text (including characters, formulas, graphics, tables, sounds, images, 3D, living).

The following is illustrated and demonstrated in connection with the embodiments:

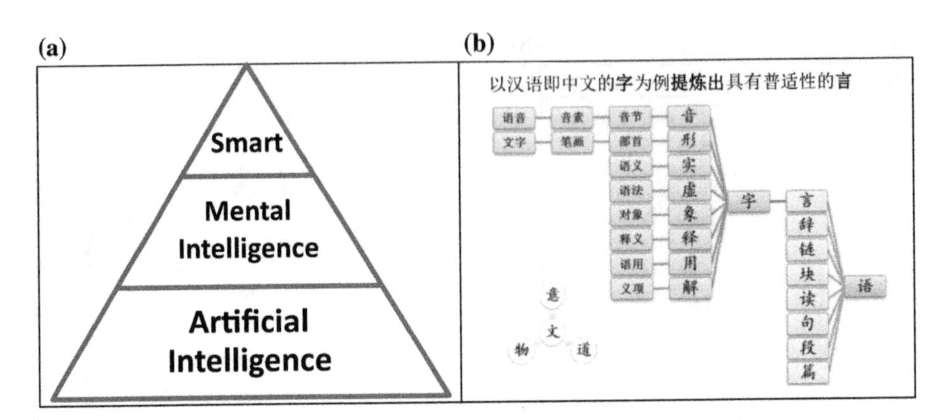

Fig. 4. God, man, machine ladder model (a) and language and speech classification (b)

It can be seen in Fig. 4a that smart (city), mental intelligence and artificial intelligence (different from ancient Greek wisdom in philosophy), converge on the connotation and extension of research object of intelligent science.

As it can be seen in Fig. 4b, Chinese as a typical natural language, we understand its own and its unambiguous formal expression, such as combination in Table 3, we can do further research and discussion. Later, in conjunction with Fig. 5a, we can also classify it.

Sequence and location of syllables, select: value-taken and confidence-building.

The so-called value-taken, here is that: around a specific goal, given a series of Chinese characters with appropriate structures. For example, in the following Table 3, two groups have been accurately displayed, and their features are clearly displayed.

The equation of string (character) formula 2.

$$Yan\,(\text{Zi, Sentence}) + Yu\,(\text{ZiZu1, Phrase; ZiZu2, Sentence}) = Word\,or\,Sentence \quad (2)$$

It can be seen in Fig. 5a that the linguistic view introduced here is essentially a broad bilingual view. Between formal language shown in Fig. 5a and formal system developed in Fig. 5b by the part of reduction theory, its example highly consistent. In other words, the formal understanding natural language here with the arithmetic language (including: machine language) formal understanding, in the same way. The type calculation of the hierarchical set is combined with the attribute set, and the bridge between the simplest single set and the most complex cluster is built up. Thus, language theory, epistemology, object theory, can be unified into the overall logic of sequence and location in twin matrix. Pure formal various functions and a variety of property functions, between each other can be in the agreed under a variety of specific

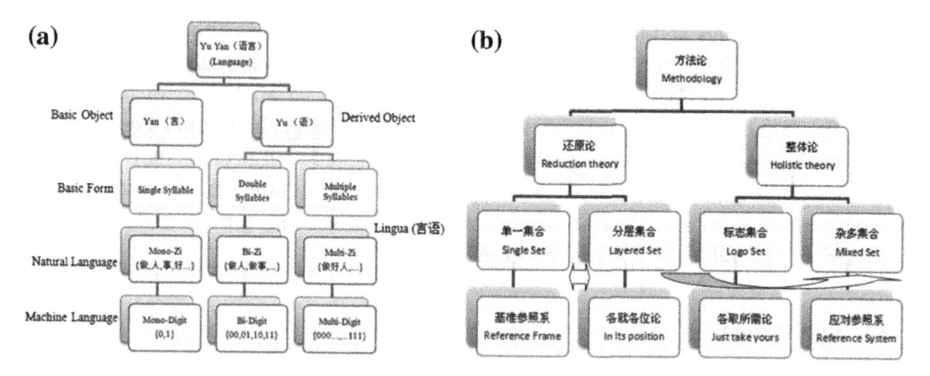

Fig. 5. Basic classification of generalized bilinguals (a) and methodology, set classification, strategy and method model (b)

constraints, the formation of the corresponding linkage function. This is the general translation, paved the way. Furthermore, not only for various forms of class functions, but also for various attribute functions (involving the content) can be done: both can be calculated, but also can statistic and can do similar measures [19].

3 Conclusion

If Global Positioning System (GPS) can help people to accurately determine the starting point and the target point and the optimal path between each other, then the language, knowledge concept, software object Global Positioning System also can help people clearly understand (interpret, translate or convert) them. In this study, the macroscopic triangular pyramid (tetrahedron) model and the microscopic table model are combined to form Global (Language, Knowledge, Software) Positioning System, and the target (language, knowledge, software) positioning effect. For example, the (eight) Yan and Yu, (three) Zhi (from wisdom to smart, mental intelligence and artificial intelligence), all can be particularly understood. Another example is the further understanding of physics, meaning and grammar (divided into three groups). With this basis, the other types of subdivision and rough (language, knowledge, software) target point of sequencing positioning can be accurate and correct. If both Figs. 2 and 4a are to look at the problem from a macro perspective, then Figs. 1b and 4b and all look at the problem in the middle of the view, while Fig. 3b looks at the problem in a microscopic point of view. So that the combination of macro and micro to do natural language understanding and formalized the expression of knowledge, can be accurate, of course, naturally can be resolved which contains a variety of ambiguities.

3.1 Result

The result is through the three groups of examples, from manual translation to machine translation and translation memory on practical view, pointed out that the fundamental way to resolve ambiguity.

The three models and their used cases tell us that each item of character or word as value-taken can be recorded, not only by computer-assisted enumeration into an electronic dictionary, but also by an enumerable corpus at all levels, in particularly through a large number of bilingual expressive knowledge bases that are precisely processed and validated for bilingual linkage. This has laid a large data environment for repeated application within the three models. Their function can be used by the majority of teachers and students to the extreme. Thus creating a basic development platform can be built for knowledge production.

3.2 Significance

Its significance is that the method can be advanced to the generalized translation and corresponding interpretation and the final practical understanding, the specific performance is that we can work to resolve various ambiguities, to ensure accurate understand, prevent and eliminate all kinds of misunderstandings,through human-computer collaboration with this method and its specific verifiable results.

The practical significance of this paper is not only that it demonstrates how a series of ambiguities between Zi (字) and Zhi (智) are gradually resolved by the three models of macro and micro (and its practical application methods and tools), but also in the process of argumentation. The basic views and basic approaches adopted have universal roles, values and meanings. For the author's innovative ideas and innovative methods of the system argument, has been beyond the scope of this article, interested in it, can refer to the author's related works.

Do they all together can be classified as Tao trying to characterize the object form and the subject content or idea? (To be further explored)

References

1. Quiroga-Clare, C.: Language ambiguity: a curse and a blessing. Translation Journal (2013)
2. Allen, J.: Natural Language Understanding. Pearson, London (1995)
3. Understanding Interpretation and Translation. Interpretation or Translation, October 2014
4. Jurafsky, D.: The Language of Food: A Linguist Reads the Menu, October 2015
5. Manning, C.D., Schütze, H.: Foundations of Statistical Natural Language Processing, 1st edn. The MIT Press, Cambridge (1999)
6. Goldberg, Y.: A primer on neural network models for natural language processing. J. Artif. Intell. Res. **57**, 345–420 (2016)
7. Motta, E., Rajan, T., Eisenstadt, M.: Methodology and tool for knowledge acquisition. Stud. Comput. Sci. Artif. Intell. **5**, 297–322 (1989)
8. Allenby, B.: Why it's a mistake to compare A.I. with human intelligence? The Citizen's Guide to the Future, 11 April 2016
9. Sejnowski, T., Crick Chair, F.: Deep learning: artificial intelligence meets human intelligence. Clark Center Auditorium, 23 May 2017
10. Smolensky, P.: Harmony in linguistic cognition. Cogn. Sci. **30**(5), 779–801 (2006)
11. Zou, X.: Paradox existing in the translation pyramid model. X Mind, August 2011
12. Zou, X.: From translation pyramid model to bilingual butterfly model. Sciencenet (2017)

13. Zou, X.: Value-taking and confidence-building of language. In: AAAS Annual Meeting (2012)
14. Zou, X., Zou, S.: New mission of contemporary Chinese universities: cultural inheritance and innovation based on Chinese thinking and bilingual processing. J. Nanjing Univ. Sci. Technol. (Soc. Sci. Ed.) (05) (2012)
15. Zou, X., Zou, S.: The relationship between words and language - based on the distinction between language and speech. In: The Fourth Session of the West International Symposium on Philosophy of Philosophy Abstracts (2014)
16. Zou, X., Zou, S.: Two types of formalization strategy. J. Comput. Appl. Softw. (09) (2013)
17. Zou, X., Zou, S.: Bilingual information processing method and principle. J. Comput. Appl. Softw. (11) (2015)
18. Zou, X.: Fundamental law of information: proved by double matrices on numbers and characters. AAAS Annual Meeting (2017)
19. Zou, X.: Characteristics of information and its scientific research. In: IS4SI-2017 » Session Conference FIS: The Seventh International Conference on the Foundations of Information Science (2017)

Exploration on Causal Law of Understanding and Fusion Linking of Natural Language

Peihong Huang(✉)

Putian Branch of Agriculture Bank of China, Putian, China
Huangpeihong6488@sohu.com

Abstract. To research the causation law of natural language understanding from its cognitive model is a kind of creative method which has been being developed recently. In this paper, the research situation of natural language understanding and interconnection are introduced, and furthermore we prospect its potential future affect to other areas.

Keywords: Natural language understanding · Perception · Interconnection · Prospection

1 Introduction

As we know humankind has a phenomenon which is interconnection. For example, if A is B's brother, and B is C's brother, we can know immediately A is C's brother. However, if apparent knowledge is not sufficient, machine wouldn't know it. The capability of interconnection makes people to comprehend from inference. This is because of capability of understanding. It is just because of the use of understanding that humankind can solve mathematic problems from simple to complex, can speak from babbling to creation of masterpiece which is rich of literary grace.

However, in the area of NLP (Natural Language Processing), the main way of NLP is based on statistics. Nevertheless, when facing even further language translation and problem solving, the method based on statistics is not a magic bullet. On the contrary, method based on understanding when doing natural language processing can do it at ease.

Different from psychology concepts, the theory of natural language understanding [1, 2] defines the perception as the minimum element of feeling, and uses perception as the basic concept of the physical theory of understanding. As for the qualitative aspect of perception, since it is abstract, it needn't to be fully analogized at current stage. Instead, just extract the qualitative invariant part of perception and identify this part with symbols. This method can achieve the same effect as that can be gained from fully analogized implementation. We define perception element as: p.

Perception: so called perception, is the unit of consciousness, which is the minimum element which can be felt and which is also the minimum element of meaning, the qualitative invariant part of which is identified as: p.

Z. Shi et al. (Eds.): ICIS 2017, IFIP AICT 510, pp. 344–350, 2017.
DOI: 10.1007/978-3-319-68121-4_37

2 Understanding of Logic and Physical Model of Interconnection

Paper [1, 2] presented that types of understanding include not only understanding of meaning but also understanding of logic. ***Understanding*** is the sense of certainty when the stimulus matches the set of perceptual patterns within the cognitive system, and the formula is:

$$b_t = u(x) = w(P^x_g, m^x) \tag{1}$$

Among them, the external stimulation is x, let an understanding of stimulation x to be $u(x)$, the corresponding perceptual pattern is m^x and the *perceptual subset* is to be P^x_g, the *assembly perception set* is to be P_g, $P^x_g \subset P_g$, the matching between x and m and its confidence function are to be w, the certainty feeling is b_t.

The stimulation material consists of various parts, the understanding of it includes all parts of the understanding. Paper [1] gives a formula of the comprehensive understanding. To stimulation x, let the perception produced by it to be p, the comprehensive understanding of x is to be $u_c(x)$; Set a stimulus x containing n sets of *assembly perceptions* and n sets of *perceptual patterns*, the corresponding *assembly perception set* and *perceptual patterns* are respectively $(P^x_g)_i$ and $(M^x)_i$ $(i = 1, 2, \ldots, n)$, then comprehensive understanding of the x is:

$$u_c(x) = \prod_{i=1}^{n} u((P^x_g)_i) = \prod_{i=1}^{n} w((P^x_g)_i, (M^x)_i) \tag{2}$$

In order to comprehensively and systematically reveal the physical law of natural language understanding, paper [2] presented the reliability and integrity theorems of the axiom system which is essential to pragmatic implication inference.

The important theorem about pragmatic implication presented and proved in Paper [2] is also based on the foundation of the integrity theorem, and the inference of pragmatic implication is very important to the semantic computing, it provides the formalization operating procedures to semantic computing. The pragmatic implication of a sentence in the corresponding context is:

$$(p^S_h)_j = (M^r - \bigcup_{i=1, i \neq j}^{n} (p^s)_i - (p^s_s)_j) \cup (p^s_t)_j \tag{3}$$

Among them, an understood context contains a collection of sentences $S = \bigcup_{j=1}^{n} s_j$, and the associated rule set $M^r \subseteq S_c$. The pragmatic meaning of sentence s is P^S_h, the pragmatic meaning of sentence s_j is $(P^s_h)_j$, its sentence meaning is $(P^s)_j$, its literal meaning is $(P^s_s)_j$, and the deduced semantic is as $(P^s_t)_j$. From the expression, the pragmatic implication is a deduction value in the sentence group context G. By taking a certain value of G, it is possible to determine a pragmatic extrapolation value in a context.

2.1 Understanding of Logic is the Prerequisite of Interconnection

There is a close relationship between understanding of logic and interconnection. In paper [3], we reveal how to implement reasoning understanding (understanding of logic) according to perception semantics dictionary within semantics sequence correspondent to language, and furthermore acquire new understandable belief. The results of these steps also further verify the theory of natural language understanding we found.

Table 1 is one of the experimental examples from paper [3], which is a sentence of a famous Chinese ancient poem of Tang Dynasty, by poet Li Bai. This example assume that the machine already has words: *in front of the bed*, *bright moon*, *light*, and has the perception semantic value, and machine can understand the semantics of these words according to the definition of understanding. Where, *Mr* is pattern set generated after context being understood, new pattern set filter function *new(Mr)* generates new pattern set *M*.

Table 1. Computing example of learning based on understanding.

Uc (*bright moon light in front of the bed,*)

=Uc (*in front of the bed*) Uc (*bright moon*) Uc (*light*) Uc (*bright moon light*) Uc (*bright moon light in front of the bed*)

↓

P1=Uc (*in front of the bed*) =Uc(P(*in front of the bed*))=1

P2=Uc (*bright moon*) =Uc (P (*bright moon*)) =1

P3=Uc (*light*) =Uc (P (*light*)) =1

=Uc (*bright moon light*) Uc (*bright moon light in front of the bed*)

↓

P4=*bright moon-*) give out light

then P5=P (*bright moon-*) give out light)

then P6=Uc (light gave from *bright moon*) =1

then P7=Uc (*bright moon* light) =1

=Uc (*bright moon light in front of the bed*)

↓

P8=*bright moon light-*) give out light to everywhere

P9=*in front of the bed*∈everywhere

then P10=*bright moon light-*) give out light *in front of the bed*

then P11=Uc (*bright moon light is given in front of the bed*) =1

then P12=Uc (*bright moon light in front of the bed*: the *light in front of bed given by bright moon*) =1

=1

Mr={P1, ..., P12}

M=new(Mr)={P(*bright moon light in front of the bed*:the *light in front of bed given by bright moon*),...}

Table 1 shows the formalization procedure of machine understanding learning of "*bright moon light in front of the bed*". It means that, is it feasible that "*bright moon*" and "*light*" are put together? The implication of "*bright moon*" shows that it can give out "*light*", then the meaning of "*bright moon light*" is the "*light*" shined by "*bright moon*", thus "*bright moon light*" is understandable. What is the comprehensive meaning of "*in front of the bed*" and "*bright moon light*"? Is it feasible that there is *"bright moon light" "in front of the bed"*? When *"bright moon"* shines in the sky, the *"light"* of it can illuminate to everywhere, thus it can illuminate to "in front of the bed". Therefore *"bright moon light in front of the bed"* is understandable. Storing after understanding is learning. Understanding is the basis and one of purposes of learning. So called reasoning of logic understanding includes deduction and induction reasoning based on perception semantics, new knowledge can only be acquired when logic understanding is implemented, thus interconnection effects can be gained.

Compared to statistics based machine learning, one of the advantages of understanding based learning approach is that its learning results have characteristics of robustness, because their learning results are fully logical. Secondly, its learning outcomes is grounded, because the result is in line with reality; thirdly, the outcomes of learning is understandable for human beings; fourthly, it is a small sample learning method. The prerequisite of using the method is that it requires careful description of common sense, and requires a comprehensive understanding. The process of comprehensive understanding is more complicated and more time-consuming than statistical method. So far, the study is still in its infancy, and many further researches are needed to do persistently.

2.2 The Physical Model of Interconnection and its Meaning

Interconnection is to fuse certain perception material and then generate new correct beliefs, thus interconnection is also known as fusion. A model of interconnection is presented in paper [2], shown in Fig. 1. In paper [2, 4], interconnection is built on the basis of understanding, machine cannot implement interconnection if there is no understanding as the basis, and the knowledge in it is only the isolated hard knowledge.

Since knowledge base system is built on the foundation of concepts, therefore, these knowledge can't be interconnected well. Differentiating, understanding and fusing concepts is fundamental roadmap to overcome these problems. To understand these concepts is useful to automatic learning and automatic usage.

In Fig. 1, When cognitive system observes Real World, it forms the perceiving Images of Real World, following-on through Instinct Mechanism of machinery understanding, judges whether it forms any function, which means to judge if it has values. If the function rule is true (i.e. $e(x) = 1$), then cognitive system differentiates and segments out the Function Rules including Concepts (C_{01}, $C_{02,\ldots}$, C_{mn})/Knowledge (K_{01}, $K_{02,\ldots}$, K_{mn}).

In cognitive system, all Concepts (C_{01}, $C_{02,\ldots}$, C_{mn})/Knowledge(K_{01}, $K_{02,\ldots}$, K_{mn}) are fusing based on Perception Semantics (PS^C_{01} PS^C_{02}......PS^C_{0n}, PS^C_{11} PS^C_{12}......PS^C_{1n}, ..., $PS^C_{m1}$$PS^C_{m2}$......$PS^C_{mn}$) and Perception Semantics of Knowledge (PS^K_{01} PS^K_{02}......PS^K_{0n}, PS^K_{11} PS^K_{12}......PS^K_{1n}, ..., PS^K_{m1} PS^K_{m2}......PS^K_{mn}). These concepts and knowledge are in fusion linking state when cognitive system is thinking or using these concepts.

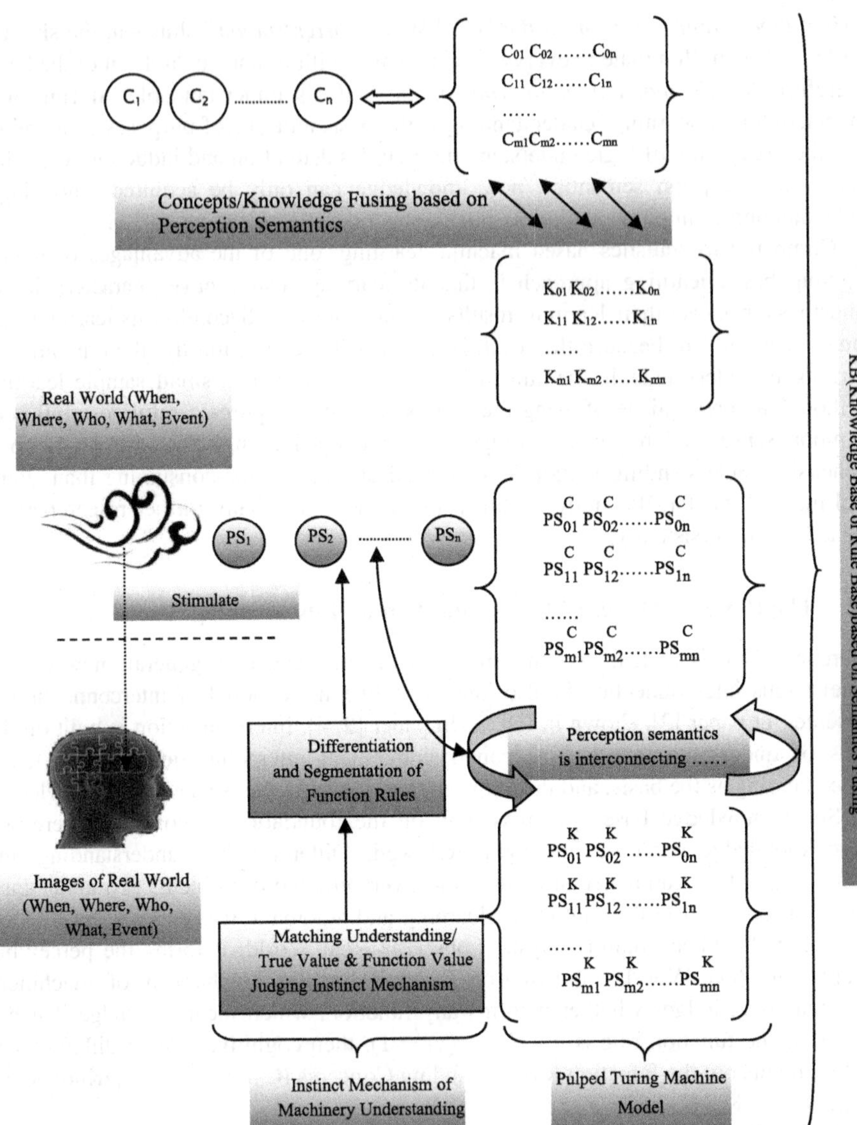

Fig. 1. Extended (pulped) Turing Machine Model processed information based on perception semantics

3 Prospection of Understanding Theory

During the course of language translation, supposing that the machine has already got some basic facts and knowledge, if the machine has the ability of interconnection, it will have the ability of analogy. It can interconnects the old material (content) again with perception elements, and can generate new beliefs according to new requirements, and thus can extract new experiences from old material constantly. This means if the machine grasps a typical interconnection, it will comprehend the whole category by analog. This process is just like paper making: Pound paper fiber and mixing fiber with water, and finally fuse them into fine pulp, then manufacture paper products meeting various kinds of requirements. Lack of ready-made and suitable knowledge is often a problem, thus machine-oriented interconnection is extremely significant.

On the other hand, since there is no limitation (restrains) of the magnitude of the concept particles of variable or status in Turing Machine Model, formalization system based on Turing Machine Model is normally local, and thus the fusion of systems is difficult. This limitation is manifested when programming. Generally a program is a formalization system. When programming, firstly the variables of various kinds of data structure are defined, and then the logical flow are defined or structured, and then variable mathematical operation or formal variation are implemented, and thus the final results are got and stored into the result variables. We can see that a program is consists of variables and limited operation instruction sequence. Program is just a formalization system which can stop within limited steps.

The particle size of variable (or status) can be changed. Size and unit can be element, or even can be big enough to the magnitude of a combined concept which is composed of multiple concepts. These concepts can be huge, and concepts of different systems are independent, thus the definition of variables also can be of huge quantity, this means human need to program huge quantity of program, and build huge quantity of formalization system, only in this way we can adapt to the change of the reality world.

After all, the limitation of Turing Machine Model is that it cannot fuse the knowledge it uses. The interconnection built on understanding will enhance the usage rate and usability of knowledge, and thus we outlook that our natural language understanding theory will impact software engineering, especially software production automation significantly.

The theory of natural language understanding can also be applied to network content security, spoken dialogue system, information retrieval, verification code recognition and voice content retrieval, etc. [5–12], and thus can incrementally fulfill the dream of The Imitation Game of Alan Turing.

4 Conclusion

Natural language understanding is a conclusion of one of the important rules about human language phenomenon. It is the same as physics that theory of natural language understanding can be summarized and extracted from the psychological and physical phenomenon. In the field of computer science, Turing Machine is its base, and the initial intention of Turing Machine is to analog thinking computationally. This is the nature of Turing Machine. Just as what is analyzed in paper [2], the interconnection ability of original Turing Machine Model is limited. Natural language understanding theory [1, 2] is an extension to Turing Machine Model. Just like that Turing machine liberate human being from cumbersome computation, we believe that natural language understanding theory will liberate human being from cumbersome analogy of thinking which results in building of infinite formalization systems for human using Turing Machine. This is just the motivation of the natural language understanding theory research.

References

1. Huang, P.H.: Formalization of natural language understanding. Comput. Eng. Sci. **29**(6), 113–116 (2007)
2. Huang, P.H.: NLU-A logic theory on machine perceiving language of humankind. J. Sichuan Ordnance **30**(1), 138–142 (2009)
3. Huang, P.H.: A method of machinery inference learning based on text explaining. J. Chongqing Commun. Inst. **29**(6), 89–91 (2009)
4. Huang, P.H.: What-why understanding effect of natural language processing. Mod. Comput. (Res. Dev.) **10**(7), 9–16 (2016)
5. Xiao, N., Zhao, E., Yan, B.: Research progress of network content security. Technol. Appl. Netw. Secur. **11**, 30–32 (2008)
6. Wu, T.: Spoken dialogue system of tourism information. Dissertation of Beijing Forestry University, Beijing, China (2015)
7. Zhang, L.: Telecom business software design based on natural language processing and speech recognition. Dissertation of University of Electronic Science and Technology, Chengdu, China (2014)
8. Jiang, T.: Design and implementation of universal demand retrieval data processing subsystem for Baidu video. Dissertation of Beijing Jiaotong University, Beijing, China (2014)
9. Zhou, H.: Overview of network content security. In: China Science and Technology Expo, vol. 2 (2012)
10. He, Z.: Research and development of natural language. Sci. J.-Electron. Vers. (MID) **5** (2016)
11. Yin, G.: Research on verification code recognition algorithm based on SVM. Dissertation of Anhui University, Anhui, China (2010)
12. Cheng, X.: Tourism website content search based on ontology. Dissertation of Nanjing University of Science and Technology, Nanjing, China (2008)

Depression Tendency Recognition Model Based on College Student's Microblog Text

Jie Qiu[1] and Junbo Gao[2(✉)]

[1] College of Information Engineering, Shanghai Maritime University, Shanghai 201306, China
qiujiel125@163.com

[2] Shanghai Maritime University, Shanghai, China
jbgao@shmtu.edu.cn

Abstract. In order to solve the issue of identifying depression tendency hidden in microblog text, a depression emotional inclination recognition model based on emotional decay factor is proposed. Make the self-rating depression scale, collect students' microblog text and ask the psychology specialist to annotate the microblog artificially. Construct the depressive emotion dictionary, and then build a depression emotion classifier based on support vector machine. Considering the continuity of depression mood swing, the mathematical model of emotional decay factor is constructed to realize the continuity of discrete emotional state. The experimental results show that the model can effectively identify the depression of user for a period of time, the recognition accuracy rate is 83.82%.

Keywords: Depression tendency recognition · Emotional decay factor · Support vector machine (SVM) · Depression emotional dictionary · Sina micro-blog

1 Introduction

Web2.0 era, micro-blog style format because the free, easy to use and a large number of users to publish their own thoughts and feel. Many researchers use microblogging text for emotional analysis research [1], but the research is mainly for a specific thing [2], such as film critics, product reviews, etc., and for the text researched on depression emotion a bit less. How to timely and effectively identify the user through the microblogging tendencies, to prevent the long-term spread of depression has become an urgent problem that to be solved.

In the world, depression is one of the most common mental that illness people are faced with [3]. With the development of social networking platform such as microblogging, many researchers use the characteristics of user network to determine their psychological depression [4]; literature [5] based on the frequency of maternal postpartum social networks, to analyze language style and establish a statistical model of maternal depression. Wang [6] and so on. The depressive patient is regarded as a node, and a graph network is constructed as the center. According to the attributes of the adjacent nodes in the network and the weight of the connection, the model is given

Z. Shi et al. (Eds.): ICIS 2017, IFIP AICT 510, pp. 351–359, 2017.
DOI: 10.1007/978-3-319-68121-4_38

to calculate the depression [7]. The method of using brain-imaging study starting at the resting of the changes of brain function in patients with depression [8]. In terms of time users send microblogging, number of fans and a number of concerns such as depression and to analyze the situation of the user.

Existing studies have focused on the microblogging network and external factors to analyze user behavior characteristics of depressive tendencies situation, ignoring the role of micro-blog text Tech in terms of expression. In this paper, we will study the tendency of depression in the students' microblogging text, which integrate the emotional decay factor and time factor into the change of emotion, and describe the fluctuation of the depressive mood of the students more effectively and effectively. Hospital staff to identify patients with depression and provide aid.

2 Related Work

2.1 Depression

Depression is a common psychological disease and the cause is very complex, which the researchers on the pathogenesis of depression put forward many theoretical hypotheses [9]. Psychological and medical researchers also proposed a variety of Depression Diagnostic Scale, which provides an important experimental basis for the relevant practice. Zung [10] proposed depression self-rating scale with a high degree of operability and adaptability; many medical institutions also use this scale to measure the degree of depression in patients. It divides depression into four categories according to the score, [20, 41] indicates normal, [42, 49] indicates mild depression, [50, 57] indicates moderate depression, [58, 80] indicates severe depression. This article produces an online depression self-rating scale (https://sojump.com/jq/9743549.aspx) to obtain a score for each student's depression.

3 Depression Emotional Tendency Recognition Model

In this paper, the depression and emotional model is shown in Fig. 1.

The model uses the support vector machine algorithm to construct the depressed emotion classifier, and uses the pretreatment training sample training classifier to obtain the reliable depression emotion recognizer. The identification module that the decay factor into emotions and time factor or the discrete continuous depressive emotional state. The emotional chart obtained by Chart efficient user who determines the depression for a period of time.

3.1 Preprocessing Module

Before the microblogging text is used to identify the depression tendency, the acquired raw data must be converted into a data form that the computer can understand.

In order to solve the high dimensionality of feature space in text categorization, this paper uses CHI for feature selection [11], and then calculates the weight using the classical TF-IDF method [12] in the field of text processing. In order to solve the

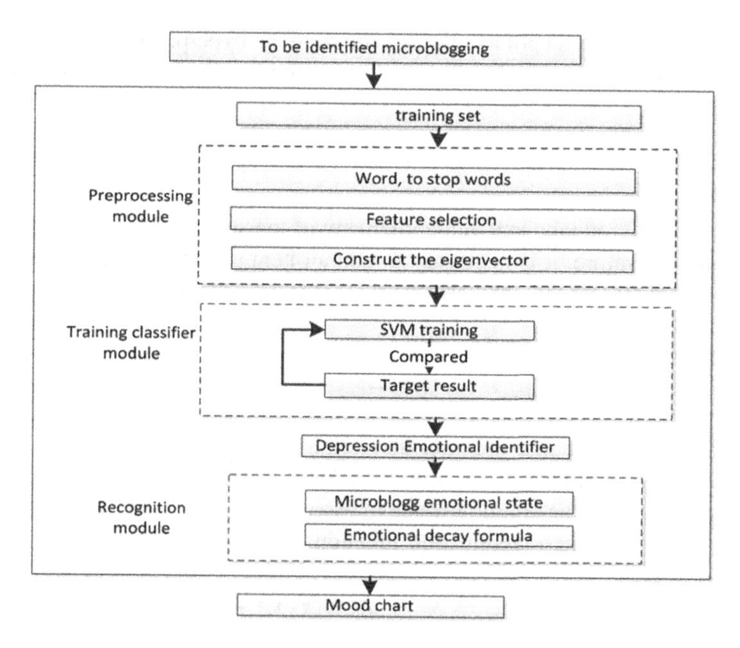

Fig. 1. Depression emotional tendency identification model

sparseness problem of microblogging feature, we use (L T: W) to represent each microblogging, where L is the label of each microblogging, T is the characteristic item, and W is the weight of the characteristic item. Such as: "I really love you, close your eyes, that I can forget, but shed tears, or did not deceive themselves," get five feature words uttered that word feature selection, are "cheat", "Love", "Tears", "close your eyes", "shed". Therefore, this sentence can be expressed as "1.0 28: 0.4528 39: 0.2295 49: 0.3215 862: 0.5811 1832: 0.54878", where 1.0 is the label, 28 is the characteristic word "cheat" index number and 0.4528 is the weight of the characteristic word.

3.2 The Construction of Emotion Classifier

Based on SVM, the depressive emotion classifier is constructed. Firstly, the nonlinear mapping function is constructed to convert from the original microblogging data x to the high dimensional space H. Then, the linear classifier is used to classify the microblogging emotion in the high dimensional space H. The Gaussian kernel function is used to linearize the nonlinear problem, and the Gaussian kernel function is shown in Eq. (1).

$$K(x \cdot x_i) = \exp^{\frac{-\|x - x_i\|^2}{2\delta^2}} \tag{1}$$

In the data samples may be time-linear high-dimensional space, SVM obtains the optimal hyperplane by calculating the minimization. At the same time, taking into account the number of data points may deviate from the hyperplane, it is slack variables introduced in the existing conditions based on the constraints, constraints as shown in Eq. (2).

$$y_i(w^T x_i + b) \geq 1 - \xi_i \quad \mathrm{i} = 1, 2, \cdots, n \tag{2}$$

At this point if you take any value, then any of the super-plane are in line with the requirements. Therefore, it is necessary to add an item after the original minimization function, so that all and the smallest, the new objective function as formula (3).

$$\min\left\{\frac{1}{2}\|w\|^2 + \mathrm{C}\sum\nolimits_{i=1}^{n} \xi_i\right\} \quad 0 \leq \alpha_i \leq \mathrm{C} \tag{3}$$

where C is a parameter that controls the weight between the two above-mentioned objective function. Further, the constraint is added to the objective function, Lagrange function is configured, the solution was expressed constraints corresponding Lagrange multiplier. Corresponding classification function as shown in Eq. (4).

$$f(x) = \sum\nolimits_{i=1}^{n} \alpha_i y_i K(x \cdot x_i) + b \tag{4}$$

where x is the text of the microblog to be classified, x_i is the support vector for the text, and y_i is the corresponding classification of x_i. When $f(x) > 0$, there x has a tendency to depression; $f(x) \leq 0$ is normal.

3.3 Mathematical Model of Depression and Emotional Decay

3.3.1 Basic Assumptions

Depression emotion classifier can identify whether each of these two micro-blog text depressive emotional state, and cannot effectively portray the evolution of these two emotional states. As a thoughtful individual, the person's depression is constantly changing over time, depression decay before giving formula, we first proposed the following basic assumptions.

1. The uncertainty of a person is not a depressed patient; this depression only assumes that individuals within any a period of time tend to be fluctuating mood.
2. An initial state is assumed that any normal individuals are emotional state, represented by state c. In this experiment, the state is the result of emotion classifier, there are two states, where c = 1 for the depression tendency state and c = 0 for the normal emotional state. The state of a microblogging can only explain the individual's emotional state, and cannot describe the degree of individual state, the corresponding state value.
3. The state definition of a value greater than 0 indicates predisposition to depression, state 0 indicates a value less than pleasant state, emotional state value equal to 0 indicates an equilibrium state. Moreover, the larger the state, the higher status individuals tend to some degree bias. The happy state is a subset of the normal state.

3.3.2 Depression Emotional Decay Formula

The structural decay formula is (5).

$$f(t) = f(t-1) + (-1)^n e^{-\lambda t} \tag{5}$$

where $f(t)$ represents time, the micro-blog corresponding depressive emotional state value; $f(t-1)$ is the depressive emotion state value corresponding to the microblogging text of the last time. λ is the emotional decay factor, which indicates the decay speed of the emotion, $\lambda = 0.5$, that is to say that the depression emotion is in accordance with the half-life law; the value of n is related to the state of microblogging at two adjacent time points, and according to the above basic assumption 3, the value of n is as follows (6).

$$n = \begin{cases} 2 & \forall c \in \{0,1\} \to c = 1 \\ 1 & \forall c \in \{0,1\} \to c = 0 \end{cases} \tag{6}$$

In view of the limited number of students' microblogging text, the time of microblogging has no regularity. In this paper, the time t which in the above-mentioned emotional decay mathematical model is defined as the time interval of two adjacent microblogs, then the range of t is: T = 0, 1, 2, 3 ⋯, n, and the initial state of any individual $f(t = 0) = 0$.

Basic assumptions above 2 (3) specifically explained in emotion decay formula. When two or more consecutive 0 state occurs, when the state of the next moment $c = 1$, then t_i. On the contrary, when the two consecutive or the above state one, if the next time t of the state $c = 0$, then the value of t is not set, but then the next time in turn increase. In the above two kinds of state alternation process, the value of $f(t-1)$ remains unchanged, still the state value of the last moment.

To quantitatively characterize the condition of the individual depression tendency over time, individuals taking herein mean value of the emotional state of a period of time for the metric, as shown in Eq. (7). Where $t = i$ represents the depressive state value of the i-th microblog from the i-th microblogging, and $f(t = i)$ represents the mean value of depression from the i-th microblog to the n-th microblog.

$$Avg = \frac{\sum_{t=i}^{n} f(t = i)}{n - i + 1} \tag{7}$$

Above mathematical model of emotional decay, this paper presents a mathematical model that identifies whether the individual is depressed over a period, as in Eq. (8).

$$\mathrm{E}(Avg) = \begin{cases} normal, Avg \in [-1.6, 0.2] \\ depression, Avg \in [0.2, 2] \end{cases} \tag{8}$$

Equation (8) shows that the average state of individual Yu state in the [−1.6, 0.2] interval, the emotional state is normal; depressive state of the average in the [0.2, 2] interval, there is a tendency to depression.

3.3.3 Case Analysis

Twitter test set is selected from the nickname "Fuji child package" student embodiment, with the emotional classifier on its pre-processed microblogging text classification, in view of the limited space, only select the first five microblogging content and classifier output. The emotional state presented as shown in Table 1 below. The state sequence of the output is {1, 0, 1, 0, 0, 1, 1, 0, 0, 0, 1, 0, 0, 1, 0}. According to Hypothesis 3, the state value $c = 0$ at time $t = 0$, that is. After the initial state, the $f(t = 0) = 0$ sequence is {0, 1, 0, 1, 0, 0, 1, 1, 0, 0, 0, 1, 0, 0, 1, 0}, according to the emotional depression Decay the mathematical model that get the emotional state values, as shown in Table 2 below. According to the different emotional status values at all times, you can draw depressive mood charts.

For this case, from $t = 0$ to $t = 15$, calculate the mean value of depression $Avg = \sum_{t=0}^{15} f(t)/16 = 0.7631$. According to the formula (8), it can be judged that the student is in a depressed state during that time.

Table 1. Microblogging content and emotional classifier output status

Label	Content
1	Did not expect stomach pain into the hospital
0	Meet the Tianzi Square
1	To TM computer, brush more than 200 sheets of A4 and close to the collapse of the end of the period
0	Headphones placed in the draft of Stefanie and always can cause a lot of resonance and back
0	How the end of the holiday began a little homesick

Table 2. Emotional state value corresponding to the emotional state sequence

$f(t=0) = 0$	$f(t=8) = f(t=7) + (-1)^1 e^{-0.5\times3} = 0.9957$
$f(t=1) = f(t=0) + (-1)^2 e^{-0.5} = 0.6065$	$f(t=9) = f(t=8) + (-1)^1 e^{-0.5\times4} = 0.8604$
$f(t=2) = f(t=1) + (-1)^1 e^{-0.5\times2} = 0.2387$	$f(t=10) = f(t=9) + (-1)^1 e^{-0.5\times5} = 0.7783$
$f(t=3) = f(t=2) + (-1)^2 e^{-0.5\times3} = 0.4618$	$f(t=11) = f(t=10) + (-1)^2 e^{-0.5} = 1.3848$
$f(t=4) = f(t=3) + (-1)^1 e^{-0.5\times4} = 0.3264$	$f(t=12) = f(t=11) + (-1)^1 e^{-0.5\times2} = 1.0169$
$f(t=5) = f(t=4) + (-1)^1 e^{-0.5\times5} = 0.2444$	$f(t=13) = f(t=12) + (-1)^1 e^{-0.5\times3} = 0.7938$
$f(t=6) = f(t=5) + (-1)^2 e^{-0.5} = 0.8509$	$f(t=14) = f(t=13) + (-1)^2 e^{-0.5} = 1.4003$
$f(t=7) = f(t=6) + (-1)^2 e^{-0.5\times2} = 1.2188$	$f(t=15) = f(t=14) + (-1)^1 e^{-0.5\times2} = 1.0324$

4 Experimental Results and Analysis

4.1 Data Acquisition and Annotation

From December 2016 to March 2017, data collected from 271 students' microblogging, microblogging that get 7321 text. From the scale score results, 80 people tend to depression, normal 191 people. Ratio of 3:1, were randomly selected from two types of training with examples and test students, the number of training set and test concentration shown in Table 3 below. After microblog corpus manually labeled, received a total of depression microblogging 1512 and 3786 normal microblogging. There is 1154 test set microblogging. This experiment is carried out in win10 32-bit system, using python language.

Table 3. Training set and test set number and microblogging number

	Training	Test
Depression	60	20
Normal	143	48
Total	5298	1154

4.2 Result Analysis

Experiment 1 test concentration Twitter 1154, using the depression and emotional depression without using dictionaries and dictionary emotion experiment, the results shown in Table 4 below.

Table 4. The experimental accuracy of depression and emotional dictionary

Training sample	No adopted dictionary	Adopted dictionary
2000	70.62%	80.58%
3000	73.83%	85.68%
4000	75.85%	87.36%
5298	78.45%	90.04%

As can be seen from Table 4, the artificial depiction of emotional depression that a single microblogging recognition accuracy significantly improved.

Experiment 2 uses the Depression Emotional Classifier to classify the microblogging text of each student in the test set, and obtain the intelligence status and classification accuracy of each microblogging emotion of 68 test cases. Some of the data are shown in Table 5 below.

Using the mathematical model of depressive emotion decay to draw out the emotional trend graphs of 68 test cases, limited to the limited length of the articles, the trend graphs of some test cases were randomly selected from two types of test sets, as shown in Figs. 2 and 3.

As can be seen from Figs. 2 and 3, a normal emotional state of the individual curves are generally fluctuations in the x axis, the majority of the individuals are in a state of pleasure. On the contrary, the emotional curves of individuals with depression tend to be above the x axis; two types of individuals over time curve mood swings are likely to exceed the critical emotional point $f(t) = 0$, which is also a composite of the real situation of human emotional changes.

Table 5. Different students' single microblogging classification accuracy

	Microblog name	Number of microblog	Total	Ac
Normal	盛夏SQH	15	150	0.96
	谭华要做江南小清新	9	224	0.9821
	小疯子白日梦想家	9	159	0.9811
	夜夜夜先森	7	54	0.9629
	真菌菌凹	20	85	0.8118
Depression	我是一把宝剑	3	27	0.9630
	肸肸是笑声	26	103	0.7865
	Arioster	10	58	0.8793
	图拉拉树袋熊	11	42	0.9048
	小桑meng	6	43	0.9767

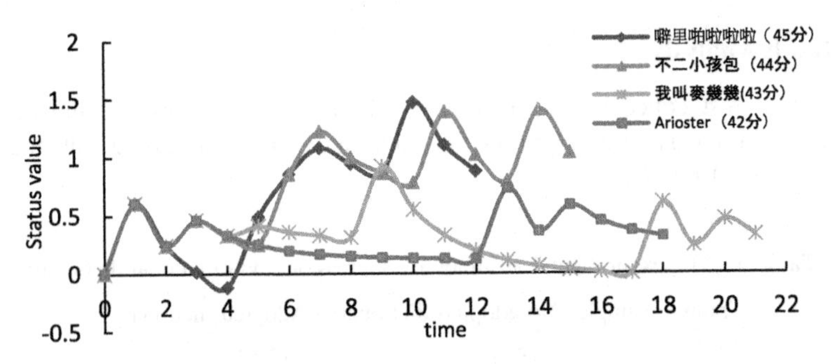

Fig. 2. The emotional trend of students with depression

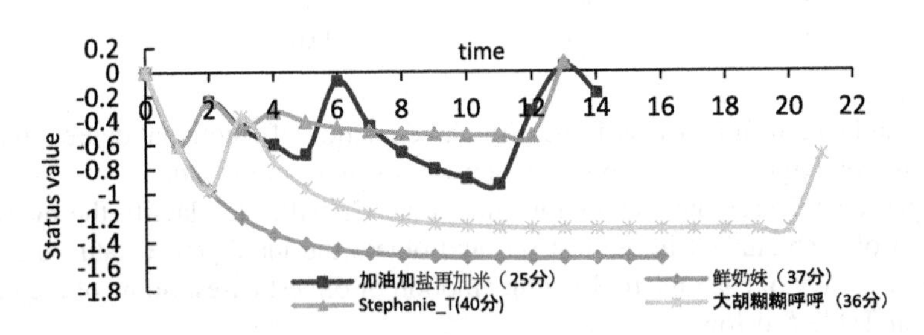

Fig. 3. Normal student's emotional chart

Experiment 3 randomly selected two periods for each test case in 68 test cases, with 15 microblogs as a period. Among them, from 20 depressed tendencies of the students which in the depression trend chart selected 40 time periods, from 48 emotional normal students of which the depression trend chart selected 96 time periods, respectively, using the formulas (7) and (8) to determine. Results are shown in Table 6. Recognition accuracy than the normal emotions of depressive tendencies rate is low, probably due to the students in the normal state that issued microblog contains words related to depression.

Table 6. Test results for the test set

	Total	Correctly identify	Accuracy
Depression tendency	40	34	85.00%
Normal mood	96	80	83.75%
Total	136	114	83.82%

5 Conclusion

Based on the microblogging text, this paper constructs a model of depression and emotion tendency recognition. The experiment shows that the model can identify the tendency of the individual in a period of time. This paper only considers the interaction of depression at adjacent time points, but does not consider the broader time interval factor. At the same time, the value of emotional decay factor is related to various factors such as environment and personality. It is only assumed that the emotional decay meets the half-life. Regularly, follow-up will be the above aspects of research, with a view to a better experimental result.

References

1. Liu, N., He, Y.: Emotional analysis for microblogging essay, pp. 24–91. Wuhan University, Wuhan (2013)
2. Liu, L., Wang, Y., Wei, H.: Fine-grained emotion analysis for product reviews. J. Comput. Appl. **35**(12), 3481–3486 (2015)
3. World Health Organization. http://www.who.int/en/
4. Youn, S.J., Trinh, N.H., Shyu, I., et al.: Using online social media, facebook, in screening for major depressive disorder among college students. Int. J. Clin. Health Psychol. **13**(1), 74–80 (2013)
5. De Choudhury, M., Counts, S., Horvitz, E.: Predicting postpartum changes in emotion and behavior via social media. In: Proceedings of the 2013 ACM Annual Conference on Human Factors in Computing Systems, pp. 3267–3276. ACM (2013)
6. Wang, X., Zhang, C., Ji, Y., et al.: A depression detection model based on sentiment analysis in micro-blog social network. In: Proceedings of the PAKDD-2013 Workshop on Data Analysis for Targeted Healthcare (2013, in press)
7. Qiu, T.-S., Dai, R.-J., Liu, Y.-J.: Study on the Relationship between depression and functional brain region based on restricted FMRI low frequency amplitude. Chin. J. Data Acquis. Process. **30**(5), 940–947 (2015)
8. Li, P., Yu, G.: Microblogging social network in the student depression identification method, pp. 17–60. Harbin Institute of Technology, Harbin (2014)
9. Huang, S., Chen, Y., Zhang, Y.: Magnolia treatment of depression and antidepressant mechanism. World J. Integr. Tradit. W. Med. (07) (2015)
10. Zung, W.W.K., Richards, C.B., Short, M.J.: Self-rating depression scale in an outpatient clinic: further validation of the SDS. Arch. Gen. Psychiatry **13**(6), 508 (1965)
11. Aizawa, A.: An information-theoretic perspective of TF–IDF measures. Inf. Process. Manag. **39**(1), 45–65 (2003)
12. Vapnik, V.: The Nature of Statistics Learning Theory. Springer, New York (1995). doi:10.1007/978-1-4757-2440-0

Intelligent Applications

Traffic Parameters Prediction Using a Three-Channel Convolutional Neural Network

Di Zang[1,2](✉), Dehai Wang[1,2], Jiujun Cheng[1,2], Keshuang Tang[3], and Xin Li[4]

[1] Department of Computer Science and Technology, Tongji University, Shanghai, China
zangdi@tongji.edu.cn
[2] The Key Laboratory of Embedded System and Service Computing, Ministry of Education, Tongji University, Shanghai, China
[3] Department of Transportation Information and Control Engineering, Tongji University, Shanghai, China
[4] Shanghai Lujie Electronic Technology Co., Ltd., Pudong, Shanghai, China

Abstract. Traffic three elements consisting of flow, speed and occupancy are very important parameters representing the traffic information. Prediction of them is a fundamental problem of Intelligent Transportation Systems (ITS). Convolutional Neural Network (CNN) has been proved to be an effective deep learning method for extracting hierarchical features from data with local correlations such as image, video. In this paper, in consideration of the spatiotemporal correlations of traffic data, we propose a CNN-based method to forecast flow, speed and occupancy simultaneously by converting raw flow, speed and occupancy (FSO) data to FSO color images. We evaluate the performance of this method and compare it with other prevailing methods for traffic prediction. Experimental results show that our method has superior performance.

Keywords: Deep learning · Convolutional Neural Network · Traffic prediction · Intelligent Transportation System

1 Introduction

Currently, real time traffic information prediction has got more and more attention of individual drivers, business sectors and governmental agencies with the increasing numbers of vehicles and the development of ITS. Flow, speed and occupancy as three elements of traffic describe the different characteristics of traffic and record the spatiotemporal evolution over a period of time. Accurate prediction of them can facilitate people's travel and reduce traffic congestion and accidents. Meanwhile, it can help traffic managers allocate traffic resources systematically and improve regulatory efficiency.

There exist the spatiotemporal correlations of traffic data due to the consecutive evolution of traffic state on the time-space dimension. It's necessary to retain and leverage inherent spatiotemporal correlations when forecasting traffic information. In addition, there are undoubtedly inner correlations between traffic parameters such as

Z. Shi et al. (Eds.): ICIS 2017, IFIP AICT 510, pp. 363–371, 2017.
DOI: 10.1007/978-3-319-68121-4_39

flow, speed and occupancy within a time unit, most of which have been explained theoretically and mathematically. Intuitively, when traffic flow is high, speed usually won't be low. Therefore, it's reasonable to take these important correlations into account in the modeling to make the prediction more robust.

Existing traffic prediction methods can be mainly divided into three categories: time-series methods, nonparametric methods and deep leaning methods.

Autoregressive moving average (ARIMA) [1] model is one of the representative time-series models, which focuses on finding the patterns of the temporal evolution formation by two steps: moving average (MA) and autoregressive (AR) considering the essential traffic information characteristics. Many variants of this model, such as subset ARIMA, seasonal ARIMA, KARIMA, ARIMAX, were proposed to improve prediction accuracy. However, these models are inept at extracting spatiotemporal feature for prediction because of ignoring spatiotemporal correlations.

Nonparametric methods were widely used because of its advantages, such as their ability to deal with multi-dimensional data, implementation flexibility. In [2], a prediction model was built by support vector regression (SVR) and optimized by particle swarm algorithm. Chang et al. proposed a dynamic multi-interval traffic flow prediction using k-nearest neighbors (KNN) [3]. In addition, various artificial neural network (ANNs) [4, 5] were designed to predict traffic information. Unfortunately, these shallow architectures failed to learn deep features.

Recently, deep learning models such as deep belief network (DBN) [6], recurrent neural network (RNN) [7], and long short-term memory (LSTM) [8] have been used for traffic prediction owe to their excellent capability of extracting complex features and generalizability and strong forecasting performance. However, these models usually put the time and space into same dimension and they violate the two-dimensional basis of spatiotemporal features. CNN has been demonstrated to have excellent performance in large-scale image-processing tasks [9]. Ma et al. [10] first applied CNN to traffic speed prediction by converting network traffic to gray images and achieved significant improvement on average accuracy.

In this paper, we propose a CNN-based method to forecast traffic information from a comprehensive perspective. We design a novel way in which quite complete spatiotemporal features can be extracted from the data by converting raw flow, speed and occupancy (FSO) data to FSO images. Furthermore, we use the features to forecast the traffic flow, speed, occupancy simultaneously. Finally, we demonstrate the performance of the proposed model and compare it with other prevailing methods.

The rest of the paper is organized as follows: In Sect. 2, converting raw FSO data to FSO images and CNN model for traffic prediction are introduced. In Sect. 3, experiments and results are detailed. Finally, conclusions are drawn in Sect. 4.

2 Methodologies

Raw FSO data with space and time dimensions can be integrated and converted to FSO color images as shown in Fig. 1. Then a CNN model based on LeNet-5 architecture is designed to learn the mapping relationship between spatiotemporal features and FSO images. Finally, we compute predictive performance indicators by mapping the learned features back to original FSO data space.

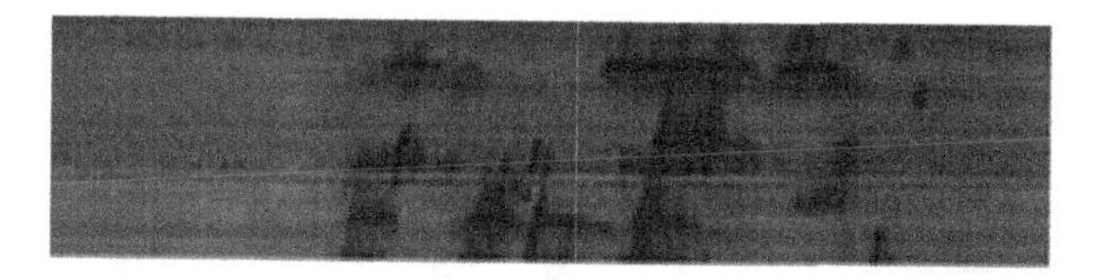

Fig. 1. FSO image converted from FSO data of a day. The area containing purple blocks corresponds to the period when the traffic volume is large i.e. day time whereas the green area corresponds to the opposite side (Color figure online).

2.1 Data Conversion

Traffic data such as flow, speed, occupancy collected by sensors such as inductive loop detectors at a certain time interval, which is usually not more than 5 min, record the evolution of traffic conditions in a particular region.

For the width of the image, time sequences are fitted linearly into the width-axis chronologically. Thus, the length of time interval determines the width of the image. For a 5-min interval, there will be 288 data sequences on the time dimension corresponding to the image width of 288 pixels. In general, short intervals like 5 s are meaningless for traffic prediction. Therefore, in most cases data sequences need to be aggregated to generate available data by dealing with several adjacent time intervals.

For the height of the image, we simply fit the number of sensors ordered spatially into the height-axis. We can also make height-axis compact and informative, in the same way as width dimension, by aggregating data from serval adjacent sensors. Notably, different traffic data may use different aggregation methods. For example, flow use accumulation while speed and occupancy use mean.

Finally, three time-space matrix generated by the method mentioned above are directly merged into FSO color images for flow as green channel, speed as red channel and occupancy as blue channel. A FSO image can be denoted as:

$$P = \begin{bmatrix} p_{11} & p_{12} & \cdots & p_{1N} \\ p_{21} & p_{22} & \cdots & p_{2N} \\ \vdots & \vdots & \ddots & \vdots \\ p_{M1} & p_{M2} & \cdots & p_{MN} \end{bmatrix} \quad (1)$$

where M is the number of sensors, N is the length of time units and pixel p_{ij} is a triple consisting of flow, speed and occupancy.

2.2 CNN Architecture

Figure 2 illustrates the proposed three-channel convolutional neural network model. In this case, traffic data of flow, speed and occupancy are respectively encoded into three channels corresponding to the red, green and blue color channels. These three traffic parameters are then learned by different convolutional kernels to generate feature maps which represent the fusion of traffic information. After the consecutive processing of convolution and pooling, the fully-connected layer maps the extracted traffic features back to the original space for final prediction as the output of the model.

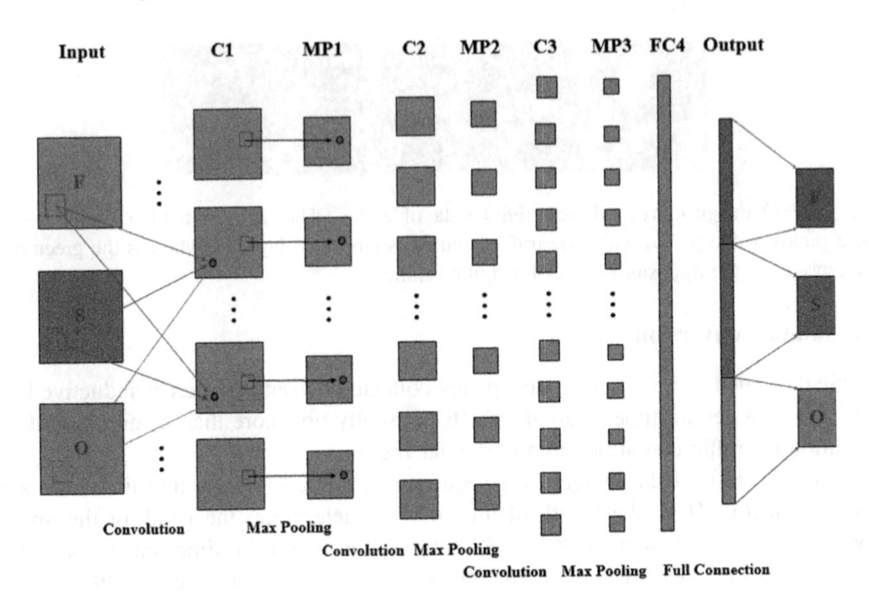

Fig. 2. Proposed CNN model for extracting traffic spatiotemporal features. C, MP and FC represent convolution, max pooling and fully-connected layer, respectively (Color figure online).

2.2.1 Input Layer and Output Layer

First, raw FSO data are converted to FSO images. However, different from CNN models for classification, for an FSO image as input, the output of the model is also a FSO image right next to it. This means that the network not only learns hierarchical features but also learns to map the extracted features back to original space, which is similar to the mechanism of the autoencoder (AE).

Given the lengths of input and output time units as τ and ε, respectively and prediction interval between input and output as π, which is set to 1 in this paper, the input of the model can be written as:

$$x^i = \begin{bmatrix} p_{1,i} & p_{1,i+1} & \cdots & p_{1,i+\tau-1} \\ p_{2,i} & p_{2,i+1} & \cdots & p_{2,i+\tau-1} \\ \vdots & \vdots & \ddots & \vdots \\ p_{M,i} & p_{M,i+1} & \cdots & p_{M,i+\tau-1} \end{bmatrix} \quad (2)$$

where i is the sample index in the range of $[1, N-\tau-\pi-\varepsilon+2]$, and M is defined as in formula (1), $p_{i,j}$ is a triple consisting of flow, speed and occupancy. Accordingly, the predicted FSO image can be written as:

$$y^i = \begin{bmatrix} p_{1,i+\tau+\pi-1} & p_{1,i+\tau+\pi} & \cdots & p_{1,i+\tau+\pi+\varepsilon-2} \\ p_{2,i+\tau+\pi-1} & p_{2,i+\tau+\pi} & \cdots & p_{2,i+\tau+\pi+\varepsilon-2} \\ \vdots & \vdots & \ddots & \vdots \\ p_{M,i+\tau+\pi-1} & p_{M,i+\tau+\pi} & \cdots & p_{M,i+\tau+\pi+\varepsilon-2} \end{bmatrix} \quad (3)$$

2.2.2 Convolutional Layer and Pooling Layer

Convolutional layer plays a major role in extracting the spatiotemporal features. The previous layer's features are convolved with learnable kernels and put through the activation function to form more complex features. Generally, a convolutional layer is usually followed by a pooling layer to reduce parameters of the model and make the learned features more robust. Rectified linear unit (ReLU) activation function and max pooling procedure are used in our model because of their respective superior performance. The output of convolutional and pooling layers can be written as:

$$x_j^l = mp\left(\varphi\left(\sum_{i=1}^{c^{l-1}} x_i^{l-1} * k_{ij}^l + b_j^l\right)\right) \tag{4}$$

where l is the index of layers and j is the index of feature maps in the lth layer, x_i^{l-1}, x_j^l, k_{ij}^l and b_j^l denote the input, output, kernels and bias of the layer, respectively. $*$ indicates convolution operation and mp denotes max pooling procedure. φ is ReLU activation function defined as:

$$\varphi(x) = max(0, x). \tag{5}$$

2.2.3 Fully-Connection Layer

Fully-connection layer provides an effective way to map the spatiotemporal features back to original FSO image space. The output of this layer as predictive value of the model can be written as:

$$\hat{y} = \sigma\left(w^l x^{l-1} + b^l\right) \tag{6}$$

where w^l and b^l are the weights and bias of the layer, respectively. $\hat{y}$ are predicted values, that is, the output of the model. σ is sigmoid activation function defined as:

$$\sigma(x) = \frac{1}{1+e^{-x}} \tag{7}$$

Finally, Mean squared errors (MSEs) are employed as loss function to measure the distance between predictions and ground-truth, which are optimized by mini-batch gradient descent (mini-batch GD) algorithm. MSE can be defined as:

$$MSE = \frac{1}{N}\sum_{i=1}^{N}\left(\hat{y}^i - y^i\right)^2 \tag{8}$$

where $\hat{y}^i$ and y^i are predicted values and true values of the ith sample, respectively, and N is the number of samples.

3 Experiments and Results

In this section, we first detail the generation of samples. Then we display the trained model from different aspects. Finally, we give the experimental results.

3.1 Data Description

The proposed method is evaluated on the data of West Yan'an Road in Shanghai, which were collected every 5 min from 35 individual inductive loops during the whole year of 2012. In fact, there are a total of 361 days of data because of the absence of data from March 20 to March 23. For practical reasons, there are inevitably some missing pieces in the data. Therefore, a proper mending method is applied to the data with spatiotemporal adjacent records. As shown in Fig. 1, the green area from 21:00 pm of previous day to 7:00 am of next day corresponds to the traffic state that there are few vehicles on the road. It is incompatible with the characteristics of CNN model and almost impossible to extract spatiotemporal features because of lacking representational and differentiable patterns when using image patches to train the network. Therefore, the data with obvious traffic patterns from 7:00 am to 21:00 pm of a day are chosen to generate samples.

For the time interval of 5 min as a time unit, there are 168 time sequences from 7:00 am to 21:00 pm. Accordingly, when converting these sequences to FSO images, the width of the images will be 168. As a result, there are 361 FSO images of $35 \times 168 \times 3$ available for intercepting image patches as samples. When the lengths of input and output are set to τ and ε, respectively, which means using 5τ-min FSO data to forecast next 5ε-min FSO data, the number of the image patches of a day will be $168 - \tau - \varepsilon + 1$. In total, there are $361 \cdot (168 - \tau - \varepsilon + 1)$ samples. First 90% of them are used as training data and the rest are used as test data.

3.2 Model Display

The model is designed based on LeNet-5 architecture. The parameters of the model as listed in Table 1 are set based on the principle that the network converges to a better solution and that the train time of the network is acceptable (Fig. 3).

Table 1. Parameters of the model.

Layer	Parameter dim	Feature dim	Parameter num
Input	–	(3, 35, 35)	–
C1	Kernel (8, 3, 5, 5)	(8, 31, 31)	608
MP1	Pooling (2, 2)	(8, 16, 16)	–
C2	Kernel (16, 8, 3, 3)	(16, 14, 14)	1168
MP2	Pooling (2, 2)	(16, 7, 7)	–
C3	Kernel (32, 16, 3, 3)	(32, 5, 5)	4640
MP3	Pooling (2, 2)	(32, 3, 3)	–
FC4	Weight (288, 105)	105	147968
Output	–	105	–

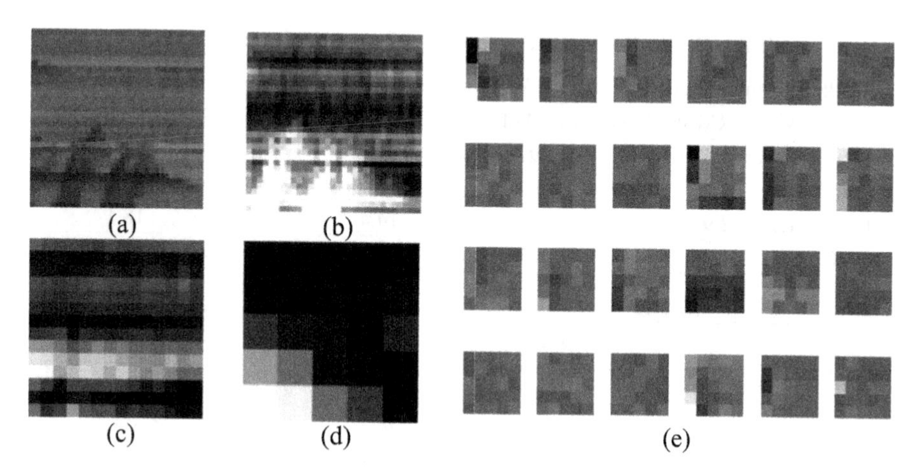

Fig. 3. (a): FSO input image. (b): Feature map of the first convolutional layer. (c): Feature map of the second convolutional layer. (d): Feature map of the third convolutional layer. Hierarchical spatiotemporal features of FSO data from simple to complex and concrete to abstract are extracted automatically through the model. (e): Kernels of the first convolutional layer. The model has learnt homogeneous kernels through training, which is consistent with the characteristics of traffic data.

3.3 Results and Comparison

We first evaluate the proposed method for short-term traffic prediction with the lengths of input and output are set to 35 and 1, respectively. This means we use 175-min FSO data to forecast next 5-min data. As mentioned in Sect. 3.1, 43211 training samples are used to train the model and 4802 test samples are used to validate the model. We compare three indexes: mean absolute errors (MAE), mean relative errors (MRE) and relative mean square errors (RMSE) with other prevailing methods for short-term traffic prediction: ANN, DBN and random work (RW). These methods are all optimized by mini-batch SGD algorithm with batch size set to 300 except RW. Results show that our model has the best average performance as shown in Table 2 (AVG indicates average value). Notably, under the same index, three channels of flow, speed and occupancy differ due to their different data fields ranging from 10 to 1017, 1.0 to 116.0 and 1.1 to 98.7, respectively.

Table 2. Prediction performance of CNN and other models on test data.

TYPE	CNN			ANN			DBN			RW		
	MAE	RMSE	MRE (%)	MAE	RMSE	MRE (%)	MAE	RMSE	MRE	MAE	RMSE	MRE (%)
F	23.8	31.7	8.3	24.1	32.0	8.6	24.3	32.9	8.4	27.5	39.1	8.2
S	3.0	4.4	5.6	3.3	4.6	6.0	3.0	4.4	5.7	3.8	6.2	8.7
O	1.7	2.9	14.3	1.8	2.9	14.9	1.8	2.9	15.4	2.4	4.4	14.3
AVG	9.5	18.6	9.4	9.7	18.7	9.8	9.7	19.2	9.8	11.2	23.0	10.4

As shown in Table 3, we further evaluate the performance of our model on forecasting longer time span when setting the input length to 35 i.e. 175 min. From the table we can see that, with the increase of time span, the prediction accuracy is decreasing but within the acceptable range. This is due to the fusion of three-channel information which can effectively filter out influences of the outliers in channels.

Table 3. Prediction performance of CNN on different time span.

TYPE	5-min prediction			15-min prediction			30-min prediction		
	MAE	RMSE	MRE (%)	MAE	RMSE	MRE (%)	MAE	RMSE	MRE (%)
F	23.8	31.7	8.3	26.1	34.9	9.0	25.8	33.9	9.2
S	3.0	4.4	5.6	3.3	4.9	6.2	3.2	4.6	6.1
O	1.7	2.9	14.3	1.8	3.0	14.4	2.1	3.2	18.1
AVG	9.5	18.6	9.4	10.4	20.4	9.9	10.4	19.8	11.2

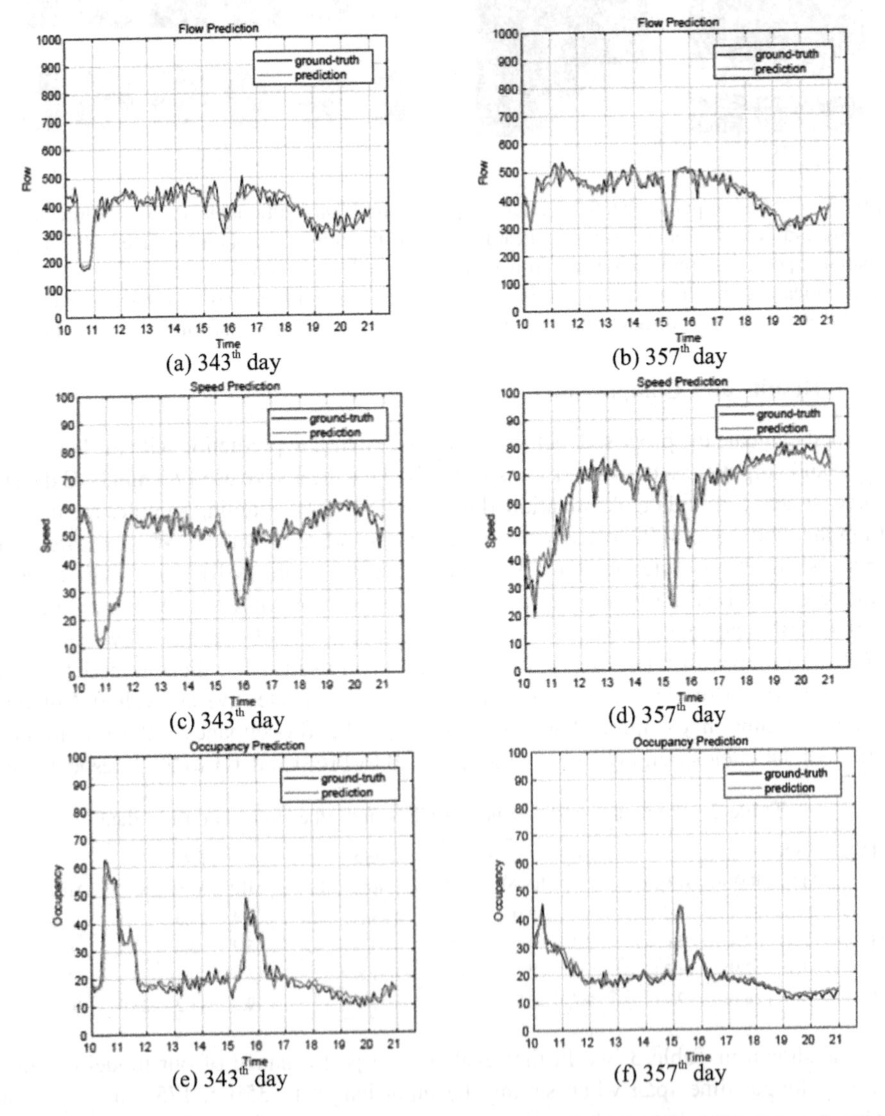

Fig. 4. (a), (c) and (e) are the flow, speed and occupancy predictions of the 343th day at the 26th loop, respectively. (b), (d) and (f) are the flow, speed and occupancy predictions of the 357th day at the 22th loop, respectively.

Finally, we present the prediction curves of the 343th day at the 26th loop and the 357th day at the 22th loop from 10:00 am to 21:00 pm as shown in Fig. 4.

4 Conclusion

In this paper, we proposed a CNN-based method to forecast traffic information. Unlike most single prediction models, traffic flow, speed and occupancy are simultaneously predicted by the model to provide more complete traffic information. Spatiotemporal traffic features of traffic data can be learned automatically by converting flow, speed, and occupancy data to color images as the input of the model.

We evaluated the proposed method on the data of West Yan'an road in Shanghai and compared it with ANN, DBN and RW methods. And results show that the proposed method is superior to others. In addition, we explored the prediction performance on different tasks of 5-min prediction, 15-min prediction and 30-min prediction and results show that proposed model has certain robustness as the prediction accuracy descends but within an acceptable range.

Acknowledgments. This work has been supported by the Fundamental Research Funds for the Central Universities of China and by National Natural Science Foundation of China under grant 61472284.

References

1. Ahmed, M.S., Cook, A.R.: Analysis of freeway traffic time-series data by using Box-Jenkins techniques. Transp. Res. Rec. **722**, 1–9 (1979)
2. Jeong, Y.S., Byon, Y.J., Castro-Neto, M.M., Easa, S.M.: Supervised weighting-online learning algorithm for short-term traffic flow prediction. IEEE Trans. Intell. Transp. Syst. **14**(4), 1700–1707 (2013)
3. Chang, H., Lee, Y., Yoon, B., Baek, S.: Dynamic near-term traffic flow prediction: system oriented approach based on past experiences. IET Intell. Transp. Syst. **6**(3), 292–305 (2012)
4. Chan, K.Y., Dillon, T.S., Singh, J., Chang, E.: Neural-network-based models for short-term traffic flow forecasting using a hybrid exponential smoothing and Levenberg–Marquardt algorithm. IEEE Trans. Intell. Transp. Syst. **13**(2), 644–654 (2012)
5. Kumar, K., Parida, M., Katiyar, V.K.: Short term traffic flow prediction for a non urban highway using artificial neural network. Proc. Soc. Behav. Sci. **104**, 755–764 (2013)
6. Huang, W., Song, G., Hong, H., Xie, K.: Deep architecture for traffic flow prediction: deep belief networks with multitask learning. IEEE Trans. Intell. Transp. Syst. **15**(5), 2191–2201 (2014)
7. Ma, X., Yu, H., Wang, Y., Wang, Y.: Large-scale transportation network congestion evolution prediction using deep learning theory. PLoS ONE **10**(3), e0119044 (2015)
8. Tian, Y., Pan, L.: Predicting short-term traffic flow by long short-term memory recurrent neural network. In: 2015 IEEE International Conference on Smart City/SocialCom/SustainCom (SmartCity), pp. 153–158. IEEE (2015)
9. Krizhevsky, A., Sutskever, I., Hinton, G.E.: ImageNet classification with deep convolutional neural networks. In: Advances in Neural Information Processing Systems (2012)
10. Ma, X., Dai, Z., He, Z., Ma, J., Wang, Y., Wang, Y.: Learning traffic as images: a deep convolution neural network for large-scale transportation network speed prediction. Sensors **17**(4), 818 (2017)

Research of the Evaluation Index System of Green Port Based on Analysis Approach of Attribute Coordinate

Xueyan Duan[1(✉)], Xiaolin Xu[2], and Jiali Feng[3]

[1] Department of Economic and Management, Shanghai Polytechnic University, Shanghai 201209, China
xyduan@sspu.edu.cn

[2] Department of Electronics and Information, Shanghai Polytechnic University, Shanghai 201209, China
xlxu@sspu.edu.cn

[3] Information Engineering College, Shanghai Maritime University, Shanghai 201306, China
jlfeng@189.cn

Abstract. With the coming of low-carbon era the concept of green port becomes more and more important due to the occurrence of abnormal global environmental change. Founded on the previous literatures, the Analysis Approach of Attribute Coordinate is used to analyze evaluation of green port. An index system including three attributes and ten factors is put forwarded. The quantification methods of different indicators are also proposed according to the different property and scoring methods of indicators. The green port evaluation index system can be used to evaluate the port's green operation performance. It is also a good instrument to construct a green port operation for decision makers of port organizations.

Keywords: Green port · Analysis Approach of Attribute Coordinate · Index system · Quantification methods

1 Introduction and Literature Review

Port is an important infrastructure and strategic resource in the economic and social development. In the low-carbon economy era Green Port development has become the important field of sustainable development in the world transportation today. How to boost port service capabilities, as well as achieve green sustainable development, has important strategic significance for each port. During green port development, science evaluation and policy guidance is the key point.

In recent years, the domestic and foreign scholars have carried out a large number of research on green port evaluation.

Stone [1] studies environmental pollution problems in port construction.

Knight [2], Brooke [3] and Peris-Moraa and Orejas [4] build the evaluation index system of port based on the concept of green and ecological.

Z. Shi et al. (Eds.): ICIS 2017, IFIP AICT 510, pp. 372–378, 2017.
DOI: 10.1007/978-3-319-68121-4_40

Shao et al. [5] establishes ecological port evaluation index system based on Driver-Pressure-State-Impact-Response (DPSIR) model.

Ge [6] builds a comprehensive evaluation index system of ecological port.

Ling [7] introduces the evaluation index system of green port taking Shanghai port as an example.

Essence [8] analyses factors and performance of green port and put forward an index system including five dimensions and thirteen factors based on Fuzzy AHP model.

Ouyang [9] discusses the connotation and features of green and low-carbon port and establishes an evaluation index system including four categories of indexes.

However existing studies somewhat have some difficulties in indicator scoring and do not take the experts' psychological preferences and evaluation process into account. The Analysis Approach of Attribute Coordinate can make up for this deficiency. Therefore, this article attempts to construct a green port evaluation index system based on the Analysis Approach of Attribute Coordinate in order to provide research ideas for related research.

2 Analysis Approach of Attribute Coordinate

According to the Attribute Theory, everything that exists objectively has two kinds of stipulation: quality and quantity. Therefore, the attributes of things also have two characteristic values of quantity and quality. The sensory feature extraction system is a converter that transforms the quantitative characteristics x of sensory attributes $a(o)$ into their corresponding quality characteristics $p(x, o)$. The Attribute Theory calls this process Qualitative Mapping (QM). Conversely, it is Converse Qualitative Mapping (CQM) (see as Fig. 1). The definition of QM and CQM can be extended from 1 dimension to n dimension.

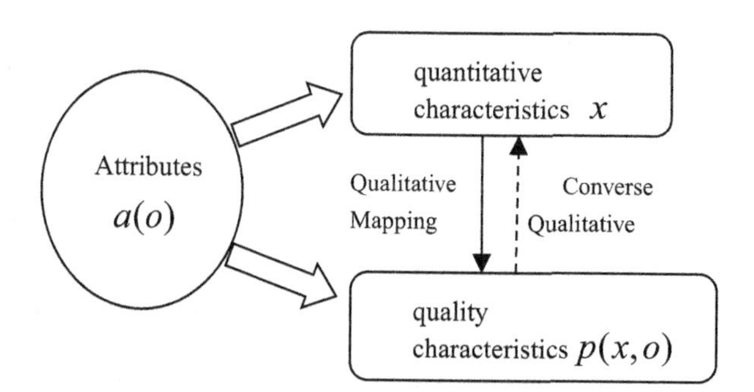

Fig. 1. QM and CQM model of attribute theory

For human assessment and decision making, different decision attributes have different decision weights (or preferences). An evaluation and decision model called Analysis Approach of Attribute Coordinate is proposed [11]. The Analysis Approach of Attribute Coordinate reflects the psychological standard of the decision-maker and the process of evaluating. It considers the evaluation criteria for nonlinear problems. The nonlinear change of judgment criterion is tracked dynamically by human-computer interaction in order to make appropriate evaluation decisions. It is more consistent with the nonlinear characteristics of human decision-making process.

Meanwhile the Analysis Approach of Attribute Coordinate defines a mental weight which is easier to express psychological preference of decision makers. The meaning of mental weight is the score of corresponding decision attributes when total score equals a certain value. The psychological weight is directly linked to the psychological preference of the decision maker, and the distribution is the most reasonable or satisfactory when corresponding to the mental weight of the decision maker.

In recent years the Analysis Approach of Attribute Coordinate is applied to the emergency decision-making evaluation system of college entrance examination [12], the evaluation of third-party logistics enterprises [13], the evaluation of Chinese Software Enterprises [14] and the evaluation of bullwhip effect in supply chain [15]. These applications have achieved very good results.

3 Attribute Characteristics of Green Port

Green port means the organic combination of port development and ecological environment protection. Through rational utilization of resources, ports can achieve low energy consumption and low pollution, and achieve harmonious and unified development between people and ports and the environment.

On the one hand, ports need to the scientific integration of existing resources and to ensure the improvement of port development speed; on the other hand, ports also need to pay attention to the quality and efficiency of development, low resource consumption, less environmental pollution, growth advantages and scale effect of the road of sustainable development.

As a systematic concept, green port can deepen its understanding of the three attributes of economy, society and environment (see as Fig. 2).

(1) Economic attribute: The economic attribute is the basic index in the evaluation index system of green port, which can fully reflect the economic development level of a port. Green ports also seek an increase in economic returns. The characteristics of port economic benefits can be drawn from the indexes of port GDP, profit and tax ratio, input output ratio, etc.
(2) Society attribute: On the premise of ensuring proper economic development social attribute of port is another important criterion for measuring green ports. The port should be able to play an important role in promoting regional social development, promoting the employment of the population in the port area and improving the quality of the population near the port area.
(3) Environment attribute: In essence, the development of green port is resource-efficient and environment-friendly. Resource-efficiency involves conservation of energy, land, coastline and materials, while the environment-friendliness involves objects such as atmosphere, water, sound and ecology.

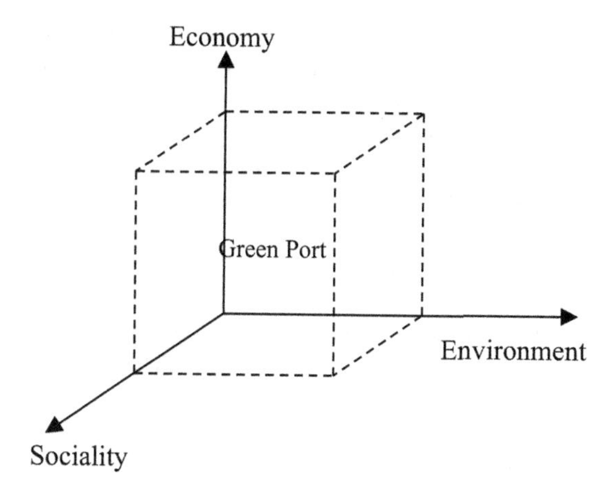

Fig. 2. Three attributes of green port

4 Evaluation Index System of Green Port

According to the principles of the index system and the objectives of this study, we establish a recursive hierarchy evaluation index system of attribute coordinate. Evaluation index system of green port is set up including target layer, criterion layer and indicator level. At criterion layer we divide green port system into three dimensions which are economic benefit, social efficiency and ecological environment. Each dimension separately includes several specific indicators. In all there are 10 evaluation indicators.

The index system of green port is shown in Table 1.

Table 1. The evaluation index system of green port

Green port (U)	Economical benefit (u1)	Port GDP (u11) Investment profit rate (u12) Input-output ratio (u13)
	Social efficiency (u2)	Cargo throughput (u21) Passenger throughput (u22) Port employment (u23)
	Ecological environment (u3)	Qualified rate of wastewater discharge (u31) Air quality qualification rate in harbor area (u32) Qualified rate of noise pollution in port area (u33) Green degree of port area (u34)

5 Indicator Quantification Method

In order to carry out the assessment, we must first quantify the 10 indicators of evaluation index system of green port, that means to convert each index value to a standard uniform score. According to Analysis Approach of Attribute Coordinate 100 point system is adopted.

Qualitative indicators are quantified by Converse Qualitative Mapping (CQM).

Let $a(0)$ be an attribute of something (or object). x is a quantity characteristic for $a(0)$, $x \in \mathrm{X} \subseteq \mathrm{R}$. $p_i(0)$ is a quality characteristic for $a(0)$, $p_i(0) \in p_0$.

$\mathrm{A} = (\alpha_i, \beta_i) \in \Gamma$ is qualitative benchmark (or "degree") of $p_i(0)$.

So we call $\mu : X \times \Gamma \rightarrow p_0$ is the Qualitative Mapping (QM) of x based on (α_i, β_i) which can be written into: $\mu : X \times \Gamma \rightarrow \{0, 1\}_i$, making:

$$\mu(x, (\alpha_i, \beta_i)) = x \bot (\alpha_i, \beta_i) = \mu_i(x).$$

In consideration of $[\alpha_i, \beta_i]$ is a topological neighborhood with $\xi_i = (\alpha_i + \beta_i)/2$ as the center and $\delta_i = (\beta_i - \alpha_i)/2$ as the radius, the QM can also be written into: $\mu : X \times \Gamma \rightarrow \{0, 1\}_i$, making:

$$\mu(x, N(\xi_i, \delta_i)) = x \bot N(\xi_i, \delta_i) = \mu_i(x).$$

Then some qualitative mappings that can be transformed to the Converse Qualitative Mapping (CQM) can be written.

When an expert evaluates a qualitative indicator, the evaluation results are divided into four grades which correspond to four ranges of scores. "Excellent" corresponds to [90–100], "Good" corresponds to [80–90], "General" corresponds to [60–80] and "Inferior" corresponds to [0–60]. Meanwhile experts also need to give the degree of evaluation to the extent of the range.

If the evaluation results are "excellent", "good" and "general", Formula (1) is used to obtain the quantization results.

$$x(\mu_i) = \beta_i + \delta_i(\mu_i - 1) \tag{1}$$

If the evaluation results are "Inferior", Formula (2) is used to obtain the quantization results.

$$x(\mu_i) = \alpha_i + \delta_i(1 - \mu_i) \tag{2}$$

The quantitative indicators that are percentage are quantified by linear interpolation.

If the indicator are positive when the index value is bigger it is the better, Formula (3) is used to obtain the quantization results.

$$p = f(i) + \frac{f(j) - f(i)}{j - i}[x - f(i)] \tag{3}$$

If the indicator are negative when the index value is smaller it is the better, Formula (4) is used to obtain the quantization results.

$$p = f(j) + \frac{f(j) - f(i)}{j - i}[x - f(j)] \tag{4}$$

The Indicator Property and Quantification Method are shown in Table 2.

Table 2. Indicator property and quantification method

Indicator	Property	Quantification method
U11	Positive	Expert evaluation
U12	Positive	Linear interpolation method
U13	Negative	Linear interpolation method
U21	Positive	Expert evaluation
U22	Positive	Expert evaluation
U23	Positive	Expert evaluation
U31	Positive	Linear interpolation method
U32	Positive	Linear interpolation method
U33	Positive	Linear interpolation method
U34	Positive	Linear interpolation method

6 Conclusion

Ports play an important role in economic development. The evaluation of green ports is of great significance to the development of the world ports. Based on the Analysis Approach of Attribute Coordinate, economic attributes, social attributes and environmental attributes of green port is deeply analyzed. The evaluation index system of the green port is constructed and the corresponding indicator properties and quantization method are put forward. The index system of green ports can be applied to the objective evaluation of the current situation of the development of green ports, so as to guide and promote the construction and development of green ports all over the world. In the follow-up research, the index system can be used in the evaluation of specific ports, and the practical research can also be improved on the basis of the characteristics of the development trend of green ports.

References

1. Stone, J.H.: Environmental impact of a super port in the Gulf of Mexieo. Inst. Chem. Eng. Publ, Water **75**, 1217–1229 (1975)
2. Knight, K.D.: Conceptual ecological modeling and interaction matrices as environmental assessment tools in coastal planning. Water Sci. Technol. **16**(3–4), 559–567 (1983)
3. Brooke, J.: Environmental appraisal for ports and harbours. Doek Harb. Auth. **71**(820), 89–94 (1990)
4. Peris-Mora, E., Orejas, J.M.D., Subirats, A.: Development of a system of indicators for sustainable port management. Mar. Pollut. Bull. **50**(12), 1649–1660 (2005)
5. Shao, C., Ju, M., He, Y., Sun, X.: Study on index system of eco-ports based on DPSIR model. Mar. Environ. Sci. **28**(3), 333–337 (2009). (in Chinese)

6. Ge, Z., Zhou, X., Cheng, J., Chen, B., Wang, T., Wang, K.: Preliminary study on the comprehensive evaluation index system for ecological ports—Shanghai port as the example. Resour. Environ. Yangtze Basin **03**, 329–335 (2008). (in Chinese)
7. Ling, Q.: The Preliminary study of evaluation index system for the Shanghai green port. Sci. Technol. Ports **1**, 4–7 (2010). (in Chinese)
8. Chiu, R.H., Lin, L.H., Ting, S.C.: Evaluation of green port factors and performance: a fuzzy AHP analysis. Math. Probl. Eng. **5**, 1–12 (2014)
9. Ouyang, B., Wang, L., Huang, J., Gao, A.: Research and application of green and low-carbon port evaluation index system. Port Waterw. Eng. **4**, 73–80 (2015). (in Chinese)
10. Yin, Y., Zhang, Q.: Study on construction level evaluation of ecological ports based on D-S evidence theory. Logistics Technol. **35**(3), 50–54 (2016). (in Chinese)
11. Wu, Q., Feng, J., Dong, Z., Zhang, Y.: A kind of evaluation and decision model based on analysis and learning of attribute coordinate. J. Nanjing Univ. (Nat. Sci.) **39**(2), 182–188 (2003). (in Chinese)
12. Xie, X., Liu, J., Lu, Q.: Admissions decisions in the university entrance exam system. J. Guilin Eng. **21**(4), 402–406 (2001). (in Chinese)
13. Duan, X., Liu, Y., Xu, G.: Evaluation on 3PL's core competence based on method of attribute theory. J. Shanghai Marit. Univ. **27**(1), 41–43 (2006). (in Chinese)
14. Xu, G., Liu, Y., Feng, J.: Evaluation of Chinese software enterprises' core competence in attribute theory. J. Guangxi Nor. Univ. (Nat. Sci. Ed.) **24**(4), 34–37 (2006)
15. Li, J., Liu, Y., Feng, J.: Comprehensive assessment & optimized selection of supplier in scm based on attribute theory. Logistics Technol. **26**(5), 75–79 (2007). (in Chinese)

Two Stages Empty Containers Repositioning of Asia-Europe Shipping Routes Under Revenue Maximization

Hengzhen Zhang[1(✉)], Lihua Lu[2], and Xiaofeng Wang[1]

[1] College of Information Engineering, Shanghai Maritime University, Shanghai, China
{zhanghz,xfwang}@shmtu.edu.cn

[2] School of Electronic Information Engineering, Shanghai Dianji University, Shanghai, China
lulihua@sdju.edu.cn

Abstract. The problem of empty containers repositioning (ECR) is to dispatch the empty containers coupled with laden containers transportation flow between the surplus ports and the deficit ports according to the fixed schedule. Because this problem involves many parameters, variables and constraints, how to build and solve the model is always focus point. Aiming at this problem, firstly one loop itinerary of liner is divided into several stages to analyze the changes of two typical containers located in the ports and on board with the time evolving. Secondly, for attacking the difference between supply and demand empty containers in preview stage t and the ones in next stage t + 1, one dynamic region-to-region ECR model, which adopts the dynamic across-region port set to port set redistribution strategy, is proposed to reduce the various possible costs which commonly exist in existing static port-to-port ECR policy. Finally, through some deductive instances and contrast model we analyze these two models. Results show that in any case our proposed model has absolutely more advantages than the one who uses the static port-to-port ECR strategy. Moreover, the former has significant information support for making the safety stock of ports.

Keywords: Empty containers repositioning · Laden containers transportation · Across-region · Revenue maximization · Dynamic programming

1 Introduction

Containers transportation has increasingly become popular in international trade transportation activities in the last few decades. As various reasons consisting of the history, political and economy the world trades are getting more imbalanced in recent years especially the Trans-Pacific and Asia-Europe shipping routes [1]. The growth of containerized shipping has presented challenges inevitably, in particular to the management of empty containers arising from the highly imbalanced trade between continents. Some ports have accumulated many surplus empty containers, which are called the surplus ports. At the same time, other ports (deficit ports) need many empty

Z. Shi et al. (Eds.): ICIS 2017, IFIP AICT 510, pp. 379–389, 2017.
DOI: 10.1007/978-3-319-68121-4_41

containers to load the cargoes and have to lease some ones to meet the customers demand. For example, the Europe region can be regarded as surplus region while the Asia is deficient region. Under such imbalanced situation, efficiently and effectively repositioning empty containers by using the residual vessel space has become an important strategy to fortify the competitive market of liner company.

Basically, the liner shipping has fixed service route with a number of vessels deployed in fixed schedules with the weekly service frequency. Reasonable arrangement plan of empty and laden containers not only guarantee the arrival of laden ones on time but ensure to reposition empty containers as possible as reducing the operation cost. However the factors influencing the ECR policy are rather complex including the sail schedule, various costs, laden containers, vessel capacity etc. So this problem is a typical combinational optimization problem.

2 Literature Review

There have been many literatures related to the ECR. Crainic et al. [3] consider the factors of long-term leasing containers for attacking the dynamic random ECR. Cheung and Chen [4] model the ECR as a two-stage stochastic optimization model. They propose a time-space network model which is the opening the maritime ECR network modeling for discussion. Lam et al. [5] apply the actual service schedule so that the general networking techniques to the shipping industry can be developed. Bell et al. [6] focus on the assignment of laden and empty containers over a given shipping service network. Erera et al. [7] and Brouer et al. [8] confirmed the economic benefits of simultaneously considering laden and empty containers when modeling cargo allocation in a shipping network. Both above literatures do not consider the asynchronism between planning repositioning and actual repositioning. Song and Dong [9] formulate the problem of ECR for general shipping service routes based on container flow balancing. Imai et al. [10] and Meng and Wang [11] are the only two papers found in the related literatures that explicitly consider ECR decisions together with shipping network design or ship routing, in which the shipping routes are limited within a few pre-specified options. To incorporate uncertainties in the operational model, Long et al. [12] formulate a two-stage stochastic programming model with random demand, supply, ship weight, and ship capacity.

To solve long-haul journey of ECR, we firstly divide one circle journey into several stages to attack the asynchronism of the planned repositioning and actual repositioning. Through analyzing the numerical changes of empty and laden containers in ports and on board in different stages, we can dynamically redistribute the empty containers to meet the current actual demand of different ports. Then the dynamic region-to-region ECR (DRR-ECR) optimization model is built. Figure 1 shows the details of liner shipping, ports and regions, where the circle represents the port, the directed arc the two successive visited ports, and the rectangle the surplus or deficit region.

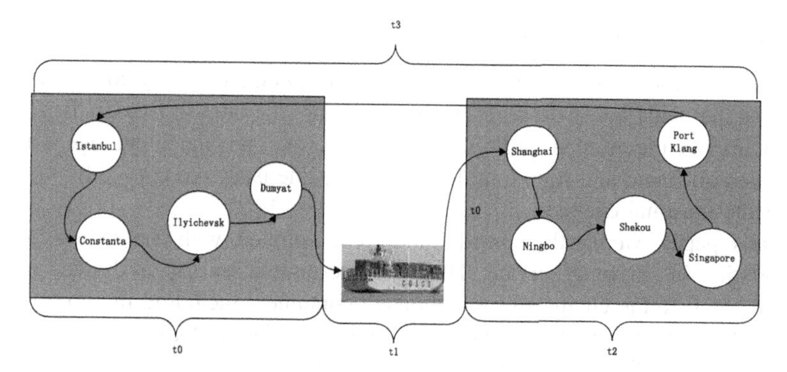

Fig. 1. Shipping line between Europe and Asia.

2.1 Phase t_0: Planned Stage

Since the Asia is the deficit region, it sends the total number of demand empty containers of all ports to the control center in stage t_0. Then each port in Europe firstly loads the laden containers to the vessel according to the control center. The number of loading empty containers of each port equals to the minimum value of demand ones of Asia, supply ones of Europe and residual vessel space in t_0. Meanwhile the number of containers in each leg of liner shipping should be less than the vessel space. In addition, the empty containers in surplus ports which do not temporarily be transported to the Asia or other ports will incur the storage costs.

2.2 Phase t_1: On Board Stage

After the liners successively visit the ports in Europe, they will sail into stage t_1. In this stage, the numbers of empty and laden containers on board are unchanged until the liners reach the first port of Asia: Shanghai. But note that the demand or supply empty containers in all ports are constantly changing with the time evolving because the other liners may visit the corresponding ports. This will cause the changes of ports states and their number of empty and laden containers.

2.3 Phase t_2: Dynamic Repositioning Stage

When the liners arrive the first port of Asia in t_2, it can observe that the actual states of ports have changed compared with the corresponding ones in t_0. It is mainly caused by the unpredictable demand/supply in ports. For the laden containers, which reach the destination ports, they will be discharged and release their space. For the empty containers, the control center will dynamically redistribute and discharge them to meet the actual demands. Moreover, the vessels also load the laden containers, which will be shipped to their corresponding destination ports in Europe.

2.4 Phase t_3: Return Trip Stage

The liners visit each port in Asia and discharge the laden and empty containers as well as loading the laden ones whose destination ports are the ones in Europe. In general, it does not involve the problem of ECR since the Asia region is the deficit one. The above four phases go round and round until the shipping line is altered caused by the tactical level or the seasonal change.

In this paper, we firstly consider the ECR coupled with the laden containers transportation in shipping service. Then the one circulation trip is divided into four phases to analyze the changes of empty/laden containers including the ports and vessels. Finally a maximized revenue optimization model: DRR-ECR is proposed to dynamically solve the empty/laden containers transportation problem.

The remainder of this paper is organized as follows. In the next section, the ECR and laden containers transportation problem are formulated. In the subsequent section, it presents the experimental results. Finally, conclusions are drawn and future researches are derived.

3 Problem Formulation

3.1 Problem Description

In this paper suppose that all the proposed models are subject to the following assumptions. The schedules of services are given and fixed in advance. In the different stages, the numbers of deficit/surplus empty containers in all ports have been dealt with by the branch lines and they refer to the integrate ones. In t_0, the number of surplus/deficit empty containers in their corresponding regions is also known a prior according to the history data of ports. There is no limit on the number of leasing containers for each port in any moment. All the containers are measured in TEU. The demand empty containers must be satisfied. All costs are calculated in dollars.

3.2 Parameters

To simplify the narrative, the following notations are introduced to formulate the containers transportation problems.

P: set of ports within the scope of research, $P = PE \cup PA$.
$PE/\ PA$: set of Europe/Asia ports within the scope of research.
i: port identifier in Europe region, $i \in \{1, 2, \ldots, |PE|\}$.
j: ports identifier in Asia region, $j \in \{1, 2, \ldots, |PA|\}$.
S: super port which refers to the liner; t_0, t_1, t_2, t_3: time stage.
$D_{j,t}$: demand empty containers in j and t, $t \in \{t_0, t_1, t_2, t_3\}$.
$S_{i,t}$: supply empty containers in i and t, $t \in \{t_0, t_1, t_2, t_3\}$.
$LadenTrans_{ij,t}$: number of transportation laden containers from i to j in t.
$CE_i^l/CL_i^l/CL_i^u/CE_i^r/CE_i^s/$: loading/discharging/loading/leasing/storage cost of unit empty container in i (Unit: \$/container/time).
CE_{ij}^{tr}: transportation cost of unit empty containers from i to j (Unit: \$/container).

CE^{tr}_{iS}: transportation cost of unit empty containers from i to S.
CL^{tr}_{ij}: transportation cost of unit laden containers from i to j.
PL^{tr}_{ij}: profit of unit laden container from i to j.
Capacity: the vessel capacity.

3.3 Decision Variables

According to the problem, we give the decision variables as follows.

$trLaden_{ij,t_0}$: number of transportation laden containers from i to j in t_0
$trLaden_{ji,t_3}$: number of transportation laden containers from j to i in t_3.
$trEmp_{iS,t_0}$: number of transportation empty containers from i to super port S.
$trEmp_{Sj,t_2}$: distribution number of empty containers from the liner to j in t_2.
lA_{j,t_2}: number of leasing empty containers in Asia port j and in t_2.
sE_{i,t_0}: number of storage empty containers in Europe port i and in t_0.
sA_{i,t_2}: number of storage empty containers in Asia port i and in t_2.

3.4 Objective Function

The objective function of our paper is to maximize the total profit minus the various costs. The formal model is described in (1), which is subject to constraints (2)–(12).

maximize

$$\sum_i \sum_j trLaden_{ij,t_0} * PL^{tr}_{ij} + \sum_j \sum_i trLaden_{ji,t_3} PL^{tr}_{ji} - \sum_i \sum_j trLaden_{ij,t_0} * (CL^{tr}_{ij} + CL^{l}_{i} + CL^{u}_{j})$$
$$- \sum_j \sum_i trLaden_{ji,t_3} * (CL^{tr}_{ji} + CL^{l}_{j} + CL^{u}_{i}) - \sum_i trEmp_{iS,t_0} * (CE^{tr}_{iS} + CE^{l}_{i})$$
$$- \sum_i sE_{i,t_0} * CE^{s}_{i} - \sum_j trEmp_{Sj,t_2} * (trEmp_{Sj} + CE^{l}_{j}) - \sum_j sA_{j,t_2} * CE^{s}_{j} - \sum_j lA_{j,t_2} * CE^{l}_{j}; \quad (1)$$

$$trEmp^{tr}_{iS,t_0} \le S_{i,t_0}; \quad (2)$$

$$\sum_i trEmp^{tr}_{iS,t_0} \le \min(\sum_i S_{i,t_0}, \sum_j D_{j,t_0}); \quad (3)$$

$$\sum_i trEmp^{tr}_{iS,t_0} = \sum_j trEmp_{Sj,t_2}; \quad (4)$$

$$sE_{i,t_0} = S_{i,t_0} - trEmp_{iS,t_0}; \quad (5)$$

$$\sum_j sA_{j,t_2} = \sum_i trEmp^{tr}_{iS,t_0} - \sum_j trEmp_{Sj,t_2} \quad (6)$$

$$lA_{j,t_2} = \max((D_{j,t_2} - trEmp^{tr}_{Sj,t_2}), 0) \quad (7)$$

$$\sum_i \sum_j trLaden_{ij,t_0}^{tr} \leq Capactity \tag{8}$$

$$\sum_j \sum_i trLaden_{ji,t_3}^{tr} \leq Capactity \tag{9}$$

$$\sum_i trEmp_{iS,t_0}^{tr} \leq Capactity - \sum_i \sum_j trLaden_{ij,t_0}^{tr} \tag{10}$$

$$trLaden_{ij,t_0}^{tr} \leq LadenTrans_{ij,t_0} \tag{11}$$

$$trLaden_{ji,t_3}^{tr} \leq \sum_j \sum_i LadenTrans_{ji,t3} \tag{12}$$

The first two terms in (1) represent the profits between Europe and Asia. The next two terms describe the corresponding transportation costs, loading costs and discharging costs. The fifth and sixth terms show various costs in Europe region. The last three terms explain the cost of redistribution empty containers from liner to the ports of Asia, storage cost of additional ones and leasing cost of deficit ones.

4 Experimental Results and Analysis

4.1 Dataset

To evaluate DRR-ECR, one case, which simulates the shipping company business, is conducted in this section. Assume that the surplus region has 5 ports and the deficit region 4 ports. The number of transportation laden containers from surplus region to deficit one or in the opposite direction are randomly ranged from 100 to 200. Both two directional laden containers flows have the same profits which is random generated within [650, 1500]. The laden containers transportation costs are changed between 100 and 150 since they are different among ports. Suppose that they discount 20%–30% compared with the transportation costs of laden containers and are within [20, 50]. The supply empty containers in surplus region are from 150 to 400 and the demand ones in deficit ports [150, 200] in t_0 as the imbalance exists. The actual demand empty containers in deficit region in t_3 are the fluctuation with $\pm 5\%$ based on the supply ones in surplus region in t_0, where the positive value represents the increasing rate and the negative one the decreasing rate.

To test the performances of DRR-ECR, many instances with varying characteristics are generated. The entire set of instances is divided into some subsets so that different cost parameters influencing the total cost can be tested. We refer to the data generated with the above method as base instance. The other instances are deduced from it. Table 1 show the details for each subset, where instances I11–I15 are changed in the transportation cost, instances I21–I25 the demand of Europe and instances I31–I35 the leasing cost. The symbol "-" represents the corresponding values are the same as the base instance.

Table 1. Characteristics of the instances.

Instance subset Base instance	Transportation cost	In t_3 demand of Europe	Leasing cost of Asia
I11	+5%	-	-
I12	+10%	-	-
I13	+20%	-	-
I14	+50%	-	-
I15	+70%	-	-
I21	-	[−10%, 10%]	-
I22	-	[±10%, ±20%]	-
I23	-	[±20%, ±50%]	-
I24	-	[±50%, ±80%]	-
I25	-	[±80%, ±100%]	-
I31	-		[−10%, ±10%]
I32	-		[±10%, ±20%]
I33	-		[±20%, ±50%]
I34	-		[±50%, ±80%]
I35	-		[±80%, ±100%]

4.2 Compared Model

To discuss the DRR-ECR, a static port-to-port ECR model (SPP-ECR) is also designed. Its objective function and constraints conditions are given as (13)–(21).

$$\begin{aligned} &maximize \\ &\sum_i \sum_j trLaden_{ij,t_0} * PL_{ij}^{tr} + \sum_j \sum_i trLaden_{ji,t_3} PL_{ji}^{tr} - \sum_i \sum_j trLaden_{ij,t_0} * (CL_{ij}^{tr} + CL_i^l + CL_j^u) \\ &- \sum_j \sum_i trLaden_{ji,t_3} * (CL_{ji}^{tr} + CL_j^l + CL_i^u) - \sum_i \sum_j trEmp_{ij,t_0} * (CE_{ij}^{tr} + CE_i^l + CE_j^u) \\ &- \sum_j (D_{j,t_0} - \sum_i trEmp_{ij}) * CE_j^r - \sum_i (S_{i,t_0} - \sum_j trEmp_{ij}) * CE_i^S \end{aligned} \tag{13}$$

$$\sum_j trEmp_{ij,t_0}^{tr} \le S_{i,t_0}; i \in PE, j \in PA \tag{14}$$

$$\sum_i trEmp_{ij,t_0}^{tr} \le D_{j,t_0}; j \in PA \tag{15}$$

$$\sum_i \sum_j trEmp_{ij,t_0}^{tr} \le \min(\sum_i S_{i,t_0}, \sum_j D_{j,t_0}) \tag{16}$$

$$\sum_i \sum_j trLaden_{ij,t_0} \le Capactity; \tag{17}$$

$$\sum_j \sum_i trLaden_{ji,t_3} \leq Capactity; \quad (18)$$

$$\sum_i \sum_j trEmp_{ij,t_0} \leq Capactity - \sum_i \sum_j trLaden_{ij,t_0}; \quad (19)$$

$$\sum_i \sum_j trLaden^{tr}_{ij,t_0} \leq \sum_i \sum_j LadenTrans_{ij,t_0}; \quad (20)$$

$$\sum_j \sum_i trLaden^{tr}_{ji,t_3} \leq \sum_j \sum_i LadenTrans_{ji,t_3}; \quad (21)$$

We use CPLEX 12.6 to solve these mathematical models. For each instance, we run 10 replications and achieve the average value. Two indicators are employed to analyze them. One is the values of objective functions. Another is the decreasing rates of total cost for 8 extracted instances which are the combination of four parameters.

4.3 Objective Function

Table 2 gives the details when various costs fluctuate in deficit region. In the third line of Table 2, the cost, total profit and net profit of the base instance are first shown in order to compare with other instances.

Table 2. Cost, total profit and net profit when the transportation cost, the demand of deficit region and leasing cost in deficit region fluctuate.

Base instance	SPP-ECR			DRR-ECR		
	Cost	Total profit	Net profit	Cost	Total profit	Net profit
	994,815	6,600,735	5,605,920	989,564	6,600,735	5,611,171
111	1,032,460	6,600,735	5,568,275	1,027,218	6,600,735	5,573,517
112	1,071,520	6,600,735	5,529,215	1,066,248	6,600,735	5,534,487
113	1,146,427	6,600,735	5,454,308	1,141,327	6,600,735	5,459,408
114	1,375,110	6,600,735	5,225,625	1,369,806	6,600,735	5,230,929
115	1,525,477	6,600,735	5,075,258	1,520,157	6,600,735	5,080,578
121	996,682	6,600,735	5,604,053	990,687	6,600,735	5,610,048
122	1,002,424	6,600,735	5,598,311	990,179	6,600,735	5,610,556
123	1,009,839	6,600,735	5,590,896	987,787	6,600,735	5,612,948
124	1,044,780	6,600,735	5,555,955	997,082	6,600,735	5,603,653
125	1,061,906	6,600,735	5,538,829	997,191	6,600,735	5,603,544
131	994,832	6,600,735	5,605,903	989,564	6,600,735	5,611,171
132	994,757	6,600,735	5,605,978	989,564	6,600,735	5,611,171
133	994,878	6,600,735	5,605,857	989,564	6,600,735	5,611,171
134	994,896	6,600,735	5,605,839	989,564	6,600,735	5,611,171
135	994,800	6,600,735	5,605,935	989,564	6,600,735	5,611,171

As we do not change the numbers of laden containers transportation in the whole operation and they are less than the vessel space, all the laden containers can be transportation from Europe to Asia. The instances I11–I35 have the same total profits in column 3 and 6. And it is no difficulty to observe that the DRR-ECR is always superior to the SPP-ECR with no relationship with any cost parameter. The reason is mainly that both the storage costs and leasing costs in DRR-ECR are lower than the ones of ECR-SPP because we use the dynamic redistribution strategy in t_3.

4.4 The Cost, Total Profit and Net Profit Under Different Circumstances

Considering with the actual numbers of ECR, which is limited by the residual space of liners excluding the laden containers, the supply ones of surplus region in t_0, and the demand ones of deficit region in t_0 and t_2, we divide the whole process into two stages. The first stage is planned stage and the second stage the implementation stage. We combine with these parameters and deduce eight instances shown as Table 3 to explain above models, where the RTEC refers to the real transportation empty containers and the RC the residual capacity. For example, the $S_{surplus,t_0} > D_{deficit,t_0}$ means the whole supply in surplus region is larger than the total demands ones in deficit region. The $S_{deficit,t_0} > RC$ describes that the whole demand of all ports in deficit region is larger than the residual capacity. Table 4 describes the details. From it, we can discover that the RTEC always equals to the minimum value of RC, the whole supply empty containers of surplus region in t_0, and the whole demand ones of deficit region in t_0.

Compared with other instances, both I45 and I46 have higher costs reducing, because the RTEC, which equals to the surplus empty containers of surplus region in t_0, fluctuates wildly than the demand ones of deficit region in t_2. So, for SPP-ECR it directly leads to that both the storage cost in deficit region and the transportation cost increase. However, for DRR-ECR the dynamic redistribution policy among the ports of deficit region brings down the storage cost to some extent.

Table 3. Eight combined instances according to the whole supply empty containers in surplus region in t_0, the demand ones in deficit region in t_0 and in t_2.

	Stage 1		Stage 2
I41	$S_{surplus,t_0} > D_{deficit,t_0}$	$D_{deficit,t_0} > RC$	$RTEC > D_{deficit,t_2}$
I42	-	-	$RTEC < D_{deficit,t_2}$
I43	-	$D_{deficit,t_0} < RC$	$RTEC > D_{deficit,t_2}$
I44	-	-	$RTEC < D_{deficit,t_2}$
I45	$S_{surplus,t_0} < D_{deficit,t_0}$	$D_{deficit,t_0} > RC$	$RTEC > D_{deficit,t_2}$
I46	-	-	$RTEC < D_{deficit,t_2}$
I47	-	$D_{deficit,t_0} < RC$	$RTEC > D_{deficit,t_2}$
I48	-	-	$RTEC < D_{deficit,t_2}$

Table 4. the percentage of each cost in the total operation cost of these two models.

	DRR-ECR			SPP-ECR			Cost decreasing rate
	Cost	Profit	Net profit	Cost	Profit	Net profit	
141	990,687	6,600,735	5,610,048	996,682	6,600,735	5,604,053	0.6015%
142	988,517	6,600,735	5,612,218	996,942	6,600,735	5,603,793	0.8451%
143	980,567	6,600,735	5,620,168	994,718	6,600,735	5,606,017	1.4246%
144	1,008,971	6,600,735	5,591,764	1,015,455	6,600,735	5,585,280	0.6385%
145	945,474	6,600,735	5,655,261	1,007,152	6,600,735	5,593,583	6.124%
146	960,920	6,600,735	5,639,815	1,023,915	6,600,735	5,576,820	6.1524%
147	951,774	6,600,735	5,648,961	970,763	6,600,735	5,629,972	1.9561%
148	971,646	6,600,735	5,629,089	990,840	6,600,735	5,609,895	1.9371%

5 Conclusion and Future Work

This paper presents a dynamic across-region ECR model: DRR-ECR, in which we use the region-region redistribution strategy to reduce the possible increasing costs which commonly exist in tradition static port-port ECR policy. To evaluate these models, we compare them through some deductive instances. Moreover, the DRR-ECR has significant impact on the safety stock of ports.

In this paper, our model considers the numbers of supply/demand empty containers given in advance. This cannot match the practice. So, one of the extensions is to develop the uncertainties. Another possible future work is to study by combining the problem of ECR with the tactical level decision even the strategic level decision. If we take the two level decisions or three level decisions into account, the meta-heuristic algorithms have to be involved to face this complex problem.

References

1. UNCTAD: Review of maritime transport. United Nation Publication (2015)
2. Álvarez, J.F.: Joint routing and deployment of a fleet of container vessels. Marit. Econ. Logist. **11**(2), 186–208 (2009)
3. Crainic, T.G., Gendreau, M., Dejax, P.: Dynamic and stochastic models for the allocation of empty containers. Oper. Res. **41**, 102–126 (1993)
4. Cheung, R.K., Chen, C.-Y.: Two-stage stochastic network model and solution methods for the dynamic empty container allocation problem. Transp. Sci. **32**, 142–162 (1998)
5. Lam, S.-W., Lee, L.-H., Tang, L.-C.: An approximate dynamic programming approach for the empty container allocation problem. Transp. Res. Part C **15**(4), 265–277 (2007)
6. Bell, M.G.H., Liu, X., Angeloudis, P., Fonzone, A., Hosseinloo, S.H.: A frequency-based maritime container assignment model. Transp. Res. Part E **45**(8), 1152–1161 (2011)
7. Erera, A.L., Morales, J.C., Savelsbergh, M.: Global intermodal tank container management for the chemical industry. Transp. Res. Part E **41**(6), 551–566 (2005)
8. Brouer, B.D., Pisinger, D., Spoorendonk, S.: Liner shipping cargo allocation with repositioning of empty containers. INFOR **49**(2), 109–124 (2011)

9. Song, D.-P., Dong, J.-X.: Flow balancing-based empty container repositioning in typical shipping service routes. Marit. Econ. Logist. **13**(1), 61–77 (2011)
10. Imai, A., Shintani, K., Papadimitriou, S.: Multi-port vs. hub-and-spoke port calls by containerships. Transp. Res. Part E **45**(5), 740–757 (2009)
11. Meng, Q., Wang, S.: Liner shipping service network design with empty container repositioning. Transp. Res. Part E Logist. Transp. Rev. **47**(5), 695–708 (2011)
12. Long, Y., Lee, L.H., Chew, E.P.: The sample average approximation method for empty container repositioning with uncertainties. Eur. J. Oper. Res. **222**, 65–75 (2012)

Speed Optimization of UAV Vehicle Tracking Algorithm

Yanming Xu[1(✉)], Wei Pei[2], Yongying Zhu[3], Mingyu Lu[1], and Lei Wu[1]

[1] Information Science and Technology College, Dalian Maritime University, Dalian, Liaoning Province, China
itrec_ming@hotmail.com

[2] Environmental Science and Engineering College, Dalian Maritime University, Dalian, Liaoning Province, China
peiweidl@gmail.com

[3] Ocean and Civil Engineering Department, Dalian Ocean University, Dalian, Liaoning Province, China

Abstract. An improved method for vehicle tracking in UAV-captured video is introduced in this paper, this method based on GPU (Graphics Processing Units) acceleration which have efficient parallel computing powers. Today, vehicle tracking in UAV videos has many meaningful applications in computer vision, such as traffic management and illegal vehicle tracking. However, the increased computational complexity makes it difficult to be used in real-time tracking system. To overcome the limitation of tracking speed and meet the highly intensive calculation required, we perform image processing with GPUs and adopt an optimized structured output support vector machine (SVM) to online learning. The proposed approach is evaluated in UAV videos taken from DJI Matrice 100 drone and the corresponding conclusions are received.

Keywords: Vehicle tracking · Speed optimization · GPU · UAV

1 Introduction

With the rapid development of unmanned aerial vehicle technology (UAV), the detection and tracking of moving vehicles in UAV videos has been paid more attention. In military, UAV can be used in casualty rescue, battlefield investigation and surveillance, medicine material transport, damage assessment and so on [1, 2], etc. UAV object tracking has also been applied to many civilian areas. Traffic police department can conduct the identification of illegal vehicle behavior in a remote distance which greatly saves money and material resources [3]. Including Forest fire, marine pollution monitoring and other highly destructive events can also be prevented by UAV detection. The most widespread use of agricultural UAV is medicine spraying [4]. Drone object tracking is a significant research which involves many disciplines such as artificial intelligence, image processing, machine learning.

However, target tracking of UAV remains as a challenge because of the relatively lower video resolution, non-uniform illumination, object occlusion, shape deformation,

Z. Shi et al. (Eds.): ICIS 2017, IFIP AICT 510, pp. 390–399, 2017.
DOI: 10.1007/978-3-319-68121-4_42

irregular rotation, small-sized target, drone vibration and so on [5]. In order to track a moving vehicle with a UAV, the following points should be implemented in the tracking algorithm. Firstly, the accuracy of the result should be guaranteed in object tracking. Algorithm should have the ability to identify the tracking target from a complex, noise-filled background. Secondly, the tracking algorithm is supposed to keep a certain degree of robustness when the moving target is partially or completely obscured. It means that an automatic re-detection process platform should be established to keep tracking the object. Lastly, the most important thing one has to take into consideration is the processing speed. Videos transferred from a real UAV camera ought to be processed at high speed. Many works have been done to reduce the processing time, but the processing consumption increases exponentially with a significant improvement in UAV video resolution. A single CPU processing has been unable to meet the requirements for speed. With the rapid development of GPU hardware and its programming model, using the GPU's massive high-speed parallel processing mechanism to accelerate the machine vision algorithm has become an inevitable trend. This paper strives to improve the processing speed under the premise of high precision on an improved tracking algorithm [6] based on Struck [7] by GPU acceleration.

2 Related Works

Large quantities of algorithms have been proposed in recent years, but they all have significant limitations. G. Mattyus pointed out that the rapid movement of the UAV camera which resulting in the distribution of samples can be used to express characteristics are too little, so by normalizing the European distance to separate the front and back points, and then get the moving target binary image [8]. But the calculations is too complex to lead to a slow tracking process. This problem can be solved by CUDA Parallel operation as described in this paper. Based on background difference, an improved method of moving object detection and tracking is proposed, and the target is identified with the background model to construct the deviation among different image frames [9]. Considering the relationship between the target height and the shadow scale, GV.Reilly performed shadow removal and target tracking by combining wavelet features with SVM features in 2013 [10]. The background subtraction was used to overcome the difficulties of scene constantly shifts [11, 12], etc. In order to achieve higher accuracy and get a relatively good tracking performance, many algorithms involve large feature vectors which bring a significant impact on the speed of the tracker.

The purpose of this paper is to accelerate the tracking process By CUDA computation. CUDA is a general purpose parallel computing platform and programming model that it leverages the parallel compute engine in NVIDIA GPUs to solve many complex computational problems in a more efficient way. Using a machine learning method to choose the best sparse matrix representation model on GPU [13], the target tracking algorithm was implemented on the CUDA platform with the GPU's ability of parallel computing [14, 15], etc. Pei [6] proposed an improved tracking method based on Struck [7], the extended affine transformation motion estimation method had been adopted to automatically adjust the tracking window's scale and rotation. Compared with the original Struck, the tracking accuracy has been enhanced, however, the number of frames processed per second have been dropped. The meaningful contribution in this paper is that all those tracking models are ported to the GPU and keep detection and tracking via CUDA parallel calculations.

3 System Achievement

In this paper, GPU's powerful parallel computing capabilities are exploited to accelerate the speed of improved tracking algorithm. We use Gaussian/Haar-like combination to reach an acceptable compromise between processing speed and tracking accuracy. An online structured output SVM learning framework has been used to incorporate multiple types of image features and kernels. In this section, some key points will be introduced briefly.

3.1 CUDA Programming Model

According to CUDA's model architecture, the task of the CPU is to perform serial operation such as data transmission and logic operations. Originally graphics data processing consumed lots of resources now is handled by GPU parallel computing. CUDA not only completes the traditional image rendering work, but also achieves good performance of complex operation. A complete CUDA program includes the serial processing program in host (CPU) and the device's (GPU) parallel processing code. Code which runs on the GPU is called the CUDA kernel function. To complete respective subtasks of each model, the CUDA threads run on a physically separate device (GPU) which plays as a medium to the host running the C code. The host code will calls the parallel code running on the device side when needed [16].

3.2 Haar-Like Feature Extraction

Papageorgiou first proposed Haar-like features and applied it to face description [19]. Pixel-based face detection algorithm has a high computational cost, but the Haar-Like feature based on block reduces processing costs. This paper uses six kinds of Haar-like features (feature template), as shown in Fig. 1. The reason why we choose this six features is it basically cover all the features. The feature value is that the sum of black pixels gray value subtracts the sum of white pixels gray value.

The main idea of integral images is to form a rectangular region from the starting point to each point and let the sum of pixels value in each rectangular region as an element of an array stored in memory. When you need to calculate the pixels value of a region, you can index the array elements, Instead of recalculating the pixels of this region, thus the computation will be speeded up. Integral image algorithm is a matrix representation method which can describe the global image information. The principle is that the element value $ii(x, y)$ at location (x, y) in integral image is the sum of all pixels above from the upper left corner to (x, y) direction in original image f. The recurrence formulae are as follows.

$$s(x,y) = s(x,y-1) + i(x,y). \tag{1}$$

$$ii(x,y) = ii(x-1,y) + s(x,y). \tag{2}$$

Where $s(x,y)$ is the cumulative row sum with the initialization of $s(x,-1) = 0$. $ii(-1,y) = 0$ and $ii(x,y)$ is an integral image. Scanning the image and when the image reaches the lower right corner of the pixel, the integral image is over.

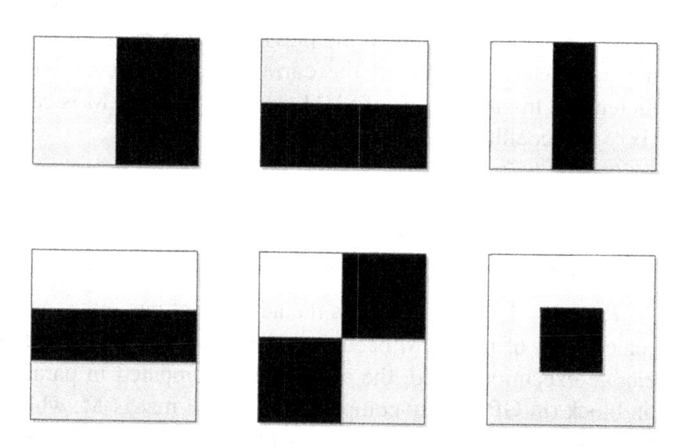

Fig. 1. Six box features template for extracting Haar-like feature.

3.3 Structured Output SVM Optimization

Support vector machine (SVM) is a statistical learning theory based on Structural Risk Minimization Principle (SRM). The goal of SVM is to determine the optimal hyperplane of feature space partition and the core idea of SVM is to maximize the classification of marginal. Structured output SVM tracking is used to directly estimate the change in target position between frames by learning a predictive function $f : X \to Y$, so the output space is a combination of the deformation of Y, instead of traditional binary labels -1 and $+1$. In this method, the sample (x,y) as a whole, y is the change function of target optimal position, the target position at the time of t is $P_t = P^{\circ}_{t-1}y_t$. So the framework used to learn a predictive function f can be obtained by introducing the discriminant function $F : \mathrm{X} \times \mathrm{Y} \to \mathrm{R}$. The result can be predicted according to (3).

$$y_t = f\left(x_t^{p_{t-1}}\right) = \underset{y \in Y}{\operatorname{argmax}}\, F\left(x_t^{p_{t-1}}, y\right). \tag{3}$$

In order to update the predictive function online, we mark every new tracking position $(x_t^{p_t}, y^0)$. The discriminant function F measures the compatibility between (x,y) samples and chooses the best score in all samples. Equation (3) can be further expressed as $\mathrm{F}(x,y) = \langle w, \Phi(x,y) \rangle$ where $\theta(x,y)$ is a joint kernel projection. By minimizing the convex objective function (4), it can be learned in a new framework from a set of pairs $\{(x_1,y_1),\ldots,(x_n,y_n)\}$.

$$\min_{w} \frac{1}{2}||w||^2 + C\sum_{i=1}^{n} \xi_i. \tag{4}$$

$$\text{s.t. } \forall i : \xi_i \geq 0 \forall i, \forall y \neq y_i : \langle w, \delta\Phi_i(y)\rangle \geq \Delta(y_i, y) - \xi_i$$
$$\delta\Phi_i(y) = \Phi(x_i, y_i) - \Phi(x_i, y).$$

The original Struck algorithm contains two sets of data in the support vectors machine (vector x). One is the features of the previous frame reserved support vectors, the other is the parameters that record the current support vector and their corresponding coefficients β. In the analysis of SVM algorithm, matrix M is calculated from vector x, matrix y and coefficient β.

$$\text{M} = \sum_{i=1}^{m}\sum_{j=1}^{n} x_i x_j K(y_i, y_j, \beta)k. \tag{5}$$

Where $x_{i,j} \in R^2$, i, j $= 1, 2, \ldots,$ m, n. k is the adjustment parameter and K is kernel function. The calculation of matrix M occupies a large amount of computation time in CPU. To accelerate execution speed, the algorithm is computed in parallel, and each kernel thread in block (in GPU) only computes the row of matrix M, which means the calculation of initial matrix M is split into n subtasks and each subtask is executed asynchronously. The formula of each subtask is defined as follows.

$$\text{M} = \sum_{i=1}^{m} x_i^2 K(y_i, y_i, \beta). \tag{6}$$

This computational framework design is based on CUDA programming model and every kernel function runs a number of threads which are executed in parallel. In kernel programming, kernel thread only executes a simple cycle to improve the parallel procession of the algorithm.

4 Experimental Results

Hardware Environment: a standard PC equipped with 4G Memory, Dual-core Intel(R) Core(TM) i3-3220, 3.3 GHz, GTX 750 TI graphics card with 640 CUDA cores, DJI Matrice 100. Software Environment: Windows 7, Microsoft Visual Studio 2010, CUDA 6.5, OpenCV2.4.10.

It is important to collect a representative dataset for a comprehensive and impartial performance evaluation of the tracking algorithm. The groundbreaking in this work is the VTB dataset presented in which contains 100 sequences from recent literatures, and it is the most influential visual target tracking dataset. Another influential visual target tracking dataset is the VOT dataset. In this section we will use the VTB dataset with different frame sizes and video captured with UAV flying over traffic intersections to perform the tracking test. This part outlines the results of experiments which carried out during this research.

4.1 Evaluation of VTB Test Set

Object tracking experiments can be divided into two parts. In this part, five VTB image sequences are selected at random and manually marked for the ground truth which contains 2926 image frames. This experiment provides a comparison between the processing speed on GPU with CUDA and CPU for object tracking, and the resolution of picked image is identified below the figure. The record is measured once every 10 frames, 20 frames or 50 frames according to different sequence. Take the limitation of the length into consideration, we demonstrated the detailed results of the RedTeam public dataset to test the improved tracking algorithm, the other datasets only give the final statistical results.

We compare the processing speed of original CPU-only algorithm with the speed of GPU-accelerated code. The GPU-accelerated algorithm contains the image features calculation, dual SVM, learning/adaptation and multiple-kernel learning. As shown in Fig. 2, it provides details about the timing of processing the frames (the frame interval is 50 every record) of the proposed approach run on the video such as Fig. 3. The overall runtime of processing in VTB dataset are observed between the CPU and GPU versions. The results show that the speed of algorithm processes a video stream in GPU platform achieves a great effect than running on original CPU-only algorithm which is running about 2500–3000 ms every 10 frames, and the speed running on CPU platform is unstable which changes intensely.

Figure 4 and Table 1 show the results of performance comparison of tracking algorithm execution on CPU and GPU platform. In CarDark dataset, the speed boost obtained due to using a massive parallel CUDA programming model changes from 3.04 fps to 33.99 fps which achieved nearly 11 times speedup per second in comparison with CPU implementation. From the Fig. 4 we can see that the average FPS from the CPU implementation is below four frames per second which means the overall execution time will become greater as time goes on.

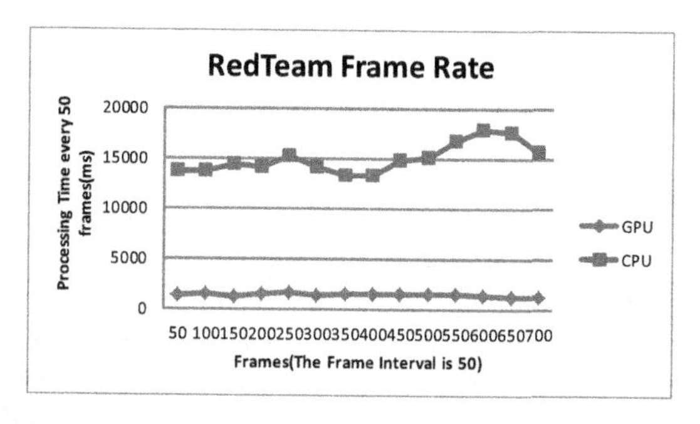

Fig. 2. Comparison of calculation performance on CPU and GPU in RedTeam dataset at 352×340 resolution.

Fig. 3. RedTeam dataset tracking.

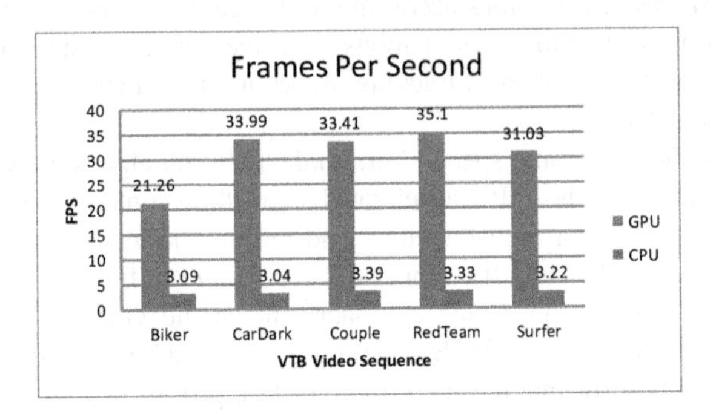

Fig. 4. FPS test of RedTeam.

Table 1. The statistics table for acceleration factor.

Dataset	FPS in CPU	FPS in GPU	Acceleration factor
Bike	3.09	21.26	5.88
CarDark	3.04	33.99	10.18
Couple	3.39	33.41	8.86
RedTeam	3.33	35.1	9.54
Surfer	3.22	31.03	8.64

Next, the tracking performance will be valued on RedTeam dataset (as shown in Fig. 3) using two popular evaluation criteria: center location error (CLE) and overlap ratio (VOR). In order to reduce the accident, each video sequence is processed twice and picked the RedTeam sequence to show the performance in Figs. 5 and 6. The improved ratio between the original algorithm on CPU and improved algorithm on GPU in terms of center location error and overlap rate can be seen from the figure. In Fig. 5, the average CLE of original code is about 8.5 but the average CLE on GPU is only 2.4 which means the better capability of vehicle tracking in our method. In Fig. 6, the improved Struck algorithm is 31.63% higher than original Struck algorithm in average tracking overlap ratio, and the average high score of VOR shows a good adaptability to complex environment.

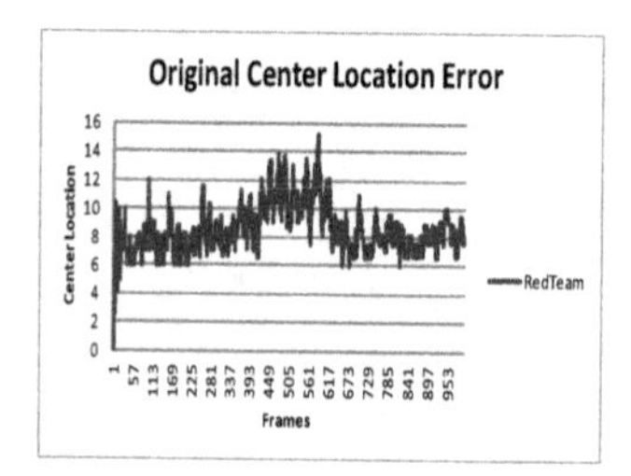

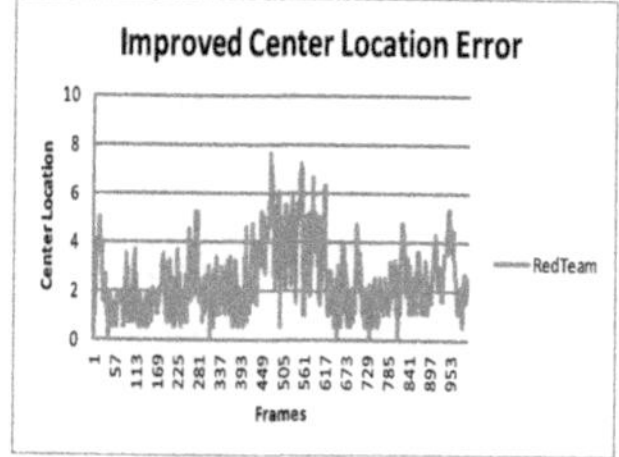

Fig. 5. Quantitative comparison of CPU and GPU versions in CLE on the RedTeam video.

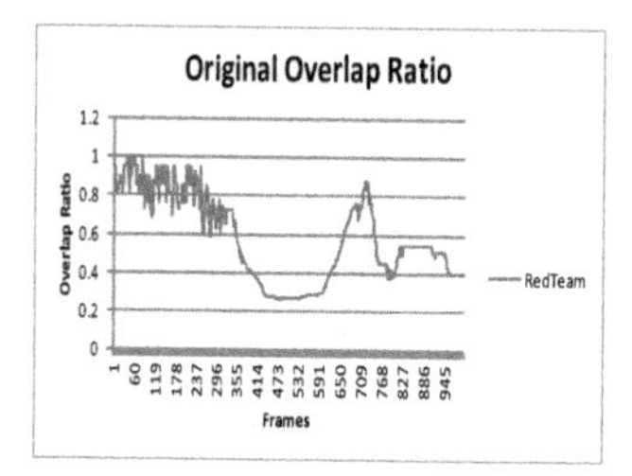

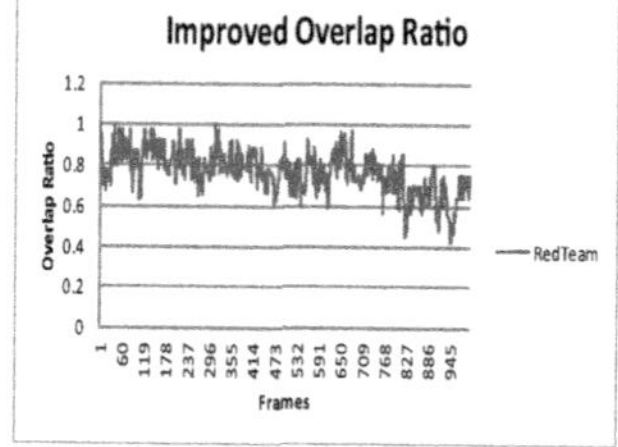

Fig. 6. Quantitative comparison of CPU and GPU versions in VOR on the RedTeam video.

4.2 Evaluation of UAV Test Set

In this section, three scenes are captured by the DJI X3 camera with UAV in Dalian, Chinese and handled as follows: scene one contains the UAV shooting multi-targets. Scene two consists of the UAV overlooking a white car on a campus road and scene three includes the UAV overlooking a truck from a moving perspective. The purpose of the experiments was to compare the performance of the improved algorithm on CPU and GPU platforms on the same PC. The precision rate is not considered in this experiment, the time to process the video is what we care about.

In order to get a reasonable comparison, images with different resolutions are used to test. Figure 8 shows the comparison of FPS (frames per second) running on CPU and GPU with different resolutions. From Fig. 8 we can see that the implementation of GPU algorithm has achieved a very significant acceleration effect, lower resolution of image means faster processing speed on GPU. The FPS of original algorithm on CPU platforms improves little despite the resolution of image varies greatly. The average speed of tracker using CUDA parallel optimization is about 4.5–5 times FPS in comparison with CPU implementation. Figure 7 is the tracking performance in different UAV testing sequences.

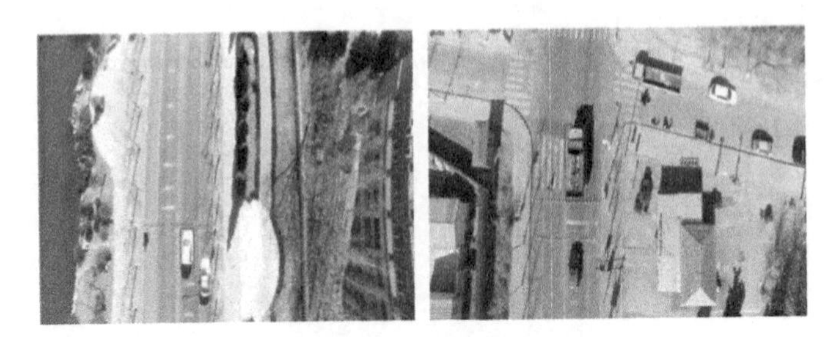

Fig. 7. Vehicle tracking in UAV videos.

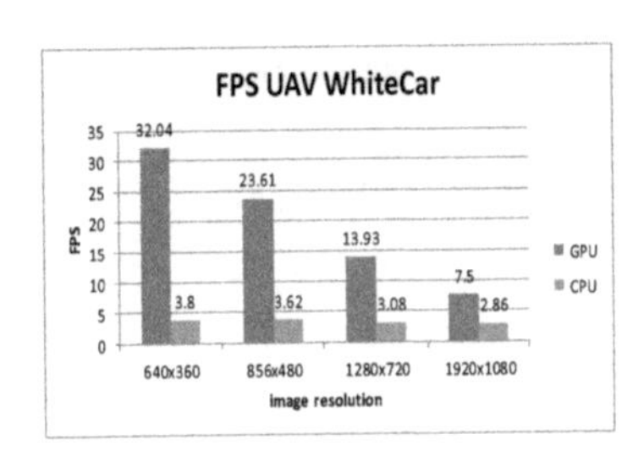

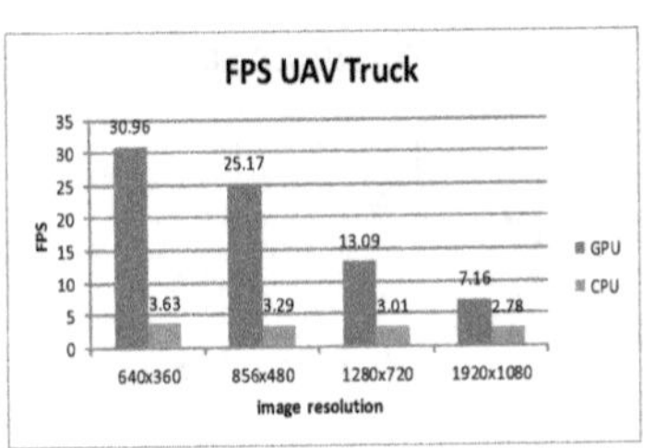

Fig. 8. FPS test at different resolutions of white car and truck dataset.

5 Conclusions

In this paper, structured output support vector machine and haar-like features has been used to improve the original Struck tracking method and accelerate the overall runtime with the implementation of improved algorithm on CUDA. The algorithm processes a video sequence with a frame rate close to 31 frames per second, and it is significant excellent than CPU. This method has been tested on UAV videos captured by DJI M100 UAV, and the experimental results show that the proposed algorithm is feasible to track and identify moving vehicles. In the future, the program should be optimized to better integrate into the CUDA programming architecture to achieve a better performance.

Acknowledgments. The work was supported by the National Natural Science Foundation of China (No. 61001158, 61272369 and 61370070), Liaoning Provincial Natural Science Foundation of China (Grant No. 2014025003), Scientific Research Fund of Liaoning Provincial Education Department (Grant No. L2012270), Specialized Research Fund for the Doctoral Program of Higher Education (SRFDP no 200801511027), and Fundamental Research Funds for the Central Universities (Grant No. 3132015084).

References

1. Callam, A.: Drone wars: armed unmanned aerial vehicles. Int. Aff. Rev. **18** (2015)
2. Springer, P.J.: Military Robots and Drones: A Reference Handbook. ABC-CLIO, Santa Barbara (2013)

3. Yu, Q., Medioni, G.: Motion pattern interpretation and detection for tracking moving vehicles in airborne video. In: Computer Vision and Pattern Recognition, pp. 2671–2678 (2009)
4. Zhang, C., Kovacs, J.M.: The application of small unmanned aerial systems for precision agriculture: a review. Precision Agric. **13**(6), 693–712 (2012). Computer Vision and Pattern Recognition
5. Fu, C., Duan, R., Dogan, K., Erdal, K.: Onboard robust visual tracking for UAVs using a reliable global-local object model. Sensors **16**(9), 1406 (2016)
6. Pei, W., Zhu, Y., Zuo, X.: A multi-scale vehicle tracking algorithm based on structured output SVM. In: International Congress on Image & Signal Processing, pp. 1–6 (2015)
7. Hare, S., Saffari, A., Torr, P.: Struck: structured output tracking with kernels. In: International Conference on Computer Vision, pp. 263–270 (2011)
8. Máttyus, G., Benedek, C., Szirányi, T.: Multi target tracking on aerial videos. In: ISPRS Workshop on Modeling of Optical Airborne & Space Borne Sensors, pp. 1–7 (2010)
9. Ramakrishnan, V., Kethsy, A., Devishree, J.: A survey on vehicle detection techniques in aerial surveillance. Int. J. Comput. Appl. **55**(18), 43–47 (2012)
10. Reilly, V., Solmaz, B., Shah, M.: Shadow casting out of plane (SCOOP) candidates for human and vehicle detection in aerial imagery. Int. J. Comput. Vis. **101**(2), 350–366 (2013)
11. Uzkent, B., Hoffman, M.J., Vodacek, A.: Real-time vehicle tracking in aerial video using hyperspectral features. In: Computer Vision & Attern Recognition Workshops, pp. 36–44 (2016)
12. Chen, B.J., Medioni, G.: Motion propagation detection association for multi-target tracking in wide area aerial surveillance. In: Advanced Video and Signal Based Surveillance, pp. 1–6 (2015)
13. Benatia, A., Ji, W., Wang, Y., Shi, F.: Sparse matrix format selection with multiclass SVM for SpMV on GPU. In: International Conference on Parallel Processing, pp. 496–505 (2016)
14. Zorski, W., Sklodowski, P.: Object tracking and recognition using massively parallel processing with CUDA. In: International Conference on Methods & Models in Automation & Robotics, pp. 977–982 (2015)
15. Rao, G.M.: GPU based video tracking system. In: IEEE Tenth International Conference on Semantic Computing, pp. 170–171 (2016)
16. Luebke, D.: Scalable parallel programming for high-performance scientific computing. In: IEEE International Symposium on Biomedical Imaging: from Nano to Macro, pp. 836–838 (2008)
17. Papageorgiou, C., Poggio, T.: A trainable system for object detection. Int. J. Comput. Vis. **38**(1), 15–33 (2000)

Application of Ant Colony Optimization in Cloud Computing Load Balancing

Zheng-tao Wu(✉)

School of Information Engineering, Shanghai Maritime University, Shanghai 201306, China
wuzhengtao520@126.com

Abstract. In cloud computing, because all the tasks are submitted and requested by the users, it is the key to improve the quality of service whether all tasks are scheduled to every host to achieve the shortest completion time for all tasks. In this paper, the basic ant colony algorithm is modified appropriately to enable task scheduling and load balancing. The simulation results show that the ant colony algorithm can effectively obtain the global optimal solution and realize the shortest task completion time.

Keywords: Cloud computing · Load balance · Ant colony optimization

1 Introduction

Cloud computing is a new business and computing model which is proposed in recent years. Thousands of hosts are linked together over the network to form an ultra large pool of resources. Google corporation is the first to use cloud computing service, its scale has reached millions of machines, and other companies such as Microsoft, Amazon, Aliyun and so on, also reached the scale of several hundred thousand units [1]. Depending on the type of service, cloud computing is divided into three layers, namely, IaaS (infrastructure as a service), PaaS (platform as a service), and SaaS (software as a service). No matter which layer, there is the need to balance the task of scheduling to the host, that is, load balancing. However, due to the large number of computers in the cloud computing platform, the complexity of the structure and the difference of the allocation of resources, it is difficult to achieve load balancing for such large-scale, heterogeneous and non-centralized task scheduling. The unbalanced load will lead to the decrease of system efficiency and throughput, and the instability of the daily operation, which will seriously affect the quality of service of cloud computing [2].

Ant Colony Optimization (ACO) is a kind of bionic intelligent algorithm to find the optimal solution, simulating the foraging behavior of ants. The wisdom of a single ant is not high, but it achieves an optimal population effect by sharing information among groups. Ant colony algorithm is a kind of positive feedback algorithm, which has good robustness and adaptability, and is easy to combine with other algorithms [3]. Therefore, ant colony algorithm can be used in load balancing.

The ant colony system was firstly proposed by Italy scholars Dorigo, Maniezzo and others in 1990s. When they studied ants foraging, they found that the behavior of a

Z. Shi et al. (Eds.): ICIS 2017, IFIP AICT 510, pp. 400–408, 2017.
DOI: 10.1007/978-3-319-68121-4_43

single ant was relatively simple, but the whole ant colony could show some intelligent behavior. For example, an ant colony can find the shortest path to a food source in different environments. This is because ants in the ant colony can transfer information through some kind of information mechanism. Further studies have shown that ants release a substance called "pheromones" on their path.

The ants in the colony can apperceive pheromones, they will follow the path with high concentration pheromone, and every ant will leave the pheromones on the road, which forms a kind of positive feekback mechanism. Thus, after a period of time, the whole ant colony will arrive at the food source along the shortest path.

The basic idea of applying ant colony algorithm to solve the optimization problem is to use the path of the ant to represent the feasible solution of the problem to be optimized. All the paths of the whole ant colony form the solution space of the optimization problem. The shorter path ants release more pheromones, and as time goes on, the pheromone concentration on the shorter path increases gradually, and the number of ants in the path is increasing. Finally, the whole ant will concentrate on the best path under the positive feedback. At this time, the corresponding solution is the optimal solution of the problem to be optimized.

Researchers based on the study of ACO use traveling salesman problem as a test of performance, not others. In this paper, the basic ACO is modified accordingly, using cloudsim platform as simulation experiment, and design, modify the corresponding parameters, in order to design a load balancing algorithm.

2 Mathematic Model of Task Scheduled

ACO is a new bionic intelligent algorithm. The first ant guides and influences the selection of the following ants by means of pheromone, so as to achieve the goal of cooperation among ants. The application of ant colony algorithm in the task scheduling of cloud computing is studied in this paper.

In a data center of cloud computing, there are n virtual hosts, they are set as $Vm_1, Vm_2, Vm_3, \cdots, Vm_n$. Each virtual host has a different MIPS (millions of instructions per second). At the same time, there are m tasks that need to be scheduled to work on the n virtual hosts. The total amount of time spent in these tasks is:

$$\mathrm{T}_{total} = \max(T_1, T_2, \cdots, T_n) \tag{1}$$

where T_i indicates that the i virtual host completes the time allocated to all of its tasks, calculated by formula (2) and formula (3).

$$\mathrm{t}_j = \frac{MI_j}{MIPS} \tag{2}$$

$$\mathrm{T}_i = \sum_{j=1}^{k} t_j \tag{3}$$

In the above two formulas, suppose that the performance of platform I virtual host is MIPS, and there are k tasks allocated on it, the number of instructions corresponding to each task is MI_j (Million Instructions), so the time for each task is t_j.

3 Ant Colony Optimization

3.1 Theory of Ant Colony Optimization

After finding the food, the ants will randomly explore the way from home to food, and leave on the road with pheromones that are inversely proportional to the length of the journey. Subsequent ants will be based on the concentration of pheromones, select the nearest road, so as to achieve the group to find the best path of the results. If the shortest path is interrupted, the ants will randomly explore and converge quickly to the new shortest path.

The basic ACO in the TSP (traveling salesman problem) on the state as follows: at time 0, all the ants and n city will initialize m ants randomly in a city as a starting point, each edge between the city has an initial pheromone concentration $\tau_{ij}(0)$. Each ant places its city as the first element of the tabu list. Each ant then selects the next city to move, based on the probability function of the two parameter (α, β) (formula (6)), moving from the city I to the city J, and fill out the corresponding tabu list. After a cycle of N times, all ants go through all the cities and fill out the tabu list. After that, calculate the path length of each ant L_k, and then update corresponding pheromone $\Delta\tau_{ij}^{k}$ according the formula (4). Save the shortest path that ants walk through, and get the shortest path $minL_k$ after one traversal, and empty all the tabu tables and start the next traversal loop. Repeat the process until the number of iterations reaches the required number of iterations (N_C), get the shortest path $minN_C$ in the global situation.

$$\tau_{ij}(\mathrm{t}+1) = \rho\tau_{ij}(\mathrm{t}) + \sum \Delta\tau_{ij}^{k} \tag{4}$$

where $\tau_{ij}(\mathrm{t}+1)$ is the updated pheromone concentration, ρ is the rate of pheromone emission, $\sum \Delta\tau_{ij}^{k}$ is the sum of pheromones left by all ants.

$$\Delta\tau_{ij}^{k} = \begin{cases} \frac{Q}{L_k} \; \textit{through the road} \\ 0 \; \textit{not through the road} \end{cases} \tag{5}$$

where Q is a constant, L_k is the path length of ants.

The probability function of the first k ant moving from city I to city J is:

$$p_{ij}^{k} = \begin{cases} \frac{(\tau_{ij})^{\alpha}(\eta_{ij})^{\beta}}{\sum_{k}(\tau_{ij})^{\alpha}(\eta_{ij})^{\beta}}, & j \in J(i) \\ 0, & other \end{cases} \tag{6}$$

where $J(i)$ means In the tabu list of K, the J city has not passed yet.

In summary, Fig. 1 shows the flow chart using ACO to solve the TSP problem.

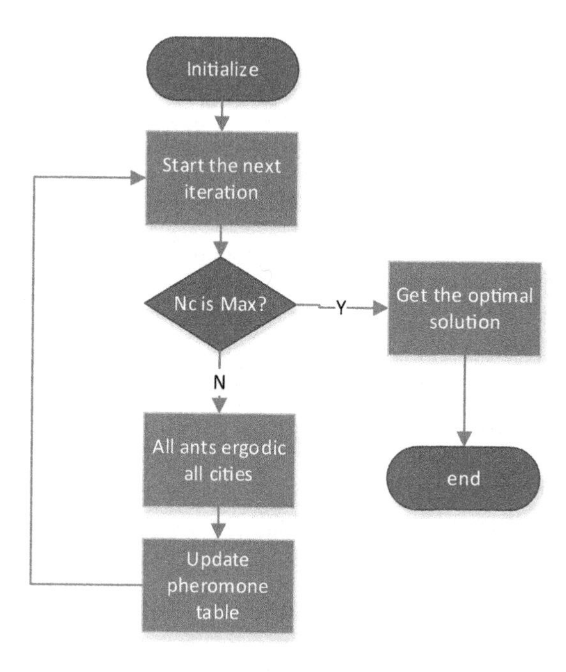

Fig. 1. ACO solve TSP problem

3.2 Realization of Ant Colony Optimization

In the solution of the shortest time, m ants were placed in the initial position, according to the rules of a certain probability choice, every ant can arrange Vm for next task, until all tasks completed, and then pheromone increment table will be updated based on the total time. After all ants have finished their work, the best solution is selected from all candidate allocation and the historical optimal solution, and it is recorded, then the homologous pheromone table is updated. Then the pheromone is corrected, simulating the ants releasing pheromone on the optimal path and the natural volatilization of pheromone, leading the ant to find the global optimal solution. Figure 2 shows the flow chart of the algorithm implementation.

(1) calculation rules of pheromone increment table

After each walk, each ant leaves a certain amount of pheromones according to the length of the path. The more ants there are in the same path, the more pheromones they leave behind at the same time. The pheromone increment table records the size of the pheromone emitted by the ants after each cycle. In this paper, the pheromone increment with Δ_{ij}^{k}, Δ_{ij}^{k} calculation rules such as formula (7) for:

$$\Delta_{ij}^{k} = \begin{cases} \frac{Q}{L}, & j \in Vm_i \\ 0, & j \notin Vm_i \end{cases} \tag{7}$$

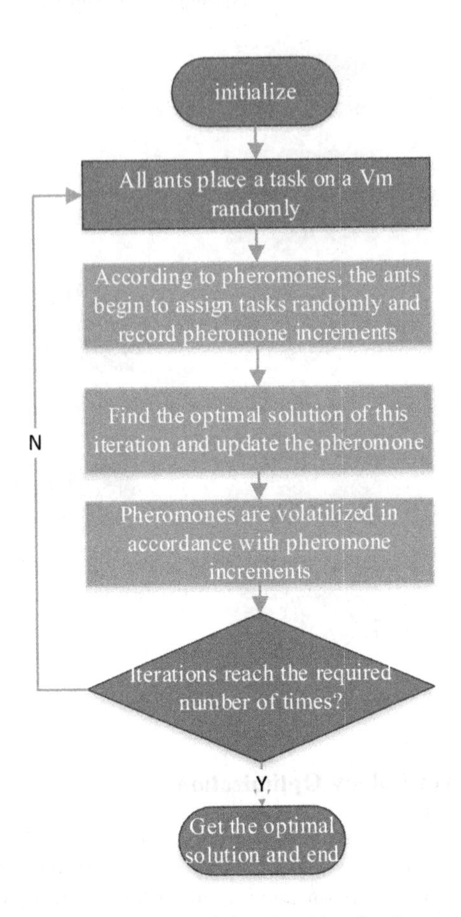

Fig. 2. The flow chart of the algorithm implementation

In the formula, the Q is expressed as a constant, and L is the total time after which the current ant is assigned the task. Each ant will leave pheromones on its path, and in the end, the global pheromone increment is the sum of the pheromone left by all the ants.

(2) calculation rule of task transfer probability

When the ant chooses the virtual host for the next task, it is determined by the pheromone left before it and a heuristic information. The probability formula for assigning the task J to the virtual host I is shown in formula (6), where τ_{ij} is pheromone concentration, η_{ij} is the heuristic information. The formula of η_{ij} is given by formula (8) and (9):

$$\eta_{ij} = \frac{1}{D_{ij}} \tag{8}$$

$$D_{ij} = \frac{Task_i}{MIPS_i} + \frac{L_j}{BW_i} \tag{9}$$

where $Task_i$ represents the total number of tasks allocated on Table 1 Vm, $MIPS_i$ is the execution speed of the CPU for this Vm, L_j is the length of the task to be assigned, and BW_i is the interface bandwidth of this Vm, which is a constant in this paper. α, β are control parameters that control the relative importance of pheromone and heuristic information.

(3) Amendment rules for pheromone tables

When all ants completing one iteration and getting an optimal solution, pheromone table is needed to have a comprehensive revision according to the pheromone emitted by ants and the phenomenon that simulates the natural evaporation of residual pheromones, the correction rule is such as formula (10):

$$\tau_{ij}(t+1) = (1-\rho)\tau_{ij}(t) + \sum \Delta_{ij}^{k} \tag{10}$$

where $\tau_{ij}(t+1)$ is the pheromone table after correction, $\tau_{ij}(t)$ is pheromone table before correction, ρ is the control parameter, indicates the degree of evaporation of pheromones, $\sum \Delta_{ij}^{k}$ is the sum of pheromone increments left by all ants. The purpose of pheromone correction is to assign as many pheromones as possible to the optimal allocation mode, and better guide the selection of ants. The rule will not only store the pheromones left by the ants, but will also evaporate them properly.

3.3 Algorithm Parameter Design

According to the above principle and various rules, this paper has designed and stipulated a series of parameters. In the formula (7), the value of the constant Q will affect the left ant pheromone on global pheromone influence, it will weaken the influence of a better path on later ants if Q is too small, otherwise it will faster convergence which can easily fall into local optimal solution. According to the task length and the MIPS value of Vm, repeated experiments were conducted, and finally 100 was selected. In the formula (6) and (10), α, β, ρ are control parameter, where α, β respectively indicate the importance of pheromone and heuristic information, and ρ is the volatilization rate of pheromones. Studies have shone that when α is 1, the resulting optimum is better than the other. When value of β is from 1 to 5, the optimal solution quality will gradually become better, and when β is more then 5, the quality of the solution can be reduced, therefore, 5 is assigned to β. P value changes between 0.3–0.9, the solution changed little, when its value is 0.5, the quality of solution is the best, ρ is suggested to value of 0.5 [5]. In formula (9), MIPS is randomly chosen in 100–300, and BW is the interface bandwidth. In this paper, the default value of simulation software is taken. The task length of all tasks is also random, with a range of 10000–50000.

4 Simulation and Results

This paper uses cloudsim3.0 simulation platform to do simulation experiments. By inheriting the DatacenterBroker class, we extend the DatacenterBrokerACO class to ACO, and compile the Ant class and ACO class. We can regulate single ant activity and ant colony activity separately.

4.1 Introduction of Cloudsim

Cloudsim is a cloud computing simulation platform developed by Melbourne University in Australia on the basis of Gridsim. CloudSim has several versions, and has now grown to version 3 of CloudSim. CloudSim 3 uses a layered architecture. Simulation layer provide a support for the configuration and simulation of virtual data center environment simulation for cloud computing, including virtual machines, memory, capacity, and bandwidth interface. This layer is used to study strategy of virtual machine to the host distribution, and the implementation of virtual machine scheduling function expansion core. The top layer is the user code layer, which provides basic entities such as hosts, applications, virtual machines, user data and application types, and agent scheduling policies. CloudSim 3 provides virtualization engines that build and manage multiple virtualized tasks on data center nodes and flexibly switch between time sharing and spatial sharing policies when virtualized services are allocated. Cloudsim 3 is open source, users can expand the open source code by expanding the interface, realizing their own scheduling strategy, and test and verify the corresponding function based on specific environment and configuration.

4.2 Result of Simulation

Before the simulation starts, the paper specifies some parameters in the ant colony algorithm, as shown in Table 1:

Table 1. Parameters of ACO

Q	α	β	ρ
100	1.0	5.0	0.5

In order to ensure the heterogeneity of all virtual hosts, Java is used to randomly specify 20 different MIPS hosts. The range of MIPS is 100–300, and the value of MIPS is shown in Table 2:

Table 2. The MIPS of 20 Vm

VmId	1	2	3	4	5
Mips	228	299	108	204	152
VmId	6	7	8	9	10
Mips	275	202	278	109	195
VmId	11	12	13	14	15
Mips	208	294	196	143	140
VmId	16	17	18	19	20
Mips	209	292	194	212	255

At the same time, 100 tasks with different task lengths are generated randomly, and the length is 10000–50000.

In order to determine the optimal solution, the number of ants is taken as half of the number of virtual hosts, 1000 iterations, and the final total running time is 813.1402877697841. Task assignment is shown in Table 3:

Table 3. Task allocation

VmId	Task id							
0	27	73	97	50	16	15		
1	72	0	56	58	68	94	42	14
2	9	84	43					
3	3	74	46	54	13	17		
4	11	49	86	24				
5	90	88	98	93	47	8		
6	7	71	62	30				
7	63	59	22	37	55	95	87	
8	38	70	77					
9	51	89	75	10				
10	52	36	91	18				
11	85	44	53	92	45	12	19	
12	33	61	99					
13	29	39	83	35				
14	32	78	4					
15	28	66	1	26	20			
16	6	21	64	67	40	82		
17	31	34	57	80	81			
18	23	65	96	76	2	60	48	
19	25	69	79	5	41			

5 Conclusion

In this paper, a method of load balancing for Cloud Computing Based on ant colony algorithm is proposed, and simulation experiments are carried out. The method guarantees that all virtual hosts can receive task assignments as evenly as possible, and that the total time of task completion is as short as possible. It can be seen that ACO has the following characteristics in load balancing:

As the ants continue to spread pheromones to the environment, new information will soon be updated to the environment, with dynamic characteristics.

Each ant makes its own choice according to the environment, and also sends out its pheromones. Therefore, ACO has distributed features.

Ants share environmental information and influence environment at the same time, so ACO has the characteristics of cooperative operation.

These characteristics are consistent with the super large scale, heterogeneous and realtime characteristics of cloud computing, so ACO is important for load balancing.

Although ACO can solve the problem of load balancing, but also has the following problems need further study:

Load balancing is a multi-dimensional and multi constrained QoS problem. In this paper, the shortest time problem in load balancing is discussed, as well as the minimum transfer and the economic principle.

ACO needs to initialize some parameters, and the choice of parameters has an important influence on the performance and the final results. Therefore, in the cloud computing environment, the setting of various parameters needs further study.

The search time of ACO is too long, and the convergence speed is slow. How to combine ACO with neural network and particle swarm algorithm to improve the performance and efficiency of the algorithm in order to meet the requirements of response time and response speed of cloud computing.

References

1. Liu, P.: Cloud Computing, vol. 3. Publishing House of Electronics Industry, Beijing (2015)
2. Li, F.: Research on Load Balancing Scheduling Algorithm of Cloud Computing Resource Based on Ant Colony Algorithm. Yunnan University, Yunnan (2013)
3. Zhang, N.: Research on Task Scheduling of Cloud Computing Based on Adaptive ant Colony Algorithm. Dalian University of Technology, Dalian (2015)
4. Kong, L., Zhang, X., Chen, J.: Basic ant colony algorithm and its improvement. J. Beihua Univ. **5**(6), 572–574 (2004)
5. Ye, Z., Zheng, Z.: The parameters alpha, beta in the ant colony algorithm, with an example of TSP problem. J. Wuhan Univ. **29**(7), 597–601 (2004)
6. Wang, X.: Cloudsim, cloud computing simulation tools, research and application. Micro Comput. Appl. **29**(8), 59–61 (2013)
7. Tian, Q.: An abstract of ant colony optimization. J. Shijiazhuang Vocat. Technol. Inst. **16**(6), 37–38 (2004)
8. Zhang, H., Luo, X.: Status and prospect of ant colony optimization algorithm. Hum.: Inf. Control **33**(3), 318–324 (2004)

Beam Bridge Health Monitoring Algorithm Based on Gray Correlation Analysis

Jianguo Huang[1(✉)], Lang Sun[2], and Hu Meng[2]

[1] Center for Information Technology, Hefei Preschool Education College, Hefei, China
71067566@qq.com
[2] HEFEI City Cloud Data Center Co., Ltd., Hefei, China

Abstract. Bridge construction investment is huge and the service cycle is long. During the service cycle, the bridge structure not only beared the load effect caused by fatigue damage, but also effected by the natural environment and human damage. Beam bridge is the most kind of bridge built on the highway and had a long-term service in China. The main beam of beam bridge is the main load-bearing component. Real-time evaluation of main beam's health degree will greatly improve the safety of highway transportation. Through the rapid assessment of the main beam of the bridge, it can not only directly reflect whether the deflection of the main beam is beyond the dangerous range and the overall condition of the main beam, but also observe the long-term variation rule of the main beam. The current assessment algorithm only stays in the monitoring of whether the deflection of the main beam is beyond the dangerous range, without a complete assessment combined with massive historical data. Based on the theory of Gray Correlation Analysis and combined with the real - time data and historical data of bridge monitoring, we calculate the statistical indicator and morphological indicator of the main beam quickly, and evaluate the comprehensive health indicator of the bridge according to the technical specification in this paper.

Keywords: Beam bridge · Bridge health assessment · Gray Correlation Analysis

1 Introduction

In recent years, with the continuous development of computer technology, communication technology, embedded sensors and other technologies, the use of computer systems for automatic health monitoring has become the main method of bridge monitoring. The linear evaluation of bridge main beam is an important indicator which can reflect the safety of the bridge. Through the linear monitoring of the main beam of the bridge, it can not only directly reflect whether the deflection of the main beam is beyond the dangerous range and the working condition of the main beam in the operating state, but also observe the long-term variation rule of the main beam. And the linear evaluation of the main beam of the bridge is of great significance to the bridge bearing capacity detection and the bridge earthquake disaster mitigation.

Z. Shi et al. (Eds.): ICIS 2017, IFIP AICT 510, pp. 409–416, 2017.
DOI: 10.1007/978-3-319-68121-4_44

The linear evaluation of bridge main beam theory is still in the exploratory stage and without unified evaluation method in our country. The method remains on the level of monitoring whether the original data of the main beam deflection exceeds the risk threshold. There is no standardized assessment of the bridge data.

Gray system theory is a discipline with good theoretical research and application value. Gray Correlation Analysis plays a significant role in theoretical research and application study, which is one of the important parts of Gray system theory. Gray Correlation Analysis is a method of analyzing and determining the degree of correlation between factors or factors on the main behavior of the system through the calculation of gray relational degree. The calculation of Gray Correlation Grade is the basement and an important tool of Gray Correlation Analysis. Therefor the establishment and improvement of the Gray Correlation Model is an important discussion topic of the Gray Correlation Analysis. Gray Correlation Grade theories now catch the attention of the scholars at home and abroad, and become an important branch of future developments and researches of the Gray system study area. Applying the Gray Correlation Grade theory to evaluate the morphological characteristics of the main beam can provide reliable analysis theory support for bridge monitoring.

In this paper, a health assessment algorithm for the main beam of beam bridge based on Gray Correlation Grade is proposed, which can solve the shortcomings that evaluation cannot be synthetically in existing technologies.

2 Health Monitoring Algorithm

The algorithm mainly includes the following steps:

Step 1: Define the evaluation strategy

Define the range of the outliers and the reference time range of the main beam deflection data.

Step 2: Obtain the original data of deflection

Read original data obtained by the i sensors installed on the main beam in real time, and store them as arrays in memory. The array is defined as $rawdata_i$ *{{id0, time, value}... {idn, time, value}}*.

Step 3: Filter the deflection data

Classify the scrolling raw data based on the evaluation strategy, dividing the array $rawdata_i$ into the array of reference values $refdata_i$ and the array to be evaluated $candata_i$.

The value of the attribute *value* in the array $rawdata_i$ is filtered based on the range of the outliers of the main beam deflection data, and the arrays in the range of the outliers are discarded and the arrays in the range of the outliers are kept in the array $rawdata_i$.

The value of the attribute *time* in the array $rawdata_i$ is filtered based on the reference time range of the main beam deflection data, and the arrays in the time range of the reference value are divided into the array of reference values $refdata_i$, and the array is not within the time range of the reference values $candata_i$.

Step 4: Calculate reference values

Get the reference mean value a_{i0} and the reference variance value b_{i0} the reference displacement value d_{i0} and the reference numerical sequence L_0 from $refdata_i$.

Calculate the base mean a_{i0} as follows:

$$a_{i0} = \frac{refdata_{i0} + refdata_{i1} + \ldots + refdata_{in}}{n} \tag{1}$$

In formula (1), i is the number of measured points, $refdata_{in}$ is the n-th measured value of the i-th measure.

Calculate the benchmark variance b_{i0} as follows:

$$b_{i0} = \frac{1}{n}\sum_{j=1}^{n}\left|\frac{refdata_{ij} - refdata_{i0}}{refdata_{i0}}\right| \tag{2}$$

In formula (2), i is the number of measured points, $refdata_{ij}$ is the j-th measured value of the i-th measuring point.

Calculate the reference displacement value d_{i0} as follows:

$$d_{i0} = \frac{|refdata_{i0} - refdata_{i1}| + |refdata_{i1} - refdata_{i2}| + \ldots + \left|refdata_{i(n-1)} - refdata_{in}\right|}{n} \tag{3}$$

In formula (3), i is the number of measured points, $refdata_{in}$ is the n-th measured value of the i-th measure.

Calculate the displacement monitoring data $L_{0.}$ The baseline sequence is calculated as follows:

$$L_0 = \{a_{00}, a_{10}, \ldots\ldots, a_{i0}\} \tag{4}$$

In formula (4), d_{00} is the reference mean of the first measurement point, a_{i0} is the reference mean of the i-th measurement point.

Step 5: Evaluate the health characteristics of the main beam indicators

According to the health data of the main beam, calculate the variance variation coefficient, the maximum variance variation coefficient, the maximum displacement change coefficient, the ride coefficient and the curvature change coefficient.

First, calculate the mean change coefficient. Calculate the mean value of the deflection monitoring point along the main beam distribution, and define the mean change coefficient Δa as follows:

$$\Delta a = \frac{1}{n}\sum_{i=1}^{n}|candata_i - a_{i0}| \tag{5}$$

In formula (5), n is point numbers, $candata_i$ is the measured value of the i-th measuring point, a_{io} is the reference mean of the i-th measurement point.

Then calculate the variance change coefficient. Calculate the variance change of the overall deflection of the data at the deflection monitoring point along the main beam, and define the variance change coefficient Δb as follows:

$$\Delta b = \frac{1}{n}\sum_{i=1}^{n}\left|\frac{b_i - b_{i0}}{b_{i0}}\right| \tag{6}$$

In formula (6), Δb is the variance change coefficient of the deflection point, b_i is the measured variance of the i-th measuring point, b_{io} is the reference variance.

Then calculate the maximum variance change coefficient. Define the maximum variance change coefficient Δc as follows:

$$\Delta c = \max\left|\frac{b_i - b_{i0}}{b_{i0}}\right| \tag{7}$$

In formula (7), Δc is the maximum variance change coefficient, b_i is the measured variance of the i-th measuring point, b_{io} is the reference variance.

Then calculate the maximum displacement change coefficient. Define the maximum displacement change coefficient Δd as follows:

$$\Delta d = \max\left|\frac{d_i - d_{i0}}{d_{i0}}\right| \tag{8}$$

In formula (8), d_i is the measured displacement of the i-th measuring point, d_{io} is the reference displacement value.

According to the theory of Gray Correlation Analysis, the ride coefficient $\bar{p}$ is calculated, and the sequence of the j-th of the monitoring points is defined as L_j

$$L_j = \{\text{candata}_j(1), candata_j(2), candata_j(3), \ldots\ldots, candata_j(i)\} \tag{9}$$

Define the correlation coefficient as p_j, the formula is as follows:

$$p_j = r(L_0, L_j) = \frac{1}{i-1}\sum_{k=1}^{i-1}\left(\frac{1}{1 + \left|\frac{L_0(k+1) - L_0(k)}{L_0(k+1)} - \frac{L_j(k+1) - L_j(k)}{L_j(k+1)}\right|}\right), \quad k = 1, 2, 3\ldots, i-1 \tag{10}$$

In formula (10), $L_O\ (k + 1)$ is the $k + 1$ value of the reference numerical sequence, $L_O\ (k)$ is the k-th value of the reference numerical sequence, $L_j\ (k + 1)$ is the $k + 1$ value of the data sequence to be evaluated, $L_j\ (k)$ is the k-th value of the data sequence to be evaluated.

Then normalize the degree of correlation. Take the p_j average as a ride coefficient, define smoothness coefficient of $\bar{p}$ as follows:

$$\bar{p} = \frac{\sum_{n=0}^{j} p_n}{j} \tag{11}$$

In formula (11), j is the number of sampling points, p_j is the correlation coefficient obtained for the vector corresponding to the j-th sample point, $\bar{p}$ is the ride coefficient.

Then calculate the curvature change coefficient. Through the reference value data to obtain the reference deflection deformation curvature s_{i0}, define the data to be evaluated deflection deformation curvature s_i as follows:

$$s_i = d_i'' = \frac{d_{i+1} - 2d_i + d_{i-1}}{\Delta x^2} \tag{12}$$

In formula (12), s_i is the deformation curvature of the i-th measuring point, d_{i+1}, d_i, d_{i-1} are the i + 1 measuring point, i point, i − 1 measuring point deflection deformation value, Δx is the distance between two adjacent measuring points.

The mean value of the difference of the vector is taken as the variation coefficient of curvature, and the variation coefficient of curvature is defined as Δs, and its formula is as follows:

$$\Delta s = \frac{1}{n} \sum_{i=1}^{n} \left| \frac{s_i - s_{i0}}{s_{i0}} \right| \tag{13}$$

The data are scored based on the scoring rules and the weight table. Calculate the health status score F, the formula is as follows:

$$F = S(\Delta a)w_a + S(\Delta b)w_b + S(\Delta c)w_c + S(\Delta d)w_d + S(\bar{p})w_p + S(\Delta s)w_s \tag{14}$$

In formula (14), F is the total score of linear evaluation, full score is 100 points, $S(\Delta a)$ is the mean change coefficient score value, w_a is the mean change coefficient weight, $S(\Delta b)$ is the variance change coefficient score value, w_b is the variance change coefficient weight, $S(\Delta c)$ is the maximum variance change coefficient w_c is the maximum variance change coefficient weight, $S(\Delta \bar{p})$ is the ride factor coefficient score value, w_p is the ride coefficient weight, $S(\Delta s)$ is the curvature change coefficient score value, w_s is the curvature change coefficient weight.

3 Experiment and Result

The experiment data were randomly selected from July 4 to July 18, 2016, a total of 9 days data of Wohe Bridge in Anhui Province for analysis. The data acquisition frequency is 1 Hz. After statistical analysis, calculate the maximum displacement of the monitoring points as shown in Table 1, the displacement mean as shown in Table 2.

Table 1. The maximum displacement monitoring table (mm)

Date	Monitoring point1	Monitoring point2	Monitoring point3	Monitoring point4	Monitoring point5
2016-7-4	−1.879	−1.602	−0.249	−20.081	−16.15
2016-7-5	−2.085	−1.56	0.176	−9.999	−16.153
2016-7-6	−1.841	−1.404	0.038	−17.71	−15.89
2016-7-13	−2.048	−1.406	−0.272	−21.094	−16.003
2016-7-14	−82.111	−58.193	−58.564	−60.877	−60.38
2016-7-15	−2.438	−1.727	−0.599	−20.436	−16.458
2016-7-16	−2.415	−1.708	−0.668	−9.998	−16.384
2016-7-17	−82.114	−58.196	−58.567	2.048	−60.348
2016-7-18	−2.143	−1.483	−0.221	−8.959	−15.842

Table 2. The mean displacement monitoring table (mm)

Date	Monitoring point1	Monitoring point2	Monitoring point3	Monitoring point4	Monitoring point5
2016-7-4	−1.81	−1.54	−0.14	−18.43	−16.01
2016-7-5	−1.80	−1.36	0.02	−13.83	−15.76
2016-7-6	−1.82	−1.39	0.02	−17.25	−15.85
2016-7-13	−1.98	−1.39	−0.24	−20.04	−15.96
2016-7-14	−2.01	−1.38	−0.13	−13.97	−15.86
2016-7-15	−2.25	−1.63	−0.47	−19.40	−16.26
2016-7-16	−2.21	−1.65	−0.50	−15.05	−16.22
2016-7-17	−2.11	−1.46	−0.24	−7.12	−15.78
2016-7-18	−2.13	−1.47	−0.21	−8.63	−15.82

With the data of Tables 1 and 2, the mean result were drawn line chart of the main beam under permanent load condition as shown in Fig. 1.

As you can see from Fig. 1, there was a large fluctuation of displacement at No. 4 monitoring point at 80 m of the bridge which is under permanent load condition during the monitoring period. So we initially suspected there was something wrong of this bridge. Through the calculation of basic grey relational degree algorithm, the real health data of the bridge was obtained after analysis, the evaluation results are listed in Table 3.

According to the calculated values, the line chart of the bridge can be evaluated. Referring to 《Rules for inspection and assessment of bearing capacity of Highway Bridges》 (2003) and 《Technical code for highway maintenance》 (JTGH11-2004), the technical state of bridge is divided into five categories. The first class is intact and in good state; the second class is in good state; the third class is in poor state; the fourth class is in bad state; the fifth class is in dangerous state. The specific classification criteria are shown in the Table 4.

From Table 4, this bridge's Technical condition evaluation is Grade III, a poor state. It needs a maintenance work.

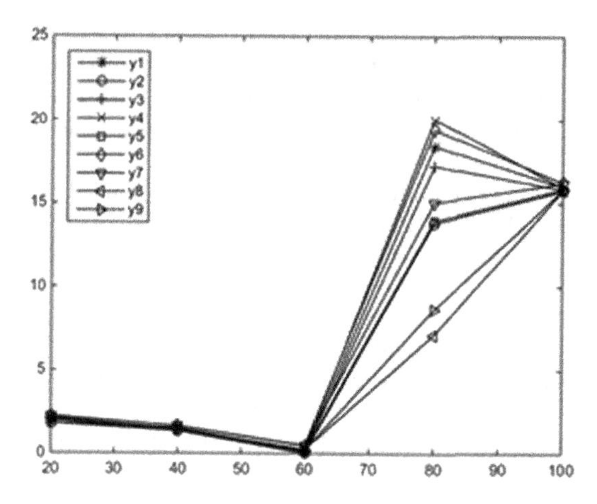

Fig. 1. Line chart of main beam

Table 3. Linear evaluation results table

Ref	Δa	Δb	ai0	Curv
Def1	−2.11	0.70	−2.11	2.00
Def2	−1.54	0.50	−1.54	2.00
Def3	−0.34	0.53	−0.34	2.00
Def4	−16.45	3.65	−16.45	3.00
Def5	−16.10	0.43	−16.10	2.00

Table 4. Technical condition evaluation

Grade	I	II	III	IV	V
Value	80–100	60–80	40–60	20–40	0–20

4 Conclusion

The evaluation algorithm proposed in this paper can make the bridge supervisors quick to judge the overall health status of the bridge and the damage status of the main beams compared with the existing technology. This method makes the monitoring data of the bridge have the characteristics of standardization and high efficiency, and improves the robustness and ease of use of the system, and has high engineering application value.

References

1. Lee, Y.-L., Pan, J., Hathaway, R.B., Barkey, M.E.: Fatigue Testing and Analysis (Theory and Practice). Elsevier, Amsterdam (2005)
2. Stuart, L.M.: Data Analysis for Science and Engineering. Wily, Hoboken (1975)
3. Liu, S.: Gray System Theory and Application. Science Press, October 2004. (in Chinese)

4. Yi, T., Li, H., Gu, M.: Recent research and application of GPS based technology for bridge health monitoring. J. Sci. China **153**(10), 2597–2610 (2010). (in Chinese)
5. Ang, Y., Li, Z.: Parameter estimation and test of fatigue stress probability distribution function of large span bridge based on monitoring samples. Mod. Traffic Technol. **5**(6), 17–21 (2008)
6. Morris, A.J.: Foundations of Structural Optimization: A Unified Approach. Wiley Inter Science Publication, Hoboken (1982)
7. Fox, R., Kapoor, M.: Rate of change of eigenvalues and eigenvectors. AIAA J. **6**, 2421–2429 (1968)
8. Moller, P.W., Friberg, O.: Updating large finite element models in structural dynamics. AIAA J. **36**(10), 1861–1868 (1998)
9. Zhao, J., De Wolf, J.T.: Sensitivity study for vibrational parameters used in damage detection. J. Struct. Eng. **125**(4), 410–416 (1999)
10. Pmdey, A.K., Biswas, M.: Damage detection in structures using changes in flexibility. J. Sound Vib. **169**(1), 3–17 (1994)

Designing an Optimal Water Quality Monitoring Network

Xiaohui Zhu[1,3], Yong Yue[1(✉)], Yixin Zhang[2], Prudence W.H. Wong[3], and Jianhong Tan[4]

[1] Department of Computer Science and Software Engineering, Xi'an-Jiaotong Liverpool University, Suzhou 215123, Jiangsu Province, People's Republic of China
{xiaohui.zhu,yong.yue}@xjtlu.edu.cn

[2] Research Institute of New-Type Urbanization, Huai'an 223005, Jiangsu Province, People's Republic of China
yixin.zhang@xjtlu.edu.cn

[3] Department of Computer Science, University of Liverpool, Liverpool L69 3BX, UK
pwong@liverpool.ac.uk

[4] Jiangsu Province Hydrology and Water Resources Investigation Bureau Suzhou Branch, Suzhou 215011, Jiangsu Province, People's Republic of China
jianhong_tan@qq.com

Abstract. The optimal design of water quality monitoring network can improve the monitoring performance. In addition, it can reduce the redundant monitoring locations and save the investment and costs for building and operating the monitoring system. This paper modifies the original Multi-Objective Particle Swarm Optimization (MOPSO) to optimize the design of water quality monitoring network based on three optimization objectives: minimum pollution detection time, maximum pollution detection probability and maximum centrality of monitoring locations. We develop a new initialization procedure as well as a discrete velocity and position updating function to optimize the design of water quality monitoring network. The Storm Water Management Model (SWMM) is used to model a hypothetical river network which was studied in the literature for comparative analysis of our work. We simulate pollution events in SWMM to obtain all the pollution detection time for all the potential monitoring locations. Experimental results show that the modified MOPSO can obtain steady Pareto frontiers and better optimal deployment solutions than genetic algorithm (GA).

Keywords: Optimal water quality monitoring network · Multi-objective optimization · SWMM · Closeness centrality

1 Introduction

River systems play a crucial role in the sustainable development of a community. However, industry and living activities are creating more and more pollutants to freshwater sources. It is estimated that 40 billion dollars are lost each year in China due

Z. Shi et al. (Eds.): ICIS 2017, IFIP AICT 510, pp. 417–425, 2017.
DOI: 10.1007/978-3-319-68121-4_45

to freshwater pollution events [1]. Water quality monitoring has become one of the routine efforts for environmental protection all over the world. However, monitoring water quality remains a very complex process due to the large number of factors to consider such as monitoring locations, selection of water quality parameters, monitoring frequency and identification of monitoring objectives [2]. The costs of building and operating an automatic monitoring station are also very high (about 500,000–600,000 dollars per station for construction and 14,000 dollars per year for operating and maintaining). Planning and optimizing water quality monitoring networks have been addressed since 1940s. Dozens of papers have been published on this subject [3–6].

In this paper, we proposed a modified MOPSO algorithm to optimize the design of water quality monitoring network based on three optimization objectives: maximum pollution detection probability, minimum pollution detection time and maximum centrality of a monitoring network. Two discrete functions were developed to calculate particle's velocity and update particle's position respectively. Experimental results show that our algorithm can get steady Pareto frontiers and obtain better optimal deployment solutions than genetic algorithm (GA).

2 Methodology

2.1 Hypothetical River Network

To compare our study results with the achievements given by the literature (Ouyang et al. [6]; Telci et al. [7, 11]), we use the same hypothetical river network as Fig. 1 shows, there are 6 inlet locations (1, 3, 5, 8, 10, 11), 5 intermediate locations (2, 4, 6, 7, 9) and 1 outlet location (12) in the river network. We assume that a pollution event can occur at any location randomly with the same amount of pollutant spilling and there is only one pollution event at each time. We simulate the water flows for 24 h from 00:00 to 23:59 with a steady water flow of 10 ft^3/s for each inlet location. The pollution event occurs at 10:00 and lasts for 1 h during the simulation. We also assume that the pollutant concentration is 10 mg/L. The remaining characteristics of the river network is shown in Table 1, which is the same as Telci used [7].

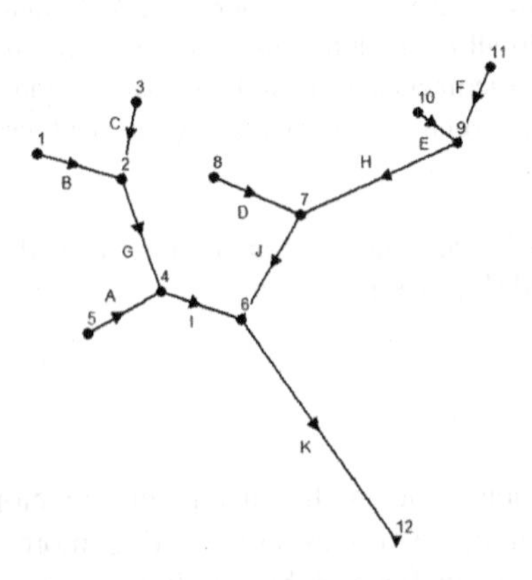

Fig. 1. Hypothetical river network

Table 1. Hydraulic characteristics of the river network

Catchment	Width (ft)	Channel's slope	Manning's coefficient	Flow rate (ft^3/s)
A	10	0.0001	0.02	10
B	10	0.0001	0.02	10
C	10	0.0001	0.02	10
D	10	0.0001	0.02	10
E	10	0.0001	0.02	10
F	10	0.0001	0.02	10
G	10	0.0001	0.02	20
H	10	0.0001	0.02	20
I	10	0.0001	0.02	30
J	10	0.0001	0.02	30
K	10	0.0001	0.02	60

2.2 Hydraulic Simulations

The Storm Water Management Model (SWMM) is a dynamic rainfall-runoff simulation model used for single event or long-term (continuous) simulation of runoff quantity and quality from primarily urban areas [8]. It is widely used for dynamically simulating storm water runoff and drainage systems in urban areas. Here we use SWMM to simulate the hydraulic model, pollution events and pollutants transport along the river system.

We set the pollution detection threshold to 0.01 mg/L and run hydraulic simulations in SWMM. Table 2 shows the simulation results of pollution detection time for each potential monitoring location. The "_" in Table 2 represents an infinite value, which means the pollution event cannot be successfully detected at a monitoring location.

Table 2. Pollution detection time at each potential locations

Locations	Pollution detection time and probability at each location											
	1	2	3	4	5	6	7	8	9	10	11	12
1	0	27	_	81	_	118	_	_	_	_	_	198
2	_	0	_	40	_	75	_	_	_	_	_	152
3	_	27	0	81	_	118	_	_	_	_	_	198
4	_	_	_	0	_	23	_	_	_	_	_	96
5	_	_	_	28	0	62	_	_	_	_	_	139
6	_	_	_	_	_	0	_	_	_	_	_	62
7	_	_	_	_	_	38	0	_	_	_	_	113
8	_	_	_	_	_	79	27	0	_	_	_	157
9	_	_	_	_	_	111	57	_	0	_	_	190
10	_	_	_	_	_	133	78	_	10	0	_	213
11	_	_	_	_	_	156	99	_	27	_	0	236
12	_	_	_	_	_	_	_	_	_	_	_	0

2.3 Optimization Objectives

(a) Minimum Pollution Detection Time and Maximum Pollution Detection Probability

Telci et al. [7] proposed a real-time optimal monitoring network design in river networks. They designed two optimization objectives of minimum pollution detection time and maximum pollution detection. A genetic algorithm (GA) was used to optimize the water quality monitoring network according to these two optimization objectives. Here we also use the same two objectives in our algorithm. The detailed definition of these two optimization objectives can be found in [7].

(b) Maximum Centrality of Monitoring Network

We argue that in practical environment, different potential monitoring locations may have different monitoring priorities, which should be considered when we design a monitoring network. In river network simulations, graph theory and network analysis are usually used to model river systems. The centrality is one of the most important indicators in graph theory. It can identify the importance of vertexes in a graph or network. Here we use the closeness centrality as an evaluation criterion of location priority. The closeness centrality for each potential monitoring location is described as Formula 1.

$$\mathrm{C(i)} = \left[\frac{\sum_{j=1}^{m} d(i,j)}{m-1}\right]^{-1} \tag{1}$$

where m is the total number of potential monitoring locations, $d(i, j)$ is the length from location i to location j. The total closeness centrality for a potential deployment solution S_k is shown in Formula 2.

$$\mathrm{C}(S_K) = \sum_{i=1}^{n} C(S_{ki}) \tag{2}$$

where n is the number of monitoring devices deployed in a river network, S_{ki} is a monitoring location in deployment solution S_k.

The third optimization objective is to maximize the total closeness centrality for all deployment solutions, which is shown in Formula 3.

$$C(S) = \max\{C(S_1), C(S_2), \ldots, C(S_T)\} \tag{3}$$

where T is the total number of potential deployment solutions.

2.4 MOPSO Algorithm

Assume that we will deploy n monitoring devices in a river system out of m potential monitoring locations ($n \leq m$). It is easy to know that the total number of potential deployment combinations T is:

$$T = C_m^n \tag{4}$$

We can find from Formula 4 that when we increase the value of m and/or n, the number of deployment solutions will be also increased exponentially, which is too large to calculate using enumeration search methods within a reasonable time. In addition, these optimization objectives normally conflict with each other, which means we aim to find some good trade-off solutions among multi objectives [9, 10].

Multi-objective Particle Swarm Optimization (MOPSO) is one of the popular evolution algorithms used in recent year. Coello et al. [10] compared MOPSO against three state-of-the art multi-objective evolutionary algorithms of Nondominated Sorting Genetic Algorithm II (NSGA-II), Pareto Archived Evolution Strategy (PAES) and Microgenetic Algorithm for Multi-Objective Optimization (MicroGA) using 5 different test functions. Experimental results show that MOPSO has a highly competitive performance and can be considered a viable alternative to solve multi-objective optimization problems with low computational time. Here we use MOPSO to design an optimal water quality monitoring network. The velocity and position of particles during the computing iteration are updated by the following equations:

$$V_i(t+1) = w\ V_i(t) + c_1 r_1 (pbest(i,t) - P_i(t)) + c_2 r_2 (gbest(t) - P(t)) \tag{5}$$

$$P_i(t+1) = P_i(t) + V_i(t+1) \tag{6}$$

where V denotes the particle's velocity, w is an inertia weight constant, r_1 and r_2 are uniformly distributed random variables within range [0, 1], $pbest(i, t)$ is the best position that the particle i has had, $gbest(t)$ is the best position in all current particles, and c_1 and c_2 are positive constant coefficients for acceleration.

The classical MOPSO is a powerful algorithm to get global optimal results for continuous definition domains. However, it cannot be applied to discrete problems directly. In this paper, we redesign the initialization procedure and the velocity and position calculation function to optimize this discrete issue.

(a) Particle Design and Swarm Initialization

Assume we select n locations to deploy water quality monitoring devices out of m potential monitoring locations in a river network ($n \leq m$). Each potential monitoring location is named from 1 to m respectively resulting in a location set $S = \{1, 2, 3, \ldots, m\}$. Each particle in a swarm denotes a deployment solution with n monitoring locations. Therefore, each particle has n positions and each position represents a monitoring location in set S. As a result, particle P can be defined as a vector with n elements shown in Formula 7.

$$\begin{gathered} P = [p_1 p_2 \ldots p_i \ldots p_n] \\ subject\ to\ n \leq m\ \&\ 1 \leq p_i \leq m \end{gathered} \tag{7}$$

Assume we create k initial particles, we use a random integer function to initialize n positions and velocities for each particle in MOPSO. The swarm initialization

```
procedure INITIALIZATION (Integer k)
for i=1 to k do
   particle(i).position = [ ];
   particle(i).velocity = [ ];
   for j=1 to n do
      particle(i).position(j) = randomi(1, m);
      particle(i).velocity(j) = 0;
   end for
end for
end procedure
```

Fig. 2. Pseudocode of MOPSO initialization

```
procedure VEL_POS_UPDATING (int k)
MaxVel = round ((m − 1)/10);
for i=1 to k do
    for j=1 to n do
        particle(i).velocity(j) = round(w * particle(i).velocity(j)
        +c1 * r1 * (particle(i).best.position(j) − particle(i).position(j))
        +c2 * r2 * (reph.position(j) − particle(i).position(j)));
        particle(i).velocity(j) = min(max(particle(i).velocity(j), −MaxVel), +MaxVel);
        particle(i).position(j) = particle(i).position(j) + particle(i).velocity(j);
        if particle(i).position(j) < 1 or particle(i).position(j) > m then
            particle(i).velocity(j).flag = −particle(i).velocity(j).flag;
            particle(i).position(j) = min(max(particle(i).position(j), 1), m);
        end if
    end for
end for
end procedure
```

Fig. 3. Pseudocode of velocity and position updating

procedure is as Fig. 2 shows. The position value for all the particles is constrained between 1 and m. All the velocities are initialized to 0.

(b) Velocity and Position Updating

Equations 5 and 6 show that the original velocity and position in MOPSO are both real values. However, we use integers to denote monitoring locations in a particle. Figure 3 shows a procedure to update particle's velocity and position. We let *MaxVel* be a maximum velocity during calculation. A *round* function is used to calculate a new integer value of velocity for each particle based on current *gbest* and *pbest* particles. Because the new velocity may be out of the boundary of [−*MaxVel*, +*MaxVel*], we use *max* and *min* functions to restrict the velocity scope. The detailed velocity updating process is shown in Fig. 4.

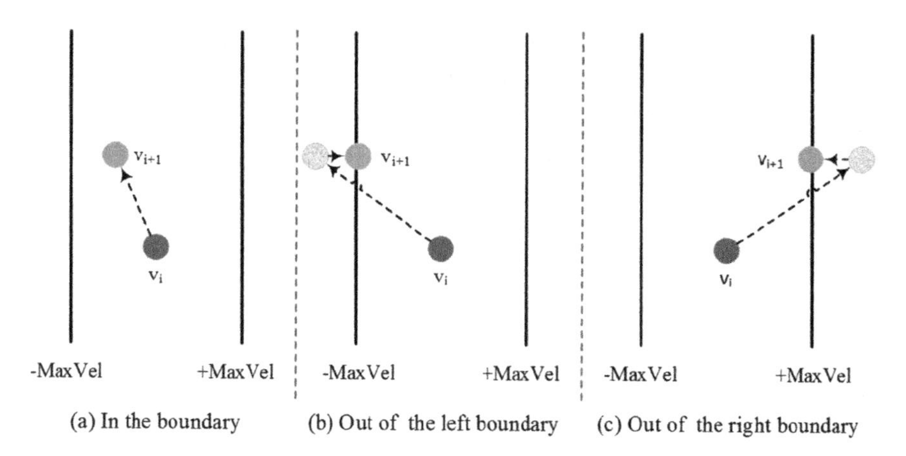

Fig. 4. Velocity updating process

3 Simulations and Analysis

We run the MOPSO algorithm several times based on data in Table 2 and get several Pareto frontiers as Fig. 5 shows. The simulation results show that these Pareto frontiers are almost the same except for few (one or two) Pareto frontier particles. It means that the simulation results of MOPSO algorithm are quite steady and we can use it to design an optimal water quality monitoring network. Table 3 shows the optimal deployment solutions based on the Pareto frontier particles.

We can find from Table 3 that if we deploy 3 monitoring devices at locations 6, 9 and 12, we can detect all the potential pollution events and the pollution detection time is only 45.8 min. The centrality of this monitoring network is 0.0414. If we deploy monitoring devices at locations 6, 7 and 12, potential pollution events can still be all detected. However, the pollution detection time is increased to 54.8 min. If the monitoring devices are deployed at locations 4, 6 and 9. The pollution detection time is significantly decreased to 34.9 min while the detection probability is slightly decreased

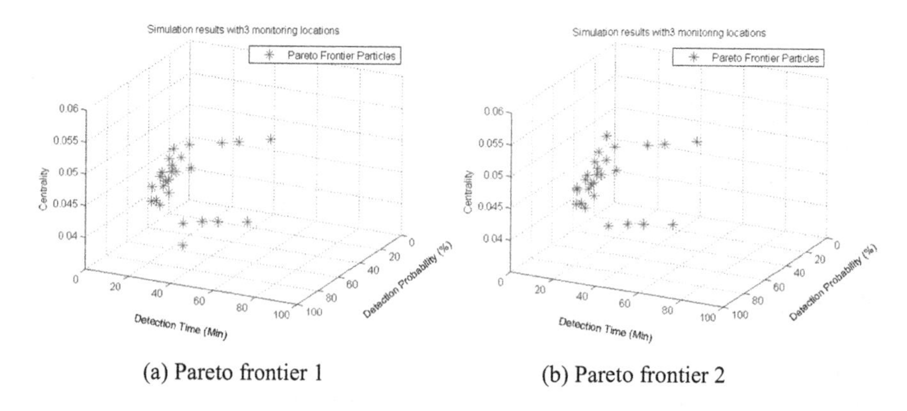

Fig. 5. Hypothetical river network

Table 3. Optimal deployment solutions

Monitoring locations	Detection time (min)	Detection probability	Centrality
6, 9, 12	45.8	100%	0.0414
6, 7, 12	54.8	100%	0.0455
4, 6, 9	34.9	91.7%	0.0500
2, 6, 7	36.4	91.7%	0.0514
4,6,7	44.6	91.7%	0.0561
4, 7, 9	29.4	83.3%	0.0487
2, 4, 7	34.3	83.3%	0.0505
2, 7, 9	14.8	66.7%	0.0451
2, 4, 9	14.9	66.7%	0.0455
2, 5, 9	13	58.3%	0.0420
5, 7, 9	10.7	50%	0.0447
7, 8, 9	7.4	41.7%	0.0444
2, 4, 5	10.8	41.7%	0.0466
5, 9, 11	2.5	33.3%	0.0379
2, 3, 5	6.75	33.3%	0.0401
5, 8, 10	0	25%	0.0399

to 91.7%, which is also the second highest pollution detection probability solution in our deployment solutions and we can get a better centrality of 0.05. However, the second highest pollution detection probability in Telci's paper is only 83%.

4 Conclusions and Future Work

We presented a novel method based on a modified MOPSO algorithm to design an optimal water quality monitoring network with three optimization objectives of minimum pollution detection time, maximum pollution detection probability, maximum closeness centrality. Results show that the modified MOPSO can get a better optimal deployment solutions than GA.

In the future, this novel approach will be applied to a real case of water quality monitoring network. Further research is planned to explore the feasibility of redesigning the velocity and position calculation procedure to avoid same positions in a particle, which can further improve the computing performance.

Acknowledgements. This work was partly supported by the Natural Science Foundation of Jiangsu Province (BK20151245), the Natural Science Foundation of Huai'an City (HAG2015007), the Natural Science Foundation of Nantong City (MS12016048).

References

1. Yang, Y.: How to look at the GDP receiving much attention. Revolution **12**, 26–32 (2015)
2. Behmel, S., Damour, M., Ludwig, R., et al.: Water quality monitoring strategies - a review and future perspectives. Sci. Total Environ. **571**, 1312–1329 (2016)

3. Park, S.Y., Choi, J.H., Wang, S., et al.: Design of a water quality monitoring network in a large river system using the genetic algorithm. Ecol. Modell. **199**(3), 289–297 (2006)
4. Chilundo, M., Kelderman, P.: Design of a water quality monitoring network for the Limpopo River Basin in Mozambique. Phys. Chem. Earth, Parts A/B/C **33**(8), 655–665 (2008)
5. Chang, C.-L., Lin, Y.-T.: A water quality monitoring network design using fuzzy theory and multiple criteria analysis. Environ. Monit. Assess. **186**(10), 6459–6469 (2014)
6. Ouyang, H.T., Yu, H., Lu, C.H., Luo, Y.H.: Design optimization of river sampling network using genetic algorithms. J. Water Resour. Plan. Manag. **134**(1), 83–87 (2008)
7. Telci, I.T., Nam, K., Guan, J., Aral, M.M.: Real time optimal monitoring network design in river networks. In: World Environmental and Water Resources Congress 2008: Ahupua'A, pp. 1–10 (2008)
8. Rossman, L.A.: Storm water management model user's manual, version 5.0. National Risk Management Research Laboratory, Office of Research and Development, US Environmental Protection Agency, Cincinnati (2010)
9. Reyes-Sierra, M., Coello, C.C.: Multi-objective particle swarm optimizers: a survey of the state-of-the-art. Int. J. Comput. Intell. Res. **2**(3), 287–308 (2006)
10. Coello, C.A., Lechuga, M.: MOPSO: a proposal for multiple objective particle swarm optimization. In: Congress on Evolutionary Computation, pp. 1051–1056 (2002)
11. Telci, I.T., Nam, K., Guan, J., Aral, M.M.: Optimal water quality monitoring network design for river systems. J. Environ. Manag. **90**(10), 2987–2998 (2009)

Ship Identification Based on Ship Blade Noise

Haitao Qi(✉) and Zhijing Xu

College of Information Engineering, Shanghai Maritime University,
Shanghai, China
932392658@qq.com

Abstract. In this paper, we propose a new method about ship identification, which is based on paddle noise for classification and recognition. Blade noise is the feature that is used to identify ships, blade noise is collected by digital hydrophone, then the signal is converted to spectrum. We extract features based on the difference of the pixel in the picture. What's more, the softmax classification function is used to training blade noise samples by combining with CDBN technology. The maximal output is the sample that satisfies the minimum cost function. The results of our experiment show that the method can be applied for ships identification by combining CDBN, spectrum and softmax functions, the classification recognition rate increases with the increase of the convolution kernel.

Keywords: Blade noise · Digital hydrophone · Softmax · CDBN

1 Introduction

In recent years, more and more ship identification methods are forming, because of the development of pattern recognition and artificial intelligence. For example: wavelet analysis theory, support vector machine (SVM) [1], neural network and deep Learning model [2]. Of course, there are other research methods, such as AIS and XAIRCRAFT XCope UAV vision system and Ship identification based on ship blade noise [3].

The traditional ship identification generally uses the theory of wavelet analysis to analyze the characteristics and characteristics of ship noise. The wavelet packet technique is used to separate the characteristics of different frequency bands, and the maximum and outband energy is extracted in the wavelet band. This method is cumbersome and complex, and SVM is valid for small samples. In the neural network of ship identification, the DEMON spectrum is extracted by spectral estimation method, and then BP neural network is used to classify and identify, but the feature extraction can not represent the ship noise well, which is not conducive to the actual operation.

At present, the deep learning technology development is rapid, it can forms a more abstract high-level representation of attribute categories or characteristics by combining low-level features. Convolutional Deep Belief Networks (CDBN) [4] is a method of deep learning, it has been successfully applied in recognition tasks, such as handwritten character recognition, face recognition and pedestrian detection. The experiment of the results show that CDBN model can extract features effectively.

In this paper, the noise of different types of ship blades are simulated by setting the number and size of the blades. The traditional neural network and the deep learning

Z. Shi et al. (Eds.): ICIS 2017, IFIP AICT 510, pp. 426–434, 2017.
DOI: 10.1007/978-3-319-68121-4_46

model are used respectively. Speedboat model's blades can produce different noise signals and simulate the cavitation noise [5] by using different blades, and then signals are processed by the convolutional belief neural network (CDBN).

2 Noise Collection of Ship Blade Signal

Ship radiation noise is mainly composed of propeller noise, mechanical noise and hydrodynamic noise. During the high speed of the ship, the noise generated by the propeller blades is divided into vacuosity and non-voiding noise. Vacuum noise is the main noise in ship sailing.

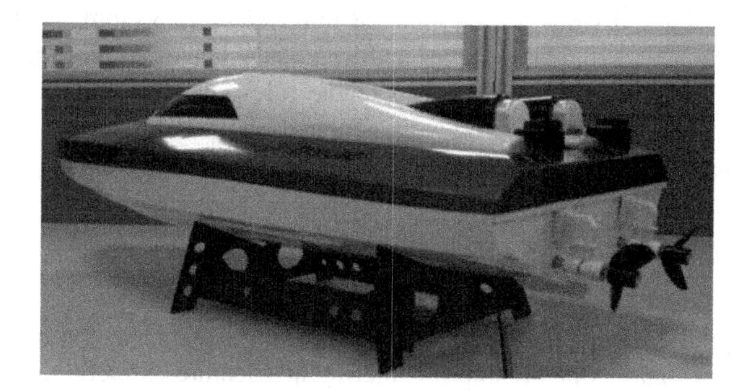

Fig. 1. Speedboat model

As the experimental conditions are limited, this experiment can not collect propeller noise signal, the signals are simulated sampling in still water by a speedboat model, speedboat model is shown in Fig. 1. The experiment simulated different ships By changing the type and size of the propeller blades constantly. Noise of the software control interface is shown in Fig. 2.

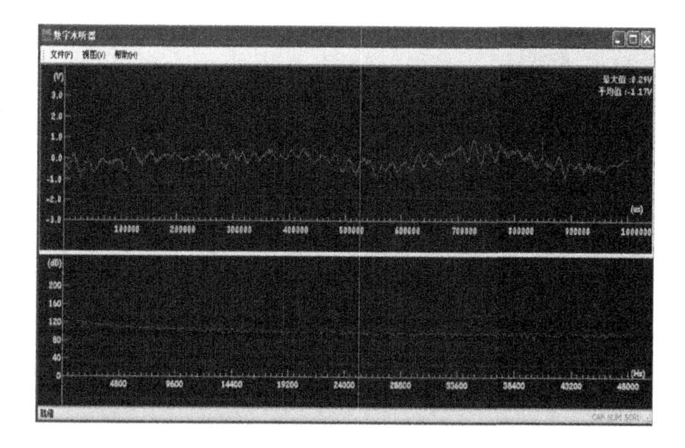

Fig. 2. Software control interface

Table 1. Number of different types of blades collected signal

Size and model	24 mm	28 mm	32 mm	36 mm	40 mm	44 mm
Two blades	150	150	150	150	150	150
Three blades	150	150	150	150	150	150
Four blades	150	150	150	150	150	150
Irregular four blades	150					
Irregular seven blades	150					

The ship blade noise signal can be divided through the number of blades and the blade size of the propeller. In this experiment, five different blades of the propeller: 2 leaves, 3 leaves, 4 leaves, irregular 4 leaves, irregular 7 leaves. The rules of the ship propeller blades are divided to 24 mm, 28 mm, 32 mm, 36 mm, 40 mm, 44 mm according to the size; Irregular 4 leaves and irregular 7 leaves are not required in size. The signals of irregular blades are added to the rules signal to enhance the generalization ability of the deep neural network. Table 1 shows the number of signals collected for different types of propeller blades.

3 Blade Noise Spectrum

DHP8501 digital hydrophone collects the blade noise signal in the form of binary. Since the binary data is not suitable for filtering and feature extraction, the file format must be converted into a wav format, it is convenient to feature extraction and deep learning, so the binary file is converted to wav format files.

The sound signal is subjected to a windowing function and then subjected to Fourier transform. The resulting amplitude is represented by a spectrum with different shades of color, which is the phonogram of the sound signal we have obtained. In this paper, the blade noise signal is processed in the form of a spectrum similar to the speech signal

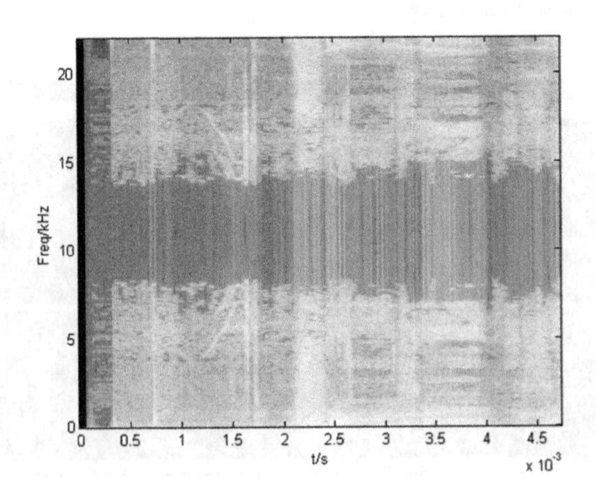

Fig. 3. Impeller noise spectrum

processed in the form of speech spectrum. From the nature of the spectrum, we can see that the two-dimensional spectrum expresses the three-dimensional characteristics of the sound signal, which is the time characteristic of the horizontal axis, the frequency characteristic of the vertical axis and the energy characteristic of the gray scale in the spectrum.

Taking the pulse signal of the speedboat model as an example, this signal collects one of the 20 kinds of blade noise signals of the speedboat model. The number of blades is 4 leaves, the blade size is 32 mm, and the time of blade noise signal is 30 s. The wav format sound signal is represented by a spectrum, as shown in Fig. 3.

4 Dimension Reduction and Classification Processing

4.1 Probability of the Largest Pool Layer

In order to solve the problem that the feature value is too large to fitted for the classifier, the pool layer is added to the classifier. Pooling generally uses the method of aggregate statistics to find the characteristic mean for different positions [6]. The pooling process not only reduces the dimensionality of the feature but also enhances the generalization ability of the deep model, which can effectively prevent overfitting.

In the deep learning model [7], there are three types of pooling, which are named average pooling, maximum pooling and random pooling. The average pooling is the average of the numbers in the pooled domain. The maximum pooling is the characteristic point of the neighborhood. The random pool is probability assigns a variable to sampling point and sub-sampled by probability. Convolutional Restricted Boltzmann Machines (CRBM) [8] uses random pooling.

As can be seen from Fig. 4, the hidden layer and the pooling layer have the same number of units, it set as k group according to the mentioned above, the total of pool layer binary is $N_P \times N_P$. Defining a binary interval $B_\alpha \triangleq \{(i,j)\}$: h_{ij} belong to α. The area block B_α of the hidden layer is fastened to the cell of the pooled layer. Once the hidden layer of cell area block is activated, which is equals to the corresponding cell block is activated.

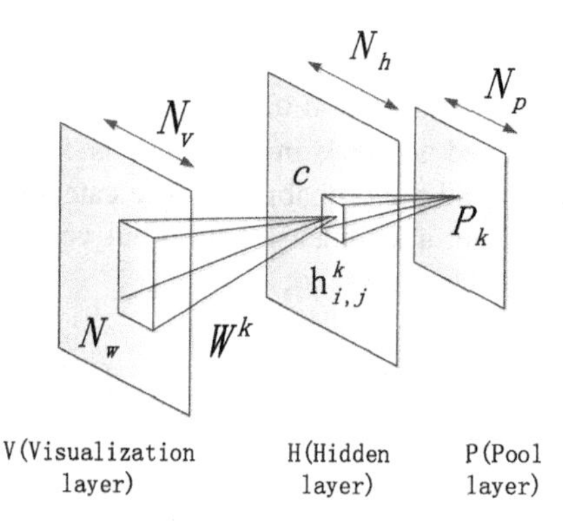

Fig. 4. Added the pool layer to conditional restricted boltsman machine (CRBM)

Then, the maximum pooled CRBM energy function is:

$$E(v,h) = -\sum_{k}\sum_{i,j}(h_{i,j}^{k}(\tilde{W}^{k} * v)_{i,j} + b_{k}h_{i,j}^{k}) - c\sum_{i,j}v_{i,j} \tag{1}$$

The corresponding conditional probability is:

$$P(h_{i,j}^{k} = 1 \mid v) = \frac{\exp(I(h_{i,j}^{k}))}{1 + \sum_{(i',j')\in B_{\alpha}}\exp(I(h_{i',j'}^{k}))} \tag{2}$$

$$P(p_{\alpha}^{k} = 0 \mid v) = \frac{1}{1 + \sum_{(i',j')\in B_{\alpha}}\exp(I(h_{i',j'}^{k}))} \tag{3}$$

In the above equation, $-I(h_{i,j}^{k})$ is the energy increment expression of the activation function, the form follow as:

$$I(h_{i,j}^{k}) \triangleq b_{k} + (\tilde{W}^{k} * v)_{i,j} \tag{4}$$

4.2 Softmax Classification Function

The softmax classifier is a logistic regression function classifier that can be classified as multiple categories, which is a commonly used classifier for the deep learning classification model. The basic principle is to calculate the probability value of all independent samples, the probability value which meets the minimum sample regard as the optimal sample output, so as to achieve the effect of classification. Sigmoid is the activation function of the softmax classifier:

$$g(z) = \frac{1}{1 + e^{-z}} \tag{5}$$

the way of mathematical description is the same as below. Supposing there are m training sample: $\{(\mathrm{x}^{(1)}, \mathrm{y}^{(1)}), \ldots, (\mathrm{x}^{(m)}, \mathrm{y}^{(m)})\}$ For sample x, there is always a corresponding k-dimensional vector y as classification label. Then for $\{(\mathrm{x}^{(1)}, \mathrm{y}^{(1)}), \ldots, (\mathrm{x}^{(m)}, \mathrm{y}^{(m)})\}$ with $\mathrm{y}^{(\mathrm{m})} \in \{1, 2, \ldots, k\}$, the dimension of x is $\mathrm{n}+1$, $\theta = \{\mathrm{W_{ij}}, \mathrm{b_j}\}$, then the probability that the class is judged to be k in the input is x is the $p(y = j|x)$, and the probability is chosen among all the probabilities. The category is the optimal solution. The K input classifier will get a k-dimensional output vector, and the sum of all the elements of the vector is 1. Output set to:

$$h_{\theta}(x^{(i)}) = \begin{pmatrix} p(y^{(i)} = 1 \mid x^{(i)}; \theta) \\ p(y^{(i)} = 2 \mid x^{(i)}; \theta) \\ \vdots \\ p(y^{(i)} = k \mid x^{(i)}; \theta) \end{pmatrix} = \frac{1}{Z}\begin{pmatrix} \exp(\theta_{1}^{T}(x^{(i)})) \\ \exp(\theta_{2}^{T}(x^{(i)})) \\ \vdots \\ \exp(\theta_{k}^{T}(x^{(i)})) \end{pmatrix} \tag{6}$$

Where the model parameter is $\theta_1, \theta_2, \cdots \theta_k$ and the normalized function is $Z = \sum_{j=1}^{k} \exp\left(\theta_1^T \left(X^{(i)}\right)\right)$. Now solving the probabilistic problem is transformed into solving the model parameters. In order to solve the model function, the cost function of the regression model is introduced. If the parameter makes the value of the cost function reach the minimum, the parameter is the solution. The cost function is set to:

$$J(\theta) - \frac{1}{m}\sum_{i=1}^{m}\sum_{j=1}^{k} 1\{y^{(i)} = j\} \lg \frac{\exp(\theta_1^T(x^{(i)}))}{\sum_{j=1}^{k} \exp(\theta_1^T(x^{(i)}))} \tag{7}$$

For solving the cost function, the usual method is to find the partial derivative of the cost function, that is $j(\theta)$ for each parameter θ_j partial derivative, so that the partial derivative is 0, solving its value, and reversing to solve the sample parameter probability value.

5 Experimental Setup and Analysis of Results

5.1 Spectral Classification and Recognition Model

Using the web tutorial to produce the generated blade noise map into a data set similar to CIFAR-10.

Taking 30,000 samples from the data set and dividing the sample into two cases:

In first case, 10,000 samples will be divided into two groups, one group selected 8,000 samples as a training sample, leaving the 2000 samples as a test sample.

In second case, 20,000 samples will be divided into two groups, one group selected 16,000 samples as a training sample, leaving 4000 samples as a test sample.

First, preprocessed the data set: the first step is to whiten the data, reduce the correlation of adjacent pixels, so as to reduce the correlation between the features, and between the characteristics has the same variance. The second step, the data is divided into several copies, the number of each sample is 100. In this experiment, the CDBN structure is set as: the input layer is $32 \times 32 \times 3$; the convolution layer is $7 \times 7 \times 10$; the pool layer is $2 \times 2 \times 2$; and the second layer structure is set as: the convolution layer is $7 \times 7 \times 12$; the Pool layer is $2 \times 2 \times 2$. In this experiment, the number of layers and convolution kernel, and the size of the convolution kernel are not specifically prescribed.

5.2 Classification and Recognition on Samples

In this experiment, the classification recognition rate was tested separately, in the first case tested by 10,000 samples, in the second case tested by 20,000 samples. The network structure of CDBN is set as: the input layer is $32 \times 32 \times 3$; the convolution layer is $7 \times 7 \times 10$; the pool layer is $2 \times 2 \times 2$; and the second layer structure is set as: the convolution layer is $7 \times 7 \times 12$; the pool layer is $2 \times 2 \times 2$. The specific parameters of the network are: learning rate is 0.05, the number of iterations is 100 times. According to the above data to get the spectrum CDBN model classification recognition rate as shown in Table 2:

Table 2. The number of samples and the impact of ICA and ASSASE on the classification recognition rate

The way	The first case	The second case
Classification recognition rate	65.2%	68.2%

As can be seen from the table, in general, the sample classification recognition rate is not very high, and the number of samples is high and the classification rate of the model is higher.

5.3 Recognition Rate About CDBN Network Structure

Since the recognition rate of the spectrum increases with the increase of the test sample, so in the test CDBN network structure on the impact of the experiment, only the second group of samples were tested. In this experiment, the experimental data are set as follows: The experiment divided 30,000 samples into two groups, 22,500 as training samples, and 7,500 as test samples. CDBN is a single layer structure, the input layer is $32 \times 32 \times 3$, the pool layer is $2 \times 2 \times 2$. Learning rate is 0.05, the number of iterations are 200 times. The experimental results are shown in Fig. 5 and Table 3.

It can be seen from Fig. 5 that as the number of convolution nuclei is from less to more, the classification recognition is increased firstly, and the classification recognition rate is almost constant at a certain stage. In the case of a certain number, the classification recognition rate is higher. The classification of the kernel increases the classification rate, but the classification recognition rate does not fluctuate in the case of a large number.

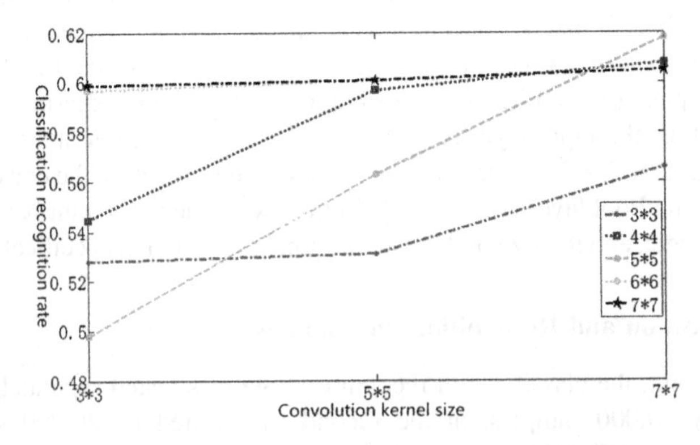

Fig. 5. Influence of network structure on classification recognition rate

Table 3. The influence of CDBN network structure on classification recognition rate

Single layer structure: A layer of convolution + A layer of pooling		
Convolution count	Convolution kernel size	Classification recognition rate
3	3 × 3	52.8%
	5 × 5	52.9%
	7 × 7	56.0%
4	3 × 3	54.6%
	5 × 5	59.6%
	7 × 7	60.3%
5	3 × 3	49.2%
	5 × 5	55.6%
	7 × 7	61.8%
6	3 × 3	59.8%
	5 × 5	60.0%
	7 × 7	60.3%
7	3 × 3	60.0%
	5 × 5	60.3%
	7 × 7	60.7%

6 Conclusions

In this paper, we provide a new method of ship identification, which is based on CDBN technology to carry out ship classification and identification. In the experiment, we demonstrated the possibility of CDBN, spectrum and softmax to identify ships. Experimental results on identification rate of ship show that recognition rate of ship will increase when increasing the samples of the training; the number of convolution kernel is different, the recognition rate is different, too. Correspondingly, when we increase the number of convolution kernel, the recognition rate will increase. Therefore, if we want to improve the recognition rate of ship, we must increase the number of training samples and convolution kernel.

Acknowledgments. This study was funded by the National Natural Science Foundation of China (Grant No. 61404083).

References

1. Suykens, J.A.K., Vandewalle, J.: Least squares support vector machine classifiers. Neural Process. Lett. **9**(3), 293–300 (1999)
2. Tang, Y.: Deep learning using linear support vector machines. arXiv preprint arXiv:1306.0239 (2013)
3. Can, G., Akbaş, C.E., Çetin, A.E.: Recognition of vessel acoustic signatures using non-linear teager energy based features. In: 2016 International Workshop on Computational Intelligence for Multimedia Understanding (IWCIM), pp. 1–5. IEEE (2016)

4. Lee, H., Pham, P., Largman, Y., et al.: Unsupervised feature learning for audio classification using convolutional deep belief networks. In: Advances in Neural Information Processing Systems, pp. 1096–1104 (2009)
5. Harrison, M.: An experimental study of single bubble cavitation noise. J. Acoust. Soc. Am. **24**(6), 776–782 (1952)
6. Adhikari, S.P., Yang, C., Kim, H., et al.: Construction of CNN template via learning with random weight change algorithm. In: CNNA 2016 (2016)
7. Sainath, T.N., Kingsbury, B., Saon, G., et al.: Deep convolutional neural networks for large-scale speech tasks. Neural Netw. **64**, 39–48 (2015)
8. Krizhevsky, A., Hinton, G.E.: Factored 3-way restricted boltzmann machines for modeling natural images. In: International Conference on Artificial Intelligence and Statistics, pp. 621–628 (2010)

A Composite Weight Based Access Network Selection Algorithm in Marine Internet

Liang Zhou[1,2](✉) and Shengming Jiang[1]

[1] College of Information Engineering, Shanghai Maritime University, Shanghai 201306, China
[2] Network and Information Center, Shanghai Municipal Educational Examinations Authority, Shanghai 200433, China
lzhou@shmeea.edu.cn

Abstract. The 21st century is the ocean century, and ocean informatization is one of the key requirements. Marine Internet (MI) is the basic infrastructure of ocean informatization, so developing the MI technology is of great significance. There are several access networks maybe available in a MI system, and a network user has to choose the best one to support the application. The selection scheme needs to consider may factors in order to make a smart decision, which is a multi-index decision problem. These factors include user preferences, business types, and the performance as well as cost of alternative access networks. Combining with the Analytic Hierarchy Process (AHP) algorithm and entropy theory, we develop a method for the user to select access networks in a smart way. It adopts subjective and objective weights and a utility function to select the target access network. The simulation results carried out on the Exata platform show that the proposed scheme can improve the network performance with reduced cost.

Keywords: Marine internet · Access method · Analytic Hierarchy Process (AHP) · Entropy · Exata simulation

1 Introduction

The 21st century is the ocean century, and ocean informatization is one of the key requirements. Marine Internet (MI) is the basic infrastructure of ocean informatization, so developing the MI technology is of great significance In the marine Internet, there are multiple access networks maybe available for selection, which typically include shore-based networks, self-organizing ship networks, high-altitude communication platforms, satellite networks and so on [1]. Network users often need to choose one from a number of access networks to support the current application, considering the user preferences, business type and the performance as well as cost of alternative access networks and other factors.

When multiple access networks are available, a network selection is usually based on a mathematical algorithm that gives a weight to each metric. With the utility function, the comprehensive utility value of each index corresponding to each alternative network is calculated. Finally, the access network with the largest utility value is

Z. Shi et al. (Eds.): ICIS 2017, IFIP AICT 510, pp. 435–444, 2017.
DOI: 10.1007/978-3-319-68121-4_47

selected as the best solution. The problem with such kind of algorithm is that the assignment of weights is dependent on the subjective experience of the operator as mentioned in [2], which then proposed an algorithm based on the combination of subjective and objective weight to determine the weight of the network attributes, taking into account the user preferences and different business types and Quality of Service (QoS). The algorithm works well in a dense user scenario. However, the marine Internet is characterized by sparse distribution of users, which mainly consist of ships. So far, such kind of scenario has not been investigated in the literature. Therefore, this paper combines the Analytic Hierarchy Process (AHP) [3] and the entropy theory to set the subjective and objective weights respectively, and then selects the target network through the utility function.

2 Access Network Selection Algorithm Based on Compound Weights

2.1 Framework of the Proposed Algorithm

See Fig. 1.

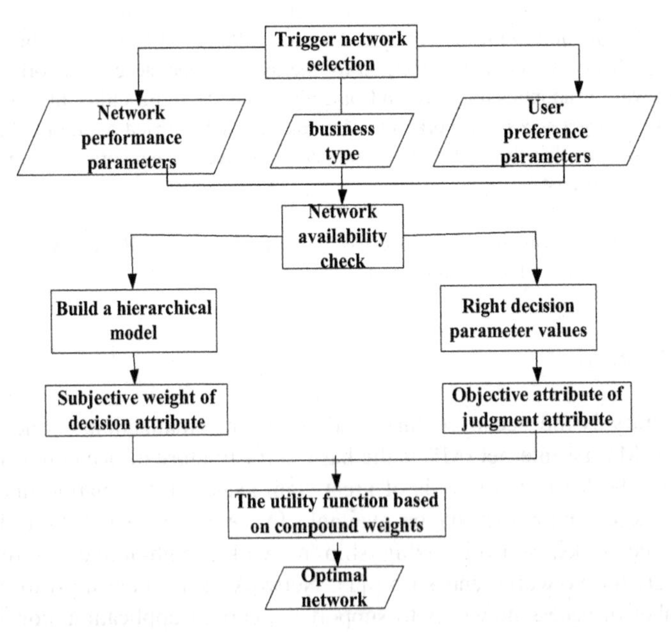

Fig. 1. Framework of the proposed network selection algorithm. It shows the framework of the algorithm. The Analytic Hierarchy Process (AHP) is an effective method to solve the complex multi-index decision problem, and the entropy weight method has objective fairness and can judge the change of the judgment index objectively [4]. Therefore, this paper combines these two algorithms to deal with the access network selection in the marine Internet. The algorithm considers the network performance requirements and user preferences, makes the algorithm have a good applicability.

2.2 Network Selection Model Based on Analytic Hierarchy Process

Many factors affect the heterogeneous network selection, and it is unrealistic to choose an optimal network without considering them comprehensively. We can focus on several common, important factors as an indicator. The users of the same business costs have different tendencies, and some hope to have high-quality services, and some with a pursuit of economic benefits. Furthermore, different types of applications have different basic requirements of network performance. For example, for marine emergency search and rescue applications, the network needs to ensure the stability of communication and timeliness. Based on the above considerations, the AHP algorithm is used to establish the hierarchical model. With consideration of the user preference in the criterion layer, the sub-criterion layer considers the application requirements of the network type.

(1) Hierarchical model of network selection based on user preference

As shown in Fig. 2, first, select the QoS level (α), the network available load (β) and the price charge (γ) as the network performance indicators, and then combine them with the user's network performance bias, QoS priority mode and price priority mode.

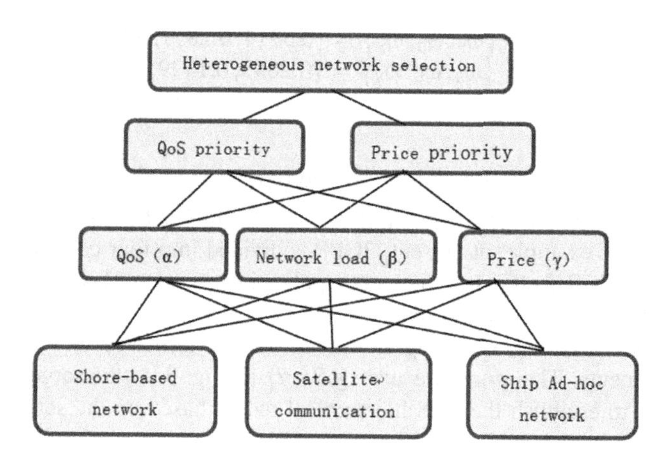

Fig. 2. The hierarchical model based on the user's preferred access network. Users in the QoS priority mode in the access network selection give priority to application requirements with an inclination to the performance of the access network an acceptable price. Under this strategy, the priority of the three network attribute indicators is as follows: QoS level > network load > price. Users in the price priority mode is more concerned about the cost of the access network, while the sensitivity of the network performance is low. Then the three network attribute indicators priority is price > QoS level > network load.

We first establish the two models of the decision matrix, set the two indicators of the importance of each in Table 1, where the values depend on the experience of the decision maker, and can be appropriately adjusted to suit the specific needs of the algorithm.

Table 1. Setting for different strategies under the decision matrix

(a) QoS priority

QoS priority	QoS level (α)	Load (β)	Price (γ)
QoS level (α)	1	2	4
Load (β)	1/2	1	2
Price (γ)	1/4	1/2	1

(b) Price priority

Price priority	QoS level (α)	Load (β)	Price (γ)
QoS level (α)	1	2	1/2
Load (β)	1/2	1	1/4
Price (γ)	2	4	1

The judgment matrix is sorted and the consistency is verified. The maximum Eigen values of the two matrices are calculated as 3, and the calculated values are all zero. Note that the settings of the matrix conform to the consistency requirements. Then, the weights of the indexes in the two modes are obtained and normalized, and the calculated values are all four decimal places.

$$\begin{cases} W_{QoS} = \{\omega_\alpha, \omega_\beta, \omega_\gamma\} = \{0.5714, 0.2857, 0.1429\} \\ W_{Price} = \{\omega_\alpha, \omega_\beta, \omega_\gamma\} = \{0.2857, 0.1429, 0.5714\} \end{cases} \quad (1)$$

(2) QoS hierarchical model based on service type

There are many types of network applications with different network service quality requirements. For example, at present, 3GPP is divided into four categories according to the different QoS features of network application: session class, streaming class, interactive class and background class service [7].

Different services corresponding to the network performance indicators have different requirements. Therefore, we use QoS (α) in Fig. 3 as the target layer of the criterion layer to establish the QoS hierarchical model based on the service type. The quasi-side layer is divided into session class (T1), streaming media class (T2), interactive class (T3), and background class (T4). The subordinate criterion layer includes four parameters: delay (α_1), jitter (α_2), packet loss rate (α_3) and rate (α_4). Take *Session class* (T1) as an example, the network performance indicators between the decision matrixes are set as shown in Table 2.

According to the AHP algorithm, the maximum Eigen values of the decision matrix corresponding to the four business types are calculated, which are 4.0735, 3.8480, 4 and 4, respectively. The four decision matrices pass the consistency test; then calculate the QoS sub-parameters based on different types of business weights as listed in Table 3.

$$W_{QoS\_i} = \{\omega_{Q\_\alpha 1}, \omega_{Q\_\alpha 2}, \omega_{Q\_\alpha 3}, \omega_{Q\_\alpha 4}\}\, i \in \{T1, T2, T3, T4\} \quad (2)$$

Among them, the corresponding weight of the parameters are shown in Table 3.

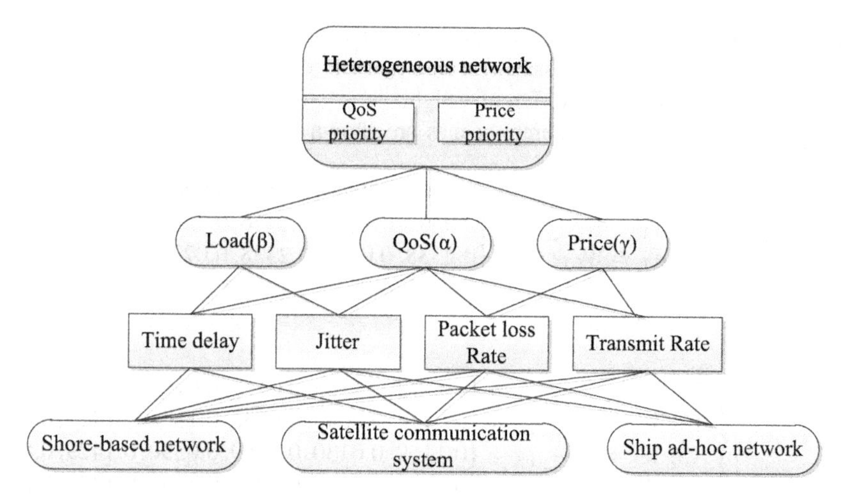

Fig. 3. Network selection hierarchy model. The total rank weights are calculated by single rank weight at each level. Formulas 1 and 2 are used to calculate the total order of each index under different user preference model, different business types, namely for the time delay (α_1), jitter (α_2), the rate of packet loss (α_3), Rate (α_4), price (γ) and subjective weight vector based on different cases: $W_1 = \left\{\omega'_{\alpha_1}, \omega'_{\alpha_2}, \omega'_{\alpha_3}, \omega'_{\alpha_4}, \omega'_{\beta}, \omega'_{\gamma}\right\}$

Table 2. Judgment matrices based on various business types

Session class - T1				
	α1	α2	α3	α4
Delay (α1)	1	1	3	7
Jitter (α2)	1	1	3	7
Packet loss rate (α3)	1/3	1/3	1	5
TxRate (α4)	1/7	1/7	1/5	1

Table 3. The QoS (α) weights of different service types

Type of service	α_1	α_2	α_3	α_4
(T1)	0.3966	0.3966	0.16	0.0468
(T2)	0.0615	0.2273	0.2583	0.4529
(T3)	0.4091	0.0455	0.4091	0.1364
(T4)	0.0601	0.0601	0.2647	0.6150

(3) Ranking the total order and calculating the subjective weight

An overall multi-level hierarchical model based on the network selection problem is drawn in Fig. 3, consisting of two modes of QoS priority and price priority target layer in the model, which represent different user preferences. QoS, available network load and price factors are the center of the rule layer, the weights corresponding to the target layer are calculated with Formula 1. In addition, the sub-rule layer under QoS criterion includes delay, jitter, packet loss rate, traffic rate and other factors, also correspond to different business types with different weights.

The weight vectors for different scenes are shown in Formulas 3 and 4.

$$\text{QoS priority}\begin{cases} \text{T1: } W_1 = W_{C1\_T1} = \{0.2266, 0.2266, 0.0914, 0.0267, 0.2857, 0.1429\} \\ \text{T2: } W_1 = W_{C1\_T2} = \{0.0351, 0.1299, 0.1476, 0.2588, 0.2857, 0.1429\} \\ \text{T3: } W_1 = W_{C1\_T3} = \{0.2338, 0.0260, 0.2338, 0.0779, 0.2857, 0.1429\} \\ \text{T4: } W_1 = W_{C1\_T4} = \{0.0343, 0.0343, 0.1512, 0.3514, 0.2857, 0.1429\} \end{cases} \tag{3}$$

$$\text{Price Priority}\begin{cases} \text{T1: } W_1 = W_{C2\_T1} = \{0.1133, 0.1133, 0.0457, 0.0134, 0.1429, 0.5714\} \\ \text{T2: } W_1 = W_{C2\_T2} = \{0.0176, 0.0649, 0.0738, 0.1294, 0.1429, 0.5714\} \\ \text{T3: } W_1 = W_{C2\_T3} = \{0.1169, 0.0130, 0.1169, 0.0390, 0.1429, 0.5714\} \\ \text{T4: } W_1 = W_{C2\_T4} = \{0.0172, 0.0172, 0.0756, 0.1757, 0.1429, 0.5714\} \end{cases} \tag{4}$$

After the layered model is built up, should calculate the weight of the single layer firstly, and calculate the total weight of the hierarchy finally. Because the decision matrix is set by the decision-maker itself, it has some limitations.

2.3 Objective Weight Based on Information Entropy

In order to eliminate the subjective unilateralism of the total weight obtained by AHP, the proposed method uses entropy weight as the objective weight of judgment. Before calculating the entropy weight, the dimensionless decision attributes are introduced, followed by the method of the dimensionless of the decision attributes.

In a decision problem with m alternative network schemes and n network metrics, the given parameter evaluation matrix is:

$$R' = \begin{pmatrix} r'_{11} & r'_{12} & \cdots & r'_{1n} \\ r'_{21} & r'_{22} & & r'_{21} \\ & \vdots & \ddots & \vdots \\ r'_{m1} & r'_{m1} & \cdots & r'_{mn} \end{pmatrix} \tag{5}$$

For some network parameter indicators, the smaller the better, such as delay; while for the other, the greater the better, such as speed. In order to describe the changing trend of two kinds of indexes in a unified way, normalization is needed.

The delay (α_1), jitter (α_2), packet loss rate (α_3), rate (α_4), available load (β) and network price (γ) are chosen as decision criteria in network parameters. Among them, for the delay (α_1), jitter (α_2), packet loss rate (α_3) and the network price (γ), the smaller the better, and its normalization is:

$$r_{ij} = \frac{\max\left(r'_{1j}, r'_{2j}, \ldots r'_{mj}\right) + \min\left(r'_{1j}, r'_{2j}, \ldots r'_{mj}\right) - r'_{1j}}{\max\left(r'_{1j}, r'_{2j}, \ldots r'_{mj}\right) + \min\left(r'_{1j}, r'_{2j}, \ldots r'_{mj}\right)} \tag{6}$$

For the rate (α_4) and the available load (β), the bigger the better, and the following formula is used for the normalization processing:

$$r_{ij} = \frac{r'_{ij}}{\max\left(r'_{1j}, r'_{2j}, \ldots r'_{mj}\right) + \min\left(r'_{1j}, r'_{2j}, \ldots r'_{mj}\right)} \tag{7}$$

After the parameter is dimensionless, the normalized evaluation matrix R is obtained:

$$R = \begin{pmatrix} r_{11} & r_{12} & \cdots & r_{1n} \\ r_{21} & r_{22} & & r_{21} \\ & \vdots & \ddots & \vdots \\ r_{m1} & r_{m1} & \cdots & r_{mn} \end{pmatrix} \tag{8}$$

Then calculate the network attributes corresponding to entropy weight, composed of objective weight vector W_2 as follows:

$$W_2 = \left\{\omega''_{\alpha 1}, \omega''_{\alpha 2}, \omega''_{\alpha 3}, \omega''_{\alpha 4}, \omega''_{\beta}, \omega''_{\gamma}\right\} \tag{9}$$

2.4 Utility Function

Selection algorithm is based on a combination of subjective and objective weights from the user perspective and network management. The user preferences and different requirements for network performance by different services are considered firstly. The hierarchical model using the AHP algorithm is built, and the subjective weight of each index is calculated. Then entropy method is used to reflect the dynamic relationship between the various indicators of competition, and reduce the error caused by subjective judgment from the objective standpoint. The selection algorithm is based on the combination of subjective and objective weights,aiming to the process of network selection more reasonably. The algorithm itself is not complicated and is beneficial to generalization. The form of compound weights is as follows:

$$\omega_j = \theta \cdot \omega'_j + (1 - \theta) \cdot \omega''_j,\ j = \alpha_1, \alpha_2, \alpha_3, \alpha_4, \beta, \gamma \tag{10}$$

$$W = \left\{\omega_{\alpha_1}, \omega_{\alpha_2}, \omega_{\alpha_3}, \omega_{\alpha_4}, \omega_{\beta}, \omega_{\gamma}\right\} \tag{11}$$

Among them $\theta \in (0, 1)$, $\theta = 0.5$ in this paper.

For a network with N candidates, evaluation matrix of the network parameter normalization is $R = (r_{ij})_{n\times 6}$, and we define the utility of each network as follows:

$$U = [u_1, u_2, \ldots u_n] = R \cdot W^T \quad (12)$$

$$u_i = \sum_j (w_j \cdot r_{ij}) \quad j = \alpha_1, \alpha_2, \alpha_3, \alpha_4, \beta, \gamma, \ i = 1, 2, \ldots n \quad (13)$$

Finally, the network with the highest utility value is chosen as the switching access network as follows:

$$\mathbf{u_{Best}} = \operatorname{argmax}(\mathbf{u_j}), \mathbf{i} = 1, 2, \ldots \mathbf{n} \quad (14)$$

3 Simulation and Analysis

In order to verify the performance of the proposed access selection method (En-AHP), we simulated En-AHP on the Exata platform, comparing with, the satellite network priority algorithm (Simple) and the unmodified AHP algorithm. The simulation scenario follows a real ship distribution with a total of 29 ships, and other infrastructure network equipment. The moving speed of ship follows the random moving model, and the velocity distribution is from 6 to 13 m/s with the moving granularity of 10 m. The simulation time of the scene is 30 s, and the application traffic type is constant bit rate (CBR). The data packet delivered by the application is 512 bytes, and the maximum number of packets is 100.

As shown in Fig. 4, where the simulation time and the size of data packet remain unchanged, as the contract rate decreases, the total number of contracts is reduced, and the throughput of the two links shows a downward trend. When the transmission interval is 0.8 s, the trend becomes flat, and the packet delivery rate decreases gradually with respect to the throughput per unit time. The throughput of En-AHP per unit time is more than *Simple* and *AHP*. The reason is the EN-AHP selects the ship ad hoc network,

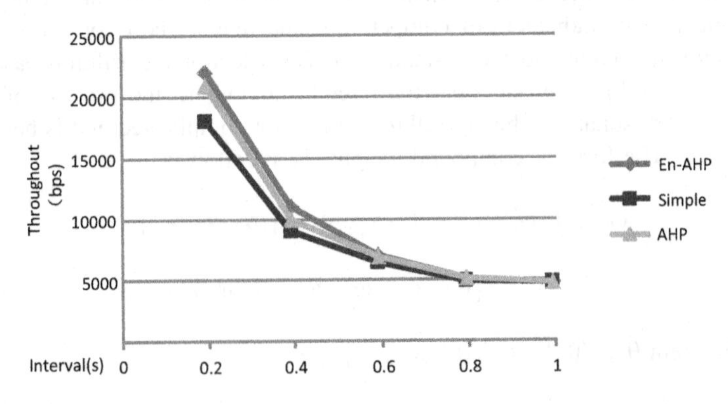

Fig. 4. Comparison of throughput. The throughput shows a downward trend mainly.

and the speed of base network in EN-AHP is faster than the Simple and the unimproved AHP, which select the satellite communication. As a result, the data processing capabilities of EN-AHP are superior to those of Simple and the unmodified AHP.

As shown in Fig. 5, the end-to-end delay of the Simple algorithm is always higher than that of EN-AHP. The primary variable is the transmission distance. In En-AHP, the data exchange between ships increases the delay, but the delay of *Simple* (satellite communication) is too large, accounting for the main proportion. With the reduce of packet sending rate, the end-to-end delay of Simple and unimproved AHP remain stable at 0.29 s basically, because for satellite, transmission distance of Simple algorithm is mainly composed of a satellite transmission path, and the characteristics of satellite communication is the delay remain stable.

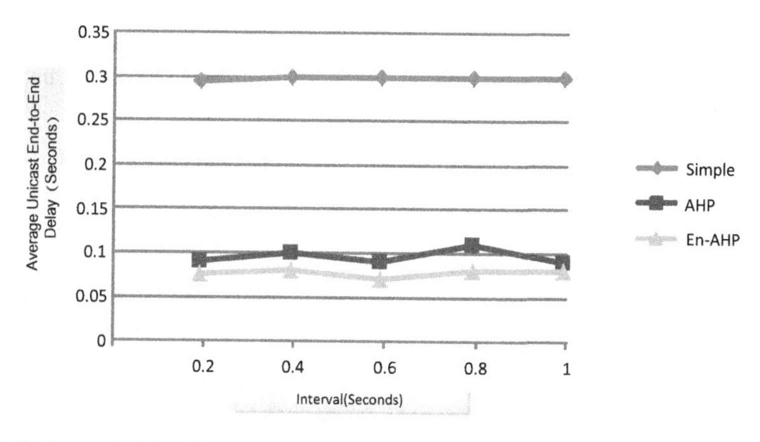

Fig. 5. End-to-end delay. There is little difference between the En-AHP and AHP algorithm.

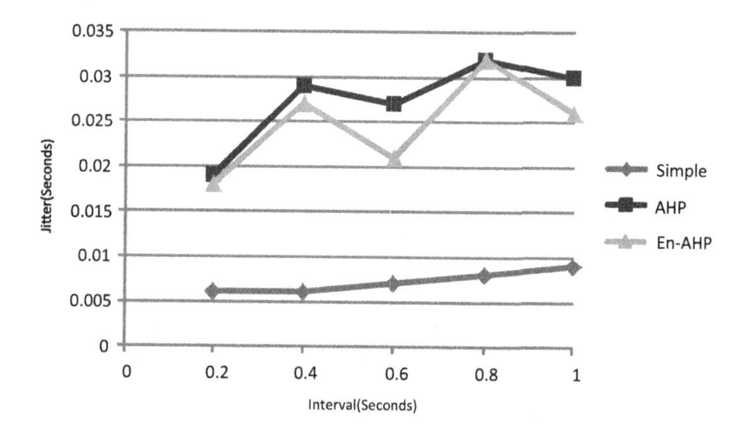

Fig. 6. Comparison of Jitter. We found that the packet loss rate of EN-AHP is always higher than that of Simple and not improved AHP. The movement of the nodes in the ship's self-organizing network causes the link state to be unstable.

As shown in Fig. 6, the Jitter of EN-AHP is always higher than Simple and unmodified AHP. Due to the number of ships is limited, the impact of link congestion is not considered. In EN-AHP, the movements of nodes in the ship self-organizing network lead to the instability of the link state, which is not as good as the satellite system. With the decrease of the packet transmission rate and the increase of the ship's moving distance, the data transmission jitter will intensify. The jitter of Simple remained stable, because the jitter of satellite communication is almost zero.

In terms of s the delay, jitter, and throughput, the benefits of EN-AHP are higher than Simple and unmodified AHP significantly.

4 Conclusion

Multiple access networks may coexist in the Marine Internet. Considering the different preferences of users and different needs of different applications, we proposes an access network selection algorithm based on the composite weights. The algorithm is simulated on Exata, and the simulation results show that the network performance of the access network using the proposed algorithm is better with lower cost.

References

1. Jiang, S.M.: On the marine internet and its potential applications for underwater inter-networking. In: 8th ACM International Conference on Underwater Networks and Systems (2013)
2. Sun, Y.B.: Research on heterogeneous network selection algorithms based on network performance and service feature perception. Nanjing University of Posts and Telecommunications (2012)
3. Yang, G., Ji, Y., Yu, K., et al.: General link layer technology in heterogeneous wireless access networks. Radio Eng. **35**(10), 19–22 (2005)
4. Jia, H.L.: Research on access selection and admission control in heterogeneous wireless networks. Zhejiang University (2007)
5. Lv, Q.F.: Research on network selection algorithm based on entropy method and network analytic hierarchy process. Inner Mongolia University (2013)
6. Varma, V.K., Ramesh, S., Wong, K.D., et al.: Mobility management in integrated UMTS/WLAN networks. In: IEEE International Conference on Communications, pp. 1048–1053 (2003)
7. Buddhikot, M., Chandranmenon, G., Han, S., et al.: Integration of 802.11 and third-generation wireless data networks. Phys. C Supercond. **1**(s 3–4), 503–512 (2003)
8. Xu, C.K., High, M.T.: Use LSA to reduce the dimensionality of the improved ART2 neural network clustering. Comput. Eng. Appl. (24), 133–138+177 (2014)
9. Zhou, Q.P., Feng, J.L.: The model of port road monitoring based on shape mapping and attribute computing network. Mod. Comput. (09), 3–8 (2013)

Online Shopping Recommendation with Bayesian Probabilistic Matrix Factorization

Jinming Wu, Zhong Liu, Guangquan Cheng(✉), Qi Wang, and Jincai Huang

Science and Technology on Information Systems Engineering Laboratory, National University of Defense Technology, Changsha, China
cgq299@nudt.edu.cn

Abstract. Recommendation system plays a crucial role in demand prediction, arousing attention from industry, business, government and academia. Widely employed in recommendation system, matrix factorization can well capture the potential relationships between users, items and latent variables. In this paper, we focus on a specific recommendation task on the large scale opinion-sharing online dataset called Epinions. We carried out recommendation experiments with the Bayesian probabilistic matrix factorization algorithm and the final results showed the superior performance in comparison to six representative recommendation algorithms. Meanwhile, the Bayesian probabilistic matrix factorization was investigated in depth and the potential advantage was explained from the model flexibility in parameters' adjustment. The findings would guide further research on applications of Bayesian probabilistic matrix factorization and inspire more researchers to contribute in this domain.

Keywords: Recommendation system · Bayesian probabilistic matrix factorization · Monte Carlo Markov chain

1 Introduction

The past several decades have witnessed the great necessity in designing novel recommendation systems, keeping pace with the fast development of information technologies. The increasing popularity of recommendation systems has influenced human life in many aspects. The potential value of recommendation is being excavated in fruitful applications, such as online social voting recommendation on social networks like Facebook, research paper recommendation for accelerating the speed of crucial scientific findings [1], multimedia recommendation for satisfying users' interest [2], citation recommendation for sorting relevant papers [3]. Interestingly, these commercial or academic applications have generalized well and at the same time brought us convenience in a personalized way. However, the demand has already confronted difficulties and challenges from both data quality and volume, requiring more novel algorithms or frameworks to deal with. A specific instance in recommendation research lies in that the ratings on items by users tend to be sparse and imbalanced, leading to

Z. Shi et al. (Eds.): ICIS 2017, IFIP AICT 510, pp. 445–451, 2017.
DOI: 10.1007/978-3-319-68121-4_48

the cold start phenomenon in the domain. On the other hand, the boom of internet allows the interactive behaviors like rating the goods bought before or feeding back the preferences of items such as foods, movies or music. In China, some famous corporations including Alibaba, Xiecheng Company have perceived the importance of feed-backing comments from customers and there is no doubt that accurate predictions on customers' preferences no matter in the future or later would bring much profits. It is obvious that the potential relationships between the users we take interests in may contribute to the recommendation in some sense and such social information involved in the model forms the representative family of recommendation algorithms as collaborative filtering.

To address these issues, we firstly make a brief survey regarding the probabilistic matrix factorization (PMF) algorithm which is proved to be effective in sparse and imbalanced recommendation dataset with the social information well exploited [4]. Distinguished from the traditional methodologies which mainly capture the information from the user-item ratings, we combined the social information with the rating circumstances for the performance promotion in our study and a superior adaptive Bayesian based model was employed in shopping recommendation.

The remainders of paper are arranged as follows. Section 2 elaborates the related developments of probabilistic matrix factorization and Bayesian based recommendation system. The experiments on the collected dataset using Bayesian probabilistic matrix factorization are described in Sect. 3 with some detailed analysis. Section 4 summarizes our work and light the direction in our future research.

2 Related Work and Employed Methodology

2.1 Review on Probabilistic Matrix Factorization

Probabilistic matrix factorization (PMF) is regarded as one of the outstanding algorithms in several recommendation competitions like Netflix Prize, compared to several collaborative filtering methodologies. The essence of the PMF is to make use of the low rank approximation and assume the potential distribution of dataset, overcoming the shortness of matrix factorization model. After both of user and item feature matrix obtained by low-rank approximating the rating matrix, we utilize the feature matrix to complete the potential missing ratings. Such model maintains well prediction accuracies in practice.

Due to the advantages of PMF, it has successfully attracted researchers' attention as well as interest. Dueck et al. proposed a probabilistic sparse matrix factorization algorithm, which well explained the uncertainty of noise in different extents [5]. Based on the matrix factorization ideology, Salakhutdinov et al. made assumption that user-feature matrix, item-feature matrix and user-item matrix were Gaussian in prior distribution, and then by modelling the latent variables, the PMF were derived in first time [6]. Multiple rating indexes were integrated into a user-item weighted matrix, assuming the latent matrix obeyed the Gaussian distribution and density probabilistic distribution conditionally correlated with user-feature and item-feature matrixes [7]. Another adaptation on the PMF was SoRec recommendation model [8], where user-feature vectors from both PMF model and confidence based matrix factorization model were combined to enhance the performance. Apart from the theoretical research on the PMF, some applications have also driven the interest of adaptation on PMF. Kernel probabilistic matrix factorization [7, 9].

2.2 A Brief Introduction to the Bayesian Probabilistic Matrix Factorization

In this section, we would focus our attention on the mechanism of Bayesian probabilistic matrix factorization (BPMF) algorithm proposed by Salakhutdinov et al. [10] and the essence of the recommendation algorithm would be elaborated.

The PMF is the basic model for the BPMF. Generally speaking, the PMF is a linear probabilistic model for solving recommendation tasks with some plausible information containing N users, M items and N * M rating matrix as R. Missing elements in R are ratings which the user never response to the item. The element r_{ij} in R records the preference score of the user i towards the item j.

We apply the low rank approximation to the rating matrix R and then obtain two matrix in low dimensions as

$$R \cong U^T * V \tag{1}$$

where the D * N matrix U and D * M matrix V are respectively user-feature matrix and item-feature matrix in low rank representations, describing the latent representations of users and items respectively.

The probabilistic graphical model for PMF can be translated in the Fig. 1.

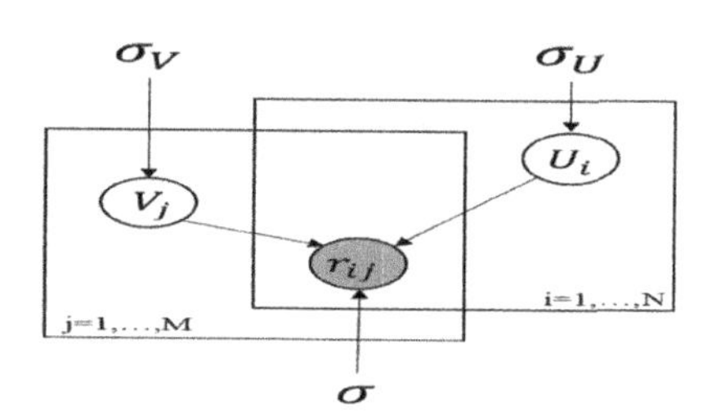

Fig. 1. The probabilistic graphical model of PMF

Assuming that the difference between the actual rating r_{ij} and the predicted rating $\tilde{r_{ij}}$ obeys the Gaussian distribution with mean 0 and variance σ, we derive

$$r_{ij} - U_i^T V_j \sim Gaussian\left(0, \sigma^2\right) \tag{2}$$

By the constant shifting, the distribution of rating r_{ij} can be represented as

$$r_{ij} \sim Gaussian\left(U_i^T V_j, \sigma^2\right) \tag{3}$$

Meanwhile, the assumption on the prior distribution on both latent user and latent item is the Gaussian distribution as follows:

$$p(U|\sigma_u^2) = \prod_{i=1}^{N} N(U_i|0, \sigma_U^2 I) \quad (4)$$

$$p(V|\sigma_V^2) = \prod_{j=1}^{M} N(V_j|0, \sigma_V^2 I) \quad (5)$$

Based on the Bayesian inference, we can acquire the loss function by minimizing the log of the posterior distribution accompanied with quadratic regularization terms.

$$E = \frac{1}{2}\sum_{i=1}^{N}\sum_{j=1}^{M} I_{ij}\left(r_{ij} - U_i^T V_j\right)^2 + \frac{\lambda_U}{2}\sum_{i=1}^{N}||U_i||_{Fro}^2 + \frac{\lambda_V}{2}\sum_{j=1}^{M}||V_j||_{Fro}^2 \quad (6)$$

where the equations follow $\lambda_U = \frac{\sigma^2}{\sigma_U^2}$ and $\lambda_V = \frac{\sigma^2}{\sigma_V^2}$.

During the past years in the research, the adaptations on the feature dimension D and prior distribution's assignment make great improvements on the capability of PMF's generalization.

For the BPMF, the U and V are assumed to follow the Gaussian distributions instead of constants.

$$p(U|\mu_U, Z_U) = \prod_{i=1}^{N} N\left(U_i|\mu_U, Z_U^{-1}\right) \quad (7)$$

$$p(V|\mu_V, Z_V) = \prod_{j=1}^{M} N\left(V_j|\mu_V, Z_V^{-1}\right) \quad (8)$$

Furthermore, the Wishart distribution W_0 is introduced as the priors for hyperparameters $\Theta_U = \{\mu_U, Z_U\}$ and $\Theta_V = \{\mu_V, Z_V\}$.

$$p(\Theta_U|\mu_0, v_0, W_0) = p(\mu_U|Z_U)p(Z_U) \quad (9)$$

$$p(\Theta_V|\mu_0, v_0, W_0) = p(\mu_V|Z_V)p(Z_V) \quad (10)$$

where we set $v_0 = D$, $\mu_0 = 0$ and $W_0 = I_D$.

In the prediction process of BPMF, the rating can be obtained by marginalizing over model parameters and hyper parameters as follows.

$$p\left(r_{ij}^*|R, \mu_0, v_0, W_0\right) = \int\int p\left(r_{ij}^*|U_i, V_j\right)p(U, V|R, \mu_U, Z_U, \mu_V, Z_V)p(\mu_U, Z_U, \mu_V, Z_V|\mu_0, v_0, W_0)d\{\mu_U, Z_U, \mu_V, Z_V\} \quad (11)$$

In our research, the Monte Carlo approximation is utilized for the computation.

$$p\left(r_{ij}^*|R, \mu_0, v_0, W_0\right) \approx \frac{1}{K}\sum_{k=1}^{K} p\left(r_{ij}^*\middle|U_i^k, V_j^k\right) \quad (12)$$

where the samples $\left\{U_i^k, V_j^k\right\}$ are generated from a Markov chain. Detailed information on MCMC's application to the BPMF refers to [10].

3 Experiments and Analysis

3.1 Evaluation Metrics and Selected Dataset

In our research, the root means square error (RMSE) is chosen as the evaluation metric. More specifically, for the rating r_{ij} which measures the user i's preference towards the item j and the predicted rating $r_{\tilde{ij}}$ derived by the recommendation system, the RMSE is computed as

$$RMSE = \left(\sum\nolimits_{i,j \in T} \left(r_{ij} - r_{\tilde{ij}}\right)^2\right)^{\frac{1}{2}} / |T| \tag{13}$$

Due to the severe penalty on inaccurate predictions using square error in RMSE, the evaluation metric is regarded as a strict but reasonable criteria.

The dataset in our experiments is from the electronic business website as Epions (www.Epinions.com). The ratings reflecting the preference of users towards items are collected, processed and prepared in user-item matrix. Another crucial information matrix describes the friend relationship between users is also provided in Table 1.

Table 1. Related information about Epinions

Total users	Total items	Density of ratings	Density of user relationships	Averaged ratings on items	Averaged friends
49289	139738	664824 (0.0097%)	487183 (0.0201%)	4.76	9.88

In comparison to the commonly used dataset like Movielens and IMBD, the Epnions behaves rather sparse in social relationships.

3.2 Results and Analysis

To verify the effectiveness of the BPMF, we selected some frequently used popular algorithms which were believed to perform well in several recommendation tasks as baselines. Specifically, they were SVD [11], NMF [12], BaselineOnly [13], SloperOne [14], KNNBaseline [13] and CoClustering [15]. In the experiments, both of the parameters λ_U and λ_V were set as 0.002 and the turns of iteration was 200. Some other parameters were selected as the default. Aiming at determining the optimal feature dimension D, we ranged the parameter from 5 to 20 at the step width 5 and the 5-fold cross validation was performed in final selection. The results were arrange in the Table 2.

Table 2. The RMSE in different D value

D	5	10	15	20
BPMF	1.2232	1.1678	0.9978	1.4471

From the table, we can notice that the value of RMSE decreased steadily in the small interval while such metric fluctuated severely when D was over 15. So 15 was chosen as the optimal value for the feature dimension.

A persuasive metric is the k-fold evaluation to average the testing results, in which the k-1 folds are prepared for model fitting with the last fold for testing and such manipulation is performed k times in one turn by ranging all of k − 1 folds.

Furthermore, we carried out experiments using the above-mentioned seven algorithms in 10 independent turns and provided the results in average. The fold partition ranged from 1 to 5 with the step width 1. The Table 3 recorded the averaged RMSE in ten turns using various algorithms in different folds and it revealed the superior performance of BPMF with the lowest RMSE.

Table 3. Averaged k-fold RMSE in different algorithms

	Fold 1	Fold 2	Fold 3	Fold 4	Fold 5
NMF	1.1168	1.1186	1.1173	1.1153	1.1151
SVD	1.0914	1.0941	1.0969	1.0926	1.0935
BaselineOnly	1.1084	1.1086	1.108	1.1094	1.1091
SlopeOne	1.1062	1.1061	1.1067	1.1082	1.1057
KNNBaseline	1.0963	1.0942	1.0967	1.0936	1.0933
CoClustering	1.1155	1.1169	1.1163	1.1142	1.1128
BPMF	0.9959	0.9955	0.9951	0.9952	0.9954

To further validate the outstanding performance of BPMF in statistics, the paired t-test was introduced to test the results of 10 independent turns with seven algorithms in the case when the 5-fold RMSE was chosen as the metric. Specifically, the 10 results in each turn obtained by BPMF were compared with those derived by NFM, SVD, BaselineOnly, SlopeOne, KNNBaseline and Coclustering respectively with the help of SPSS. The testing results indicated the BPMF outperformed others significantly. And the compared results were mapped into three element tuple as win\tie\loss. Then the counts of win\tie\loss in 10 turns with some baseline algorithm were summarized in the Table 4 as follows.

Table 4. Win\Tie\Loss statistical results in comparison experiments

	NMF	SVD	BaselineOnly	SlopeOne	KNNBaseline	CoClustering
BPMF	9\1\0	8\1\1	10\0\0	9\0\1	8\2\0	10\0\0

The potential reason for obtaining satisfying results using BPMF was that the Bayesian framework was more flexible in searching optimal hyper parameters and the model maintained high complexity to fit the dataset of high volume.

4 Conclusions

In the end, the Bayesian probabilistic matrix factorization was employed in online item recommendation backgrounded in the opinion-sharing platform. The BPMF showed its superiority in recommendation task, deriving the lowest RMSE in comparison to six representative algorithms. The conclusions were supported by the paired t-test in ten independent experiments and the persuasive results motivated us to research in depth in other applications. In the future, we would design more variant matrix factorization algorithms incorporating the other social information to promote the performance in recommendations.

References

1. Beel, J., Gipp, B., Langer, S., Breitinger, C.: Research-paper recommender systems: a literature survey. Int. J. Digital Libr. **17**, 1–34 (2016)
2. Lee, W.P., Tseng, G.Y.: Incorporating contextual information and collaborative filtering methods for multimedia recommendation in a mobile environment. Multimedia Tools Appl. 1–21 (2015)
3. Liu, H., Kong, X., Bai, X., Wang, W., Bekele, T.M., Xia, F.: Context-based collaborative filtering for citation recommendation. IEEE Access **3**, 1695–1703 (2015)
4. Fuchs, M., Zanker, M.: Multi-criteria ratings for recommender systems: an empirical analysis in the tourism domain. In: Lecture Notes in Business Information Processing 123, 100–111 (2012)
5. Dueck, D., Frey, B.J.: Probabilistic sparse matrix factorization. Technical report PSI, University of Toronto (2004)
6. Salakhutdinov, R., Mnih, A.: Probabilistic matrix factorization. In: International Conference on Neural Information Processing Systems, pp. 1257–1264 (2007)
7. Yang, X., Guo, Y., Liu, Y., Steck, H.: A survey of collaborative filtering based social recommender systems. Comput. Commun. **41**, 1–10 (2014)
8. Ma, H., Yang, H., Lyu, M.R., King, I.: SoRec: social recommendation using probabilistic matrix factorization. In: ACM Conference on Information and Knowledge Management, CIKM 2008, Napa Valley, California, USA, pp. 931–940, October 2008
9. Zhou, T., Shan, H., Banerjee, A., Sapiro, G.: Kernelized Probabilistic Matrix Factorization: Exploiting Graphs and Side Information. SDM (2012)
10. Salakhutdinov, R., Mnih, A.: Bayesian probabilistic matrix factorization using Markov chain Monte Carlo. In: International Conference on Machine Learning, pp. 880–887 (2008)
11. Kim, J.K., Cho, Y.H.: Using web usage mining and SVD to improve e-commerce recommendation quality. In: Lee, J., Barley, M. (eds.) PRIMA 2003. LNCS, vol. 2891, pp. 86–97. Springer, Heidelberg (2003). doi:10.1007/978-3-540-39896-7_8
12. Luo, X., Zhou, M., Xia, Y., Zhu, Q.: An efficient non-negative matrix-factorization-based approach to collaborative filtering for recommender systems. IEEE Trans. Ind. Inform. **10**, 1273–1284 (2014)
13. Koren, Y.: Factor in the neighbors: scalable and accurate collaborative filtering. ACM Trans. Knowl. Disc. Data **4**, 1–24 (2010)
14. Sun, L., Jingjiao, L.I.: SlopeOne collaborative filtering recommendation algorithm based on dynamic k-nearest-neighborhood. J. Front. Comput. Sci. Technol. **5**, 857–864 (2011)
15. George, T., Merugu, S.: A scalable collaborative filtering framework based on co-clustering. In: IEEE International Conference on Data Mining, p. 4 (2005)

Author Index

Printed in the United States
By Bookmasters